Applications of Nonlinear Fiber Optics

Second Edition

Applications of Nonlinear Fiber Optics

Second Edition

GOVIND P. AGRAWAL

The Institute of Optics
University of Rochester
Rochester, New York

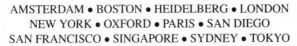

AMSTERDAM • BOSTON • HEIDELBERG • LONDON
NEW YORK • OXFORD • PARIS • SAN DIEGO
SAN FRANCISCO • SINGAPORE • SYDNEY • TOKYO

ELSEVIER   Academic Press is an imprint of Elsevier

Academic Press is an imprint of Elsevier
30 Corporate Drive, Suite 400, Burlington, MA 01803, USA
525 B Street, Suite 1900, San Diego, California 92101-4495, USA
84 Theobald's Road, London WC1X 8RR, UK

Library of Congress Cataloging-in-Publication Data
Application submitted.

British Library Cataloguing-in-Publication Data
A catalogue record for this book is available from the British Library.

ISBN: 978-0-12-374302-2

For information on all Academic Press publications
visit our Web site at www.books.elsevier.com

Transferred to Digital Printing, 2010

Printed and bound in Great Britain by
CPI Antony Rowe, Chippenham and Eastbourne

In the memory of my parents and

for Anne, Sipra, Caroline, and Claire

Contents

Preface   **xiii**

1   Fiber Gratings   **1**
  1.1   Basic Concepts   1
     1.1.1   Bragg Diffraction   2
     1.1.2   Photosensitivity   3
  1.2   Fabrication Techniques   4
     1.2.1   Single-Beam Internal Technique   4
     1.2.2   Dual-Beam Holographic Technique   5
     1.2.3   Phase-Mask Technique   7
     1.2.4   Point-by-Point Fabrication Technique   8
     1.2.5   Technique Based on Ultrashort Optical Pulses   9
  1.3   Grating Characteristics   10
     1.3.1   Coupled-Mode Equations   11
     1.3.2   CW Solution in the Linear Case   13
     1.3.3   Photonic Bandgap   14
     1.3.4   Grating as an Optical Filter   16
     1.3.5   Experimental Verification   17
  1.4   CW Nonlinear Effects   20
     1.4.1   Nonlinear Dispersion Curves   20
     1.4.2   Optical Bistability   22
  1.5   Modulation Instability   24
     1.5.1   Linear Stability Analysis   24
     1.5.2   Effective NLS Equation   26
     1.5.3   Experimental Results   28
  1.6   Nonlinear Pulse Propagation   28
     1.6.1   Bragg Solitons   29
     1.6.2   Relation to NLS Solitons   30
     1.6.3   Formation of Bragg Solitons   31
     1.6.4   Nonlinear Switching   33
     1.6.5   Effects of Birefringence   37
  1.7   Related Periodic Structures   39
     1.7.1   Long-Period Gratings   39
     1.7.2   Nonuniform Bragg Gratings   40

     1.7.3   Transient and Dynmaic Fiber Gratings 44

 Problems 47

 References 48

2 Fiber Couplers   **54**

 2.1   Coupler Characteristics 54

     2.1.1   Coupled-Mode Equations 55

     2.1.2   Low-Power Optical Beams 57

     2.1.3   Linear Pulse Switching 60

 2.2   Nonlinear Effects 61

     2.2.1   Quasi-CW Switching 62

     2.2.2   Experimental Results 64

     2.2.3   Nonlinear Supermodes 66

     2.2.4   Modulation Instability 67

 2.3   Ultrashort Pulse Propagation 70

     2.3.1   Nonlinear Switching of Optical Pulses 71

     2.3.2   Variational Approach 73

     2.3.3   Coupler-Paired Solitons 76

     2.3.4   Higher-Order Effects 78

 2.4   Other Types of Couplers 81

     2.4.1   Asymmetric Couplers 81

     2.4.2   Active Couplers 83

     2.4.3   Grating-Assisted Couplers 85

     2.4.4   Birefringent Couplers 87

 2.5   Fibers with Multiple Cores 88

     2.5.1   Dual-Core Photonic Crystal Fibers 89

     2.5.2   Multicore Fiber Arrays 91

 Problems 94

 References 95

3 Fiber Interferometers   **100**

 3.1   Fabry–Perot and Ring Resonators 100

     3.1.1   Transmission Resonances 101

     3.1.2   Optical Bistability 103

     3.1.3   Nonlinear Dynamics and Chaos 105

     3.1.4   Modulation Instability 106

     3.1.5   Ultrafast Nonlinear Effects 108

 3.2   Sagnac Interferometers 110

     3.2.1   Nonlinear Transmission 110

     3.2.2   Nonlinear Switching 111

     3.2.3   Applications 116

 3.3   Mach–Zehnder Interferometers 120

     3.3.1   Nonlinear Characteristics 120

     3.3.2   Applications 123

 3.4   Michelson Interferometers 124

 Problems 125

References 126

4   Fiber Amplifiers   **131**
   4.1   Basic Concepts 131
      4.1.1   Pumping and Gain Coefficient 132
      4.1.2   Amplifier Gain and Bandwidth 133
      4.1.3   Amplifier Noise 135
   4.2   Erbium-Doped Fiber Amplifiers 137
      4.2.1   Gain Spectrum 137
      4.2.2   Amplifier Gain 139
      4.2.3   Amplifier Noise 142
   4.3   Dispersive and Nonlinear Effects 143
      4.3.1   Maxwell–Bloch Equations 144
      4.3.2   Ginzburg–Landau Equation 145
   4.4   Modulation Instability 147
      4.4.1   Distributed Amplification 147
      4.4.2   Periodic Lumped Amplification 148
      4.4.3   Noise Amplification 150
   4.5   Optical Solitons 152
      4.5.1   Properties of Autosolitons 152
      4.5.2   Maxwell–Bloch Solitons 155
   4.6   Pulse Amplification 157
      4.6.1   Anomalous-Dispersion Regime 158
      4.6.2   Normal-Dispersion Regime 160
      4.6.3   Higher-Order Effects 164
   4.7   Fiber-Optic Raman Amplifiers 168
      4.7.1   Pulse Amplification through Raman Gain 169
      4.7.2   Self-Similar Evolution and Similariton Formation 170
   Problems 172
   References 173

5   Fiber Lasers   **179**
   5.1   Basic Concepts 179
      5.1.1   Pumping and Optical Gain 180
      5.1.2   Cavity Design 181
      5.1.3   Laser Threshold and Output Power 183
   5.2   CW Fiber Lasers 185
      5.2.1   Nd-Doped Fiber Lasers 185
      5.2.2   Yb-Doped Fiber Lasers 187
      5.2.3   Erbium-Doped Fiber Lasers 190
      5.2.4   DFB Fiber Lasers 192
      5.2.5   Self-Pulsing and Chaos 195
   5.3   Short-Pulse Fiber Lasers 197
      5.3.1   Q-Switched Fiber Lasers 197
      5.3.2   Physics of Mode Locking 200
      5.3.3   Active Mode Locking 201

      5.3.4   Harmonic Mode Locking 204
5.4   Passive Mode Locking 210
      5.4.1   Saturable Absorbers 210
      5.4.2   Nonlinear Fiber-Loop Mirrors 213
      5.4.3   Nonlinear Polarization Rotation 216
      5.4.4   Hybrid Mode Locking 219
      5.4.5   Other Mode-Locking Techniques 221
5.5   Role of Fiber Nonlinearity and Dispersion 226
      5.5.1   Saturable-Absorber Mode Locking 226
      5.5.2   Additive-Pulse Mode Locking 227
      5.5.3   Spectral Sidebands and Pulse Width 228
      5.5.4   Phase Locking and Soliton Collisions 231
      5.5.5   Polarization Effects 233
Problems 235
References 236

6 Pulse Compression   **245**
6.1   Physical Mechanism 245
6.2   Grating-Fiber Compressors 247
      6.2.1   Grating Pair 248
      6.2.2   Optimum Compressor Design 250
      6.2.3   Practical Limitations 253
      6.2.4   Experimental Results 254
6.3   Soliton-Effect Compressors 259
      6.3.1   Compressor Optimization 259
      6.3.2   Experimental Results 261
      6.3.3   Higher-Order Nonlinear Effects 263
6.4   Fiber Bragg Gratings 266
      6.4.1   Gratings as a Compact Dispersive Element 267
      6.4.2   Grating-Induced Nonlinear Chirp 268
      6.4.3   Bragg-Soliton Compression 270
6.5   Chirped-Pulse Amplification 271
      6.5.1   Chirped Fiber Gratings 272
      6.5.2   Photonic Crystal Fibers 274
6.6   Dispersion-Managed Fibers 276
      6.6.1   Dispersion-Decreasing Fibers 277
      6.6.2   Comblike Dispersion Profiles 280
6.7   Other Compression Techniques 283
      6.7.1   Cross-Phase Modulation 283
      6.7.2   Gain Switching in Semiconductor Lasers 286
      6.7.3   Optical Amplifiers 288
      6.7.4   Fiber-Loop Mirrors and Other Devices 290
Problems 292
References 293

7 Fiber-Optic Communications   301

  7.1 System Basics 301

      7.1.1 Loss Management 302

      7.1.2 Dispersion Management 304

  7.2 Impact of Fiber Nonlinearities 305

      7.2.1 Stimulated Brillouin Scattering 306

      7.2.2 Stimulated Raman Scattering 308

      7.2.3 Self-Phase Modulation 311

      7.2.4 Cross-Phase Modulation 315

      7.2.5 Four-Wave Mixing 318

  7.3 Solitons in Optical Fibers 322

      7.3.1 Properties of Optical Solitons 322

      7.3.2 Loss-Managed Solitons 325

      7.3.3 Dispersion-Managed Solitons 327

      7.3.4 Timing Jitter 331

  7.4 Pseudo-Linear Lightwave Systems 336

      7.4.1 Intrachannel Nonlinear Effects 336

      7.4.2 Intrachannel XPM 338

      7.4.3 Intrachannel FWM 339

  Problems 341

  References 342

8 Optical Signal Processing   349

  8.1 Wavelength Conversion 349

      8.1.1 XPM-Based Wavelength Converters 350

      8.1.2 FWM-Based Wavelength Converters 355

  8.2 Ultrafast Optical Switching 360

      8.2.1 XPM-Based Sagnac-Loop Switches 360

      8.2.2 Polarization-Discriminating Switches 363

      8.2.3 FWM-Based Ultrafast Switches 365

  8.3 Applications of Time-Domain Switching 368

      8.3.1 Channel Demultiplexing 368

      8.3.2 Data-Format Conversion 373

      8.3.3 All-Optical Sampling 375

  8.4 Optical Regenerators 377

      8.4.1 SPM-Based Regenerators 377

      8.4.2 FWM-Based Regenerators 383

      8.4.3 Regeneration of DPSK Signals 385

      8.4.4 Optical 3R Regenerators 387

  Problems 390

  References 391

9 Highly Nonlinear Fibers   397
   9.1 Microstructured Fibers 397
        9.1.1 Design and Fabrication 398
        9.1.2 Nonlinear and Dispersive Properties 399
   9.2 Wavelength Shifting and Tuning 403
        9.2.1 Raman-Induced Frequency Shifts 404
        9.2.2 Four-Wave Mixing 410
   9.3 Supercontinuum Generation 414
        9.3.1 Multichannel Telecommunication Sources 415
        9.3.2 Nonlinear Spectroscopy 416
        9.3.3 Optical Coherence Tomography 420
        9.3.4 Optical Frequency Metrology 424
   9.4 Photonic Bandgap Fibers 431
        9.4.1 Properties of Hollow-Core PCFs 432
        9.4.2 Applications of Air-Core PCFs 434
        9.4.3 PCFs with Fluid-Filled Cores 436
   Problems 439
   References 440

10 Quantum Applications   447
   10.1 Quantum Theory of Pulse Propagation 447
         10.1.1 Quantum Nonlinear Schrödinger Equation 448
         10.1.2 Quantum Theory of Self-Phase Modulation 449
         10.1.3 Generalized NLS Equation 451
         10.1.4 Quantum Solitons 452
   10.2 Squeezing of Quantum Noise 454
         10.2.1 Physics behind Quadrature Squeezing 454
         10.2.2 FWM-Induced Quadrature Squeezing 455
         10.2.3 SPM-Induced Quadrature Squeezing 457
         10.2.4 SPM-Induced Amplitude Squeezing 461
         10.2.5 Polarization Squeezing 466
   10.3 Quantum Nondemolition Schemes 468
         10.3.1 QND Measurements through Soliton Collisions . . . 468
         10.3.2 QND Measurements through Spectral Filtering . . . 470
   10.4 Quantum Entanglement 472
         10.4.1 Photon-Pair Generation 472
         10.4.2 Polarization Entanglement 477
         10.4.3 Time-Bin Entanglement 481
         10.4.4 Continuous-Variable Entanglement 482
   10.5 Quantum Cryptography 485
   Problems 487
   References 488

A Acronyms   493

Index   495

Preface

Since the publication of the first edition of my book *Nonlinear Fiber Optics* in 1989, this field has virtually exploded. During the 1990s, a major factor behind such a sustained growth was the advent of fiber amplifiers and lasers, made by doping silica fibers with rare-earth materials such as erbium and ytterbium. Erbium-doped fiber amplifiers revolutionized the design of fiber-optic communication systems, including those making use of optical solitons, whose very existence stems from the presence of nonlinear effects in optical fibers. Optical amplifiers permit propagation of lightwave signals over thousands of kilometers as they can compensate for all losses encountered by the signal in the optical domain. At the same time, fiber amplifiers enable the use of massive wavelength-division multiplexing, a technique that by 1999 led to the development of lightwave systems with capacities exceeding 1 Tb/s. Nonlinear fiber optics plays an important role in the design of such high-capacity lightwave systems. In fact, an understanding of various nonlinear effects occurring inside optical fibers is almost a prerequisite for a lightwave-system designer.

Starting around 2000, a new development occurred in the field of *nonlinear fiber optics* that changed the focus of research and led to a number of advances and novel applications in recent years. Several kinds of new fibers, classified as highly nonlinear fibers, have been developed. They are referred to with names such as *microstructured fibers, holey fibers*, or *photonic crystal fibers*, and share the common property that a relatively narrow core is surrounded by a cladding containing a large number of air holes. The nonlinear effects are enhanced dramatically in such fibers. In fact, with a proper design of microstructured fibers, some nonlinear effects can be observed even when the fiber is only a few centimeters long. The dispersive properties of such fibers also are quite different compared with those of conventional fibers, developed mainly for telecommunication applications. Because of these changes, microstructured fibers exhibit a variety of novel nonlinear effects that are finding application in the fields as diverse as optical coherence tomography and high-precision frequency metrology.

The fourth edition of *Nonlinear Fiber Optics*, published in 2007, has been updated to include recent developments related to the advent of highly nonlinear fibers. However, it deals mostly with the fundamental aspects of this exciting field. Since 2001, the applications of nonlinear fiber optics have been covered in a companion book that also required updating. This second edition of *Applications of Nonlinear Fiber Optics* fills this need. It has been expanded considerably to include the new research material published over the last seven years or so. It retains most of the material that appeared in the first edition.

The first three chapters deal with three important fiber-optic components—fiber-based gratings, couplers, and interferometers—that serve as the building blocks of lightwave technology. In view of the enormous impact of rare-earth-doped fibers, amplifiers and lasers made by using such fibers are covered in Chapters 4 and 5. Chapter 6 deals with the pulse-compression techniques. Chapters 7 and 8 has been revised extensively to make room for the new material. The former is devoted to fiber-optic communication systems, but Chapter 8 now focuses on the ultrafast signal processing techniques that make use of nonlinear phenomena in optical fibers. Last two chapters, Chapters 9 and 10, are entirely new. Chapter 9 focuses on the applications of highly nonlinear fibers in areas ranging from wavelength laser tuning and nonlinear spectroscopy to biomedical imaging and frequency metrology. Chapter 10 is devoted to the applications of nonlinear fiber optics in the emerging technologies that make use of quantum-mechanical effects. Examples of such technologies include quantum cryptography, quantum computing, and quantum communications.

This volume should serve well the needs of the scientific community interested in such diverse fields as ultrafast phenomena, high-power fiber amplifiers and lasers, optical communications, ultrafast signal processing, and quantum information. The potential readership is likely to consist of senior undergraduate students, graduate students enrolled in the M.S. and Ph.D. programs, engineers and technicians involved with the telecommunication and laser industry, and scientists working in the fields of optical communications and quantum information. Some universities may opt to offer a high-level graduate course devoted solely to nonlinear fiber optics. The problems provided at the end of each chapter should be useful to instructors of such a course.

Many individuals have contributed to the completion of this book either directly or indirectly. I am thankful to all of them, especially to my students, whose curiosity led to several improvements. Some of my colleagues also helped me in preparing this book. I thank Prof. J. H. Eberly, Prof. A. N. Pinto, and Dr. S. Lukishova for reading the chapter on quantum applications and making helpful suggestions. I am grateful to many readers for their feedback. Last, but not least, I thank my wife, Anne, and my daughters, Sipra, Caroline, and Claire, for understanding why I needed to spend many weekends on the book instead of spending time with them.

Govind P. Agrawal

Rochester, NY

December 2007

Chapter 1

Fiber Gratings

Silica fibers can change their optical properties permanently when they are exposed to intense radiation from a laser operating in the blue or ultraviolet spectral region. This photosensitive effect can be used to induce periodic changes in the refractive index along the fiber length, resulting in the formation of an intracore Bragg grating. Fiber gratings can be designed to operate over a wide range of wavelengths extending from the ultraviolet to the infrared region. The wavelength region near 1.5 μm is of particular interest because of its relevance to fiber-optic communication systems. In this chapter on fiber gratings, the emphasis is on the role of the nonlinear effects. Sections 1.1 and 1.2 discuss the physical mechanism responsible for photosensitivity and various techniques used to make fiber gratings. The coupled-mode theory is described in Section 1.3, where the concept of the photonic bandgap is also introduced. Section 1.4 is devoted to the nonlinear effects occurring under continuous-wave (CW) conditions. The phenomenon of modulation instability is discussed in Section 1.5. The focus of Section 1.6 is on propagation of optical pulses through a fiber grating with emphasis on optical solitons. The phenomenon of nonlinear switching is also covered in this section. Section 1.7 is devoted to related fiber-based periodic structures such as long-period, chirped, sampled, transient, and dynamic gratings together with their applications.

1.1 Basic Concepts

Diffraction gratings constitute a standard optical component and are used routinely in various optical instruments such as a spectrometer. The underlying principle was discovered more than 200 years ago [1]. From a practical standpoint, a diffraction grating is defined as any optical element capable of imposing a periodic variation in the amplitude or phase of light incident on it. Clearly, an optical medium whose refractive index varies periodically acts as a grating since it imposes a periodic variation of phase when light propagates through it. Such gratings are called *index gratings*.

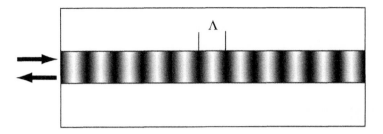

Figure 1.1: A fiber grating. Dark and light shaded regions within the fiber core show periodic variations of the refractive index.

1.1.1   Bragg Diffraction

The diffraction theory of gratings shows that when light is incident at an angle θ_i (measured with respect to the planes of constant refractive index), it is diffracted at an angle θ_r such that [1]

$$\sin \theta_i - \sin \theta_r = m\lambda/(\bar{n}\Lambda), \tag{1.1.1}$$

where Λ is the grating period, $\lambda/\bar{n}$ is the wavelength of light inside the medium with an average refractive index $\bar{n}$, and m is the order of Bragg diffraction. This condition can be thought of as a phase-matching condition, similar to that occurring in the case of Brillouin scattering or four-wave mixing [2] and can be written as

$$\mathbf{k}_i - \mathbf{k}_d = m\mathbf{k}_g, \tag{1.1.2}$$

where $\mathbf{k}_i$ and $\mathbf{k}_d$ are the wave vectors associated with the incident and diffracted light. The grating wave vector $\mathbf{k}_g$ has magnitude $2\pi/\Lambda$ and points in the direction in which the refractive index of the medium is changing in a periodic manner.

In the case of single-mode fibers, all three vectors lie along the fiber axis. As a result, $\mathbf{k}_d = -\mathbf{k}_i$ and the diffracted light propagates backward. Thus, as shown schematically in Figure 1.1, a fiber grating acts as a reflector for a specific wavelength of light for which the phase-matching condition is satisfied. In terms of the angles appearing in Eq. (1.1.1), $\theta_i = \pi/2$ and $\theta_r = -\pi/2$. If $m = 1$, the period of the grating is related to the vacuum wavelength as $\lambda = 2\bar{n}\Lambda$. This condition is known as the *Bragg condition*, and gratings satisfying it are referred to as *Bragg gratings*. Physically, the Bragg condition ensures that weak reflections occurring throughout the grating add up in phase to produce a strong reflection at the input end. For a fiber grating reflecting light in the wavelength region near 1.5 μm, the grating period $\Lambda \approx 0.5$ μm.

Bragg gratings inside optical fibers were first formed in 1978 by irradiating a germanium-doped silica fiber for a few minutes with an intense argon-ion laser beam [3]. The grating period was fixed by the argon-ion laser wavelength, and the grating reflected light only within a narrow region around that wavelength. It was realized that the 4% reflection occurring at the two fiber–air interfaces created a standing-wave pattern such that more of the laser light was absorbed in the bright regions. As a result, the glass structure changed in such a way that the refractive index increased permanently in the bright regions. Although this phenomenon attracted some attention during the next 10 years [4]–[16], it was not until 1989 that fiber gratings became a topic of intense investigation, fueled partly by the observation of second-harmonic generation in photosensitive fibers. The impetus for this resurgence of interest was provided by a 1989 paper in which a side-exposed holographic technique was used to make fiber gratings with controllable period [17].

Because of its relevance to fiber-optic communication systems, the holographic technique was quickly adopted to produce fiber gratings in the wavelength region near 1.55 μm [18]. Considerable work was done during the early 1990s to understand the physical mechanism behind photosensitivity of fibers and to develop techniques that were capable of making large changes in the refractive index [19]–[47]. By 1995, fiber gratings were available commercially, and by 1997 they became a standard component of lightwave technology. Soon after, several books devoted entirely to fiber gratings appeared, focusing on applications related to fiber sensors and fiber-optic communication systems [48]–[50].

1.1.2 Photosensitivity

There is considerable evidence that the photosensitivity of optical fibers is due to defect formation inside the core of Ge-doped silica (SiO_2) fibers [29]–[31]. In practice, the core of a silica fiber is often doped with germania (GeO_2) to increase its refractive index and introduce an index step at the core-cladding interface. The Ge concentration is typically 3–5% but may exceed 15% in some cases.

The presence of Ge atoms in the fiber core leads to formation of oxygen-deficient bonds (such as Si–Ge, Si–Si, and Ge–Ge bonds), which act as defects in the silica matrix [48]. The most common defect is the GeO defect. It forms a defect band with an energy gap of about 5 eV (energy required to break the bond). Single-photon absorption of 244-nm radiation from an excimer laser (or two-photon absorption of 488-nm light from an argon-ion laser) breaks these defect bonds and creates GeE′ centers. Extra electrons associated with GeE′ centers are free to move within the glass matrix until they are trapped at hole-defect sites to form the color centers known as Ge(1) and Ge(2). Such modifications in the glass structure change the absorption spectrum $\alpha(\omega)$. However, changes in the absorption also affect the refractive index since $\Delta\alpha$ and Δn are related through the Kramers–Kronig relation [51]:

$$\Delta n(\omega') = \frac{c}{\pi} \int_0^\infty \frac{\Delta\alpha(\omega)d\omega}{\omega^2 - \omega'^2}. \tag{1.1.3}$$

Even though absorption modifications occur mainly in the ultraviolet region, the refractive index can change even in the visible or infrared region. Moreover, as index changes occur only in the regions of fiber core where the ultraviolet light is absorbed, a periodic intensity pattern is transformed into an index grating. Typically, index change Δn is $\sim 10^{-4}$ in the 1.3- to 1.6-μm wavelength range but can exceed 0.001 in fibers with high Ge concentration [34].

The presence of GeO defects is crucial for photosensitivity to occur in optical fibers. However, standard telecommunication fibers rarely have more than 3% of Ge atoms in their core, resulting in relatively small index changes. The use of other

dopants, such as phosphorus, boron, and aluminum, can enhance the photosensitivity (and the amount of index change) to some extent, but these dopants also tend to
increase fiber losses. It was discovered in the early 1990s that the amount of index
change induced by ultraviolet absorption can be enhanced by two orders of magnitude ($\Delta n > 0.01$) by soaking the fiber in hydrogen gas at high pressures (200 atm)
and room temperature [39]. The density of Ge–Si oxygen-deficient bonds increases
in hydrogen-soaked fibers because hydrogen can recombine with oxygen atoms. Once
hydrogenated, the fiber needs to be stored at low temperature to maintain its photosensitivity. However, gratings made in such fibers remain intact over relatively long
periods of time, if they are stabilized using a suitable annealing technique [52]–[56].
Hydrogen soaking is commonly used for making fiber gratings.

Because of the stability issue associated with hydrogen soaking, a technique, known
as ultraviolet (UV) hypersensitization, has been employed in recent years [57]–[59].
An alternative method, known as *OH flooding*, is also used for this purpose. In this approach [60], the hydrogen-soaked fiber is heated rapidly to a temperature near 1000°C
before it is exposed to UV radiation. The resulting out-gassing of hydrogen creates
a flood of OH ions and leads to a considerable increase in the fiber photosensitivity. A comparative study of different techniques revealed that the UV-induced index
changes were indeed more stable in the hypersensitized and OH-flooded fibers [61]. It
should be stressed that understanding of the exact physical mechanism behind photosensitivity is far from complete, and more than one mechanism may be involved [57].
Localized heating can also affect the formation of a grating. For instance, damage
tracks were seen in fibers with a strong grating (index change >0.001) when the grating was examined under an optical microscope [34]; these tracks were due to localized
heating to several thousand degrees of the core region, where ultraviolet light was most
strongly absorbed. At such high temperatures the local structure of amorphous silica
can change considerably because of melting.

1.2   Fabrication Techniques

Fiber gratings can be made by using several techniques, each having its own merits
[48]–[50]. This section discusses briefly four major techniques, used commonly for
making fiber gratings: the single-beam internal technique, the dual-beam holographic
technique, the phase-mask technique, and the point-by-point fabrication technique.
The use of ultrashort optical pulses for grating fabrication is covered in the last subsection.

1.2.1   Single-Beam Internal Technique

In this technique, used in the original 1978 experiment [3], a single laser beam, often
obtained from an argon-ion laser operating in a single mode near 488 nm, is launched
into a germanium-doped silica fiber. The light reflected from the near end of the fiber
is then monitored. The reflectivity is initially about 4%, as expected for a fiber–air
interface. However, it gradually begins to increase with time and can exceed 90% after
a few minutes when the Bragg grating is completely formed [5]. Figure 1.2 shows

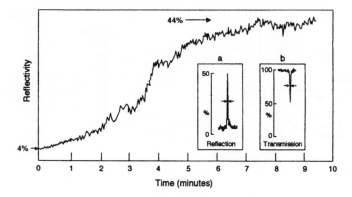

Figure 1.2: Increase in reflectivity with time during grating formation. Insets show the reflection and transmission spectra of the grating. (From Ref. [3]; ©1978 AIP.)

the increase in reflectivity with time, observed in the 1978 experiment for a 1-m-long fiber having a numerical aperture of 0.1 and a core diameter of 2.5 μm. Measured reflectivity of 44% after 8 minutes of exposure implies more than 80% reflectivity of the Bragg grating when coupling losses are accounted for.

Grating formation is initiated by the light reflected from the far end of the fiber and propagating in the backward direction. The two counterpropagating waves interfere and create a standing-wave pattern with periodicity $\lambda/2\bar{n}$, where λ is the laser wavelength and $\bar{n}$ is the mode index at that wavelength. The refractive index of silica is modified locally in the regions of high intensity, resulting in a periodic index variation along the fiber length. Even though the index grating is quite weak initially (4% far-end reflectivity), it reinforces itself through a kind of runaway process. Since the grating period is exactly the same as the standing-wave period, the Bragg condition is satisfied for the laser wavelength. As a result, some forward-traveling light is reflected backward through distributed feedback, which strengthens the grating, which in turn increases feedback. The process stops when the photoinduced index change saturates. Optical fibers with an intracore Bragg grating act as a narrowband reflection filter. The two insets in Figure 1.2 show the measured reflection and transmission spectra of such a fiber grating. The full width at half maximum (FWHM) of these spectra is only about 200 MHz.

A disadvantage of the single-beam internal method is that the grating can be used only near the wavelength of the laser used to make it. Since Ge-doped silica fibers exhibit little photosensitivity at wavelengths longer than 0.5 μm, such gratings cannot be used in the 1.3- to 1.6-μm wavelength region that is important for optical communications. A dual-beam holographic technique, discussed next, solves this problem.

1.2.2   Dual-Beam Holographic Technique

The dual-beam holographic technique, shown schematically in Figure 1.3, makes use of an external interferometric scheme similar to that used for holography. Two optical beams, obtained from the same laser (operating in the ultraviolet region) and making

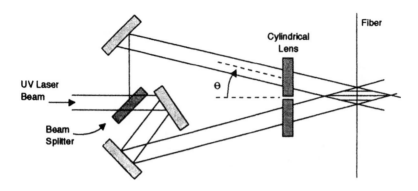

Figure 1.3: The dual-beam holographic technique.

an angle 2θ are made to interfere at the exposed core of an optical fiber [17]. A cylindrical lens is used to expand the beam along the fiber length. Similar to the single-beam scheme, the interference pattern creates an index grating. However, the grating period Λ is related to the ultraviolet laser wavelength λ_{uv} and the angle 2θ made by the two interfering beams through the simple relation

$$\Lambda = \lambda_{uv}/(2\sin\theta). \tag{1.2.1}$$

The most important feature of the holographic technique is that the grating period Λ can be varied over a wide range by simply adjusting the angle θ (see Figure 1.3). The wavelength λ at which the grating reflects light is related to Λ as $\lambda = 2\bar{n}\Lambda$. Since λ can be significantly larger than λ_{uv}, Bragg gratings operating in the visible or infrared region can be fabricated by the dual-beam holographic method even when λ_{uv} is in the ultraviolet region. In a 1989 experiment, Bragg gratings reflecting 580-nm light were made by exposing the 4.4-mm-long core region of a photosensitive fiber for 5 minutes with 244-nm ultraviolet radiation [17]. Reflectivity measurements indicated that the refractive index changes were $\sim 10^{-5}$ in the bright regions of the interference pattern. Bragg gratings formed by the dual-beam holographic technique were stable and remained unchanged even when the fiber was heated to 500°C.

Because of their practical importance, Bragg gratings operating in the 1.55-μm region were made in 1990 [18]. Since then, several variations of the basic technique have been used to make such gratings in a practical manner. An inherent problem for the dual-beam holographic technique is that it requires an ultraviolet laser with excellent temporal and spatial coherence. Excimer lasers commonly used for this purpose have relatively poor beam quality and require special care to maintain the interference pattern over the fiber core over a duration of several minutes.

It turns out that high-reflectivity fiber gratings can be written by using a single excimer laser pulse (with typical duration of 20 ns) if the pulse energy is large enough [32]–[34]. Extensive measurements on gratings made by this technique indicate a thresholdlike phenomenon near a pulse energy level of about 35 mJ [34]. For lower pulse energies, the grating is relatively weak since index changes are only about 10^{-5}. By contrast, index changes of about 10^{-3} are possible for pulse energies above 40 mJ.

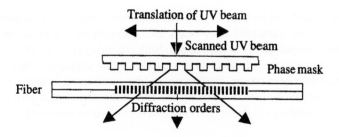

Figure 1.4: A phase-mask interferometer used for making fiber gratings. (From Ref. [48]; ©1999 Academic Press.)

Bragg gratings with nearly 100% reflectivity have been made by using a single 40-mJ pulse at the 248-nm wavelength. The gratings remained stable at temperatures as high as 800°C. A short exposure time has an added advantage. The typical rate at which a fiber is drawn from a preform is about 1 m/s. Since the fiber moves only 20 nm in 20 ns and since displacement is a small fraction of the grating period Λ, a grating can be written during the drawing stage, while the fiber is being pulled and before it is sleeved [35]. This feature makes the single-pulse holographic technique quite useful from a practical standpoint.

1.2.3   Phase-Mask Technique

This nonholographic technique uses a photolithographic process commonly employed for fabrication of integrated electronic circuits. The basic idea is to use a phase mask with a periodicity related to the grating period [36]. The phase mask acts as a master grating that is transferred to the fiber using a suitable method. In one realization of this technique [37], the phase mask was made on a quartz substrate on which a patterned layer of chromium was deposited using electron-beam lithography in combination with reactive-ion etching. Phase variations induced in the 242-nm radiation passing through the phase mask translate into a periodic intensity pattern similar to that produced by the holographic technique. The photosensitivity of the fiber converts intensity variations into an index grating of the same periodicity as that of the phase mask.

The chief advantage of the phase-mask method is that the demands on the temporal and spatial coherence of the ultraviolet beam are much less stringent because of the noninterferometric nature of the technique. In fact, even a nonlaser source such as an ultraviolet lamp can be used. Furthermore, the phase-mask technique allows fabrication of fiber gratings with a variable period (chirped gratings) and can be used to tailor the periodic index profile along the grating length. It is also possible to vary the Bragg wavelength over some range for a fixed mask periodicity by using a converging or diverging wavefront during the photolithographic process [41]. On the other hand, the quality of fiber grating (length, uniformity, etc.) depends completely on the master phase mask, and all imperfections are reproduced precisely. Nonetheless, gratings with 5-mm length and 94% reflectivity were made in 1993, showing the potential of this technique [37].

The phase mask can be used to form an interferometer using the geometry shown in Figure 1.4. The ultraviolet laser beam falls normally on the phase mask and is diffracted into several beams in the Raman–Nath scattering regime. The zeroth-order beam (direct transmission) is blocked or canceled by an appropriate technique. The two first-order diffracted beams interfere on the fiber surface and form a periodic intensity pattern. The grating period is exactly one half of the phase-mask period. In effect, the phase mask produces both the reference and object beams required for holographic recording.

There are several advantages of using a phase-mask interferometer. It is insensitive to the lateral translation of the incident laser beam and tolerant of any beam-pointing instability. Relatively long fiber gratings can be made by moving two side mirrors while maintaining their mutual separation. In fact, the two mirrors can be replaced by a single silica block that reflects the two beams internally through total internal reflection, resulting in a compact and stable interferometer [48]. The length of the grating formed inside the fiber core is limited by the size and optical quality of the silica block.

Long gratings can be formed by scanning the phase mask or translating the optical fiber itself such that different parts of the optical fiber are exposed to the two interfering beams. In this way, multiple short gratings are formed in succession in the same fiber. Any discontinuity or overlap between the two neighboring gratings, resulting from positional inaccuracies, leads to the so-called stitching errors (also called *phase errors*) that can affect the quality of the whole grating substantially if left uncontrolled. Nevertheless, this technique was used in 1993 to produce a 5-cm-long grating [42]. By 1996, gratings longer than 1 meter have been made with success [62] by employing techniques that minimize phase errors [63].

1.2.4 Point-by-Point Fabrication Technique

This nonholographic scanning technique bypasses the need for a master phase mask and fabricates the grating directly on the fiber, period by period, by exposing short sections of width w to a single high-energy pulse [19]. The fiber is translated by a distance $\Lambda - w$ before the next pulse arrives, resulting in a periodic index pattern such that only a fraction w/Λ in each period has a higher refractive index. The method is referred to as *point-by-point fabrication* since a grating is fabricated period by period even though the period Λ is typically below 1 μm. The technique works by focusing the spot size of the ultraviolet laser beam so tightly that only a short section of width w is exposed to it. Typically, w is chosen to be $\Lambda/2$ although it could be a different fraction if so desired.

This technique has a few practical limitations. First, only short fiber gratings (<1 cm) are typically produced because of the time-consuming nature of the point-to-point fabrication method. Second, it is hard to control the movement of a translation stage accurately enough to make this scheme practical for long gratings. Third, it is not easy to focus the laser beam to a small spot size that is only a fraction of the grating period. Recall that the period of a first-order grating is about 0.5 μm at 1.55 μm and becomes even smaller at shorter wavelengths. For this reason, the technique was first demonstrated in 1993 by making a 360-μm-long, third-order grating with a 1.59-μm period [38]. The third-order grating still reflected about 70% of the incident 1.55-μm light. From a fundamental standpoint, an optical beam can be focused to a spot size as small as the wavelength. Thus, the 248-nm laser commonly used in grating fabrication should be able to provide a first-order grating in the wavelength range from 1.3 to 1.6 μm with proper focusing optics similar to that used for fabrication of integrated circuits.

The point-by-point fabrication method is quite suitable for long-period gratings in which the grating period exceeds 10 μm and even can be longer than 100 μm, depending on the application [64]–[66]. Such gratings can be used for mode conversion (power transfer from one mode to another) or polarization conversion (power transfer between two orthogonally polarized modes). Their filtering characteristics have been used for flattening the gain profile of erbium-doped fiber amplifiers. Long-period gratings are covered in Section 1.7.1.

1.2.5 Technique Based on Ultrashort Optical Pulses

In recent years, femtosecond pulses have been used to change the refractive index of glass locally and to fabricate planar waveguides within a bulk medium [67]–[72]. The same technique can be used for making fiber gratings. Femtosecond pulses from a Ti:sapphire laser operating in the 800-nm regime were used as early as 1999 [73]–[75]. Two distinct mechanisms can lead to index changes when such lasers are used [76]. In the so-called type-I gratings, index changes are of reversible nature. In contrast, permanent index changes occur in type-II gratings because of multiphoton ionization and plasma formation when the peak power of pulses exceeds the self-focusing threshold. The second type of gratings can be written using energetic femtosecond pulses that illuminate an especially made phase mask [74]. They were observed to be stable at temperatures of up to 1000°C in the sense that the magnitude of index change created by the 800-nm femtosecond pulses remained unchanged over hundreds of hours [75].

In an alternative approach, infrared radiation is first converted into the UV region through harmonic generation, before using it for grating fabrication. In this case, photon energy exceeds 4 eV, and the absorption of single photons can create large index changes. As a result, the energy fluence required for forming the grating is reduced considerably [77]–[79]. In practice, one can employ either 264-nm pulses, obtained from fourth harmonic of a femtosecond Nd:glass laser, or 267-nm pulses using third harmonic of a Ti:sapphire laser. In both cases, index changes $>10^{-3}$ have been realized. Figure 1.5 shows the experimental results obtained when 264-nm pulses of 0.2-nJ energy (pulse width 260 fs) were employed for illuminating a phase mask and forming a 3-mm-long Bragg grating [77]. The left part shows the measured UV-induced change in the refractive index of fiber core as a function of incident energy fluence for (a) a hydrogen-soaked fiber and (b) a hydrogen-free fiber. The transmission spectra of three fiber gratings are shown for fluence values that correspond to the maximum fluence level for the three peak intensities. The topmost spectrum implies a peak reflectivity level of $>99.9\%$ at the Bragg wavelength and corresponds to a UV-induced change in the refractive index of about 2×10^{-3}. This value was lower for the fiber that was not soaked in hydrogen, but it could be made to exceed 10^{-3} by increasing both the peak-intensity and fluence levels of UV pulses. Similar results were obtained when

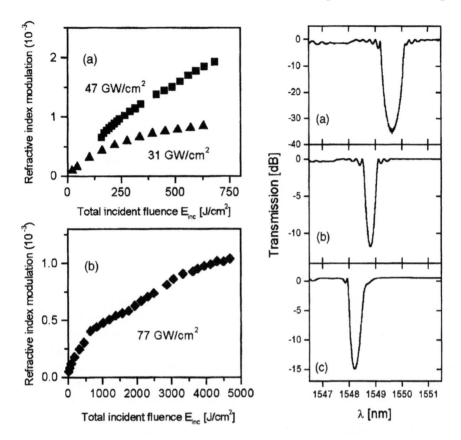

Figure 1.5: Left: Index change Δn as a function of incident energy fluence for (a) a hydrogen-soaked fiber and (b) a hydrogen-free fiber. Right: Transmission spectra of three fiber gratings for fluence values that correspond to the maximum fluence level at peak intensities of (a) 47 GW/cm^2, (b) 31 GW/cm^2, and (c) 77 GW/cm^2. (From Ref. [77]; ©2003 OSA.)

267-nm pulses, obtained through third harmonic of a 800-nm Ti:sapphire laser, were employed [78]. Gratings formed with this method are of type-I type in the sense that the magnitude of index change decreases with annealing at high temperatures [79].

1.3 Grating Characteristics

Two approaches have been used to study how a Bragg grating affects wave propagation in optical fibers. In one approach, Bloch formalism, used commonly for describing motion of electrons in semiconductors, is applied to Bragg gratings [80]. In another, forward- and backward-propagating waves are treated independently, and the Bragg grating provides a coupling between them. This method, known as the *coupled-mode theory*, has been used with considerable success in several contexts. In this section, we derive the nonlinear coupled-mode equations and use them to discuss propagation of low-intensity CW light through a Bragg grating. We also introduce the concept of photonic bandgap and use it to show that a Bragg grating introduces a large amount of dispersion.

1.3.1 Coupled-Mode Equations

Wave propagation in a linear periodic medium has been studied extensively using coupled-mode theory [81]–[83]. This theory has been applied to distributed-feedback (DFB) semiconductor lasers [84], among other things. In the case of optical fibers, we need to include both the nonlinear nature and the periodic variation of the refractive index by using

$$\tilde{n}(\omega, z) = \bar{n}(\omega) + n_2|E|^2 + \delta n_g(z), \tag{1.3.1}$$

where n_2 is the nonlinear parameter and $\delta n_g(z)$ accounts for periodic index variations inside the grating. The coupled-mode theory can be generalized to include the fiber nonlinearity since the nonlinear index change $n_2|E|^2$ in Eq. (1.3.1) is so small that it can be treated as a perturbation [85].

The starting point consists of solving Maxwell's equations with the refractive index given in Eq. (1.3.1). However, as discussed in Section 2.3 of Ref. [2], if the nonlinear effects are relatively weak, we can work in the frequency domain and solve the Helmholtz equation

$$\nabla^2 \tilde{E} + \tilde{n}^2(\omega, z)\omega^2/c^2 \tilde{E} = 0, \tag{1.3.2}$$

where $\tilde{E}$ denotes the Fourier transform of the electric field with respect to time.

Noting that $\tilde{n}$ is a periodic function of z, it is useful to expand $\delta n_g(z)$ in a Fourier series as

$$\delta n_g(z) = \sum_{m=-\infty}^{\infty} \delta n_m \exp[2\pi i m(z/\Lambda)]. \tag{1.3.3}$$

Since both the forward- and backward-propagating waves should be included, $\tilde{E}$ in Eq. (1.3.2) is of the form

$$\tilde{E}(\mathbf{r}, \omega) = F(x, y)[\tilde{A}_f(z, \omega)\exp(i\beta_B z) + \tilde{A}_b(z, \omega)\exp(-i\beta_B z)], \tag{1.3.4}$$

where $\beta_B = \pi/\Lambda$ is the Bragg wave number for a first-order grating. It is related to the Bragg wavelength through the Bragg condition $\lambda_B = 2\bar{n}\Lambda$ and can be used to define the Bragg frequency as $\omega_B = \pi c/(\bar{n}\Lambda)$. Transverse variations for the two counterpropagating waves are governed by the same modal distribution $F(x, y)$ in a single-mode fiber.

Using Eqs. (1.3.1)–(1.3.4), assuming $\tilde{A}_f$ and $\tilde{A}_b$ vary slowly with z and keeping only the nearly phase-matched terms, the frequency-domain coupled-mode equations become [81]–[83]

$$\frac{\partial \tilde{A}_f}{\partial z} = i[\delta(\omega) + \Delta\beta]\tilde{A}_f + i\kappa\tilde{A}_b, \tag{1.3.5}$$

$$-\frac{\partial \tilde{A}_b}{\partial z} = i[\delta(\omega) + \Delta\beta]\tilde{A}_b + i\kappa\tilde{A}_f, \tag{1.3.6}$$

where δ, a measure of detuning from the Bragg frequency, is defined as

$$\delta(\omega) = (\bar{n}/c)(\omega - \omega_B) \equiv \beta(\omega) - \beta_B. \tag{1.3.7}$$

The nonlinear effects in the coupled-mode eqautions are included through $\Delta\beta$. The coupling coefficient κ governs the grating-induced coupling between the forward and backward waves. For a first-order grating, κ is given by

$$\kappa = \frac{k_0 \iint_{-\infty}^{\infty} \delta n_1 |F(x,y)|^2 \, dx \, dy}{\iint_{-\infty}^{\infty} |F(x,y)|^2 \, dx \, dy}. \tag{1.3.8}$$

In this general form, κ can include transverse variations of δn_g occurring when the photoinduced index change is not uniform over the core area. For a transversely uniform grating $\kappa = 2\pi \delta n_1 / \lambda$, as can be inferred from Eq. (1.3.8) by taking δn_1 as constant and using $k_0 = 2\pi/\lambda$. For a sinusoidal grating of the form $\delta n_g = n_a \cos(2\pi z/\Lambda)$, $\delta n_1 = n_a/2$ and the coupling coefficient is given by $\kappa = \pi n_a/\lambda$.

Equations (1.3.5) and (1.3.6) can be converted to time domain by following the procedure outlined in Section 2.3 of Ref. [2]. We assume that the total electric field can be written as

$$E(\mathbf{r},t) = \tfrac{1}{2} F(x,y)[A_f(z,t)e^{i\beta_B z} + A_b(z,t)e^{-i\beta_B z}]e^{-i\omega_0 t} + \text{c.c.}, \tag{1.3.9}$$

where ω_0 is the frequency at which the pulse spectrum is centered. We expand $\beta(\omega)$ in Eq. (1.3.7) in a Taylor series as

$$\beta(\omega) = \beta_0 + (\omega - \omega_0)\beta_1 + \tfrac{1}{2}(\omega - \omega_0)^2 \beta_2 + \tfrac{1}{6}(\omega - \omega_0)^3 \beta_3 + \cdots \tag{1.3.10}$$

and retain terms up to second order in $\omega - \omega_0$. The resulting equations are converted into the time domain by replacing $\omega - \omega_0$ with the differential operator $i(\partial/\partial t)$. The resulting time-domain coupled-mode equations have the form

$$\frac{\partial A_f}{\partial z} + \beta_1 \frac{\partial A_f}{\partial t} + \frac{i\beta_2}{2} \frac{\partial^2 A_f}{\partial t^2} + \frac{\alpha}{2} A_f = i\delta A_f + i\kappa A_b + i\gamma(|A_f|^2 + 2|A_b|^2)A_f, \tag{1.3.11}$$

$$-\frac{\partial A_b}{\partial z} + \beta_1 \frac{\partial A_b}{\partial t} + \frac{i\beta_2}{2} \frac{\partial^2 A_b}{\partial t^2} + \frac{\alpha}{2} A_b = i\delta A_b + i\kappa A_f + i\gamma(|A_b|^2 + 2|A_f|^2)A_b, \tag{1.3.12}$$

where δ in Eq. (1.3.7) is evaluated at $\omega = \omega_0$ and becomes $\delta = (\omega_0 - \omega_B)/v_g$. In fact, the δ term can be eliminated from the coupled-mode equations if ω_0 is replaced by ω_B in Eq. (1.3.9). The other parameters have their traditional meaning. Specifically, $\beta_1 \equiv 1/v_g$ is related inversely to the group velocity, β_2 governs the group-velocity dispersion (GVD), and the nonlinear parameter γ is related to n_2 as $\gamma = n_2 \omega_0/(cA_{\text{eff}})$, where A_{eff} is the effective mode area (see Ref. [2]).

The nonlinear terms in the time-domain coupled-mode equations contain the contributions of both self-phase modulation (SPM) and cross-phase modulation (XPM). In fact, the coupled-mode equations are similar to and should be compared with Eqs. (7.1.15) and (7.1.16) of Ref. [2], which govern the propagation of two copropagating waves inside optical fibers. The two major differences are (i) the negative sign appearing in front of the $\partial A_b/\partial z$ term in Eq. (1.3.11) because of backward propagation of A_b and (ii) the presence of linear coupling between the counterpropagating waves governed by the parameter κ. Both these differences change the character of wave propagation profoundly. Before discussing the general case, it is instructive to consider the case in which the nonlinear effects are so weak that the fiber acts as a linear medium.

1.3.2 CW Solution in the Linear Case

In this section, we focus on the linear case in which the nonlinear effects are negligible. When the SPM and XPM terms are neglected in Eqs. (1.3.11) and (1.3.12), the resulting linear equations can be solved easily in the Fourier domain. In fact, we can use Eqs. (1.3.5) and (1.3.6). These frequency-domain coupled-mode equations include GVD to all orders. After setting the nonlinear contribution $\Delta\beta$ to zero, we obtain

$$\frac{\partial \tilde{A}_f}{\partial z} = i\delta\tilde{A}_f + i\kappa\tilde{A}_b, \tag{1.3.13}$$

$$-\frac{\partial \tilde{A}_b}{\partial z} = i\delta\tilde{A}_b + i\kappa\tilde{A}_f, \tag{1.3.14}$$

where $\delta(\omega)$ is given in Eq. (1.3.7).

A general solution of these linear equations takes the form

$$\tilde{A}_f(z) = A_1 \exp(iqz) + A_2 \exp(-iqz), \tag{1.3.15}$$

$$\tilde{A}_b(z) = B_1 \exp(iqz) + B_2 \exp(-iqz), \tag{1.3.16}$$

where q is to be determined. The constants A_1, A_2, B_1, and B_2 are interdependent and satisfy the following four relations:

$$(q-\delta)A_1 = \kappa B_1, \qquad (q+\delta)B_1 = -\kappa A_1, \tag{1.3.17}$$

$$(q-\delta)B_2 = \kappa A_2, \qquad (q+\delta)A_2 = -\kappa B_2. \tag{1.3.18}$$

These equations are satisfied for nonzero values of A_1, A_2, B_1, and B_2 if the possible values of q obey the *dispersion relation*

$$q(\omega) = \pm\sqrt{\delta^2(\omega) - \kappa^2}. \tag{1.3.19}$$

This equation is of paramount importance for gratings. Its implications will become clear soon.

One can eliminate A_2 and B_1 by using Eqs. (1.3.15)–(1.3.18) and write the general solution in terms of an effective reflection coefficient $r(q)$ as

$$\tilde{A}_f(z) = A_1 \exp(iqz) + r(q)B_2 \exp(-iqz), \tag{1.3.20}$$

$$\tilde{A}_b(z) = B_2 \exp(-iqz) + r(q)A_1 \exp(iqz), \tag{1.3.21}$$

where

$$r(q) = \frac{q-\delta}{\kappa} = -\frac{\kappa}{q+\delta}. \tag{1.3.22}$$

The q dependence of r and the dispersion relation (1.3.19) indicate that both the magnitude and the phase of backward reflection depend on the frequency ω. The sign ambiguity in Eq. (1.3.19) can be resolved by choosing the sign of q such that $|r(q)| < 1$.

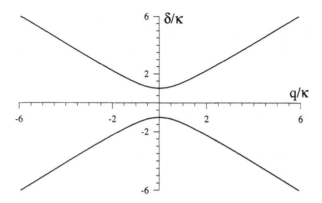

Figure 1.6: Dispersion curves showing variation of δ with q and the existence of the photonic bandgap for a fiber grating.

1.3.3 Photonic Bandgap

The dispersion relation of Bragg gratings exhibits an important property seen clearly in Figure 1.6, where Eq. (1.3.19) is plotted. If the frequency detuning δ of the incident light falls in the range $-\kappa < \delta < \kappa$, q becomes purely imaginary. Most of the incident field is reflected in that case since the grating does not support a propagating wave. The range $|\delta| \leq \kappa$ is referred to as the *photonic bandgap*, in analogy with the electronic energy bands occurring in crystals. It is also called the *stop band*, since light stops transmitting through the grating when its frequency falls within the photonic bandgap.

Consider now what happens to an optical pulse propagating inside a fiber grating such that its carrier frequency ω_0 lies outside the stop band but remains close to a band edge. It follows from Eqs. (1.3.4) and (1.3.15) that the effective propagation constant of the forward-propagating wave is $\beta_e(\omega) = \beta_B + q(\omega)$, where $q(\omega)$ is given by Eq. (1.3.19). The frequency dependence of β_e indicates that a grating exhibits dispersive effects even if it was fabricated in a nondispersive medium. In optical fibers, grating-induced dispersion adds to the material and waveguide dispersions. In fact, the contribution of grating dominates among all sources responsible for dispersion. To see this more clearly, we expand β_e in a Taylor series, in a way similar to Eq. (1.3.10), around the carrier frequency ω_0 of the pulse. The result is given by

$$\beta_e(\omega) = \beta_0^g + (\omega - \omega_0)\beta_1^g + \tfrac{1}{2}(\omega - \omega_0)^2\beta_2^g + \tfrac{1}{6}(\omega - \omega_0)^3\beta_3^g + \cdots, \qquad (1.3.23)$$

where β_m^g $(m = 1, 2, \ldots)$ is defined as

$$\beta_m^g = \frac{d^m q}{d\omega^m} \approx \left(\frac{1}{v_g}\right)^m \frac{d^m q}{d\delta^m}, \qquad (1.3.24)$$

and the derivatives are evaluated at $\omega = \omega_0$. The superscript g denotes that the dispersive effects have their origin in the grating. In Eq. (1.3.24), v_g is the group velocity of pulse in the absence of the grating ($\kappa = 0$). It occurs naturally when the frequency dependence of $\bar{n}$ is taken into account in Eq. (1.3.7). Dispersion of v_g is neglected in Eq. (1.3.24) but can be included easily.

Consider first the group velocity of the pulse inside the grating. Using $V_G = 1/\beta_1^g$ and Eq. (1.3.24), it is given by

$$V_G = \pm v_g \sqrt{1 - \kappa^2/\delta^2}, \tag{1.3.25}$$

where the choice of $\pm$ signs depends on whether the pulse is moving in the forward or backward direction. Far from the band edges ($|\delta| \gg \kappa$), the optical pulse is unaffected by the grating and travels at the group velocity expected in the absence of the grating. However, as $|\delta|$ approaches κ, the group velocity decreases and becomes zero at the two edges of the stop band, where $|\delta| = \kappa$. Therefore, close to the photonic bandgap, an optical pulse experiences considerable slowing down inside a fiber grating. As an example, its speed is reduced by 50% when $|\delta|/\kappa \approx 1.18$.

Second- and third-order dispersive properties of the grating are governed by β_2^g and β_3^g, respectively. Using Eq. (1.3.24) together with the dispersion relation, these parameters are given by

$$\beta_2^g = -\frac{\mathrm{sgn}(\delta)\kappa^2/v_g^2}{(\delta^2 - \kappa^2)^{3/2}}, \qquad \beta_3^g = \frac{3|\delta|\kappa^2/v_g^3}{(\delta^2 - \kappa^2)^{5/2}}. \tag{1.3.26}$$

The grating-induced GVD, governed by the parameter β_2^g, depends on the sign of detuning δ. Figure 1.7 shows how β_2^g and β_3^g vary with δ for three gratings for which κ is in the range of 1 to 10 cm^{-1}. The GVD is anomalous on the upper branch of the dispersion curve in Figure 1.6, where δ is positive and the carrier frequency exceeds the Bragg frequency. In contrast, GVD becomes normal ($\beta_2^g > 0$) on the lower branch of the dispersion curve, where δ is negative and the carrier frequency is smaller than the Bragg frequency. The third-order dispersion remains positive on both branches of the dispersion curve. Also note that both β_2^g and β_3^g become infinitely large at the two edges of the stop band.

The dispersive properties of a fiber grating are quite different than those of a uniform fiber. First, β_2^g changes sign on the two sides of the stop band centered at the Bragg wavelength, whose location is easily controlled and can be in any region of the optical spectrum. This is in sharp contrast with the behavior of β_2 in standard silica fibers, where β_2 changes sign at the zero-dispersion wavelength occurring near 1.3 μm. Second, β_2^g is anomalous on the shorter wavelength side of the stop band, whereas β_2 in conventional fibers becomes anomalous for wavelengths longer than the zero-dispersion wavelength. Third, the magnitude of β_2^g exceeds that of β_2 by a large factor. Figure 1.7 shows that $|\beta_2^g|$ can easily exceed 10^7 ps^2/km for a fiber grating, whereas β_2 is ~10 ps^2/km for standard fibers. This feature can be used for dispersion compensation [86]. Typically, a 10-cm-long grating can compensate the GVD acquired over fiber lengths of 50 km or more. Chirped gratings, discussed in Section 1.7.2, can provide even more dispersion when the wavelength of incident signal falls inside the stop band, although they reflect the dispersion-compensated signal [87].

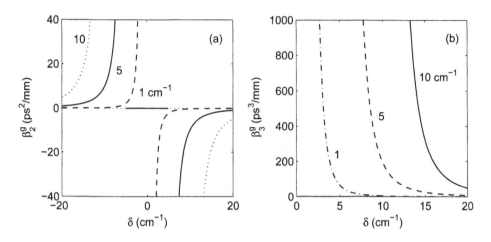

Figure 1.7: Second- and third-order dispersion parameters of a fiber grating as a function of detuning δ for three values of the coupling coefficient κ.

1.3.4   Grating as an Optical Filter

What happens to optical pulses incident on a fiber grating depends very much on the location of the pulse spectrum with respect to the stop band associated with the grating. If the pulse spectrum falls entirely within the stop band, the entire pulse is reflected by the grating. On the other hand, if a part of the pulse spectrum lies outside the stop band, only that part is transmitted through the grating. The shape of the reflected and transmitted pulses in this case becomes quite different than that of the incident pulse because of the splitting of the spectrum and the dispersive properties of the fiber grating. If the peak power of input pulses is small enough that nonlinear effects remain negligible, we can first calculate the reflection and transmission coefficients for each spectral component. The shape of the transmitted and reflected pulses is then obtained by integrating over the spectrum of the incident pulse. Considerable distortion can occur when the pulse spectrum is either wider than the stop band or when it lies in the vicinity of a stop-band edge.

The reflection and transmission coefficients can be calculated by using Eqs. (1.3.20) and (1.3.21) with the appropriate boundary conditions. Consider a grating of length L and assume that light is incident only at the front end, located at $z = 0$. The reflection coefficient is then given by

$$r_g = \frac{\tilde{A}_b(0)}{\tilde{A}_f(0)} = \frac{B_2 + r(q)A_1}{A_1 + r(q)B_2}. \tag{1.3.27}$$

If we use the boundary condition $\tilde{A}_b(L) = 0$ in Eq. (1.3.21), we find

$$B_2 = -r(q)A_1 \exp(2iqL). \tag{1.3.28}$$

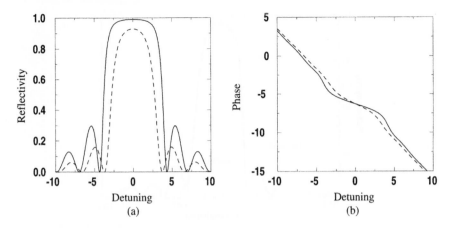

Figure 1.8: (a) Reflectivity $|r_g|^2$ and (b) the phase of r_g plotted as a function of detuning δ for $\kappa L = 2$ (dashed curves) and $\kappa L = 3$ (solid curves).

Using $r(q)$ from Eq. (1.3.22) with this value of B_2 in Eq. (1.3.27), we obtain

$$r_g = \frac{i\kappa \sin(qL)}{q\cos(qL) - i\delta \sin(qL)}. \tag{1.3.29}$$

The transmission coefficient t_g can be obtained in a similar manner. The frequency dependence of r_g and t_g governs the filtering action of a fiber grating.

Figure 1.8 shows the reflectivity $|r_g|^2$ and the phase of r_g as a function of detuning δ for two values of κL. The grating reflectivity within the stop band approaches 100% for $\kappa L = 3$ or larger. Maximum reflectivity occurs at the center of the stop band and, by setting $\delta = 0$ in Eq. (1.3.29), is given by

$$R_{\text{max}} = |r_g|^2 = \tanh^2(\kappa L). \tag{1.3.30}$$

For $\kappa L = 2$, $R_{\text{max}} = 0.93$. The condition $\kappa L > 2$ with $\kappa = 2\pi \delta n_1 / \lambda$ can be used to estimate the grating length required for high reflectivity inside the stop band. For $\delta n_1 \approx 10^{-4}$ and $\lambda = 1.55 \ \mu\text{m}$, L should exceed 5 mm to yield $\kappa L > 2$. These requirements are easily met in practice. Indeed, reflectivities in excess of 99% were achieved for a grating length of 1.5 cm [34].

1.3.5 Experimental Verification

The coupled-mode theory has been quite successful in explaining the observed features of fiber gratings. As an example, Figure 1.9 shows the measured reflectivity spectrum for a Bragg grating operating near 1.3 μm [33]. The fitted curve was calculated using Eq. (1.3.29). The 94% peak reflectivity indicates $\kappa L \approx 2$ for this grating. The stop band is about 1.7-nm wide. These measured values were used to deduce a grating length of 0.84 mm and an index change of 1.2×10^{-3}. The coupled-mode theory explains the observed reflection and transmission spectra of fiber gratings quite well.

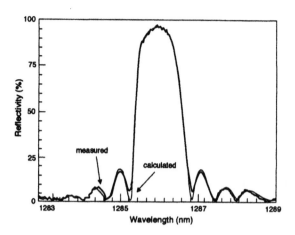

Figure 1.9: Measured and calculated reflectivity spectra for a fiber grating operating at wavelengths near 1.3 μm. (From Ref. [33]); ©1993 IEE.)

From a practical standpoint, an undesirable feature seen in Figures 1.8 and 1.9 is the presence of multiple sidebands located on each side of the stop band. These sidebands originate from weak reflections occurring at the two grating ends where the refractive index changes suddenly compared to its value outside the grating region. Even though the change in refractive index is typically less than 1%, the reflections at the two grating ends form a Fabry–Perot cavity with its own wavelength-dependent transmission. An *apodization technique* is commonly used to remove the sidebands seen in Figures 1.8 and 1.9 [48]. In this technique, the intensity of the ultraviolet laser beam used to form the grating is made nonuniform in such a way that the intensity drops to zero gradually near the two grating ends.

Figure 1.10(a) shows schematically how the refractive index varies along the length of an apodized fiber grating. In a transition region of width L_t near the grating ends, the value of the coupling coefficient κ increases from zero to its maximum value. These buffer zones can suppress the sidebands almost completely, resulting in fiber gratings with practically useful filter characteristics. Figure 1.10(b) shows the measured reflectivity spectrum of a 7.5-cm-long apodized fiber grating, made with the scanning phase-mask technique. The reflectivity exceeds 90% within the stop band, about 0.17-nm wide and centered at the Bragg wavelength of 1.053 μm, chosen to coincide with the wavelength of an Nd:YLF laser [88]. From the stop-band width, the coupling coefficient κ is estimated to be about 7 cm^{-1}. Note the sharp drop in reflectivity at both edges of the stop band and a complete absence of sidebands.

The same apodized fiber grating was used to investigate the dispersive properties in the vicinity of a stop-band edge by transmitting 80-ps pulses (with a nearly Gaussian shape) through it [88]. Figure 1.11 shows changes in (a) the pulse width and (b) the transit time during pulse transmission as a function of the detuning δ from the Bragg wavelength. For positive values of δ, grating-induced GVD is anomalous on the upper branch of the dispersion curve. The most interesting feature is the increase in the arrival time observed as the laser is tuned close to the stop-band edge because of a

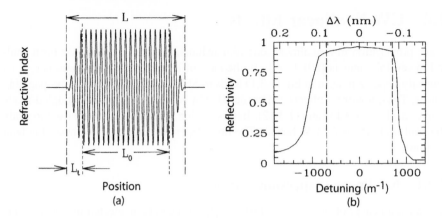

Figure 1.10: (a) Schematic variation of refractive index and (b) measured reflectivity spectrum for an apodized fiber grating. (From Ref. [88]; ©1999 OSA.)

reduced group velocity. Doubling of the arrival time for δ close to 900 m^{-1} shows that the pulse speed was only 50% of that expected in the absence of the grating. This result is in agreement with the prediction of coupled-mode theory in Eq. (1.3.25).

Changes in the pulse width seen in Figure 1.11 can be attributed mostly to the grating-induced GVD effects governed by Eq. (1.3.26). The large broadening observed near the stop-band edge is due to an increase in $|\beta_2^g|$. Slight compression near $\delta = 1200$ m^{-1} is due to a small amount of SPM that chirps the pulse. Indeed, it was necessary to include the γ term in Eqs. (1.3.11) and (1.3.12) to fit the experimental data. The nonlinear effects became quite significant at high power levels. We turn to this issue next.

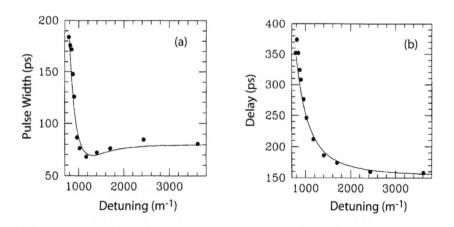

Figure 1.11: (a) Measured pulse width (FWHM) of 80-ps input pulses and (b) their arrival time as a function of detuning δ for an apodized 7.5-cm-long fiber grating. Solid lines show the prediction of the coupled-mode theory. (From Ref. [88]; ©1999 OSA.)

1.4   CW Nonlinear Effects

Wave propagation in a nonlinear, one-dimensional, periodic medium has been studied in several contexts [89]–[109]. In the case of a fiber grating, the presence of an intensity-dependent term in Eq. (1.3.1) leads to SPM and XPM of counterpropagating waves. These nonlinear effects can be included by solving the nonlinear coupled-mode equations, Eqs. (1.3.11) and (1.3.12). In this section, these equations are used to study the nonlinear effects for CW beams. The time-dependent effects are discussed in later sections.

1.4.1   Nonlinear Dispersion Curves

In almost all cases of practical interest, the β_2 term can be neglected in Eqs. (1.3.11) and (1.3.12). For typical grating lengths (<1 m), the loss term can also be neglected by setting $\alpha = 0$. The nonlinear coupled-mode equations then take the following form:

$$i\frac{\partial A_f}{\partial z} + \frac{i}{v_g}\frac{\partial A_f}{\partial t} + \delta A_f + \kappa A_b + \gamma(|A_f|^2 + 2|A_b|^2)A_f = 0, \qquad (1.4.1)$$

$$-i\frac{\partial A_b}{\partial z} + \frac{i}{v_g}\frac{\partial A_b}{\partial t} + \delta A_b + \kappa A_f + \gamma(|A_b|^2 + 2|A_f|^2)A_b = 0, \qquad (1.4.2)$$

where $v_g = 1/\beta_1$ and is the group velocity far from the stop band of the grating. These equations exhibit many interesting nonlinear effects. We begin by considering the CW solution of Eqs. (1.4.1) and (1.4.2) imposing no boundary conditions. Even though this is unrealistic from a practical standpoint, the resulting dispersion curves provide considerable physical insight. Note that all grating-induced dispersive effects are included in these equations through the κ term.

To solve Eqs. (1.4.1) and (1.4.2) in the CW limit, we neglect the time-derivative term and assume the following form for the solution:

$$A_f = u_f \exp(iqz), \qquad A_b = u_b \exp(iqz), \qquad (1.4.3)$$

where u_f and u_b remain constant along the grating length. By introducing a parameter $f = u_b/u_f$ that describes how the total power $P_0 = u_f^2 + u_b^2$ is divided between the forward- and backward-propagating waves, u_f and u_b can be written as

$$u_f = \sqrt{\frac{P_0}{1+f^2}}, \qquad u_b = \sqrt{\frac{P_0}{1+f^2}}\,f. \qquad (1.4.4)$$

The parameter f can be positive or negative. For values of $|f| > 1$, the backward wave dominates. By using Eqs. (1.4.1)–(1.4.4), q and δ are found to depend on f as

$$q = -\frac{\kappa(1-f^2)}{2f} - \frac{\gamma P_0}{2}\frac{1-f^2}{1+f^2}, \qquad \delta = -\frac{\kappa(1+f^2)}{2f} - \frac{3\gamma P_0}{2}. \qquad (1.4.5)$$

To understand the physical meaning of Eq. (1.4.5), let us first consider the low-power case so that nonlinear effects are negligible. If we set $\gamma = 0$ in Eq. (1.4.5),

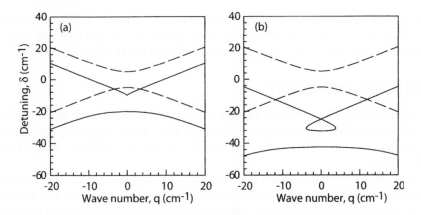

Figure 1.12: Nonlinear dispersion curves showing variation of δ with q for (a) $\gamma P_0/\kappa = 2$ and (b) $\gamma P_0/\kappa = 5$, when $\kappa = 5$ cm^{-1}. Dashed curves show the linear case ($\gamma = 0$).

it is easy to show that $q^2 = \delta^2 - \kappa^2$. This is precisely the dispersion relation (1.3.19) obtained previously. As f changes, q and δ trace the dispersion curves shown in Figure 1.6. In fact, $f < 0$ on the upper branch while positive values of f belong to the lower branch. The two edges of the stop band occur at $f = \pm 1$. From a practical standpoint, the detuning δ of the CW beam from the Bragg frequency determines the value of f, which in turn fixes the values of q from Eq. (1.4.5). The group velocity inside the grating also depends on f and is given by

$$V_G = v_g \frac{d\delta}{dq} = v_g \left(\frac{1-f^2}{1+f^2}\right). \tag{1.4.6}$$

As expected, V_G becomes zero at the edges of the stop band corresponding to $f = \pm 1$. Note that V_G becomes negative for $|f| > 1$. This is not surprising if we note that the backward-propagating wave is more intense in that case. The speed of light is reduced considerably as the CW-beam frequency approaches an edge of the stop band. As an example, it reduces by 50% when f^2 equals 1/3 or 3.

Equation (1.4.5) can be used to find how the dispersion curves are affected by the fiber nonlinearity. Figure 1.12 shows such curves at two power levels. The nonlinear effects change the upper branch of the dispersion curve qualitatively, leading to the formation a loop beyond a critical power level. This critical value of P_0 can be found by looking for the value of f at which q becomes zero while $|f| \neq 1$. From Eq. (1.4.5), we find that this can occur when

$$f \equiv f_c = -(\gamma P_0/2\kappa) + \sqrt{(\gamma P_0/2\kappa)^2 - 1}. \tag{1.4.7}$$

Therefore, a loop is formed only on the upper branch, where $f < 0$. Moreover, it can form only when the total power $P_0 > P_c$, where $P_c = 2\kappa/\gamma$. Physically, an increase in the mode index through the nonlinear term in Eq. (1.3.1) increases the Bragg wavelength and shifts the stop band toward lower frequencies. Since the amount of shift depends on the total power P_0, light at a frequency close to the edge of the upper branch can be shifted out of resonance with changes in its power. If the nonlinear parameter γ were negative (self-defocusing medium with $n_2 < 0$), the loop will form on the lower branch in Figure 1.12, as is also evident from Eq. (1.4.7).

1.4.2 Optical Bistability

The simple CW solution given in Eq. (1.4.3) is modified considerably when boundary conditions are introduced at the two grating ends. For a finite-size grating, the simplest manifestation of the nonlinear effects occurs through optical bistability, first predicted in 1979 [89].

Consider a CW beam incident at one end of the grating and ask how the fiber nonlinearity would affect its transmission through the grating. It is clear that both the beam intensity and its wavelength with respect to the stop band plays an important role. Mathematically, we should solve Eqs. (1.4.1) and (1.4.2) after imposing the appropriate boundary conditions at $z = 0$ and $z = L$. These equations are similar to those found in Section 6.3 of Ref. [2] and can be solved in terms of the elliptic functions by using the same technique used there [89]. Using $A_j = \sqrt{P_j}\exp(i\phi_j)$ and separating the real and imaginary parts, Eqs. (1.4.1) and (1.4.2) lead to the following three equations:

$$\frac{dP_f}{dz} = 2\kappa\sqrt{P_f P_b}\,\sin\psi, \tag{1.4.8}$$

$$\frac{dP_b}{dz} = 2\kappa\sqrt{P_f P_b}\,\sin\psi, \tag{1.4.9}$$

$$\frac{d\psi}{dz} = 2\delta + 3\gamma(P_f + P_b) + \frac{P_f + P_b}{(P_f P_b)^{1/2}}\kappa\cos\psi, \tag{1.4.10}$$

where ψ represents the phase difference $\phi_f - \phi_b$.

It turns out that the preceding equations have the following two constants of motion [104]:

$$P_f(z) - P_b(z) = P_t, \qquad \sqrt{P_f P_b}\cos\psi + (2\delta + 3\gamma P_b)P_f/(2\kappa) = \Gamma_0, \tag{1.4.11}$$

where P_t is the transmitted power and Γ_0 is a constant. Using them, we can derive a differential equation for P_f that can be solved in terms of the elliptic functions. The use of the boundary condition $P_b(0) = 0$ then allows us to obtain an implicit relation for the transmitted power P_t at $z = L$ as a function of the incident power P_i for a grating of finite length L. The reader should consult Ref. [101] for further details.

Figure 1.13 shows the transmitted versus incident power, both normalized to a critical power $P_{cr} = 4/(3\gamma L)$, for several values of detuning within the stop band by taking $\kappa L = 2$. The S-shaped curves are well known in the context of optical bistability occurring when a nonlinear medium is placed inside a cavity [110]. In fact, the middle branch of these curves with negative slope is unstable, and the transmitted power exhibits hysteresis, as indicated by the arrows on the solid curve. At low powers, transmittivity is small, as expected from the linear theory since the nonlinear effects are relatively weak. However, above a certain input power, most of the incident power is transmitted. Switching from a low-to-high transmission state can be understood qualitatively by noting that the effective detuning δ in Eqs. (1.4.1) and (1.4.2) becomes

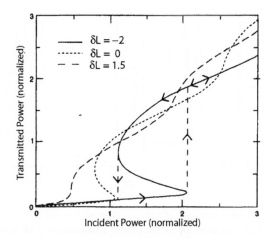

Figure 1.13: Transmitted versus incident powers for three values of detuning within the stop band. (From Ref. [89]; ©1979 AIP.)

power dependent because of the nonlinear contribution to the refractive index in Eq. (1.3.1). Thus, light that is mostly reflected at low powers, because its wavelength is inside the stop band, may tune itself out of the stop band and get transmitted when the nonlinear index change becomes large enough.

The observation of optical bistability in fiber gratings is hampered by the large switching power required ($P_0 > P_{cr} > 1$ kW). It turns out that the switching power can be reduced by a factor of 100 or more by introducing a $\pi/2$ phase shift in the middle of the fiber grating. Such gratings are called $\lambda/4$-shifted or phase-shifted gratings since a distance of $\lambda/4$ (half grating period) corresponds to a $\pi/2$ phase shift. They are used routinely for making DFB semiconductor lasers [84]. Their use for fiber gratings was suggested in 1994 [111]. The $\pi/2$ phase shift opens a narrow transmission window within the stop band of the grating. Figure 1.14(a) compares the transmission spectra for the uniform and phase-shifted gratings at low powers. At high powers, the central peak bends to the left, as seen in the traces in Figure 1.14(b). This bending leads to low-threshold optical switching in phase-shifted fiber gratings [104]. The elliptic-function solution of uniform gratings can be used to construct the multivalued solution for a $\lambda/4$-shifted grating [105]. The presence of a phase-shifted region lowers the switching power considerably.

The bistable switching does not always lead to a constant output power when a CW beam is transmitted through a grating. As early as 1982, numerical solutions of Eqs. (1.4.1) and (1.4.2) showed that transmitted power can become not only periodic but also chaotic under certain conditions [90]. In physical terms, portions of the upper branch in Figure 1.13 become unstable under certain conditions. As a result, the output becomes periodic or chaotic once the beam intensity exceeds the switching threshold. This behavior has been observed experimentally and is discussed in Section 1.6. In the following section, we turn to another instability that occurs even when the CW beam is tuned outside the stop band and does not exhibit optical bistability.

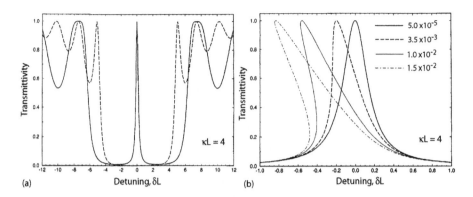

Figure 1.14: (a) Transmission spectrum of a fiber grating with (solid curve) and without (dashed curve) a $\pi/2$ phase shift. (b) Bending of the central transmission peak with increasing input power normalized to the critical power. (From Ref. [104]; ©1995 OSA.)

1.5 Modulation Instability

The stability issue is of paramount importance and must be addressed for the CW solutions obtained in the previous section. Similar to the situation discussed in Section 5.1 of Ref. [2], modulation instability can destabilize the steady-state solution and produce periodic output, even when a CW beam is incident on one end of the fiber grating [112]–[118]. Moreover, the repetition rate of pulse trains generated through modulation instability can be tuned over a large range because of large GVD changes occurring with the detuning δ.

1.5.1 Linear Stability Analysis

For simplicity, we discuss modulation instability using the CW solution given in Eqs. (1.4.3) and (1.4.4) and obtained without imposing the boundary conditions at the grating ends. Following the usual approach [2], we perturb the steady state slightly as

$$A_f = (u_f + a_f)e^{iqz}, \qquad A_b = (u_b + a_b)e^{iqz}, \qquad (1.5.1)$$

and linearize Eqs. (1.4.1) and (1.4.2), assuming that the perturbations a_f and a_b are small. The resulting equations are [117]

$$i\frac{\partial a_f}{\partial z} + \frac{i}{v_g}\frac{\partial a_f}{\partial t} + \kappa a_b - \kappa f a_f + \Gamma[(a_f + a_f^*) + 2f(a_b + a_b^*)] = 0, \quad (1.5.2)$$

$$-i\frac{\partial a_b}{\partial z} + \frac{i}{v_g}\frac{\partial a_b}{\partial t} + \kappa a_f - \frac{\kappa}{f}a_b + \Gamma[2f(a_f + a_f^*) + f^2(a_b + a_b^*)] = 0, \quad (1.5.3)$$

where $\Gamma = \gamma P_0/(1 + f^2)$ is an effective nonlinear parameter.

The preceding set of two linear coupled equations can be solved by assuming a plane-wave solution of the form

$$a_j = c_j \exp[i(Kz - \Omega t)] + d_j \exp[-i(Kz + \Omega t)], \qquad (1.5.4)$$

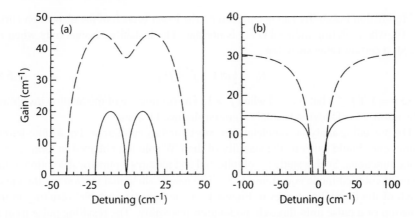

Figure 1.15: Gain spectra of modulation instability in the (a) anomalous- and (b) normal-GVD regions of a fiber grating ($f = \pm 0.5$) at two power levels corresponding to $\Gamma/\kappa = 0.5$ and 2.

where the subscript $j = f$ or b. From Eqs. (1.5.2)–(1.5.4), we obtain a set of four homogeneous equations satisfied by c_j and d_j. This set has a nontrivial solution only when the 4×4 determinant formed by the coefficients matrix vanishes. This condition leads to the the the following fourth-order polynomial:

$$(s^2 - K^2)^2 - 2\kappa^2(s^2 - K^2) - \kappa^2 f^2(s + K)^2$$
$$- \kappa^2 f^{-2}(s - K)^2 - 4\kappa\Gamma f(s^2 - 3K^2) = 0, \qquad (1.5.5)$$

where we have introduced a spatial frequency as $s = \Omega/v_g$.

The four roots of the polynomial in Eq. (1.5.5) determine the stability of the CW solution. However, a tricky issue must be resolved first. Equation (1.5.5) is a fourth-order polynomial in both s and K. The question is, which one determines the gain associated with modulation instability? In the case of the uniform-index fibers discussed in Section 5.1 of Ref. [2], the gain g is related to the imaginary part of K as light propagates only in the forward direction. In a fiber grating, light travels both forward and backward simultaneously, and it is time that moves forward for both of them. As a result, Eq. (1.5.5) should be viewed as a fourth-order polynomial in s whose roots depend on K. The gain of modulation instability is obtained using $g = 2\text{Im}(s_m)$, where s_m is the root with the largest imaginary part.

The root analysis of the polynomial in Eq. (1.5.5) leads to several interesting conclusions [117]. Figure 1.15 shows the gain spectra of modulation instability in the anomalous- and normal-GVD regions, corresponding to upper and lower branches of the dispersion curves, for two values of Γ/κ. In the anomalous-GVD case and at relatively low powers ($\Gamma < \kappa$), the gain spectrum is similar to that found for uniform-index fibers. As shown later in this section, the nonlinear coupled-mode equations reduce to a nonlinear Schrödinger (NLS) equation when $\Gamma \ll \kappa$. At high values of P_0 such that $\Gamma > \kappa$, the gain exists even at $s = 0$, as seen in Figure 1.15(a) for $\Gamma/\kappa = 2$. Thus, the CW solution becomes unstable at high power levels even to zero-frequency (dc) fluctuations.

Modulation instability can occur even on the lower branch of the dispersion curve ($f > 0$), where grating-induced GVD is normal. The instability occurs only when P_0 exceeds a certain value such that

$$P_0 > \tfrac{1}{2}\kappa(1+f^2)^2 f^p, \tag{1.5.6}$$

where $p = 1$ if $f \leq 1$ but $p = -3$ when $f > 1$. The occurrence of modulation instability in the normal-GVD region is solely a grating-induced feature.

The preceding analysis completely ignores boundary conditions. For a finite-length grating, one should examine the stability of the CW solution obtained in terms of the elliptic functions. Such a study is complicated and requires a numerical solution to the nonlinear coupled-mode equations [113]. The results show that portions of the upper branch of the bistability curves in Figure 1.13 can become unstable, resulting in the formation of a pulse train through modulation instability. The resulting pulse train is not necessarily periodic and, under certain conditions, can exhibit period doubling and optical chaos.

1.5.2   Effective NLS Equation

The similarity of the gain spectrum in Figure 1.15 with that occurring in uniform-index fibers indicates that, at not-too-high power levels, the nonlinear coupled-mode equations predict features that coincide with those found for the NLS equation. Indeed, under certain conditions, Eqs. (1.4.3) and (1.4.4) can be reduced formally to an effective NLS equation [119]–[123]. A multiple-scale method is commonly used to prove this equivalence; details can be found in Ref. [101].

The analysis used to reduce the nonlinear coupled-mode equations to an effective NLS equation makes use of the Bloch formalism, well known in solid-state physics. Even in the absence of nonlinear effects, the eigenfunctions associated with the photonic bands, corresponding to the dispersion relation $q^2 = \delta^2 - \kappa^2$, are not A_f and A_b but the Bloch waves formed by a linear combination of A_f and A_b. If this basis is used for the nonlinear problem, Eqs. (1.4.3) and (1.4.4) reduce to an effective NLS equation, provided two conditions are met. First, the peak intensity of the pulse is small enough that the nonlinear index change $n_2 I_0$ in Eq. (1.3.1) is much smaller than the maximum value of δn_g. This condition is equivalent to requiring that $\gamma P_0 \ll \kappa$ or $\kappa L_{\mathrm{NL}} \gg 1$, where $L_{\mathrm{NL}} = (\gamma P_0)^{-1}$ is the nonlinear length. This requirement is easy to satisfy in practice, even at peak intensity levels as high as 100 GW/cm^2. Second, the third-order dispersion β_3^g induced by the grating should be negligible.

When the preceding two conditions are satisfied, pulse propagation in a fiber grating is governed by the following NLS equation [117]:

$$\frac{i}{v_g}\frac{\partial U}{\partial t} - \frac{(1-v^2)^{3/2}}{\mathrm{sgn}(f)2\kappa}\frac{\partial^2 U}{\partial \zeta^2} + \frac{1}{2}(3-v^2)\gamma |U|^2 U, \tag{1.5.7}$$

where $\zeta = z - V_G t$. We have introduced a speed-reduction factor related to the parameter f through Eq. (1.4.6) as

$$v = \frac{V_G}{v_g} = \frac{1-f^2}{1+f^2} = \pm\sqrt{1 - \kappa^2/\delta^2}. \tag{1.5.8}$$

The group velocity decreases by the factor v close to an edge of the stop band and vanishes at the two edges ($v = 0$) corresponding to $f = \pm 1$. The reason the first term is a time derivative, rather than the z derivative, was discussed earlier. It can also be understood from a physical standpoint if we note that the variable U does not correspond to the amplitude of the forward- or backward-propagating wave but represents the amplitude of the envelope associated with the Bloch wave formed by a superposition of A_f and A_b.

Equation (1.5.7) has been written for the case in which the contribution of A_f dominates ($|f| < 1$) so that the entire Bloch-wave envelope is propagating forward at the reduced group velocity V_G. With this in mind, we introduce $z = V_G t$ as the distance traveled by the envelope and account for changes in its shape through a local time variable defined as $T = t - z/V_G$. Equation (1.5.8) can then be written in the following standard form of the NLS equation [2]:

$$i\frac{\partial U}{\partial z} - \frac{\beta_2^g}{2}\frac{\partial^2 U}{\partial T^2} + \gamma_g|U|^2 U = 0, \tag{1.5.9}$$

where the effective GVD parameter β_2^g and the nonlinear parameter γ_g are defined as

$$\beta_2^g = \frac{(1-v^2)^{3/2}}{\text{sgn}(f)v_g^2\kappa v^3}, \qquad \gamma_g = \left(\frac{3-v^2}{2v}\right)\gamma. \tag{1.5.10}$$

Using Eq. (1.5.8), the GVD parameter β_2^g can be shown to be the same as in Eq. (1.3.26).

Several features of Eq. (1.5.9) are noteworthy when this equation is compared with the standard NLS equation. First, the variable U represents the amplitude of the envelope associated with the Bloch wave formed by a superposition of A_f and A_b. Second, the parameters β_2^g and γ_g are not constants but depend on the speed-reduction factor v. Both increase as v decreases and become infinite at the edges of the stop band, where $v = 0$. Clearly, Eq. (1.5.9) is not valid at that point. However, it remains valid close to but outside of the stop band. Far from the stop band ($v \to 1$), β_2^g becomes quite small (<1 ps^2/km for typical values of κ). One should then include fiber GVD and replace β_2^g by β_2. Noting that $\gamma_g = \gamma$ when $v = 1$, Eq. (1.5.9) reduces to the standard NLS equation, and U corresponds to the forward-wave amplitude since no backward wave is generated under such conditions.

Before we can use Eq. (1.5.9) for predicting the modulation-instability gain and the frequency at which the gain peaks, we need to know the total power P_0 inside the grating when a CW beam with power P_{in} is incident at the input end of the grating located at $z = 0$. This is a complicated issue for apodized fiber gratings, because κ is not constant in the transition or buffer zone. However, observing that the nonlinear coupled-mode equations require $|A_f^2| - |A_b^2|$ to remain constant along the grating, one finds that the total power P_0 inside the grating is enhanced by a factor $1/v$ [124]. The predictions of Eq. (1.5.9) are in agreement with the modulation-instability analysis based on the nonlinear coupled-mode equations as long as $\gamma P_0 \ll \kappa$ [117]. The NLS equation provides a shortcut to understanding the temporal dynamics in gratings within its regime of validity.

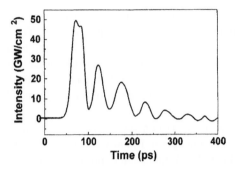

Figure 1.16: Transmitted pulse shape when 100-ps pulses with a peak intensity of 25 GW/cm^2 are propagated through a 6-cm-long fiber grating. (From Ref. [124]; ©1998 Elsevier.)

1.5.3 Experimental Results

Modulation instability implies that an intense CW beam may get converted into a pulse train if has passed through a fiber grating. The experimental observation of this phenomenon is difficult when a CW beam is used because the required input power is too large to be realistic. For this reason, experiments often use short optical pulses whose width is chosen to be much larger than the modulation period. In a 1996 experiment, 100-ps pulses—obtained from a Q-switched, mode-locked Nd:YLF laser operating close to 1.053 μm—were used, and it was found that each pulse was transformed into two shorter pulses at the grating output [116]. The grating was only 3.5-cm long in this experiment and did not allow substantial growth of modulation instability.

In a 1998 experiment, a 6-cm-long fiber grating was used with a value of $\kappa = 12$ cm^{-1} [124]. Figure 1.16 shows transmitted pulse shapes when 100-ps pulses were propagated through this grating. The peak intensity of the input Gaussian pulse is 25 GW/cm^2. Its central frequency is tuned close to but outside the stop band such that the grating provides anomalous GVD (upper branch of the dispersion curve). At lower power levels, the pulse is compressed because of the combination of GVD and SPM that leads to soliton-effect compression (discussed in Chapter 6). At the 25 GW/cm^2 power level, the transmitted pulse exhibits a multipeak structure that can be interpreted as a pulse train generated through modulation instability. This interpretation is supported by the observation that the repetition rate (spacing between two neighboring pulses) changes with the laser wavelength (equivalent to changing the detuning parameter δ), as expected from the theory of modulation instability.

1.6 Nonlinear Pulse Propagation

It is well known that modulation instability often indicates the possibility of soliton formation [2]. In the case of Bragg gratings, it is closely related to a new kind of solitons referred to as *Bragg solitons* or *grating solitons*. Such solitons were first discovered in 1987 in the context of periodic structures known as superlattices [92] and were called *gap solitons*, since they existed only inside the stop band. Later, a much larger class of Bragg solitons was identified by solving Eqs. (1.4.1) and (1.4.2) analytically [125]–[127].

The advent of fiber gratings during the 1990s provided an incentive for studying propagation of short optical pulses in such gratings [128]–[142]. The peak intensities required to observe the nonlinear effects are quite high (typically >10 GW/cm^2) for Bragg gratings made in silica fibers because of a short interaction length (typically <10 cm) and a low value of the nonlinear parameter n_2. The use of chalcogenide glass fibers for making gratings can reduce required peak intensities by a factor of 100 or more because of the high values of n_2 in such glasses [143].

1.6.1 Bragg Solitons

It was noted in 1989 that the coupled-mode equations, Eqs. (1.4.1) and (1.4.2), become identical to the well-known massive Thirring model [144] if the SPM term is set to zero. The massive Thirring model of quantum field theory is known to be integrable by the inverse scattering method [145]–[147]. When the SPM term is included, the coupled-mode equations become nonintegrable, and solitons do not exist in a strict mathematical sense. However, shape-preserving solitary waves can be obtained through a suitable transformation of the soliton supported by the massive Thirring model. These solitary waves correspond to the following solution [126]:

$$A_f(z,t) = a_+\text{sech}(\zeta - i\psi/2)e^{i\theta}, \tag{1.6.1}$$

$$A_b(z,t) = a_-\text{sech}(\zeta + i\psi/2)e^{i\theta}, \tag{1.6.2}$$

where

$$a_\pm = \pm\left(\frac{1\pm v}{1\mp v}\right)^{1/4}\sqrt{\frac{\kappa(1-v^2)}{\gamma(3-v^2)}}\sin\psi, \quad \zeta = \frac{z-V_Gt}{\sqrt{1-v^2}}\kappa\sin\psi, \tag{1.6.3}$$

$$\theta = \frac{v(z-V_Gt)}{\sqrt{1-v^2}}\kappa\cos\psi - \frac{4v}{3-v^2}\tan^{-1}[|\cot(\psi/2)|\coth(\zeta)]. \tag{1.6.4}$$

This solution represents a two-parameter family of Bragg solitons. The parameter v lies in the range $-1 < v < 1$ and the parameter ψ can be chosen anywhere in the range $0 < \psi < \pi$. The specific case $\psi = \pi/2$ corresponds to the center of the stop band [125]. Physically, Bragg solitons represent specific combinations of counterpropagating waves that pair in such a way that they move at the same but reduced speed ($V_G = vv_g$). Since v can be negative, the soliton can move forward or backward. The soliton width T_s is also related to the parameters v and ψ and is given by

$$T_s = \sqrt{1-v^2}/(\kappa V_G\sin\psi). \tag{1.6.5}$$

One can understand the reduced speed of a Bragg soliton by noting that the counterpropagating waves form a single entity that moves at a common speed. The relative amplitudes of the two waves participating in soliton formation determine the soliton speed. If A_f dominates, the soliton moves in the forward direction but at a reduced speed. The opposite happens when A_b is larger. In the case of equal amplitudes, the soliton does not move at all since V_G becomes zero. This case corresponds to the stationary gap solitons predicted in the context of superlattices [92]. In the opposite limit, in which $|v| \to 1$, Bragg solitons cease to exist since the grating becomes ineffective.

Another family of solitary waves was obtained in 1993 by searching for the shape-preserving solutions of the nonlinear coupled-mode equations [127]. Such solitary waves exist both inside and outside the stop band. They reduce to the Bragg solitons described by Eqs. (1.6.2)–(1.6.4) in some specific limits. On the lower branch of the dispersion curve where the GVD is normal, solitary waves represent dark solitons, similar to those found for constant-index fibers [2].

1.6.2   Relation to NLS Solitons

As discussed earlier, the nonlinear coupled-mode equations reduce to the NLS equation when $\gamma P_0 \ll \kappa$, where P_0 is the peak power of the pulse propagating inside the grating. Since the NLS equation is integrable by the inverse scattering method, the fundamental and higher-order solitons found with this method [2] should also exist for a fiber grating. The question then becomes how they are related to the Bragg soliton described by Eqs. (1.6.1) and (1.6.2).

To answer this question, we write the NLS equation (1.5.9) using soliton units in its standard form

$$i\frac{\partial u}{\partial \xi} + \frac{1}{2}\frac{\partial^2 u}{\partial \tau^2} + |u|^2 u = 0, \tag{1.6.6}$$

where $\xi = z/L_D$, $\tau = T/T_0$, $u = \sqrt{\gamma_g L_D}$, and $L_D = T_0^2/|\beta_2^g|$ is the dispersion length. The fundamental soliton of this equation, in its most general form, is given by (see Section 5.2 of Ref. [2])

$$u(\xi, \tau) = \eta\,\mathrm{sech}[\eta(\tau - \tau_s + \varepsilon\xi)]\exp[i(\eta^2 - \varepsilon^2)\xi/2 - i\varepsilon\tau + i\phi_s], \tag{1.6.7}$$

where η, ε, τ_s, and ϕ_s are four arbitrary parameters representing amplitude, frequency, position, and phase of the soliton, respectively. The soliton width is related inversely to the amplitude as $T_s = T_0/\eta$. The physical origin of such solitons is the same as that for conventional solitons except that the GVD is provided by the grating rather than by material dispersion.

At first sight, Eq. (1.6.7) looks quite different than the Bragg soliton described by Eqs. (1.6.2)–(1.6.4). However, one should remember that u represents the amplitude of the Bloch wave formed by superimposing A_f and A_b. If the total optical field is considered and the low-power limit ($\gamma P_0 \ll \kappa$) is taken, the Bragg soliton indeed reduces to the fundamental NLS soliton [101]. The massive Thirring model also allows for higher-order solitons [148]. One would expect them to be related to higher-order NLS solitons in the appropriate limit. It has been shown that any solution of the NLS equation (1.5.9) can be used to construct an approximate solution of the coupled-mode equations [123].

The observation that Bragg solitons are governed by an effective NLS equation in the limit $\kappa L_{NL} \gg 1$, where L_{NL} is the nonlinear length, allows us to use the concept of soliton order N and the soliton period z_0 developed in Chapter 5 of Ref. [2]. These parameters are defined as

$$N^2 = \frac{L_D}{L_{NL}} \equiv \frac{\gamma_g P_0 T_0^2}{|\beta_2^g|}, \qquad z_0 = \frac{\pi}{2} L_D \equiv \frac{\pi}{2} \frac{T_0^2}{|\beta_2^g|}. \qquad (1.6.8)$$

We need to interpret the meaning of the soliton peak power P_0 carefully since the NLS soliton represents the amplitude of the Bloch wave formed by a combination of A_f and A_b. This aspect is discussed later in this section.

An interesting issue is related to the collision of Bragg solitons. Since Bragg solitons described by Eqs. (1.6.1) and (1.6.2) are only solitary waves (because of the nonintegrablity of the underlying nonlinear coupled-mode equations), they may not survive collisions. On the other hand, the NLS solitons are guaranteed to remain unaffected by their mutual collisions. Numerical simulations based on Eqs. (1.4.1) and (1.4.2) show that Bragg solitons indeed exhibit features reminiscent of a NLS soliton in the low-power limit $\gamma P_0 \ll \kappa$ [136]. More specifically, two Bragg solitons attract or repel each other depending on their relative phase. The new feature is that the relative phase depends on the initial separation between the two solitons.

1.6.3 Formation of Bragg Solitons

Formation of Bragg solitons in fiber gratings was first observed in a 1996 experiment [128]. Since then, more careful experiments have been performed, and many features of Bragg solitons have been extracted. While comparing the experimental results with the coupled-mode theory, one needs to implement the boundary conditions properly. For example, the peak power P_0 of the Bragg soliton formed inside the grating when a pulse is launched is not the same as the input peak power P_{in}. The reason can be understood by noting that the group velocity of the pulse changes as the input pulse crosses the front end of the grating located at $z = 0$. As a result, pulse length given by $v_g T_0$ just outside the grating changes to $V_G T_0$ on crossing the interface located at $z = 0$ [80], and the pulse peak power is enhanced by the ratio v_g/V_G. Mathematically, one can use the coupled-mode equations to show that $P_0 = |A_f^2| + |A_b^2| = P_{in}/v$, where $v = V_G/v_g$ is the speed-reduction factor introduced earlier. The argument becomes more complicated for apodized fiber gratings, used often in practice, because κ is not constant in the transition region [133]. However, the same power enhancement occurs at the end of the transition region.

From a practical standpoint, one needs to know the amount of peak power P_{in} required to excite the fundamental Bragg soliton. The soliton period z_0 is another important parameter relevant for soliton formation since it sets the length scale over which optical solitons evolve. We can use Eq. (1.6.8) with $N = 1$ to estimate both of them. Using the expressions for β_2^g and γ_g from Eq. (1.5.10), the input peak power and the soliton period are given by

$$P_{in} = \frac{2(1-v^2)^{3/2}}{v(3-v^2)v_g^2 T_0^2 \kappa \gamma}, \qquad z_0 = \frac{\pi v^3 v_g^2 T_0^2 \kappa}{2(1-v^2)^{3/2}}, \qquad (1.6.9)$$

where T_0 is related to the FWHM as $T_{FWHM} \approx 1.76 T_0$. Both P_{in} and z_0 depend through v on detuning of the laser wavelength from the edge of the stop band located at $\delta = \kappa$. As $v \to 0$ near the edge, P_{in} becomes infinitely large while z_0 tends toward zero.

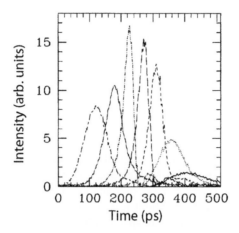

Figure 1.17: Output pulse shapes for different δ when 80-ps pulses with a peak intensity of 11 GW/cm^2 are propagated through a 7.5-cm-long fiber grating. Values of δ from left to right are 3612, 1406, 1053, 935, 847, 788, and 729 m^{-1}. (From Ref. [88]; ©1999 OSA.)

In a 1999 experiment, Bragg solitons were formed inside a 7.5-cm-long apodized fiber grating by launching 80-ps pulses, obtained from a Q-switched, mode-locked Nd:YLF laser operating at 1053 nm [88]. Figure 1.17 shows pulse shapes observed at the output end of the grating when input pulses with a peak intensity of 11 GW/cm^2 were used. The coupling coefficient κ was estimated to be 7 cm^{-1} for this grating. The detuning parameter δ was varied in the range of 7 to 36 cm^{-1} on the blue side of the stop band (anomalous GVD). The arrival time of the pulse depends on δ because of a decrease in its group velocity occurring as δ is tuned closer to the stop-band edge. This delay occurs even when nonlinear effects are negligible as shown in Figure 1.11, which was obtained under identical operating conditions, but at a much lower value of the peak intensity.

At the high peak intensities used for Figure 1.17, SPM in combination with the grating-induced anomalous GVD leads to formation of Bragg solitons. However, since both β_2^g and γ_g depend on the detuning parameter δ through ν, a Bragg soliton can form only in a limited range of δ. With this in mind, we can understand the pulse shapes seen in Figure 1.17. Detuning is so large and β_2^g is so small for the leftmost trace that the pulse acquires some chirping through SPM but its shape remains nearly unchanged. This feature can also be understood from Eq. (1.6.9), where the soliton period becomes so long as $\nu \to 1$ that nothing much happens to the pulse over a few-cm-long grating. As δ is reduced, the pulse narrows down considerably. A reduction in pulse width by a factor of 3 occurs for $\delta = 1053$ m^{-1} in Figure 1.17. This pulse narrowing is an indication that a Bragg soliton is beginning to form. However, the soliton period is still much longer than the grating length. In other words, the grating is not long enough to observe the final steady-state shape of the Bragg soliton. Finally, as the edge of the stop band is approached and δ becomes comparable to κ (rightmost solid trace), the GVD becomes so large that the pulse cannot form a soliton and becomes broader than the input pulse. This behavior is also deduced from Eq. (1.6.8), which shows that both

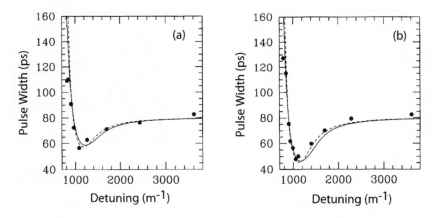

Figure 1.18: Measured pulse widths (circles) as a function of detuning for 80-ps input pulses with a peak intensity of (a) 3 GW/cm^2 and (b) 6 GW/cm^2. Predictions of the coupled-mode theory (solid line) and the effective NLS equation (dashed line) are shown for comparison. (From Ref. [88]; ©1999 OSA.)

N and z_0 tend toward zero as β_2^g tends toward infinity. A Bragg soliton can form only if $N > \frac{1}{2}$. Since the dispersion length becomes smaller than the grating length close to the stop-band edge, pulse can experience considerable broadening. This is precisely what is observed for the smallest value of δ in Figure 1.17 (solid curve).

A similar behavior was observed over a large range of pulse energies, with some evidence of the second-order soliton for input peak intensities in excess of 20 GW/cm^2 [88]. A careful comparison of the experimental data with the theory based on the nonlinear coupled-mode equations and the effective NLS equation showed that the NLS equation provides an accurate description within its regime of validity. Figure 1.18 compares the measured values of the pulse width with the two theoretical models for peak intensities of 3 and 6 GW/cm^2. The NLS equation is valid as long as $\kappa L_{NL} \gg$ 1. Using $\kappa = 7$ cm^{-1}, we estimate that the peak intensity can be as high as 50 GW/cm^2 for the NLS equation to remain valid. This is also what was found in Ref. [88].

Gap solitons that form within the stop band of a fiber grating have not been observed because of a practical difficulty: A Bragg grating reflects light whose wavelength falls inside the stop band. Stimulated Raman scattering may provide a solution to this problem, since a pump pulse, launched at a wavelength far from the stop band, can excite a "Raman gap soliton" that is trapped within the grating and propagates much more slowly than the pump pulse itself [138]. The energy of such a gap soliton leaks slowly from the grating ends, but it can survive for durations greater than 10 ns even though it is excited by pump pulses of duration 100 ps or so.

1.6.4 Nonlinear Switching

As discussed in Section 1.4.2, a fiber grating can exhibit bistable switching even when a CW beam is incident on it. However, optical pulses should be used in practice because of the high intensities required for observing SPM-induced nonlinear switching.

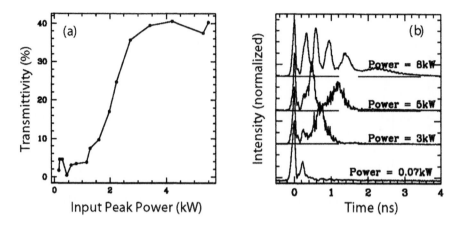

Figure 1.19: (a) Transmittivity as a function of input peak power showing nonlinear switching; (b) output pulse shapes at several peak power levels. (From Ref. [134]; ©1998 OSA.)

Even then, one needs peak-intensity values in excess of 10 GW/cm^2. For this reason, bistable switching was first observed during the 1980s using DFB semiconductor amplifiers for which large carrier-induced nonlinearities reduce the switching threshold to power levels below 1 mW [149]–[151]. Nonlinear switching in a passive grating was observed in a 1992 experiment using a semiconductor waveguide grating [98]. The nonlinear response of such gratings is not governed by the Kerr-type nonlinearity seen in Eq. (1.3.1) because of the presence of free carriers (electrons and holes) whose finite lifetime limits the nonlinear response time.

Nonlinear switching in a fiber Bragg grating was observed in 1998 in the 1.55-μm wavelength region useful for fiber-optic communications [132]. An 8-cm-long grating, with its Bragg wavelength centered near 1536 nm, was used in the experiment. It had a peak reflectivity of 98% and its stop band was only 4 GHz wide. The 3-ns input pulses were obtained by amplifying the output of a pulsed DFB semiconductor laser to power levels as high as 100 kW. Their shape was highly asymmetric because of gain saturation occurring inside the amplifier chain. The laser wavelength was inside the stop band on the short-wavelength side but was set very close to the edge (offset of about 7 pm or 0.9 GHz).

Figure 1.19(a) shows a sharp rise in the transmittivity from a few percent to 40% when the peak power of input pulses increases beyond 2 kW. Physically, the nonlinear increase in the refractive index at high powers shifts the Bragg wavelength far enough that the pulse finds itself outside the stop band and switches to the upper branch of the bistability curves seen in Figure 1.13. The pulse shapes seen in Figure 1.19(b) show what happens to the transmitted pulse. The initial spike near $t = 0$ in these traces is due to a sharp leading edge of the asymmetric input pulse and should be ignored. Multiple pulses form at the grating output, whose number depends on the input power level. At a power level of 3 kW, a single pulse is seen but the number increases to five at a power level of 8 kW. The pulse width is smallest (about 100 ps) near the leading edge of the pulse train but increases substantially for pulses near the trailing edge.

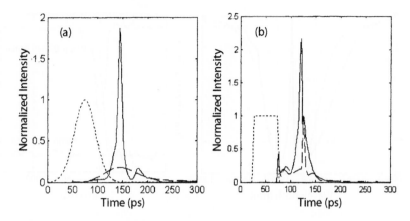

Figure 1.20: Transmitted pulse shapes in the on (solid curves) and off (dashed curves) states when 50-ps (a) Gaussian and (b) square input pulses (dotted curves) are transmitted through a fiber grating with $\kappa L = 4$. (From Ref. [139]; ©2003 IEEE.)

Several conclusions can be drawn from these results. First, the upper bistability branch in Figure 1.13 is not stable and converts the quasi-CW signal into a pulse train [90]. Second, each pulse evolves toward a constant width. Pulses near the leading edge have had enough propagation time within the grating to stabilize their widths. These pulses can be thought of as a gap soliton since they are formed even though the input signal is inside the photonic bandgap and would be completely reflected in the absence of the nonlinear effects. Third, pulses near the trailing edge are wider simply because the fiber grating is not long enough for them to evolve completely toward a gap soliton. This interpretation was supported by a later experiment in which the grating length was increased to 20 cm [137]. Six gap solitons were found to form in this grating at a peak power level of 1.8 W. The observed data were in agreement with theory based on the nonlinear coupled-mode equations.

In a 2003 study, numerical simulations were used to characterize the nonlinear switching characteristics of a wide variety of pulses of different widths and shape through uniform as well as phase-shifted fiber gratings [139]. The bistable behavior, similar to that seen in Figure 1.13 for CW beams, is realized only for pulses wider than 10 ns. The use of phase-shifted gratings reduces the switching threshold, but the on-off contrast is generally better for uniform gratings. For short pulses (width <1 ns), launched such that most of their spectrum lies inside the stop band of the grating initially, almost the entire input energy is transmitted in the form of a compressed optical pulse, once the peak-power level is large enough to switch the pulse to the "on" branch. Moreover, the shape of output pulse depends strongly on the rise and fall times associated with the pulses tails. Figure 1.20 shows the transmitted pulse shapes in the on (solid curves) and off (dashed curves) states when 50-ps Gaussian or square input pulses (dotted curves) are transmitted through a fiber grating with $\kappa L = 4$. Although nonlinear switching occurs in both cases, the pulse quality is better in the case of Gaussian pulses. In particular, a grating can be used to compress short optical pulses. In a 2005 experiment, 580-ps pulses with 1.4-kW peak power could be compressed down

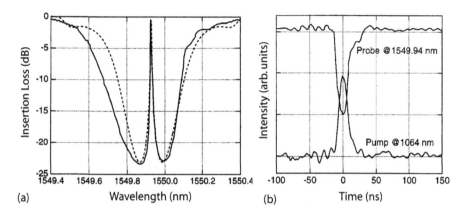

Figure 1.21: (a) Measured (solid) and calculated (dashed) transmittivity of the 2-cm-long phase-shifted fiber grating. (b) Pump and probe intensity profiles showing XPM-induced nonlinear switching. (From Ref. [154]; ©2000 IEEE.)

to 45 ps by launching them into a 10-cm-long apodized fiber grating [141].

The nonlinear switching seen in Figure 1.19 is sometimes called *SPM-induced* or *self-induced switching*, since the pulse changes the refractive index to switch itself to the high-transmission state. Clearly, another signal at a different wavelength can also induce switching of the pulse by changing the refractive index through XPM, resulting in XPM-induced switching. This phenomenon was first observed in 1990 as an increase in the transmittivity of a 514-nm signal caused by a 1064-nm pump beam [96]. The increase in transmission was less than 10% in this experiment.

It was suggested later that XPM could be used to form a "push broom" such that a weak CW beam (or a broad pulse) would be swept by a strong pump pulse and its energy piled up at the front end of the pump pulse [152]. The basic idea behind the optical push broom is quite simple. If the wavelength of the pump pulse is far from the stop band while that of the probe is close to but outside the stop band (on the lower branch of the dispersion curve), the pump travels faster than the probe. In the region where the pump and probe overlap, the XPM-induced chirp changes the probe frequency such that it moves with the leading edge of the pump pulse. As the pump pulse travels further, it sweeps more and more of the probe energy and piles it up at its leading edge. In effect, the pump acts like a push broom. At the grating output, a significant portion of the probe energy appears at the same time as the pump pulse in the form of a sharp spike because of the XPM-induced increase in the probe speed. Such a push-broom effect has been seen in a 1997 experiment [153].

In a 2000 experiment, a phase-shifted fiber grating was used to realize XPM-induced nonlinear switching at relatively low power levels [154]. In this pump–probe experiment, a 1550-nm CW laser was tuned precisely to the center of the stop band associated with the 2-cm-long grating so that it was fully transmitted by the narrow transmission peak seen in Figure 1.21(a). When a pump pulse (width $\sim$10 ns) from a Nd:YAG laser operating at 1064 nm was launched with the probe, the probe power dropped considerably over the temporal window of the pulse, as seen in Figure 1.21(b).

This switching is due to the XPM-induced shift of the probe spectrum. Even a relatively small shift moves the probe outside the narrow transmission peak, resulting in a drop in its transmissivity through the grating. The switching could be observed at a relatively low power level of 730 W because of the use of a phase-shifted fiber grating.

1.6.5 Effects of Birefringence

As discussed in Chapter 6 of Ref. [2], fiber birefringence plays an important role and affects the nonlinear phenomena considerably. Its effects should be included if Bragg gratings are made inside the core of polarization-maintaining fibers. The coupled-mode theory can be easily extended to account for fiber birefringence [155]–[158]. However, the problem becomes quite complicated, since one needs to solve a set of four coupled equations describing the evolution of two orthogonally polarized components, each containing both the forward- and backward-propagating waves. This complexity, however, leads to a rich class of nonlinear phenomena with practical applications such as optical logic gates.

From a physical viewpoint, the two orthogonally polarized components have slightly different mode indices. Since the Bragg wavelength depends on the mode index, the stop bands of the two modes have the same widths but are shifted by a small amount with respect to each other. As a result, even though both polarization components have the same wavelength (or frequency), one of them may fall inside the stop band while the other remains outside it. Moreover, as the two stop bands shift due to nonlinear index changes, the shift can be different for the two orthogonally polarized components because of the combination of the XPM and birefringence effects. It is this feature that leads to a variety of interesting nonlinear effects.

In the case of CW beams, the set of four coupled equations was solved numerically in 1994 and several birefringence-related nonlinear effects were identified [156]. One such effect is related to the onset of polarization instability discussed in Section 6.3 of Ref. [2]. The critical power at which this instability occurs is reduced considerably in the presence of a Bragg grating [159]. Nonlinear birefringence also affects Bragg solitons. In the NLS limit ($\gamma P_0 \ll \kappa$), the four equations reduce a pair of coupled NLS equations. In the case of low-birefringence fibers, the two polarization components have nearly the same group velocity, and the coupled NLS equations take the following form [155]:

$$\frac{\partial A_x}{\partial z} + \frac{i\beta_2^g}{2}\frac{\partial^2 A_x}{\partial T^2} = i\gamma_g\left(|A_x|^2 + \frac{2}{3}|A_y|^2\right)A_x + \frac{i\gamma_g}{3}A_x^*A_y^2 e^{-2i\Delta\beta z}, \quad (1.6.10)$$

$$\frac{\partial A_y}{\partial z} + \frac{i\beta_2^g}{2}\frac{\partial^2 A_y}{\partial T^2} = i\gamma_g\left(|A_y|^2 + \frac{2}{3}|A_x|^2\right)A_y + \frac{i\gamma_g}{3}A_y^*A_x^2 e^{2i\Delta\beta z}, \quad (1.6.11)$$

where $\Delta\beta \equiv \beta_{0x} - \beta_{0y}$ is related to the beat length L_B as $\Delta\beta = 2\pi/L_B$. These equations support a vector soliton with equal amplitudes, such that the peak power required for each component is only $\sqrt{3/5}$ of that required when only one component is present. Such a vector soliton is referred to as the *coupled-gap soliton* [155].

The coupled-gap soliton can be used for making an all-optical AND gate. The x and y polarized components of the input light represent bits for the gate, each bit taking a

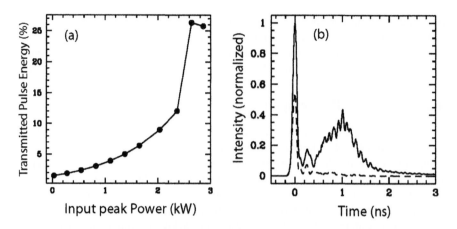

Figure 1.22: (a) Grating transmissivity as a function of input peak power showing the operation of the AND gate and (b) output pulse shapes at a peak power level of 3 kW, when only one polarization component (dashed line) or both polarization components (solid line) are incident at the input end. (From Ref. [134]; ©1998 OSA.)

value of 0 or 1 depending on whether the corresponding signal is absent or present. The AND gate requires that a pulse appears at the output only when both components are present simultaneously. This can be achieved by tuning both polarization components inside the stop band but close to the upper branch of the dispersion curve. Their combined intensity can increase the refractive index (through a combination of SPM and XPM) enough that both components are transmitted. However, if one of the components is absent at the input (0 bit), the XPM contribution vanishes and both components are reflected by the grating. This occurs simply because the coupled gap soliton forms at a lower peak power level than the Bragg soliton associated with each individual component [155].

An all-optical AND gate was realized in a 1998 experiment in which a switching contrast of 17 dB was obtained at a peak power level of 2.5 kW [131]. Figure 1.22 shows the fraction of total pulse energy transmitted (a) as a function of input peak power and the transmitted pulse shapes (b) at a peak power of 3 kW. When only one polarization component is incident at the input end, little energy is transmitted by the grating. However, when both polarization components are launched, each having the same peak power, an intense pulse is seen at the output end of the grating, in agreement with the prediction of the coupled NLS equations.

The XPM-induced coupling can be advantageous even when the two polarization components have different wavelengths. For example, it can be used to switch the transmission of a CW probe from low to high by using an orthogonally polarized short pump pulse at a wavelength far from the stop band associated with the probe [160]. In contrast with the self-induced bistable switching discussed earlier, XPM-induced bistable switching can occur for a CW probe too weak to switch itself. Furthermore, the short pump pulse switches the probe beam permanently to the high-transmission state.

1.7   Related Periodic Structures

This chapter has so far focused on uniform Bragg gratings (except for apodization) that are designed to couple the forward- and backward-propagating waves inside an optical fiber. Many variations of this simple structure exist. In this section, we consider several other kinds of gratings and discuss the nonlinear effects occurring when intense light propagates through them.

1.7.1   Long-Period Gratings

Long-period gratings are designed to couple the fundamental fiber mode to a higher-order copropagating mode [161]–[167]. In the case of a single-mode fiber, the higher-order mode propagates inside the cladding and is called a *cladding mode*. The grating period required for coupling the two copropagating modes can be calculated from Eq. (1.1.2) and is given by $\Lambda = \lambda / \Delta n$, where Δn is the difference in the refractive indices of the two modes coupled by the grating. Since $\Delta n \sim 0.01$ typically, Λ is much larger than the optical wavelength. For this reason, such gratings are called *long-period gratings*.

The coupled-mode theory of Section 1.3 can be used for long-period gratings. In fact, the resulting equations are similar to Eqs. (1.3.11) and (1.3.12) and can be written as [163]

$$\frac{\partial A_1}{\partial z} + \frac{1}{v_{g1}} \frac{\partial A_1}{\partial t} + \frac{i\beta_{21}}{2} \frac{\partial^2 A_1}{\partial t^2} = i\delta A_1 + i\kappa A_2 + i\gamma_1 (|A_1|^2 + c_1 |A_2|^2) A_1, \quad (1.7.1)$$

$$\frac{\partial A_2}{\partial z} + \frac{1}{v_{g2}} \frac{\partial A_2}{\partial t} + \frac{i\beta_{22}}{2} \frac{\partial^2 A_2}{\partial t^2} = i\delta A_2 + i\kappa A_1 + i\gamma_2 (|A_2|^2 + c_2 |A_1|^2) A_2, \quad (1.7.2)$$

where A_1 and A_2 represent the slowly varying amplitudes of the two copropagating modes coupled by the grating. A comparison of these equations with Eqs. (1.3.11) and (1.3.12) reveals several important differences. First, the two z derivatives have the same sign, since both waves travel in the forward direction. Second, the group velocities and the GVD parameters can be different for the two modes, because of their different mode indices. Third, the SPM parameters γ_j and the XPM parameters c_j are also generally different for $j = 1$ and 2. The reason is related to different spatial profiles for the two modes, resulting in different overlap factors.

In the case of low-power CW beams, both the nonlinear and the GVD effects can be neglected in Eqs. (1.7.1) and (1.7.2) by setting $\gamma_j = 0$ and $\beta_2 = 0$ ($j = 1, 2$). These equations then reduce to Eqs. (1.3.13) and (1.3.14), with the only difference being that both z derivatives have the same sign. They can be solved readily and exhibit features similar to those discussed in Section 1.3.2. When a single beam excites the A_1 mode at the fiber input, its transmission depends on its detuning δ from the Bragg wavelength and becomes quite small within the stop band centered at $\delta = 0$. The reason is easily understood by noting that the grating transfers power to the A_2 mode as light propagates inside the grating.

The nonlinear effects such as SPM and XPM can affect the amount of power transferred by changing the refractive index and shifting the Bragg wavelength toward longer wavelengths. As a result, a long-period grating should exhibit nonlinear

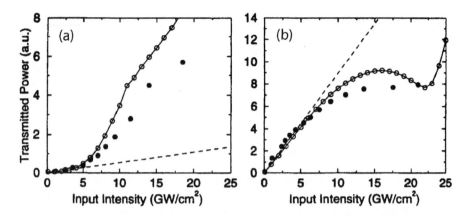

Figure 1.23: Transmitted power as a function of input peak power for (a) $\delta = 0$ and (b) -1.5 cm^{-1}. Experimental data (solid circles) are compared with coupled-mode theory (open circles). Dashed lines shows the behavior expected in the absence of nonlinear effects. (From Ref. [163]; ©1997 IEEE.)

switching. Moreover, the switching intensity is expected to be lower by a factor of $\bar{n}/\Delta n \sim 100$ compared with that required for short-period Bragg gratings. Figure 1.23 shows nonlinear changes in the transmitted power as a function of input peak intensity when 70-ps Gaussian pulses are transmitted through a 5-cm-long grating and compares the experimental data with the prediction of coupled-mode theory. Dashed lines show the linear increase in transmission expected in the absence of nonlinear effects. For $\delta = 0$ [Figure 1.23(a)], the input wavelength coincides with the Bragg wavelength, and little transmission occurs in the linear case. However, at intensity levels beyond 5 GW/cm^2, the nonlinear effects shift the Bragg wavelength enough that a significant part of the incident power is transmitted through the grating. When the input wavelength is detuned by about 5.2 nm from the Bragg wavelength ($\delta = -1.5$ cm^{-1}), the transmitted power decreases at high peak intensities, as seen in Figure 1.23(b).

Considerable pulse shaping was observed in the preceding experiment because of the use of short optical pulses. This feature can be used to advantage to compress and reshape an optical pulse. Nonlinear effects in long-period fiber gratings are likely to remain important and find practical applications.

1.7.2 Nonuniform Bragg Gratings

Both the linear and nonlinear properties of a Bragg grating can be considerably modified by introducing nonuniformities along its length [105]. Examples of such nonuniform gratings include chirped gratings, phase-shifted gratings, and superstructure gratings. The refractive index in such gratings still has the general form of Eq. (1.3.1), but its periodic part $\delta n_g(z)$ is modified to become

$$\delta n_g(z) = \delta n_1(z)\cos[2\pi z/\Lambda(z) + \phi(z)], \qquad (1.7.3)$$

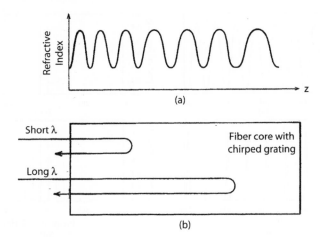

Figure 1.24: (a) Variations of refractive index in a chirped fiber grating. (b) Short and long wavelength components of a pulse are reflected at different locations within the grating because of variations in the Bragg wavelength.

where δn_1, ϕ, Λ, or a combination of them depends on z. If δn_1 varies with z, the coupling coefficient κ becomes z dependent (as in an apodized grating). In a chirped grating, the grating period Λ varies along the grating length as shown in Figure 1.24. In a phase-shifted grating, the phase ϕ is changed abruptly by $\pi/2$ in the middle of an otherwise uniform grating. It was seen in Section 1.4.2 that the use of a phase-shifted grating can reduce the switching power by a factor of 100 or more. In this section, we focus on chirped and other nonuniform gratings.

In a chirped grating, the index-modulation period $\bar{n}\Lambda$ changes along the fiber length. Since the Bragg wavelength ($\lambda_B = 2\bar{n}\Lambda$) sets the frequency at which the stop band is centered, axial variations of $\bar{n}$ or Λ translate into a shift of the stop band along the grating length. Mathematically, the parameter δ appearing in the nonlinear coupled-mode equations becomes z dependent. Typically, Λ is designed to vary linearly along the grating, resulting in $\delta(z) = \delta_0 + \delta_c z$, where δ_c is a chirp parameter. Such gratings are called linearly chirped gratings.

Chirped fiber gratings can be fabricated using several methods [48]. It is important to note that it is the optical period $\bar{n}\Lambda$ that needs to be varied along the grating length (z axis). Thus, chirping can be induced either by varying the physical grating period Λ or by changing the effective mode index $\bar{n}$ along z. In the commonly used dual-beam holographic technique, the fringe spacing of the interference pattern is made nonuniform by using dissimilar curvatures for the interfering wavefronts, resulting in Λ variations. In practice, cylindrical lenses are used in one or both arms of the interferometer. Chirped fiber gratings can also be fabricated by tilting or stretching the fiber, by using strain or temperature gradients, or by stitching together multiple uniform sections.

Chirped Bragg gratings have several important practical applications. As shown in Figure 1.24, when a pulse, with its spectrum inside the stop band, is incident on a

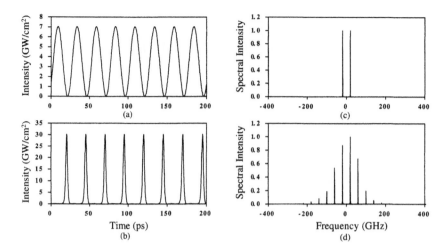

Figure 1.25: Temporal signal [(a) and (b)] and its spectrum [(c) and (d)] at the input [(a) and (c)] and output [(b) and (d)] end of a fiber grating designed with a linearly decreasing κ. (From Ref. [172]; ©1998 OSA.)

chirped grating, different spectral components of the pulse are reflected by different parts of the grating. As a result, even though the entire pulse is eventually reflected, it experiences a large amount of GVD, whose nature (normal versus anomalous) and magnitude can be controlled by the chirp. For this reason, chirped gratings are commonly used for dispersion compensation [87] and pulse compression [168]–[170]. The latter topic is discussed in Chapter 6. Chirped gratings also exhibit interesting nonlinear effects when the incident pulse is sufficiently intense. In one experiment, 80-ps pulses were propagated through a 6-cm-long grating whose linear chirp could be varied over a considerable range through a temperature gradient established along its length [171]. The reflected pulses were split into a pair of pulses by the combination of SPM and XPM for peak intensities close to 10 GW/cm^2.

When δn_1 in Eq. (1.7.3) varies with z, the coupling coefficient κ becomes nonuniform along the grating length. In practice, variations in the intensity of the ultraviolet laser beam used to make the grating translate into axial variations of κ. From a physical standpoint, since the width of the photonic bandgap is about 2κ, changes in κ result into changes in the width of the stop band along the grating length. At a fixed wavelength of input light, such local variations in κ lead to axial variations of the group velocity V_G and the GVD parameter β_2^g, as seen from Eqs. (1.3.25) and (1.3.26), respectively. In effect, the dispersion provided by the grating becomes nonuniform and varies along its length. Such gratings can have a number of applications. For example, they can be used to generate pulse trains at high repetition rates by launching the output of two CW lasers with closely spaced wavelengths.

Figure 1.25 shows the numerical results obtained by solving Eqs. (1.4.1) and (1.4.2) with $\kappa(z) = \kappa_0(1 - \kappa_1 z)$ for the case in which laser frequencies are 40 GHz apart [172]. The grating is assumed to be 70 cm long with parameters $\kappa_0 = 70$ cm^{-1} and $\delta = 160$ cm^{-1}. The parameter κ_1 is chosen such that the index changes seen at the begin-

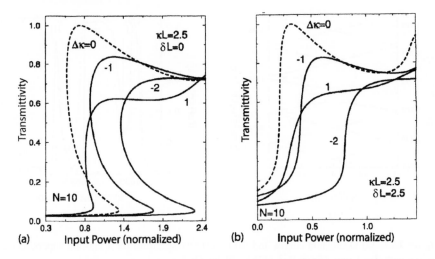

Figure 1.26: transmittivity of a CW signal as a function of its input intensity (normalized) for two values $\delta L=0$ or 2.5 when κL varies linearly over the grating length. (From Ref. [105]; ©1995 IEEE.)

ning of the grating are reduced by a factor of 5 at the end of the grating. The pulse compression can be understood by noting that the nonlinear effects (SPM and XPM) chirp the pulse and broaden its spectrum, and the GVD compresses the chirped pulse. It can also be thought of in terms of a four-wave mixing process, phase-matched by the nonlinearity, which generates multiple sidebands at the grating output, as seen in Figure 1.25.

The nonlinear switching characteristics of nonuniform gratings were first studied in 1995 using a transfer-matrix method [105]. A different technique was employed in a 2004 study, yielding the same results [173]. As an example, Figure 1.26 shows the transmittivity of a CW signal as a function of its input power [normalized to $P_c = (\gamma L)^{-1}$] for two values of detuning such that $\delta L = 0$ or 2.5. The coupling coefficient of the grating is tapered linearly such that $\kappa(z) = \kappa[1 + \Delta\kappa(z/L - \frac{1}{2})]$. In general, tapering of κ modifies the bistable characteristics and affects the power level at which the CW beam switches back and forth between the on and off states.

In another class of gratings, the grating parameters κ and δ are designed to vary periodically along the length of a grating. Such devices have double periodicity and are called *sampled* or *superstructure gratings*. They were first used in the context of DFB semiconductor lasers [174]. Fiber-based sampled gratings were made in 1994 [175]. Since then, their properties have attracted considerable attention [176]–[184]. A simple example of a superstructure grating is provided by a long grating in which constant phase-shift regions occur at periodic intervals. In practice, such a structure can be realized by placing multiple gratings next to each other with a small constant spacing between them or by blocking small regions during fabrication of a grating such that $\kappa = 0$ in the blocked regions. In a phase-sampled grating [182]–[184], the magnitude of κ remains fixed but its phase varies periodically along the grating. In

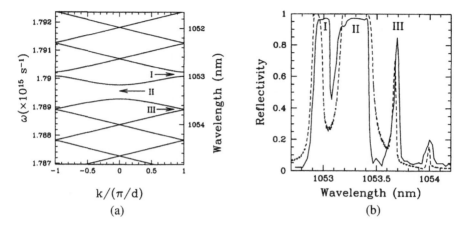

Figure 1.27: (a) Dispersion curves and (b) reflectivity spectrum for a 2.5-mm superstructure grating with $d = 1$ mm. (From Ref. [178]; ©1996 OSA.)

all cases, $\kappa(z)$ becomes a periodic function of z. It is this periodicity that modifies the stop band of a uniform grating. The period d of $\kappa(z)$ is typically ~ 1 mm. If the average index $\bar{n}$ also changes with the same period d, both δ and κ become periodic in the nonlinear coupled-mode equations.

The most striking feature of a superstructure grating is the appearance of additional photonic bandgaps on both branches of the dispersion curve seen in Figure 1.6 for a uniform grating. These bandgaps are referred to as *Rowland ghost gaps* [185]. Figure 1.27 shows the band structure of a 2.5-mm-long superstructure grating with $d = 1$ mm together with the measured reflectivity spectrum. The Rowland ghost gaps labeled I and III occur on the opposite sides of the stop band and lead to two additional reflectivity peaks. Dispersive properties near these gaps are similar to those expected near the edges of the stop band II. As a result, nonlinear effects are quite similar. In particular, Bragg solitons can form on the branch where GVD is anomalous [176]. Indirect evidence of such solitons was seen in an experiment in which a 100-ps pulse was compressed to 38 ps within the 2.5-mm-long superstructure grating when it was tuned on the high-frequency side of the Rowland ghost gap I [178]. The pulse appeared to be evolving toward a Bragg soliton, which should form if the grating were long enough. Other nonlinear effects, such as optical bistability, modulation instability, and optical switching, should also occur near Rowland ghost gaps associated with a superstructure grating. In an interesting application, a superstructure grating was used to increase the repetition rate of a 3.4-ps pulse train from 10 to 40 GHz [181]. The grating was designed to have a band structure such that it reflected every fourth spectral peak of the input spectrum.

1.7.3 Transient and Dynmaic Fiber Gratings

The fiber gratings discussed so far are of permanent nature in the sense that the periodic variations in the refractive index created during the manufacturing process may last indefinitely under normal operating conditions. In contrast, a dynamic or transient grating is formed when the index changes are induced by pumping the fiber optically and last only as long as the pump light remains incident on the fiber. Such fiber gratings were first studied in 1992 and have remained an active topic of research since then [186]–[192].

It is easy to see how a dynamic grating can be formed through optical pumping. If a CW pump beam at the carrier frequency ω_p is launched such that it propagates in the both forward and backward directions, it will form a standing wave whose intensity peaks periodically along the fiber length. The period of such a "fringe pattern" equals $\Lambda_p = \pi/\beta_p = \lambda_p/(2\bar{n})$, where $\beta_p = \bar{n}(\omega_p)\omega_p/c$ is the propagation constant, $\bar{n}(\omega_p)$ is the effective mode index, and λ_p is the pump wavelength. If such variations in the pump intensity change the material refractive index, a Bragg grating is formed. Indeed, the grating-formation technique discussed in Section 1.2.1 employs this approach to create permanent index changes.

If the fiber is doped with a rare-earth element such as erbium, the absorption or gain associated with dopants can be saturated by two counterpropagating pump beams, if their intensity is high enough to exceed the saturation intensity associated with the atomic transition. Through the Kramers–Kronig relation, such saturation helps to transfer periodic pump-intensity variations into period refractive-index variations along the fiber length, thereby creating a transient grating. In a 1992 experiment, a 12-m-long erbium-doped fiber, acting as an amplifier, was employed for this purpose [186]. The transient grating, created by launching <1 mW of pump power, exhibited 75% reflectivity. Such a grating, once created, affects the pump beams though a nonlinear process known as *two-wave mixing* [187]. It can also be used for four-wave mixing by launching a third beam whose wavelength is close enough to the pump wavelength that it falls within the grating bandwidth.

In the case of an undoped silica fiber, the intensity-dependent term in Eq. (1.3.1), whose magnitude is governed by the nonlinear parameter n_2, can transfer the periodic pump-intensity variations into refractive-index variations along the fiber length. The required pump-intensity level exceeds 10 GW/cm^2 even for a relatively weak grating with index modulations $\sim 10^{-5}$. In spite of this requirement, such a technique was used in 2002 to realize modulation instability and to form gap solitons inside a dynamic fiber grating [190].

The basic idea consists of simultaneously launching three beams at frequencies $\omega_0 - \Omega$, ω_0, and $\omega_0 + \Omega$ inside a birefringent fiber, making this scheme similar to that used for four-wave mixing. The two fields at frequencies $\omega_0 \pm \Omega$, copolarized along a principal axis of the fiber, propagate in the same direction and are launched with equal amplitudes, so that their combined intensity oscillates in the time domain at the frequency 2Ω. These periodic variations are seen by the third field, polarized orthogonal to the other two, because of an XPM-induced nonlinear coupling among the three waves. As a result, its propagation along the fiber is governed by [190]

$$\frac{\partial A}{\partial z} + \frac{i\beta_2}{2}\frac{\partial^2 A}{\partial t^2} = i\gamma|A|^2 A + 2i\kappa\cos(2\Omega t)A, \qquad (1.7.4)$$

where $\kappa = (4/3)\gamma P_g$ and P_g is the power launched into each of the two beams used

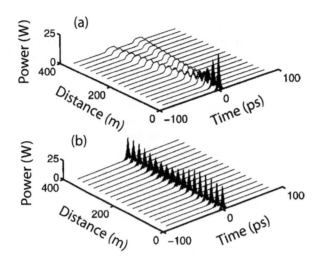

Figure 1.28: Evolution of two pulses of different wavelengths inside a fiber (a) without a grating ($\kappa = 0$) and (b) with a dynamic grating ($\kappa = 0.018$ m^{-1}). Two pulses are trapped by the grating and form a gap soliton. (From Ref. [190]; ©2002 OSA.)

to create the grating. Since periodic index variations occur in time, such a grating is called *dynamic grating*.

Equation (1.7.4) can be solved by writing its solution in the form

$$A(z,t) = A_0(z,t) + A_1(z,t)e^{i(Kz-\Omega t)} + A_2(z,t)e^{i(Kz+\Omega t)}, \qquad (1.7.5)$$

where A_1 and A_2 satisfy the following coupled-mode equations, if we assume that the amplitude of A_0 is negligibly small [190]:

$$\frac{\partial A_1}{\partial z} + \Omega\beta_2\frac{\partial A_1}{\partial t} + \frac{i\beta_2}{2}\frac{\partial^2 A_1}{\partial t^2} = i\kappa A_2 + i\gamma(|A_1|^2 + 2|A_2|^2)A_1, \qquad (1.7.6)$$

$$\frac{\partial A_2}{\partial z} - \Omega\beta_2\frac{\partial A_2}{\partial t} + \frac{i\beta_2}{2}\frac{\partial^2 A_2}{\partial t^2} = i\kappa A_1 + i\gamma(|A_2|^2 + 2|A_1|^2)A_b. \qquad (1.7.7)$$

These equations have a form similar to Eqs. (1.3.11) and (1.3.12) with the major difference being that both A_1 and A_2 propagate in the forward direction. In this sense, they are closer to Eqs. (1.7.1) and (1.7.2), obtained earlier for long-period gratings.

Equations (1.7.6) and (1.7.7) exhibit all the features associated with traditional Bragg gratings [190]. In particular, such a dynamic grating exhibits modulation instability even in the normal-GVD regime of the fiber and supports Bragg solitons. If the second-order dispersion is neglected, Eqs. (1.7.6) and (1.7.7) are found to have the following solitary-wave solution:

$$A_1(z,t) = \sqrt{\frac{\kappa}{3\gamma}}\sin\psi\,\text{sech}\left(\frac{\kappa\sin\psi}{\Omega\beta_2}t - \frac{i\psi}{2}\right)\exp(-i\kappa z\cos\psi), \qquad (1.7.8)$$

$$A_2(z,t) = -\sqrt{\frac{\kappa}{3\gamma}}\sin\psi\,\text{sech}\left(\frac{\kappa\sin\psi}{\Omega\beta_2}t + \frac{i\psi}{2}\right)\exp(-i\kappa z\cos\psi), \qquad (1.7.9)$$

where ψ can have any value in the range $0 \le \psi \le \pi$. This free parameter determines both the amplitude and width of a family of the Bragg solitons in Eqs. (1.7.8) and (1.7.9). In fact, these solitons correspond to the gap solitons discussed earlier in Section 1.6, because they are stationary in the local reference frame of a moving dynamic grating.

Physically, each such gap soliton represents a superposition of two pulses of different frequencies ($\omega_0 \pm \Omega$) that travel at different speeds in the absence of the grating. Indeed, numerical solutions of Eqs. (1.7.6) and (1.7.7) with $\kappa = 0$ show that the two pulses separate rapidly [190]. Figure 1.28(a) shows this behavior for a 400-m-long fiber using $\Omega/2\pi = 0.5$ THz and $\psi = \pi/2$. When a dynamic grating is present, the two pulses form a gap soliton that is trapped by the grating and appears stationary in the frame of the input pulse. Figure 1.28(b) shows such trapping by changing κ from 0 to 0.018 m^{-1}, while keeping all other parameters the same. A similar effect was predicted as early as 1989, and the resulting solitary waves were called *resonance solitons* [193]. The trapping mechanism is similar to that discussed in Section 6.5 of Ref. [2] in the context of vector solitons forming in a birefringent fiber.

Problems

1.1 Derive Eq. (1.1.1) from the phase-matching condition in Eq. (1.1.2).

1.2 Use Eq. (1.1.1) to find the grating period Λ for a fiber Bragg grating reflecting light near 1.55 μm. Assume $m = 1$ and $\bar{n} = 1.45$.

1.3 Describe the mechanism through which absorption of ultraviolet light produces changes in the refractive index of silica fibers.

1.4 Discuss the holographic and phase-mask techniques used to make fiber gratings. Sketch the experimental setup in each case.

1.5 Derive the nonlinear coupled-mode equations, (1.3.11) and (1.3.12), for fiber gratings starting from the Helmholtz equation (1.3.2).

1.6 What is meant by the stop band of a grating? Starting from the linear coupled-mode equations, (1.3.13) and (1.3.14), find the dispersion relation and the width of the stop band.

1.7 An optical pulse is transmitted through a fiber grating with its spectrum located close to but outside the stop band. Its energy is small enough that nonlinear effects are negligible. Derive an expression for the group velocity of the pulse.

1.8 For the previous problem, derive expressions for the second- and third-order dispersion induced by the grating. You can neglect the material and waveguide dispersion of silica fibers.

1.9 Derive an expression for the reflectivity of a fiber grating by solving the coupled-mode equations, (1.3.13) and (1.3.14). Plot it as a function of δ/κ using $\kappa L = 3$.

1.10 The coupling coefficient of an apodized grating of length L varies as $\kappa(z) = \kappa_0 \exp[-(4 - 8z/L)^{2m}]$. Solve the linear coupled-mode equations, (1.3.13) and

(1.3.14), numerically and plot the reflectivity spectrum for $m = 1, 2, 3$ as a function of δ/κ using $\kappa L = 3$.

1.11 Solve the nonlinear coupled-mode equations, (1.4.1) and (1.4.2), assuming that the powers of the forward- and backward-propagating waves are constant in time and along the grating length. Find the relative power levels when $\delta/\kappa = 1.05$ and $\gamma P_0/\kappa = 2$, where P_0 is the total power.

1.12 Use the CW solution obtained in the previous problem to discuss how the stop band of a fiber grating is affected at high power levels because of the nonlinear effects.

1.13 Perturb the CW solution of Eqs. (1.4.1) and (1.4.2) and discuss the conditions under which it may become unstable.

1.14 Develop a computer program for solving Eqs. (1.4.1) and (1.4.2) numerically and use it to reproduce the results shown in Figure 1.18.

References

[1] M. Born and E. Wolf, *Principles of Optics*, 7th ed. (Cambridge University Press, New York, 1999), Section 8.6.

[2] G. P. Agrawal, *Nonlinear Fiber Optics*, 4th ed. (Academic Press, Boston, 2007).

[3] K. O. Hill, Y. Fujii, D. C. Johnson, and B. S. Kawasaki, *Appl. Phys. Lett.* **32**, 647 (1978).

[4] B. S. Kawasaki, K. O. Hill, D. C. Johnson, and Y. Fujii, *Opt. Lett.* **3**, 66 (1978).

[5] J. Bures, J. Lapierre, and D. Pascale, *Appl. Phys. Lett.* **37**, 860 (1980).

[6] J. Lapierre, J. Bures, and D. Pascale, *Opt. Commun.* **40**, 95 (1981).

[7] D. K. W. Lam and B. K. Garside, *Appl. Opt.* **20**, 440 (1982).

[8] J. Lapierre, J. Bures, and G. Chevalier, *Opt. Lett.* **7**, 37 (1982).

[9] J. Bures, S. Lacroix, and J. Lapierre, *Appl. Opt.* **21**, 3502 (1982).

[10] D. K. W. Lam, B. K. Garside, and K. O. Hill, *Opt. Lett.* **7**, 291 (1982).

[11] M. Parent, J. Bures, S. Lacroix, and J. Lapierre, *Appl. Opt.* **24**, 354 (1985).

[12] J. Stone, *J. Appl. Phys.* **62**, 4371 (1987).

[13] C. P. Kuo, U. Österberg, C. T. Seaton, G. I. Stegeman, and K. O. Hill, *Opt. Lett.* **13**, 1032 (1988).

[14] F. P. Payne, *Electron. Lett.* **25**, 498 (1989).

[15] H. G. Park and B. Y. Park, *Electron. Lett.* **25**, 737 (1989).

[16] F. Ouellette, *Electron. Lett.* **25**, 1590 (1989).

[17] G. Meltz, W. W. Morey, and W. H. Glen, *Opt. Lett.* **14**, 823 (1989).

[18] R. Kashyap, J. R. Armitage, R. Wyatt, S. T. Davey, and D. L. Williams, *Electron. Lett.* **26**, 730 (1990).

[19] K. O. Hill, B. Malo, K. A. Vineberg, F. Bilodeau, D. C. Johnson, and I. Skinner, *Electron. Lett.* **26**, 1270 (1990).

[20] S. La Rochelle, V. Mizrahi, G. I. Stegeman, and J. Sipe, *Appl. Phys. Lett.* **57**, 747 (1990).

[21] S. La Rochelle, Y. Hibino, V. Mizrahi, and G. I. Stegeman, *Electron. Lett.* **26**, 1459 (1990).

[22] J. P. Bernardin and N. M. Lawandy, *Opt. Commun.* **79**, 194 (1990).

[23] D. P. Hand and P. St. J. Russell, *Opt. Lett.* **15**, 104 (1990).

[24] B. Malo, K. A. Vineberg, F. Bilodeau, J. Albert, D. C. Johnson, and K. O. Hill, *Opt. Lett.* **15**, 953 (1990).

[25] G. A. Ball, W. W. Morey, and J. P. Waters, *Electron. Lett.* **26**, 1829 (1990).

[26] V. Mizrahi, S. La Rochelle, and G. I. Stegeman, *Phys. Rev. A* **43**, 433 (1991).

[27] S. E. Kanellopoulos, V. A. Handerek, H. Jamshidi, and A. J. Rogers, *IEEE Photon. Technol. Lett.* **3**, 244 (1991).

[28] S. Legoubin, E. Fertein, M. Douay, P. Bernage, P. Nidy, F. Bayon, and T. Georges, *Electron. Lett.* **27**, 1945 (1991).

[29] P. St. J. Russell, L. J. Poyntz-Wright, and D. P. Hand, *Proc. SPIE* **1373**, 126 (1991).

[30] G. Meltz and W. W. Morey, *Proc. SPIE* **1516**, 185 (1991).

[31] T. E. Tsai, C. G. Askins, and E. J. Friebele, *Appl. Phys. Lett.* **61**, 390 (1992).

[32] C. G. Askins, T. E. Tsai, G. M. Williams, M. A. Puttnam, M. Bashkansky, and E. J. Friebele, *Opt. Lett.* **17**, 833 (1992).

[33] H. G. Limberger, P. Y. Fonjallaz, and R. P. Salathé, *Electron. Lett.* **29**, 47 (1993).

[34] J. L. Archambault, L. Reekie, and P. St. J. Russell, *Electron. Lett.* **29**, 28 (1993); *Electron. Lett.* **29**, 453 (1993).

[35] L. Long, J. L. Archambault, L. Reekie, P. St. J. Russell, and D. N. Payne, *Opt. Lett.* **18**, 861 (1993).

[36] K. O. Hill, B. Malo, F. Bilodeau, D. C. Johnson, and J. Albert, *Appl. Phys. Lett.* **62**, 1035 (1993).

[37] D. Z. Anderson, V. Mizrahi, T. Erdogan, and A. E. White, *Electron. Lett.* **29**, 566 (1993).

[38] B. Malo, D. C. Johnson, F. Bilodeau, J. Albert, and K. O. Hill, *Opt. Lett.* **18**, 1277 (1993).

[39] P. J. Lemaire, R. M. Atkins, V. Mizrahi, and W. A. Reed, *Electron. Lett.* **29**, 1191 (1993).

[40] L. Long, J. L. Archambault, L. Reekie, P. St. J. Russell, and D. N. Payne, *Electron. Lett.* **29**, 1577 (1993).

[41] J. D. Prohaska, E. Snitzer, S. Rishton, and V. Boegli, *Electron. Lett.* **29**, 1614 (1993).

[42] K. C. Byron, K. Sugden, T. Bricheno, and I. Bennion, *Electron. Lett.* **29**, 1659 (1993).

[43] B. Malo, K. O. Hill, F. Bilodeau, D. C. Johnson, and J. Albert, *Appl. Phys. Lett.* **62**, 1668 (1993).

[44] H. Patrick and S. L. Gilbert, *Opt. Lett.* **18**, 1484 (1993).

[45] K. O. Hill, B. Malo, F. Bilodeau, and D. C. Johnson, *Annu. Rev. Mater. Sci.* **23**, 125 (1993).

[46] S. J. Mihailor and M. C. Gower, *Electron. Lett.* **30**, 707 (1994).

[47] M. C. Farries, K. Sugden, D. C. J. Reid, I. Bennion, A. Molony, and M. J. Goodwin, *Electron. Lett.* **30**, 891 (1994).

[48] R. Kashyap, *Fiber Bragg Gratings* (Academic Press, Boston, 1999).

[49] A. Othonos and K. Kalli, *Fiber Bragg Gratings: Fundamentals and Applications in Telecommunications and Sensing* (Artech House, Boston, 1999).

[50] M. N. Zervas, *Fiber Bragg Gratings for Optical Fiber Communications* (Wiley, Hoboken, NJ, 2001).

[51] P. N. Butcher and D. Cotter, *Elements of Nonlinear Optics* (Cambridge University Press, Cambridge, UK, 1990), p. 315.

[52] T. Erdogan, V. Mizrahi, P. J. Lemaire, and D. Monroe, *J. Appl. Phys.* **76**, 73 (1994).

[53] H. Patrick, S. L. Gilbert, A. Lidgard, and M. D. Gallagher, *J. Appl. Phys.* **78**, 2940 (1995).

[54] S. R. Baker, H. N. Rourke, V. Baker, and D. Goodchild, *J. Lightwave Technol.* **15**, 1470 (1997).

[55] S. Kannan, J. Z. Y. Gus, and P. J. Lemaire, *J. Lightwave Technol.* **15**, 1478 (1997).

[56] N. K. Viswanathan and D. L. LaBrake, *J. Lightwave Technol.* **22**, 1990 (2004).

[57] J. Canning, *Opt. Fiber Technol.* **6**, 275 (2000).

[58] A. Canagasabey, J. Canning, and N. Groothoff, *Opt. Lett.* **28**, 1108 (2003).

[59] K. P. Chen, P. R. Herman, and R. Tam, *J. Lightwave Technol.* **21**, 1958 (2003).

[60] M. Fokine and W. Margulis, *Opt. Lett.* **25**, 30 (2000).

[61] M. Lancry, P. Niay, M. Douay, C. Depecker, P. Cordier, and B. Poumellec, *J. Lightwave Technol.* **24**, 1376 (2006).

[62] R. Kashyap, H.-G. Froehlich, A. Swanton, and D. J. Armes, *Electron. Lett.* **32**, 1807 (1996).

[63] J. Albert, S. Theriault, F. Bilodeau, D. C. Johnson, K. O. Hill, P. Sixt, and M. J. Rooks, *IEEE Photon. Technol. Lett.* **8**, 1334 (1996).

[64] P. St. J. Russell and D. P. Hand, *Electron. Lett.* **26**, 1840 (1990).

[65] D. C. Johnson, F. Bilodeau, B. Malo, K. O. Hill, P. G. J. Wigley, and G. I. Stegeman, *Opt. Lett.* **17**, 1635 (1992).

[66] A. M. Vangsarkar, P. J. Lemaire, J. B. Judkins, V. Bhatia, T. Erdogan, and J. Sipe, *J. Lightwave Technol.* **14**, 58 (1996).

[67] K. M. Davis, K. Miura, N. Sugimoto, and K. Hirao, *Opt. Lett.* **21**, 1729 (1996).

[68] C. B. Schaffer, A. Brodeur, J. F. Garca, and E. Mazur, *Opt. Lett.* **26**, 93 (2001).

[69] M. Will, S. Nolte, B. N. Chichkov, and A. Tünnermann, *Appl. Opt.* **41**, 4360 (2002).

[70] A. M. Streltsov and N. F. Borrelli, *J. Opt. Soc. Am. B* **19**, 2496 (2002).

[71] A. Saliminia, N. T. Nguyen, M. C. Nadeau, S. Petit, S. L. Chin, and R. Vallée, *J. Appl. Phys.* **93**, 3724 (2003).

[72] H. Zhang, S. M. Eaton, and P. R. Herman, *Opt. Express* **14**, 4826 (2006).

[73] Y. Kondo, K. Nouchi, T. Mitsuyy, M. Watanabe, P. G. Kanansky, and K. Hirao, *Opt. Lett.* **24**, 646 (1999).

[74] S. J. Mihailov, C. W. Smelser, D. Grobnic, R. B. Walker, P. Lu, H. Ding, and J. Unruh, *J. Lightwave Technol.* **22**, 94-100 (2004).

[75] C. W. Smelser, S. J. Mihailov, and D. Grobnic, *Opt. Express* **13**, 5377 (2005).

[76] L. Sudrie, M. Franco, B. Prade, and A. Mysyrowicz, *Opt. Commun.* **191**, 333 (2001).

[77] A. Dragomir, D. N. Nikogosyan, K. A. Zagorulko, P. G. Kryukov, and E. M. Dianov, *Opt. Lett.* **28**, 2171 (2003).

[78] K. A. Zagorulko, P. G. Kryukov, Yu. V. Larionov, A. A. Rybaltovsky, E. M. Dianov, S. V. Chekalin, Yu. A. Matveets, and V. O. Kompanets, *Opt. Express* **12**, 5996 (2004).

[79] S. A. Slattery, D. N. Nikogosyan, and G. Brambilla, *J. Opt. Soc. Am. B* **22**, 354 (2005).

[80] P. St. J. Russell, *J. Mod. Opt.* **38**, 1599 (1991).

[81] H. A. Haus, *Waves and Fields in Optoelectronics* (Prentice-Hall, Englewood Cliffs, NJ, 1984).

[82] D. Marcuse, *Theory of Dielectric Optical Waveguides* (Academic Press, San Diego, CA, 1991).

[83] A. Yariv, *Optical Electronics in Modern Communications*, 5th ed. (Oxford University Press, New York, 1997).

[84] G. P. Agrawal and N. K. Dutta, *Semiconductor Lasers*, 2nd ed. (Van Nostrand Reinhold, New York, 1993).

[85] B. Crosignani, A. Cutolo, and P. Di Porto, *J. Opt. Soc. Am. B* **72**, 515 (1982).

[86] N. M. Litchinitser, B. J. Eggleton, and D. B. Patterson, *J. Lightwave Technol.* **15**, 1303 (1997).

[87] G. P. Agrawal, *Lightwave Technology: Teleommunication Systems* (Wiley, Hoboken, NJ, 2005).

[88] B. J. Eggleton, C. M. de Sterke, and R. E. Slusher, *J. Opt. Soc. Am. B* **16**, 587 (1999).

[89] H. G. Winful, J. H. Marburger, and E. Garmire, *Appl. Phys. Lett.* **35**, 379 (1979).

[90] H. G. Winful and G. D. Cooperman, *Appl. Phys. Lett.* **40**, 298 (1982).

[91] H. G. Winful, *Appl. Phys. Lett.* **46**, 527 (1985).

[92] W. Chen and D. L. Mills, *Phys. Rev. Lett.* **58**, 160 (1987); *Phys. Rev. B* **36**, 6269 (1987).

[93] D. L. Mills and S. E. Trullinger, *Phys. Rev. B* **36**, 947 (1987).

[94] C. M. de Sterke and J. E. Sipe, *Phys. Rev. A* **38**, 5149 (1988); *Phys. Rev. A* **39**, 5163 (1989).

[95] C. M. de Sterke and J. E. Sipe, *J. Opt. Soc. Am. B* **6**, 1722 (1989).

[96] S. Larochelle, V. Mizrahi, and G. Stegeman, *Electron. Lett.* **26**, 1459 (1990).

[97] C. M. de Sterke and J. E. Sipe, *Phys. Rev. A* **43**, 2467 (1991).

[98] N. D. Sankey, D. F. Prelewitz, and T. G. Brown, *Appl. Phys. Lett.* **60**, 1427 (1992); *J. Appl. Phys.* **73**, 1 (1993).

[99] J. Feng, *Opt. Lett.* **18**, 1302 (1993).

[100] Y. S. Kivshar, *Phys. Rev. Lett.* **70**, 3055 (1993).

[101] C. M. de Sterke and J. E. Sipe, in *Progress in Optics*, Vol. 33, E. Wolf, Ed. (Elsevier, Amsterdam, 1994), Chap. 3.

[102] P. St. J. Russell and J. L. Archambault, *J. Phys. III France* **4**, 2471 (1994).

[103] M. Scalora, J. P. Dowling, C. M. Bowden, and M. J. Bloemer, *Opt. Lett.* **19**, 1789 (1994).

[104] S. Radic, N. George, and G. P. Agrawal, *Opt. Lett.* **19**, 1789 (1994); *J. Opt. Soc. Am. B* **12**, 671 (1995).

[105] S. Radic, N. George, and G. P. Agrawal, *IEEE J. Quantum Electron.* **31**, 1326 (1995).

[106] A. R. Champneys, B. A. Malomed, and M. J. Friedman, *Phys. Rev. Lett.* **80**, 4169 (1998).

[107] A. E. Kozhekin, G. Kurizki, and B. Malomed, *Phys. Rev. Lett.* **81**, 3647 (1998).

[108] C. Conti, G. Asanto, and S. Trillo, *Opt. Express* **3**, 389 (1998).

[109] Y. A. Logvin and V. M. Volkov, *J. Opt. Soc. Am. B* **16**, 774 (1999).

[110] H. M. Gibbs, *Optical Bistability: Controlling Light with Light* (Academic Press, San Diego, CA, 1985).

[111] G. P. Agrawal and S. Radic, *IEEE Photon. Technol. Lett.* **6**, 995 (1994).

[112] C. M. de Sterke and J. E. Sipe, *Phys. Rev. A* **42**, 2858 (1990).

[113] H. G. Winful, R. Zamir, and S. Feldman, *Appl. Phys. Lett.* **58**, 1001 (1991).

[114] A. B. Aceves, C. De Angelis, and S. Wabnitz, *Opt. Lett.* **17**, 1566 (1992).

[115] C. M. de Sterke, *Phys. Rev. A* **45**, 8252 (1992).

[116] B. J. Eggleton, C. M. de Sterke, R. E. Slusher, and J. E. Sipe, *Electron. Lett.* **32**, 2341 (1996).

[117] C. M. de Sterke, *J. Opt. Soc. Am. B* **15**, 2660 (1998).

[118] K. Porsezian, K. Senthilnathan, and S. Devipriya, *IEEE J. Quantum Electron.* **41**, 789 (2005).

[119] J. E. Sipe and H. G. Winful, *Opt. Lett.* **13**, 132 (1988).

[120] C. M. de Sterke and J. E. Sipe, *Phys. Rev. A* **42**, 550 (1990).

[121] C. M. de Sterke, D. G. Salinas, and J. E. Sipe, *Phys. Rev. E* **54**, 1969 (1996).

[122] T. Iizuka and M. Wadati, *J. Opt. Soc. Am. B* **14**, 2308 (1997).

[123] C. M. de Sterke and B. J. Eggleton, *Phys. Rev. E* **59**, 1267 (1999).

[124] B. J. Eggleton, C. M. de Sterke, A. B. Aceves, J. E. Sipe, T. A. Strasser, and R. E. Slusher, *Opt. Commun.* **149**, 267 (1998).

[125] D. N. Christodoulides and R. I. Joseph, *Phys. Rev. Lett.* **62**, 1746 (1989).

[126] A. B. Aceves and S. Wabnitz, *Phys. Lett. A* **141**, 37 (1989).

[127] J. Feng and F. K. Kneubühl, *IEEE J. Quantum Electron.* **29**, 590 (1993).

[128] B. J. Eggleton, R. R. Slusher, C. M. de Sterke, P. A. Krug, and J. E. Sipe, *Phys. Rev. Lett.* **76**, 1627 (1996).

[129] C. M. de Sterke, N. G. R. Broderick, B. J. Eggleton, and M. J. Steel. *Opt. Fiber Technol.* **2**, 253 (1996).

[130] B. J. Eggleton, C. M. de Sterke, and R. E. Slusher, *J. Opt. Soc. Am. B* **14**, 2980 (1997).

[131] D. Taverner, N. G. R. Broderick, D. J. Richardson, M. Isben, and R. I. Laming, *Opt. Lett.* **23**, 259 (1998).

[132] D. Taverner, N. G. R. Broderick, D. J. Richardson, R. I. Laming, and M. Isben, *Opt. Lett.* **23**, 328 (1998).

[133] C. M. de Sterke, *Opt. Express* **3**, 405 (1998).

[134] N. G. R. Broderick, D. Taverner, and D. J. Richardson, *Opt. Express* **3**, 447 (1998).

[135] B. J. Eggleton, G. Lenz, R. E. Slusher, and N. M. Litchinitser, *Appl. Opt.* **37**, 7055 (1998).

[136] N. M. Litchinitser, B. J. Eggleton, C. M. de Sterke, A. B. Aceves, and G. P. Agrawal, *J. Opt. Soc. Am. B* **16**, 18 (1999).

[137] N. G. R. Broderick, D. J. Richardson, and M. Isben, *Opt. Lett.* **25**, 536 (2000).

[138] V. Perlin and H. G. Winful, *Phys. Rev. A* **64**, 043804 (2001).

[139] H. Lee and G. P. Agrawal, *IEEE J. Quantum Electron.* **39**, 508 (2003).

[140] K. Senthilnathan, K. Porsezian, P. Ramesh Babu, and V. Santhanam, *IEEE J. Quantum Electron.* **39**, 1492 (2003).

[141] J. T. Mok, I. C. M. Littler, E. Tsoy, and B. J. Eggleton, *Opt. Lett.* **30**, 2457 (2005).

[142] A. Rosenthal and M. Horowitz, *Opt. Lett.* **31**, 1334 (2006).

[143] M. Asobe, *Opt. Fiber Technol.* **3**, 142 (1997).

[144] W. E. Thirring, *Ann. Phys. (NY)* **3**, 91 (1958).

[145] A. V. Mikhailov, *JETP Lett.* **23**, 320 (1976).

[146] E. A. Kuznetsov and A. V. Mikhailov, *Teor. Mat. Fiz.* **30**, 193 (1977).

[147] D. J. Kaup and A. C. Newell, *Lett. Nuovo Cimento* **20**, 325 (1977).

[148] D. David, J. Harnad, and S. Shnider, *Lett. Math. Phys.* **8**, 27 (1984).

[149] H. Kawaguchi, K. Inoue, T. Matsuoka, and K. Otsuka, *IEEE J. Quantum Electron.* **21**, 1314 (1985).

[150] M. J. Adams and R. Wyatt, *IEE Proc.* **134** 35 (1987).

[151] M. J. Adams, *Opt. Quantum Electron.* **19**, S37 (1987).

[152] C. M. de Sterke, *Opt. Lett.* **17**, 914 (1992).

[153] N. G. R. Broderick, D. Taverner, D. J. Richardson, M. Isben, and R. I. Laming, *Phys. Rev. Lett.* **79**, 4566 (1997); *Opt. Lett.* **22**, 1837 (1997).

[154] A. Melloni, M. Chinello, and M. Martinelli, *IEEE Photon. Technol. Lett.* **12**, 42 (2000).

[155] S. Lee and S. T. Ho, *Opt. Lett.* **18**, 962 (1993).

[156] W. Samir, S. J. Garth, and C. Pask, *J. Opt. Soc. Am. B* **11**, 64 (1994).

[157] W. Samir, C. Pask, and S. J. Garth, *Opt. Lett.* **19**, 338 (1994).

[158] S. Pereira and J. E. Sipe, *Opt. Express* **3**, 418 (1998).

[159] R. E. Slusher, S. Spälter, B. J. Eggleton, S. Pereira, and J. E. Sipe, *Opt. Lett.* **25**, 749 (2000).

[160] S. Broderick, *Opt. Commun.* **148**, 90 (1999).

[161] K. O. Hill, B. Malo, K. A. Vineberg, F. Bilodeau, D. C. Johnson, and I. Skinner, *Electron. Lett.* **26**, 1270 (1990).

[162] B. J. Eggleton, R. E. Slusher, J. B. Judkins, J. B. Stark, and A. M. Vengsarkar, *Opt. Lett.* **22**, 883 (1997).

[163] J. N. Kutz, B. J. Eggleton, J. B. Stark, and R. E. Slusher, *IEEE J. Sel. Topics Quantum Electron.* **3**, 1232 (1997).

[164] Y. Jeong and B. Lee, *IEEE J. Quantum Electron.* **35**, 1284 (1999).

[165] V. E. Perlin and H. G. Winful, *J. Lightwave Technol.* **18**, 329, (2000).

[166] G. W. Chern, J. F. Chang, and L. A. Wang, *J. Opt. Soc. Am. B* **19**, 1497, (2002).

[167] D. Pudo, E. C. Mägi, and B. J. Eggleton, *Opt. Express* **14**, 3763 (2006).

[168] J. A. R Williams, I. Bennion, and L. Zhang, *IEEE Photon. Technol. Lett.* **7**, 491 (1995).

[169] A. Galvanauskas, P. A. Krug, and D. Harter, *Opt. Lett.* **21**, 1049 (1996).

[170] G. Lenz, B. J. Eggleton, and N. M. Litchinitser, *J. Opt. Soc. Am. B* **15**, 715 (1998).

[171] R. E. Slusher, B. J. Eggleton, T. A. Strasser, and C. M. de Sterke, *Opt. Express* **3**, 465 (1998).

[172] N. M. Litchinitser, G. P. Agrawal, B. J. Eggleton, and G. Lenz, *Opt. Express* **3**, 411 (1998).

[173] C. S. Yang, Y. C. Chiang, and H. C. Chang, *IEEE J. Quantum Electron.* **40**, 1337 (2004).

[174] V. Jayaraman, D. Cohen, and L. Coldren, *Appl. Phys. Lett.* **60**, 2321 (1992).

[175] B. J. Eggleton, P. A. Krug, L. Poladian, and F. Ouellette, *Electron. Lett.* **30**, 1620 (1994).

[176] N. G. R. Broderick, C. M. de Sterke, and B. J. Eggleton, *Phys. Rev. E* **52**, R5788 (1995).

[177] C. M. de Sterke and N. G. R. Broderick, *Opt. Lett.* **20**, 2039 (1995).

[178] B. J. Eggleton, C. M. de Sterke, and R. E. Slusher, *Opt. Lett.* **21**, 1223 (1996).

[179] N. G. R. Broderick and C. M. de Sterke, *Phys. Rev. E* **55**, 3232 (1997).

[180] C. M. de Sterke, B. J. Eggleton, and P. A. Krug, *J. Lightwave Technol.* **15**, 1494 (1997).

[181] P. Petropoulos, M. Isben, M. N. Zervas, and D. J. Richardson, *Opt. Lett.* **25**, 521 (2000).

[182] H. Li, Y. Sheng, Y. Li, and J. E. Rothenberg, *J. Lightwave Technol.* **21**, 2074 (2003).

[183] H. Lee and G. P. Agrawal, *IEEE Photon. Technol. Lett.* **15**, 1091 (2003); *IEEE Photon. Technol. Lett.* **16**, 635 (2004).

[184] J. E. Rothenberg, H. Li, Y. Sheng, J. Popelek, and J. Zweiback, *Opt. Lett.* **31**, 1199 (2006).

[185] P. St. J. Russell, *Phys. Rev. Lett.* **56**, 596 (1986).

[186] S. Frisken, *Opt. Lett.* **17**, 1776 (1992).

[187] B. Fischer, J. L. Zyskind, J. W. Sulhoff, and D. J. DiGiovanni, *Opt. Lett.* **18**, 2108 (1993).

[188] M. D. Feuer, *IEEE Photon. Technol. Lett.* **10**, 1587 (1998).

[189] S. A. Havstad, B. Fischer, A. E. Willner, and M. G. Wickham, *Opt. Lett.* **24**, 1466 (1999).

[190] S. Pitois, M. Haelterman, and G. Millot, *J. Opt. Soc. Am. B* **19**, 782 (2002).

[191] S. Stepanov, E. Hernández, and M. Plata, *Opt. Lett.* **29**, 1327 (2004); *J. Opt. Soc. Am. B* **22**, 1161 (2005).

[192] X. Fan, Z. He, Y. Mizuno, and K. Hotate, *Opt. Express* **13**, 5756 (2005).

[193] S. Wabnitz, *Opt. Lett.* **14**, 1071 (1989).

Chapter 2

Fiber Couplers

Fiber couplers, also known as directional couplers, constitute an essential component of lightwave technology [1]. They are used routinely for a multitude of fiber-optic devices that require splitting of an optical field into two coherent but physically separated parts (and vice versa). Although most applications of fiber couplers use only their linear characteristics, nonlinear effects have been studied since 1982 and can lead to all-optical switching among other applications. This chapter is devoted to describing nonlinear optical phenomena in fiber couplers. As an introduction, linear characteristics are described first in Section 2.1 using coupled-mode theory. In Section 2.2, the nonlinear effects are considered under continuous-wave (CW) conditions, along with the phenomenon of modulation instability. Section 2.3 focuses on propagation of short optical pulses through fiber couplers, with emphasis on optical solitons and nonlinear switching. Section 2.4 extends the discussion to asymmetric, active, and birefringent couplers. Fibers with multiple cores are considered in Section 2.5.

2.1 Coupler Characteristics

Fiber couplers are four-port devices (two input and two output ports) that are used routinely for a variety of applications related to fiber optics [2]–[5]. Their function is to split coherently an optical field, incident on one of the input ports, and direct the two parts to the output ports. Since the output is directed in two directions, such devices are also referred to as *directional couplers*. They can be made using planar waveguides as well and have been studied extensively in the context of $LiNbO_3$ and semiconductor waveguides. This chapter focuses exclusively on fiber-based directional couplers.

Several techniques can be used to make fiber couplers [5]. Figure 2.1 shows schematically a fused fiber coupler in which the cores of two single-mode fibers are brought close together in a central region such that the spacing between the cores is comparable to their diameters. A dual-core fiber, designed to have two cores close to each other throughout its length, can also act as a directional coupler. In both cases, the cores are close enough that the fundamental modes propagating in each core overlap partially in the cladding region between the two cores. It will be seen in this section

54

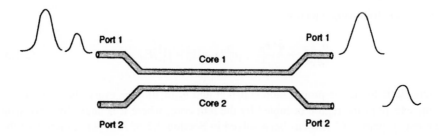

Figure 2.1: Nonlinear switching in a fiber coupler. Input pulses appear at different output ports depending on their peak powers.

that such evanescent wave coupling between the two modes can lead to the transfer of optical power from one core to another under suitable conditions. An important application of the nonlinear effects in fiber couplers consists of using them for optical switching. As shown in Figure 2.1, an optical pulse can be directed toward different output ports depending on its peak power.

Fiber couplers are called *symmetric* when their cores are identical in all respects. In general, the two cores need not be identical; such couplers are called *asymmetric*. In this section, we consider an asymmetric fiber coupler and discuss its operation using coupled-mode theory.

2.1.1   Coupled-Mode Equations

Coupled-mode theory is used commonly for directional couplers [6]–[10]. To derive the coupled-mode equations, we follow a procedure similar to that used in Section 1.3 for describing the grating-induced coupling between the counterpropagating waves inside the same core. Considering a specific frequency component at the frequency ω, we solve the Helmholtz equation:

$$\nabla^2 \tilde{\mathbf{E}} + \tilde{n}^2(x,y)k_0^2\tilde{\mathbf{E}} = 0, \tag{2.1.1}$$

where $k_0 = \omega/c = 2\pi/\lambda_0$, λ_0 is the vacuum wavelength of light, and $\tilde{\mathbf{E}}(\mathbf{r}, \omega)$ denotes the Fourier transform of the electric field $\mathbf{E}(\mathbf{r},t)$ with respect to time. The refractive index $\tilde{n}(x,y) = n_0$ everywhere in the x–y plane except in the region occupied by the two fiber cores, where it is larger by a constant amount.

The coupled-mode theory is based on the assumption that an approximate solution of Eq. (2.1.1) can be written as

$$\tilde{\mathbf{E}}(\mathbf{r}, \omega) \approx \hat{e}[\tilde{A}_1(z, \omega)F_1(x,y) + \tilde{A}_2(z, \omega)F_2(x,y)]\exp(i\beta z), \tag{2.1.2}$$

where the propagation constant β is yet to be determined. The polarization direction $\hat{e}$ of the optical field is assumed to remain unchanged during propagation. The spatial distribution $F_m(x,y)$ with $m = 1,2$ corresponds to the fundamental mode supported by the mth core in the absence of the other core. It is obtained by solving Eq. (2.1.1) and satisfies the following equation:

$$\frac{\partial^2 F_m}{\partial x^2} + \frac{\partial^2 F_m}{\partial y^2} + [n_m^2(x,y)k_0^2 - \bar{\beta}_m^2]F_m = 0, \tag{2.1.3}$$

where $\bar{\beta}_m$ is the mode-propagation constant and $n_m(x,y) = n_0$ everywhere in the x–y plane except in the region occupied by the mth core, where it is larger by a constant amount. Equation (2.1.3) has been solved in Section 2.2 of Ref. [11] in terms of the Bessel functions. The same solution applies here.

The amplitudes A_1 and A_2 vary along the coupler length because of the overlap between the two modes. To find how they evolve with z, we substitute Eq. (2.1.2) in Eq. (2.1.1), multiply the resulting equation by F_1^* or F_2^*, use Eq. (2.1.3), and integrate over the entire x–y plane. This procedure leads to the following set of two coupled-mode equations in the frequency domain:

$$\frac{d\tilde{A}_1}{dz} = i(\bar{\beta}_1 + \Delta\beta_1^{\mathrm{NL}} - \beta)\tilde{A}_1 + i\kappa_{12}\tilde{A}_2, \tag{2.1.4}$$

$$\frac{d\tilde{A}_2}{dz} = i(\bar{\beta}_2 + \Delta\beta_2^{\mathrm{NL}} - \beta)\tilde{A}_2 + i\kappa_{21}\tilde{A}_1, \tag{2.1.5}$$

where the coupling coefficient κ_{mp} and the nonlinear contribution $\Delta\beta_m^{\mathrm{NL}}$ are defined as ($m, p = 1$ or 2)

$$\kappa_{mp} = \frac{k_0^2}{2\beta} \int\!\!\int_{-\infty}^{\infty} (\tilde{n}^2 - n_p^2)F_m^*F_p \, dx \, dy, \tag{2.1.6}$$

$$\Delta\beta_m^{\mathrm{NL}} = \frac{k_0^2}{2\beta} \int\!\!\int_{-\infty}^{\infty} (\tilde{n}^2 - n_L^2)F_m^*F_m \, dx \, dy, \tag{2.1.7}$$

and n_L is the linear part of $\tilde{n}$. We have assumed that the modal distributions are normalized such that $\int\!\int_{-\infty}^{\infty} |F_m(x,y)|^2 \, dx \, dy = 1$.

The frequency-domain coupled-mode equations can be converted to the time domain following the method used in Section 1.3.1. In general, both $\bar{\beta}_m$ and κ_{mp} depend on frequency. We ignore the frequency dependence of κ_{mp} here but consider its impact on the coupler performance later (in Section 2.1.3). By expanding $\bar{\beta}_m(\omega)$ in a Taylor series around the carrier frequency ω_0 as

$$\bar{\beta}_m(\omega) = \beta_{0m} + (\omega - \omega_0)\beta_{1m} + \tfrac{1}{2}(\omega - \omega_0)^2\beta_{2m} + \cdots, \tag{2.1.8}$$

retaining terms up to second order, and replacing $\omega - \omega_0$ by a time derivative while taking the inverse Fourier transform, the time-domain coupled-mode equations can be written as [11]

$$\frac{\partial A_1}{\partial z} + \beta_{11}\frac{\partial A_1}{\partial t} + \frac{i\beta_{21}}{2}\frac{\partial^2 A_1}{\partial t^2} = i\kappa_{12}A_2 + i\delta_a A_1 + i(\gamma_1|A_1|^2 + C_{12}|A_2|^2)A_1, \tag{2.1.9}$$

$$\frac{\partial A_2}{\partial z} + \beta_{12}\frac{\partial A_2}{\partial t} + \frac{i\beta_{22}}{2}\frac{\partial^2 A_2}{\partial t^2} = i\kappa_{21}A_1 - i\delta_a A_2 + i(\gamma_2|A_2|^2 + C_{21}|A_1|^2)A_2 \tag{2.1.10}$$

where $v_{gm} \equiv 1/\beta_{1m}$ is the group velocity and β_{2m} is the group-velocity dispersion (GVD) in the mth core. We have introduced

$$\delta_a = \tfrac{1}{2}(\beta_{01} - \beta_{02}), \qquad \beta = \tfrac{1}{2}(\beta_{01} + \beta_{02}). \qquad (2.1.11)$$

The parameter δ_a is a measure of asymmetry between the two cores. The nonlinear parameters γ_m and C_{mp} ($m, p = 1$ or 2) are defined as

$$\gamma_m = n_2 k_0 \iint_{-\infty}^{\infty} |F_m|^4 \, dx \, dy, \qquad (2.1.12)$$

$$C_{mp} = 2 n_2 k_0 \iint_{-\infty}^{\infty} |F_m|^2 |F_p|^2 \, dx \, dy. \qquad (2.1.13)$$

The parameter γ_m is responsible for self-phase modulation (SPM), while the effects of cross-phase modulation (XPM) are governed by C_{mp} [11].

Equations (2.1.9) and (2.1.10) are valid under quite general conditions and include both the linear and nonlinear coupling mechanisms between the optical fields propagating inside the two cores of an asymmetric fiber coupler. They simplify considerably for a symmetric coupler with two identical cores. Using $\delta_a = 0$, $\kappa_{12} = \kappa_{21} \equiv \kappa$, and $C_{12} = C_{21} \equiv \gamma\sigma$, the coupled-mode equations for symmetric couplers become

$$\frac{\partial A_1}{\partial z} + \frac{1}{v_g} \frac{\partial A_1}{\partial t} + \frac{i\beta_2}{2} \frac{\partial^2 A_1}{\partial t^2} = i\kappa A_2 + i\gamma(|A_1|^2 + \sigma|A_2|^2)A_1, \qquad (2.1.14)$$

$$\frac{\partial A_2}{\partial z} + \frac{1}{v_g} \frac{\partial A_2}{\partial t} + \frac{i\beta_2}{2} \frac{\partial^2 A_2}{\partial t^2} = i\kappa A_1 + i\gamma(|A_2|^2 + \sigma|A_1|^2)A_2, \qquad (2.1.15)$$

where the subscript identifying a specific core has been dropped from the parameters v_g, β_2, and γ, since they have the same values for both cores. The nonlinear parameter γ can be written as $\gamma = n_2 k_0 / A_{\text{eff}}$ and is identical to that introduced in Section 2.3 of Ref. [11] for a fiber with the effective mode area A_{eff}. The XPM parameter σ is quite small in practice and can often be neglected altogether. The reason is related to the fact that the integral in Eq. (2.1.13) involves overlap between the mode intensities and is relatively small, even when the two cores are close enough that κ (involving overlap between the mode amplitudes) cannot be neglected. The coupling between A_1 and A_2 is essentially linear in that case.

2.1.2 Low-Power Optical Beams

Consider first the simplest case of a low-power CW beam incident on one of the input ports of a fiber coupler. The time-dependent terms can then be set to zero in Eqs. (2.1.9) and (2.1.10). Since the nonlinear terms are also negligible, the coupled-mode equations simplify considerably and become

$$\frac{dA_1}{dz} = i\kappa_{12} A_2 + i\delta_a A_1, \qquad (2.1.16)$$

$$\frac{dA_2}{dz} = i\kappa_{21} A_1 - i\delta_a A_2. \qquad (2.1.17)$$

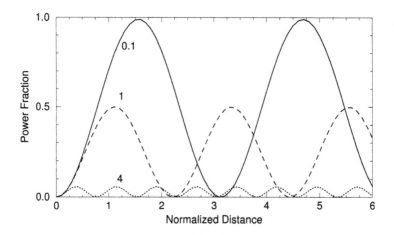

Figure 2.2: Fraction of power transferred to the second core plotted as a function of κz for three values of δ_a/κ when a CW beam is launched in one core at $z = 0$.

By differentiating Eq. (2.1.16) and eliminating dA_2/dz using Eq. (2.1.17), we obtain the following equation for A_1:

$$\frac{d^2 A_1}{dz^2} + \kappa_e^2 A_1 = 0, \tag{2.1.18}$$

where the effective coupling coefficient κ_e is defined as

$$\kappa_e = \sqrt{\kappa^2 + \delta_a^2}, \qquad \kappa = \sqrt{\kappa_{12}\kappa_{21}}. \tag{2.1.19}$$

The same harmonic-oscillator-type equation is also satisfied by A_2.

By using the boundary condition that a single CW beam is incident on one of the input ports such that $A_1(0) = A_0$ and $A_2(0) = 0$, the solution of Eqs. (2.1.16) and (2.1.17) is given by

$$A_1(z) = A_0[\cos(\kappa_e z) + i(\delta_a/\kappa_e)\sin(\kappa_e z)], \tag{2.1.20}$$

$$A_2(z) = A_0(i\kappa_{21}/\kappa_e)\sin(\kappa_e z). \tag{2.1.21}$$

Thus, even though $A_2 = 0$ initially at $z = 0$, some power is transferred to the second core as light propagates inside the fiber coupler. Figure 2.2 shows the ratio $|A_2/A_0|^2$ as a function of z for several values of δ_a/κ. In all cases, power transfer to the second core occurs in a periodic fashion. The maximum power is transferred at distances such that $\kappa_e z = m\pi/2$, where m is an integer. The shortest distance at which maximum power is transferred to the second core for the first time, called the *coupling length*, is given by $L_c = \pi/(2\kappa_e)$.

The power coming out of the two output ports of a fiber coupler depends on the coupler length L and the powers injected at the two input ends. For a symmetric coupler, the general solution of Eq. (2.1.18) can be written in a matrix form as

$$\begin{pmatrix} A_1(L) \\ A_2(L) \end{pmatrix} = \begin{pmatrix} \cos(\kappa L) & i\sin(\kappa L) \\ i\sin(\kappa L) & \cos(\kappa L) \end{pmatrix} \begin{pmatrix} A_1(0) \\ A_2(0) \end{pmatrix}. \tag{2.1.22}$$

The determinant of the 2×2 transfer matrix on the right side is unity, as it should be for a lossless coupler. Typically, only one beam is injected at the input end. The output powers, $P_1 = |A_1|^2$ and $P_2 = |A_2|^2$, are then obtained from Eq. (2.1.22) by setting $A_2(0) = 0$ and are given by

$$P_1(L) = P_0 \cos^2(\kappa L), \qquad P_2(L) = P_0 \sin^2(\kappa L), \qquad (2.1.23)$$

where $P_0 \equiv A_0^2$ is the incident power at the first input port. The coupler thus acts as a beam splitter, and the splitting ratio depends on the parameter κL.

If coupler length L is chosen such that $\kappa L = \pi/4$ or $L = L_c/2$, the power is equally divided between the two output ports. Such couplers are referred to as 50:50 or 3-dB couplers. Fiber couplers with $L = L_c$ transfer all of their input power to the second core (referred to as the *cross state*) whereas all of the launched power returns to the same core when $L = 2L_c$ (the bar state). It is important to realize that a directional coupler introduces a relative phase shift of $\pi/2$ between the two output ports, as indicated by the factor i in the off-diagonal term of the transfer matrix in Eq. (2.1.22). This phase shift plays an important role in the design of fiber interferometers (see Chapter 3).

The coupling length depends on the coupling coefficient κ, which in turn depends on the spacing d between the two cores. For a symmetric coupler, the integrals in Eq. (2.1.6) can be evaluated analytically [6]. The resulting expression is somewhat complicated as it involves the Bessel functions. The following empirical expression is useful in practice [12]:

$$\kappa = \frac{\pi V}{2k_0 n_0 a^2} \exp[-(c_0 + c_1 \bar{d} + c_2 \bar{d}^2)], \qquad (2.1.24)$$

where V is the fiber parameter, a is the core radius, and $\bar{d} \equiv d/a$ is the normalized center-to-center spacing between the two cores ($\bar{d} > 2$). The constants c_0, c_1, and c_2 depend on V as $c_0 = 5.2789 - 3.663V + 0.3841V^2$, $c_1 = -0.7769 + 1.2252V - 0.0152V^2$, and $c_2 = -0.0175 - 0.0064V - 0.0009V^2$. Equation (2.1.24) is accurate to within 1% for values of V and $\bar{d}$ in the range $1.5 \le V \le 2.5$ and $2 \le \bar{d} \le 4.5$. As an example, $\kappa \sim 1 \text{ cm}^{-1}$ for $\bar{d} = 3$, resulting in a coupling length of 1 cm or so. However, coupling length increases to 1 m or more when $\bar{d}$ exceeds 5.

One may ask whether the proximity of two cores always leads to periodic power transfer between the cores. In fact, the nature of power transfer depends on the launch conditions at the input end. The physics can be better understood by noting that, with a suitable choice of the propagation constant β in Eq. (2.1.2), the mode amplitudes $\tilde{A}_1$ and $\tilde{A}_2$ can be forced to become z independent. From Eqs. (2.1.4) and Eq. (2.1.5), this can occur when the amplitude ratio $f = \tilde{A}_2/\tilde{A}_1$ is initially such that

$$f = \frac{\beta - \bar{\beta}_1}{\kappa_{12}} = \frac{\kappa_{21}}{\beta - \bar{\beta}_2}, \qquad (2.1.25)$$

where the nonlinear contribution has been neglected. Equation (2.1.25) can be used to find the propagation constant β. Since β satisfies a quadratic equation, we find two values of β such that

$$\beta_\pm = \tfrac{1}{2}(\bar{\beta}_1 + \bar{\beta}_2) \pm \sqrt{\delta_a^2 + \kappa^2}. \qquad (2.1.26)$$

The spatial distribution corresponding to the two eigenvalues is given by

$$F_\pm(x,y) = (1 + f_\pm^2)^{-1/2}[F_1(x,y) + f_\pm F_2(x,y)], \qquad (2.1.27)$$

where $f_\pm$ is obtained from Eq. (2.1.25) using $\beta = \beta_\pm$. These two specific linear combinations of F_1 and F_2 constitute the eigenmodes of a fiber coupler (also called *supermodes*), and the eigenvalues $\beta_\pm$ correspond to their propagation constants. In the case of a symmetric coupler, $f_\pm = \pm 1$ and the eigenmodes reduce to the even and odd combinations of F_1 and F_2. When the input conditions are such that an eigenmode of the coupler is excited, no power transfer occurs between the two cores.

The periodic power transfer between the two cores, occurring when light is incident on only one core, can be understood using the preceding modal description as follows. Under such launch conditions, both supermodes of the fiber coupler are excited simultaneously. Each supermode propagates with its own propagation constant. Since β_+ and β_- are not the same, the two supermodes develop a relative phase difference on propagation. This phase difference, $\psi(z) = (\beta_+ - \beta_-)z \equiv 2\kappa_e z$, is responsible for the periodic power transfer between two cores. The situation is analogous to that occurring in birefringent fibers when linearly polarized light is launched at an angle from a principal axis. In that case, the relative phase difference between the two orthogonally polarized eigenmodes leads to periodic evolution of the state of polarization, and the role of coupling length is played by the beat length [11]. The analogy between fiber couplers and birefringent fibers turns out to be quite useful even when the nonlinear effects are included.

2.1.3   Linear Pulse Switching

In the case of low-energy optical pulses, nonlinear effects can be neglected but the effects of fiber dispersion should be included. For symmetric couplers, the coupled-mode equations, Eqs. (2.1.14) and (2.1.15), become

$$\frac{\partial A_1}{\partial z} + \frac{i\beta_2}{2}\frac{\partial^2 A_1}{\partial T^2} = i\kappa A_2, \qquad (2.1.28)$$

$$\frac{\partial A_2}{\partial z} + \frac{i\beta_2}{2}\frac{\partial^2 A_2}{\partial T^2} = i\kappa A_1, \qquad (2.1.29)$$

where $T = t - z/v_g$ is the reduced time and the parameter β_2 accounts for the effects of group-velocity dispersion (GVD) in each core of the fiber coupler.

We can introduce the dispersion length in the usual way as $L_D = T_0^2/|\beta_2|$, where T_0 is related to the pulse width. The GVD effects are negligible if the coupler length $L \ll L_D$. Since L is comparable in practice to the coupling length ($L_c = \pi/2\kappa$), GVD has no effect on couplers for which $\kappa L_D \gg 1$. Since L_D exceeds 1 km for pulses with $T_0 > 1$ ps whereas $L_c < 1$ m typically, the GVD effects become important only for ultrashort pulses ($T_0 < 0.1$ ps). If we neglect the GVD term in Eqs. (2.1.28) and (2.1.29), the resulting equations become identical to those applicable for CW beams. Thus, picosecond optical pulses should behave in the same way as CW beams. More specifically, their energy is transferred to the neighboring core periodically when such pulses are incident on one of the input ports of a fiber coupler.

This conclusion is modified if the frequency dependence of the coupling coefficient κ cannot be ignored [13]. It can be included by expanding $\kappa(\omega)$ in a Taylor series around the carrier frequency ω_0 in a way similar to Eq. (2.1.8) so that

$$\kappa(\omega) \approx \kappa_0 + (\omega - \omega_0)\kappa_1 + \frac{1}{2}(\omega - \omega_0)^2 \kappa_2, \qquad (2.1.30)$$

where $\kappa_m = d^m\kappa/d\omega^m$ is evaluated at $\omega = \omega_0$. When the frequency-domain coupled-mode equations are converted to time domain, two additional terms appear. With these terms included, Eqs. (2.1.28) and (2.1.29) become

$$\frac{\partial A_1}{\partial z} + \kappa_1\frac{\partial A_2}{\partial T} + \frac{i\beta_2}{2}\frac{\partial^2 A_1}{\partial T^2} + \frac{i\kappa_2}{2}\frac{\partial^2 A_2}{\partial T^2} = i\kappa_0 A_2, \qquad (2.1.31)$$

$$\frac{\partial A_2}{\partial z} + \kappa_1\frac{\partial A_1}{\partial T} + \frac{i\beta_2}{2}\frac{\partial^2 A_2}{\partial T^2} + \frac{i\kappa_2}{2}\frac{\partial^2 A_1}{\partial T^2} = i\kappa_0 A_1. \qquad (2.1.32)$$

In practice, the κ_2 term is negligible for pulses as short as 0.1 ps. The GVD term is also negligible if $\kappa L_D \gg 1$. Setting $\beta_2 = 0$ and $\kappa_2 = 0$, Eqs. (2.1.31) and (2.1.32) can be solved analytically to yield [13]

$$A_1(z,T) = \frac{1}{2}\left[A_0(T - \kappa_1 z)e^{i\kappa_0 z} + A_0(T + \kappa_1 z)e^{-i\kappa_0 z}\right], \qquad (2.1.33)$$

$$A_2(z,T) = \frac{1}{2}\left[A_0(T - \kappa_1 z)e^{i\kappa_0 z} - A_0(T + \kappa_1 z)e^{-i\kappa_0 z}\right], \qquad (2.1.34)$$

where $A_0(T)$ represents the shape of the input pulse at $z = 0$. When $\kappa_1 = 0$, the solution reduces to

$$A_1(z,T) = A_0(T)\cos(\kappa_0 z), \qquad A_2(z,T) = A_0(T)\sin(\kappa_0 z). \qquad (2.1.35)$$

Equation (2.1.35) shows that the pulse switches back and forth between the two cores, while maintaining its shape, when the frequency dependence of the coupling coefficient can be neglected. However, when κ_1 is not negligible, Eq. (2.1.34) shows that the pulse splits into two subpulses after a few coupling lengths, and separation between the two increases with propagation. This effect, referred to as *intermodal dispersion*, is similar in nature to polarization-mode dispersion occurring in birefringent fibers [11]. Intermodal dispersion was observed in a 1997 experiment by launching short optical pulses (width about 1 ps) in one core of a dual-core fiber with the center-to-center spacing $d \approx 4a$ [14]. The autocorrelation traces showed the evidence of pulse splitting after 1.25 m, and the subpulses separated from each other at a rate of 1.13 ps/m. The coupling length was estimated to be about 4 mm. Intermodal dispersion in fiber couplers becomes of concern only when the coupler length $L \gg L_c$ and pulse widths are ~ 1 ps or shorter. This effect is neglected in the following discussion of nonlinear effects in fiber couplers.

2.2   Nonlinear Effects

Nonlinear effects in directional couplers were studied starting in 1982 [15]–[35]. An important application of fiber couplers consists of using them for all-optical switching. Figure 2.1 showed schematically how an optical pulse can be directed toward different output ports, depending on its peak power. In this section, we focus on the quasi-CW case and consider a symmetric coupler with identical cores to simplify the discussion.

2.2.1 Quasi-CW Switching

The nonlinear coupled-mode equations for CW beams propagating inside a symmetric coupler are obtained from Eqs. (2.1.14) and (2.1.15) by neglecting the time-derivative terms. The resulting equations are

$$\frac{dA_1}{dz} = i\kappa A_2 + i\gamma(|A_1|^2 + \sigma|A_2|^2)A_1, \tag{2.2.1}$$

$$\frac{dA_2}{dz} = i\kappa A_1 + i\gamma(|A_2|^2 + \sigma|A_1|^2)A_2. \tag{2.2.2}$$

These equations are also applicable for optical pulses wide enough that the dispersion length L_D is much larger than the coupler length L (as the effects of GVD are then negligible). This is referred to as the *quasi-CW case*.

Equations (2.2.1) and (2.2.2) are similar to those studied in Section 6.3 of Ref. [11] in the context of birefringent fibers and can be solved analytically using the same technique. Introducing the powers and phases through

$$A_j = \sqrt{P_j}\exp(i\phi_j), \quad (j = 1, 2), \tag{2.2.3}$$

and defining the phase difference $\phi = \phi_1 - \phi_2$, we obtain the following set of three equations:

$$\frac{dP_1}{dz} = 2\kappa\sqrt{P_1 P_2}\sin\phi, \tag{2.2.4}$$

$$\frac{dP_2}{dz} = -2\kappa\sqrt{P_1 P_2}\sin\phi, \tag{2.2.5}$$

$$\frac{d\phi}{dz} = \frac{P_2 - P_1}{\sqrt{P_1 P_2}}\kappa\cos\phi + \frac{4\kappa}{P_c}(P_1 - P_2), \tag{2.2.6}$$

where the critical power P_c is defined as

$$P_c = 4\kappa/[\gamma(1 - \sigma)]. \tag{2.2.7}$$

The critical power level plays an important role since the solution of Eqs. (2.2.4)–(2.2.6) exhibits qualitatively different behavior depending on whether the input power exceeds P_c.

Equations (2.2.4)–(2.2.6) can be solved analytically in terms of the elliptic functions after noting that they have the following two invariants [15]:

$$P_0 = P_1 + P_2, \quad \Gamma = \sqrt{P_1 P_2}\cos\phi - 2P_1 P_2/P_c, \tag{2.2.8}$$

where P_0 is the total power in both cores. In the specific case in which all the input power is initially launched into one core of a fiber coupler, the power remaining in that core after a distance z is given by

$$P_1(z) = |A_1(z)|^2 = \tfrac{1}{2}P_0[1 + \mathrm{cn}(2\kappa z|m)], \tag{2.2.9}$$

where $\mathrm{cn}(x|m)$ is a Jacobi elliptic function with modulus $m = (P_0/P_c)^2$. The power transferred to the second core is obtained using $P_2(z) = P_0 - P_1(z)$.

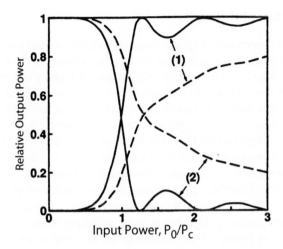

Figure 2.3: Nonlinear switching of CW beams in a fiber coupler with $\kappa L = \pi/2$. Solid lines show relative powers at the two output ports as a function of input power. Dashed lines show the coupler response in the quasi-CW case. (From Ref. [24]; ©1989 IEEE.)

In the low-power limit ($m \ll 1$), Eq. (2.2.9) reduces to the result $P_1(z) = P_0 \cos^2(\kappa z)$, as it should. Periodic transfer of the power between the two cores persists as long as $P_0 < P_c$ so that $m < 1$. However, as P_0 approaches P_c, the period begins to increase, becoming infinite when $P_0 = P_c$. The solution (2.2.9) reduces to $P_1(z) = \frac{1}{2}P_0[1 + \text{sech}(2\kappa z)]$ when $m = 1$, and at most, half of the power is transferred to the second core no matter how long the coupler is. For $P_0 > P_c$, the solution once again becomes periodic but the amount of power transferred to the second core is reduced to below 50% and becomes negligible for $P_0 \gg P_c$.

The solid lines in Figure 2.3 show the relative powers as a function of the input power at the two output ports of a coupler of length $L = L_c$. For $P_0 \ll P_c$, the launched power is transferred completely to the second core (cross state). However, the nonlinear effects influence the power transfer, and much less power is transferred to the other core (bar state) when P_0 exceeds P_c. Therefore, an optical beam can be switched from one output port to the other, depending on its input power.

The physics behind all-optical switching can be understood by noting that, when an optical beam is launched in one core of the fiber coupler, the SPM-induced phase shift is not the same in both cores because of different mode powers. As a result, even a symmetric fiber coupler behaves asymmetrically because of the nonlinear effects. The situation is, in fact, similar to that occurring in asymmetric fiber couplers where the difference in the mode-propagation constants introduces a relative phase shift between the two cores and hinders complete power transfer between them. Here, even though the linear propagation constants are the same, a relative phase shift between the two cores is introduced by SPM. At sufficiently high input powers, the phase difference—or SPM-induced detuning—becomes large enough that the input beam remains confined to the same core in which it was initially launched.

2.2.2   Experimental Results

The experimental observation of optical switching in fiber couplers using CW beams is difficult because of relatively high values of the critical power in silica fibers. We can estimate P_c from Eq. (2.2.7) using appropriate values of κ and γ and setting the XPM parameter $\sigma \approx 0$. If we use $\kappa = 1$ cm^{-1} and $\gamma = 10$ W^{-1}/km as typical values, we find that $P_c = 40$ kW. It is difficult to launch such high CW power levels without damaging silica fibers. A common practical solution is to use short optical pulses with high peak powers but wide enough that the GVD effects are not important (the quasi-CW case).

There is an obvious problem with the use of optical pulses in the quasi-CW regime. Only the central intense part of an input pulse is switched since pulse wings exhibit the low-power behavior. Therefore, a nonuniform intensity profile of optical pulses leads to distortion even when the effects of GVD are negligible. As one may expect, pulse distortion is accompanied by degradation in the switching behavior. As an example, the dashed curves in Figure 2.3 show the response of a fiber coupler to input pulses whose intensity varies as $\mathrm{sech}^2(t/T_0)$. These curves represent relative energy levels in the two cores and are obtained by integrating over the pulse shape. When compared with the case of CW beams, pulse switching is not only more gradual but also incomplete. Less than 75% of the incident peak power remains in the core in which the input pulse is launched even at peak power levels in excess of $2P_c$. This behavior restricts severely the usefulness of fiber couplers as an all-optical switch.

The results shown in Figure 2.3 do not include the effects of dispersion. As one may anticipate, the situation becomes worse in the case of normal GVD because of pulse broadening. However, the performance of fiber couplers should improve significantly for optical pulses experiencing anomalous GVD and propagating as a soliton. The reason is related to the particlelike nature of optical solitons. This topic is covered in Section 2.3.

Nonlinear effects in dual-core fiber couplers were observed starting in 1985, and a clear evidence of high-contrast optical switching had been seen by 1989 [20]–[24]. All of the experiments used short optical pulses propagating in the normal-GVD region of the fiber and, therefore, did not make use of solitons. In the 1985 experiment [20], 80-ns pulses from a frequency-doubled Nd:YAG laser ($\lambda = 0.53$ μm) were focused onto one core of a dual-core fiber. The 2.6-μm-diameter cores were separated by more than 8 μm (center-to-center spacing), resulting in a relatively small value of the coupling coefficient. Nonetheless, the transmitted power from a 18-cm-long coupler was found to increase as the launched peak power increased beyond the 100-W level. In a later experiment, the use of 50-ps pulses from a mode-locked laser provided better evidence of nonlinear switching [21].

In a 1987 experiment, 30-ps pulses from a 1.06-μm Nd:YAG laser were injected into one core of a 2-m-long dual-core fiber in which 5-μm-diameter cores were separated by 8 μm [22]. The critical power P_c was estimated to be 850 W for this coupler, and its length was about 3.8 L_c. At low input power levels, 90% of the pulse energy transferred to the neighboring core. However, the transferred energy was only 40% when the input peak power increased to about 700 W. The switching contrast improved considerably in a 1988 experiment [23] that used 100-fs pulses from a dye laser operating at 0.62 μm. The fiber coupler was only 5-mm long, consisted of two

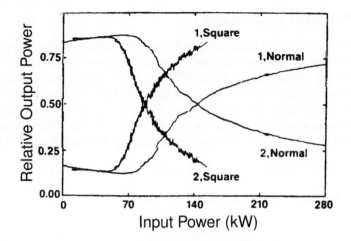

Figure 2.4: Switching data for a fiber coupler of length $L = L_c$. Relative output powers at ports 1 and 2 are shown as a function of input peak power for 90-fs bell-shaped (normal) pulses and 540-fs square-shaped pulses. (From Ref. [24]; ©1989 IEEE.)

2.8-μm-diameter cores separated by 8.4 μm, and required 32 kW of peak power for switching to occur. The measured switching characteristics were in good agreement with the theoretical prediction shown by the dashed lines in Figure 2.3. Fiber dispersion played a relatively minor role even for 100-fs pulses because of the short length of the coupler used in the experiment ($L \ll L_D$). The autocorrelation measurements showed that only the central part of the pulse underwent switching. The use of square-shaped femtosecond pulses in 1989 resulted in considerable improvement, since their use avoids the pulse breakup [24]. Figure 2.4 shows the switching characteristics measured using bell-shaped (Gaussian-like) and square-shaped pulses. Not only is the switching contrast better, the switching peak power is also lower for square pulses.

The high power levels needed for nonlinear switching in fiber couplers have hindered the use of such devices for this purpose. The switching threshold can be reduced by using fibers made with a material whose nonlinear parameter n_2 is much larger compared with that of silica. Several such materials have been used to make fiber couplers. In one case, a dye-doped polymer fiber was used [33]. Both cores of this fiber were doped with a squarylium dye and were embedded in a PMMA polymer cladding. The 6-μm-radius cores were separated by 18 μm. The coupling length was estimated to be about 1 cm. Nonlinear transmission was observed using a Q-switched, mode-locked Nd:YAG laser. In another approach, GeS_2-based chalcohalide glass was used to make the fiber [34]. The nonlinear parameter for this glass was measured to be $n_2 \approx 7.5 \times 10^{-14}$ cm^2/W, a value that is larger by more than a factor of 200 compared with that of silica. As a result, the switching threshold should also be reduced by the same factor. A third approach used a polyconjugated polymer (DPOP-PPV) to make a nonlinear directional coupler [35]. Two-photon absorption plays an important role when dye-doped or semiconductor-doped fibers are used and can affect the switching characteristics adversely.

2.2.3 Nonlinear Supermodes

An alternative approach for understanding the nonlinear effects in fiber couplers makes use of the concept of nonlinear supermodes, which represent optical fields that propagate without any change in spite of the SPM and XPM effects. Mathematically, they represent z-independent solutions (the fixed points) of Eqs. (2.2.4)–(2.2.6) and can be obtained by setting the z derivatives to zero. Here, we use an approach based on the rotation of a vector on the Poincaré sphere [18]. Let us introduce the following three real variables (in analogy with the Stokes parameters of Section 6.3 of Ref. [11]):

$$S_1 = |A_1|^2 - |A_2|^2, \qquad S_2 = 2\text{Re}(A_1 A_2^*), \qquad S_3 = 2\text{Im}(A_1 A_2^*), \qquad (2.2.10)$$

and rewrite Eqs. (2.2.1) and (2.2.2) in terms of them as

$$\frac{dS_1}{dz} = 2\kappa S_3, \qquad (2.2.11)$$

$$\frac{dS_2}{dz} = -\gamma(1 - \sigma)S_1 S_3, \qquad (2.2.12)$$

$$\frac{dS_3}{dz} = \gamma(1 - \sigma)S_1 S_2 - 2\kappa S_1. \qquad (2.2.13)$$

It can be easily verified from Eqs. (2.2.10)–(2.2.13) that $S_1^2 + S_2^2 + S_3^2 = |A_1|^2 + |A_2|^2 \equiv P_0$, where P_0 is the total power in both cores. Since P_0 is independent of z, the Stokes vector **S** with components S_1, S_2, and S_3 moves on the surface of a sphere of radius P_0 as the CW light propagates inside the fiber coupler. This sphere, known as the *Poincaré sphere*, provides a visual description of the coupler dynamics. In fact, Eqs. (2.2.11)–(2.2.13) can be written in the form of a single vector equation as

$$\frac{d\mathbf{S}}{dz} = \mathbf{W} \times \mathbf{S}, \qquad (2.2.14)$$

where $\mathbf{W} = \mathbf{W}_L + \mathbf{W}_{\text{NL}}$ such that $\mathbf{W}_L = 2\kappa\hat{y}$ and $\mathbf{W}_{\text{NL}} = \gamma(1 - \sigma)S_1\hat{x}$. Thus, the linear coupling rotates the Stokes vector **S** around the y axis while the SPM and XPM rotate it around the x axis. The combination of the two rotations determines the location of the Stokes vector on the Poincaré sphere at a given distance along the coupler length.

Figure 2.5 shows trajectories of the Stokes vector on the Poincaré sphere under three different conditions. In the low-power case, nonlinear effects can be neglected by setting $\gamma = 0$. Since $\mathbf{W}_{\text{NL}} = 0$ in that case, the Stokes vector rotates around the S_2 or y axis with an angular velocity 2κ, as seen in Figure 2.5(a). This is equivalent to the periodic solution obtained earlier. If the Stokes vector initially is oriented along the S_2 axis, it remains fixed. This can also be seen from the steady-state (z-invariant) solution of Eqs. (2.2.11)–(2.2.13) by noting that the Stokes vectors with components $(0, P_0, 0)$ and $(0, -P_0, 0)$ represent two fixed points in the linear case. These fixed points correspond to the even and odd supermodes of a fiber coupler discussed earlier.

In the nonlinear case, the behavior depends on the power level of the incident light. As long as $P_0 < P_c/2$, nonlinear effects play a minor role, and the situation is similar to the linear case, as shown in Fig. 2.5(b). At higher power levels, the motion of the Stokes vector on the Poincaré sphere becomes quite complicated since $\mathbf{W}_L$ is oriented

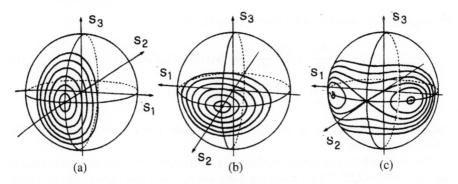

Figure 2.5: Trajectories showing motion of the Stokes vector on the Poincaré sphere. (a) Linear case; (b) $P_0 < P_c$; (c) $P_0 > P_c$. (From Ref. [18]; ©1985 AIP.)

along the y axis while $\mathbf{W}_{NL}$ is oriented along the x axis. Moreover, the nonlinear rotation of the Stokes vector along the x axis depends on the magnitude of S_1 itself. Figure 2.5(c) shows the motion of the Stokes vector when $P_0 > P_c$.

To understand the dynamical behavior, we find the fixed points of Eqs. (2.2.11)–(2.2.13) by setting the z derivatives to zero. The location and the number of fixed points depend on the optical power P_0 launched inside the fiber. More specifically, the number of fixed points changes from two to four at a critical power level, $P_0 = P_c/2$, where P_c is given in Eq. (2.2.7). For $P_0 < P_c/2$, only two fixed points, $(0, P_0, 0)$ and $(0, -P_0, 0)$, occur; these are identical to the low-power case. In contrast, when $P_0 > P_c/2$, two new fixed points emerge. The components of the Stokes vector, at the location of the new fixed points on the Poincaré sphere, are given by [18]

$$S_1 = \pm\sqrt{P_0^2 - P_c^2/4}, \quad S_2 = P_c/2, \quad S_3 = 0. \tag{2.2.15}$$

The new fixed points represent the nonlinear supermodes of a fiber coupler in the sense that, when input light excites one of these eigenmodes, the core powers do not change along the coupler length in spite of the close proximity of the two cores. Trajectories near the new fixed points are separated from those occurring near the fixed point $(0, P_0, 0)$ by a separatrix. The nonlinear switching corresponds to the transition from the low-power fixed point $(0, P_0, 0)$ to one of the new fixed points.

2.2.4 Modulation Instability

The CW solution of the coupled-mode equations (the fixed points) can become unstable in the presence of GVD in the same way as a CW beam propagating inside an optical fiber can break up into a pulse train under certain conditions. The modulation instability of couplers is similar in nature to the vector modulation instability discussed in Section 6.4 of Ref. [11] as the underlying coupled nonlinear Schrödinger (NLS) equations have the same form. This analogy is not obvious from Eqs. (2.1.14) and (2.1.15), because the XPM term is often negligible in practice. By setting $\sigma = 0$, the coupled NLS equations for a fiber coupler reduce to

$$\frac{\partial A_1}{\partial z} + \frac{i\beta_2}{2}\frac{\partial^2 A_1}{\partial T^2} = i\kappa A_2 + i\gamma|A_1|^2 A_1, \tag{2.2.16}$$

$$\frac{\partial A_2}{\partial z} + \frac{i\beta_2}{2}\frac{\partial^2 A_2}{\partial t^2} = i\kappa A_1 + i\gamma|A_2|^2 A_2, \tag{2.2.17}$$

where we have introduced, as usual, the reduced time $T = t - z/v_g$ to eliminate the group-velocity term.

The analogy between a fiber coupler and a birefringent fiber becomes quite clear if Eqs. (2.2.16) and (2.2.17) are rewritten using the even and odd supermodes of a fiber coupler. For this purpose, we introduce two new variables:

$$B_1 = (A_1 + A_2)/\sqrt{2}, \qquad B_2 = (A_1 - A_2)/\sqrt{2} \tag{2.2.18}$$

so that B_1 and B_2 correspond to the amplitudes associated with the even and odd supermodes introduced earlier [see Eq. (2.1.27)]. In terms of the new variables, Eqs. (2.2.16) and (2.2.17) can be written as

$$\frac{\partial B_1}{\partial z} + \frac{i\beta_2}{2}\frac{\partial^2 B_1}{\partial T^2} - i\kappa B_1 = \frac{i\gamma}{2}[(|B_1|^2 + 2|B_2|^2)B_1 + B_2^2 B_1^*], \tag{2.2.19}$$

$$\frac{\partial B_2}{\partial z} + \frac{i\beta_2}{2}\frac{\partial^2 B_2}{\partial T^2} + i\kappa B_2 = \frac{i\gamma}{2}[(|B_2|^2 + 2|B_1|^2)B_2 + B_1^2 B_2^*]. \tag{2.2.20}$$

The even and odd supermodes are uncoupled linearly but their phase velocities are not the same, as evident from different signs of the κ term in these two equations. Since light in the even supermode travels more slowly than that in the odd supermode, the even and odd supermodes are analogous to the light polarized along the slow and fast axes of a birefringent fiber. As seen from Eqs. (2.2.19) and (2.2.20), the coupled NLS equations written in terms of the supermodes have three nonlinear terms that correspond to SPM, XPM, and four-wave-mixing-type coupling (identical to the case of birefringent fibers).

The steady-state or CW solution of Eqs. (2.2.19) and (2.2.20) is easily obtained when the input conditions are such that either the even or the odd supermode is excited exclusively. In the case of even supermode, the CW solution is given by

$$\bar{B}_1 = \sqrt{P_0}\exp(i\theta), \qquad \bar{B}_2 = 0, \tag{2.2.21}$$

where $\theta = (\gamma P_0/2 + \kappa)z$. The solution in the case of odd supermode is obtained from Eq. (2.2.21) by changing the sign of κ and the subscripts 1 and 2. In both cases, the input power remains equally divided between the two cores, with no power exchange taking place between them. In the Poincaré sphere representation, these two CW solutions correspond to the fixed points $(0, P_0, 0)$ and $(0, -P_0, 0)$, as discussed earlier.

We follow a standard procedure [11] to examine the stability of the CW solution in Eq. (2.2.21). Assuming a time-dependent solution of the form

$$B_1 = (\sqrt{P_0} + b_1)\exp(i\theta), \qquad B_2 = b_2\exp(i\theta), \tag{2.2.22}$$

where b_1 and b_2 are small perturbations, we linearize Eqs. (2.2.19) and (2.2.20) in terms of b_1 and b_2 and obtain a set of two coupled linear equations. These equations can be solved by assuming a solution of the form

$$b_m = u_m \exp[i(K_p z - \Omega T)] + i v_m \exp[-i(K_p z - \Omega T)], \qquad (2.2.23)$$

where $m = 1$ or 2, Ω is the frequency of perturbation, and K_p is the corresponding wave number.

The four algebraic equations obtained with this technique are found to have an interesting property. The two equations for u_1 and v_1 are coupled, and so are those for u_2 and v_2. However, these two sets of two equations are not coupled. This feature simplifies the analysis considerably. The dispersion relation for the even-mode perturbation b_1 turns out to be

$$K_p^2 = \tfrac{1}{2}\beta_2 \Omega^2 (\tfrac{1}{2}\beta_2 \Omega^2 + \gamma P_0) \qquad (2.2.24)$$

and is the same dispersion relation obtained in Section 5.1 of Ref. [11], except for a factor of 2 in the last term. The features associated with modulation instability are thus identical to those found there. More specifically, no instability occurs in the case of normal GVD. When GVD is anomalous, gain curves are similar to those appearing in Figure 5.1 of Ref. [11].

The new feature for fiber couplers is that, even when CW light is launched initially into the even supermode, perturbations in the odd supermode can grow because of the coupling between the two cores. The odd-mode perturbations satisfy the dispersion relation [37]:

$$K_p^2 = (\tfrac{1}{2}\beta_2 \Omega^2 - 2\kappa)(\tfrac{1}{2}\beta_2 \Omega^2 - 2\kappa + \gamma P_0). \qquad (2.2.25)$$

The presence of κ in this equation shows that the coupling between the two cores can lead to a new kind of modulation instability in fiber couplers. Indeed, it is easy to see that K_p becomes complex under certain conditions even in the normal-GVD regime. Introducing the instability gain as $g_0 = 2\text{Im}(K_p)$, the gain is given by

$$g_0(f) = 2\kappa \sqrt{(2 - sf^2)(4p - 2 + sf^2)}, \qquad (2.2.26)$$

where $s = \text{sgn}(\beta_2)$, $f = \Omega/\Omega_c$ is the normalized frequency and $p = P_0/P_c$ is the normalized input power. They are introduced using

$$\Omega_c = \sqrt{2\kappa/|\beta_2|}, \qquad P_c = 4\kappa/\gamma. \qquad (2.2.27)$$

The critical power P_c is the same as defined earlier in Eq. (2.2.7) as $\sigma = 0$ has been assumed.

Figure 2.6 shows the gain spectra of modulation instability for both normal and anomalous GVD. In both cases, the gain exists at low frequencies, including $\Omega = 0$. This feature is similar to the polarization instability occurring in birefringent fibers (see Chapter 6 of Ref. [11]). The gain at $\Omega = 0$ occurs only when the input power exceeds $P_c/2$ ($p > 0.5$). This is related to the appearance of the two new fixed points on the Poincaré sphere (see Figure 2.5). When GVD is normal, the gain peak occurs at $\Omega = 0$ only when $p > 1$. Therefore, when the input power P_0 exceeds P_c, modulation

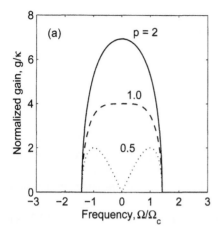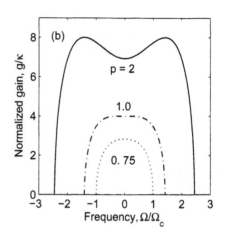

Figure 2.6: Gain spectra of modulation instability at several power levels in the cases of (a) normal and (b) anomalous dispersion.

instability is static in nature and does not lead to self-pulsing. In contrast, when the GVD is anomalous, no gain occurs until p exceeds $\frac{1}{2}$. The gain peak occurs at $\Omega = 0$ until $p = 1$ but shifts to $\Omega \neq 0$ for $p > 1$. Only for $p > 1$, CW light is converted into a pulse train whose repetition rate depends on the input power. The repetition rate is close to $\Omega_c/2\pi$ and is estimated to be ~ 1 THz for typical values of κ and β_2.

Direct experimental observation of modulation instability in fiber couplers is hampered by the fact that it is difficult to excite the even or odd supermode alone. Typically, initial conditions are such that both supermodes are excited simultaneously. Another difficulty is related to the relatively short coupler lengths used in practice ($L \sim L_c$). The growth of sidebands from noise (spontaneous modulation instability) requires the use of dual-core fibers for which $L \gg L_c$ is possible. The effects of induced modulation instability can be observed using shorter lengths since sidebands are seeded by an input signal. As an example, induced modulation instability can be used to control switching of a strong pump beam launched in one core of the coupler through a much weaker signal launched into the other core with an appropriate relative phase [19].

2.3   Ultrashort Pulse Propagation

Because of the high power levels needed for all-optical switching in fiber couplers, optical pulses are often used in practice. For short pulses, the GVD term in the coupled-mode equations can affect the switching behavior considerably, and it must be included [36]–[65]. This section first considers propagation of ultrashort optical pulses in fiber couplers by solving the coupled-mode equations numerically. A variational approach is then used to provide additional physical insight. The higher-order dispersive and nonlinear effects are included in the last subsection.

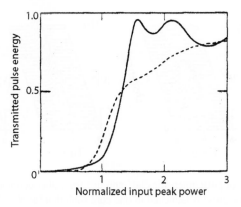

Figure 2.7: Transmitted energy as a function of $p = P_0/P_c$ in the case of solitons (solid line) and quasi-CW pulses (dashed line) for a coupler of length $L = L_c$. (From Ref. [36]; ©1988 OSA.)

2.3.1 Nonlinear Switching of Optical Pulses

To discuss pulse switching, it is useful to normalize Eqs. (2.2.16) and (2.2.17) in terms of the soliton units (see Section 5.3 of Ref. [11]) and write them as a pair of two coupled NLS equations of the form

$$i\frac{\partial u}{\partial \xi} - \frac{s}{2}\frac{\partial^2 u}{\partial \tau^2} + |u|^2 u + Kv = 0, \tag{2.3.1}$$

$$i\frac{\partial v}{\partial \xi} - \frac{s}{2}\frac{\partial^2 v}{\partial \tau^2} + |v|^2 v + Ku = 0, \tag{2.3.2}$$

where $s = \text{sgn}(\beta_2) = \pm 1$, $K = \kappa L_D$, and we have introduced the normalized variables

$$\xi = z/L_D, \quad \tau = T/T_0, \quad u = (\gamma L_D)^{1/2}A_1, \quad v = (\gamma L_D)^{1/2}A_2. \tag{2.3.3}$$

Here, $L_D = T_0^2/|\beta_2|$ is the dispersion length and T_0 is a measure of the pulse width. For $K = 0$, these equations reduce to two uncoupled NLS equations.

The coupled NLS equations, Eqs. (2.3.1) and (2.3.2), cannot be solved analytically in general. They are often solved numerically using the split-step Fourier method developed in Section 2.4 of Ref. [11]. The switching behavior depends on whether the GVD experienced by optical pulses is normal or anomalous in nature. As early as 1988, numerical simulations indicated that solitons, forming in the case of anomalous GVD, switch between the cores as an entire pulse in a manner analogous to the CW case [36]. In contrast, switching ceases to occur in the normal-dispersion regime if pulses are short enough that the dispersion length L_D becomes comparable to the coupling length L_c [38]. Soliton switching is, in fact, superior to the quasi-CW switching realized using relatively broad pulses.

Figure 2.7 compares the switching characteristics of fiber couplers in the two preceding cases. The numerical results are obtained by solving Eqs. (2.3.1) and (2.3.2) with the initial conditions

$$u(0, \tau) = N\text{sech}(\tau), \quad v(0, \tau) = 0. \tag{2.3.4}$$

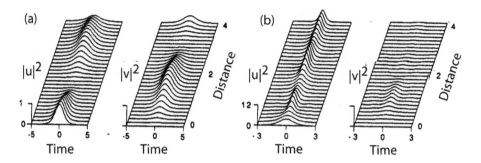

Figure 2.8: Evolution of pulse intensities $|u|^2$ and $|v|^2$ for a fiber coupler with coupling length $L_c = \pi L_D/2$. The input pulse is launched in one core such that (a) $N = 1$ or (b) $N = 2$. (From Ref. [36]; ©1988 OSA.)

The amplitude N is related to the peak power P_0 of the input pulse as $N^2 = \gamma L_d P_0 = 4Kp$, where $p = P_0/P_c$ is normalized to the CW switching power P_c. If we choose $K = 1/4$, N equals 1 when $p = 1$; that is, the input pulse propagates as a fundamental soliton when $P_0 = P_c$. The fraction of the pulse energy appearing in the core in which a soliton is initially launched is plotted as a solid line in Figure 2.7 for values of p in the range from 0 to 3. The switching behavior occurring near $p = 1$ for solitons is almost identical to that of a CW beam (compare with Figure 2.3). Since $L = L_c \sim L_D$ is required for soliton switching to occur, the input pulse width and peak power should be about 0.1 ps and 1 kW, respectively, for a 1-m-long fiber coupler. Because of relatively short propagation distances involved, higher-order effects are not likely to affect the switching behavior considerably, unless pulses become much shorter than 100 fs.

The exact value of N required for switching depends on the choice of the single parameter K appearing in Eqs. (2.3.1) and (2.3.2). As an example, when $K = 1$, $p = 1$ is realized only when $N = 2$ since $N^2 = 4Kp$. The switching behavior in this case is shown in Figure 2.8, where evolution of $|u|^2$ and $|v|^2$ along the coupler length is shown for $N = 1$ and 2. Since the first-order soliton is below the switching threshold, most of its power is transferred to the neighboring core at a distance $\xi = \pi/2$. In contrast, the second-order soliton keeps most of its power in the original core, since $p = 1$ for it. The switching threshold appears to be below $p = 1$, in contrast with the results shown in Figure 2.7. The reason can be understood by noting that a second-order soliton undergoes compression initially, resulting in higher peak powers. In fact, the pulse is compressed enough that the value of N at the output end is close to 1. The important point to note is that the entire pulse switches from one core to another. In the absence of the soliton effects, pulses are severely distorted, since only the central part is intense enough to undergo switching. Physically, this behavior is related to the fact that a fundamental soliton has the same phase over the entire pulse width in spite of SPM. A weak pulse, launched at the other input port, can also force a soliton to switch between the two output ports [39]. All-optical ultrafast logic gates have also been proposed using asymmetric fiber couplers [40].

2.3.2 Variational Approach

Particlelike switching of solitons suggests the use of a classical mechanics technique based on the Hamiltonian or Lagrangian formulation. Such an approach offers considerable physical insight [42]–[47]. The variational technique was first used in 1990 for solving Eqs. (2.3.1) and (2.3.2) approximately [43]. However, the width of solitons was assumed to remain constant in spite of changes in their amplitudes. As discussed in Chapter 5 of Ref. [11], the width and the amplitude of a soliton are related inversely when solitons evolve adiabatically. This section discusses the adiabatic case [45].

In the Lagrangian formalism, Eqs. (2.3.1) and (2.3.2) are derived from the Euler–Lagrange equation:

$$\frac{\partial}{\partial \xi}\left(\frac{\partial \mathcal{L}_d}{\partial q_\xi}\right) + \frac{\partial}{\partial \tau}\left(\frac{\partial \mathcal{L}_d}{\partial q_\tau}\right) - \frac{\partial \mathcal{L}_d}{\partial q} = 0, \tag{2.3.5}$$

where q represents u, u^*, v, or v^*; the subscripts τ and ξ denote differentiation with respect to that variable; and the Lagrangian density $\mathcal{L}_d$ is given by [43]

$$\mathcal{L}_d = \frac{i}{2}(u^* u_\xi - u u_\xi^*) + \frac{1}{2}(|u|^4 - |u_\tau|^2) + \frac{i}{2}(v^* v_\xi - v v_\xi^*)$$
$$+ \frac{1}{2}(|v|^4 - |v_\tau|^2) + K(u^* v + u v^*). \tag{2.3.6}$$

The crucial step in the variational analysis consists of choosing an appropriate functional form of the solution. In the case of adiabatic evolution, we anticipate solitons to maintain their "sech" shape even though their amplitude, width, and phase can change. We therefore assume that

$$u(\xi, \tau) = \eta_1 \text{sech}(\eta_1 \tau) e^{i\phi_1}, \qquad v(\xi, \tau) = \eta_2 \text{sech}(\eta_2 \tau) e^{i\phi_2}, \tag{2.3.7}$$

where η_j is the amplitude and ϕ_j is the phase for the soliton propagating in the jth core of the coupler. Both η_j and ϕ_j are assumed to vary with ξ. The soliton width also changes with its amplitude, as expected. Note that solitons in both cores are assumed to remain unchirped. In general, one should also include chirp variations [57].

The next step consists of integrating the Lagrangian density over τ using $\mathcal{L} = \int_{-\infty}^{\infty} \mathcal{L}_d \, d\tau$. The result is given by

$$\mathcal{L} = \frac{1}{3}(\eta_1^3 + \eta_2^3) - 2\eta_1 \frac{d\phi_1}{d\xi} - 2\eta_2 \frac{d\phi_2}{d\xi}$$
$$+ K\eta_1\eta_2 \cos(\phi_1 - \phi_2) \int_{-\infty}^{\infty} \text{sech}(\eta_1 \tau)\text{sech}(\eta_2 \tau) d\tau. \tag{2.3.8}$$

Using Eq. (2.3.8) in the Euler–Lagrange equation, we obtain a set of four ordinary differential equations for η_j and ϕ_j $(j = 1, 2)$. These equations can be simplified by noting that $\eta_1 + \eta_2 \equiv 2\eta$ is a constant of motion. Furthermore, the total phase $\phi_1 + \phi_2$ does not play a significant role since $\mathcal{L}$ depends only on the relative phase difference $\phi = \phi_1 - \phi_2$. Introducing a new dynamical variable,

$$\Delta = (\eta_1 - \eta_2)/(\eta_1 + \eta_2), \qquad |\Delta| \le 1, \tag{2.3.9}$$

the switching dynamics is governed by the equations

$$\frac{d\Delta}{dZ} = G(\Delta)\sin\phi, \qquad \frac{d\phi}{dZ} = \mu\Delta + \cos\phi\frac{dG}{d\Delta}, \tag{2.3.10}$$

where $Z = 2K\xi \equiv 2\kappa z$, $\mu = \eta^2/K$, and

$$G(\Delta) = \int_0^\infty \frac{(1-\Delta^2)dx}{\cosh^2 x + \sinh^2(x\Delta)}. \tag{2.3.11}$$

The parameter η is related to the total energy Q of both solitons as

$$Q = \int_{-\infty}^\infty (|u|^2 + |v|^2)d\tau = 2(\eta_1 + \eta_2) \equiv 4\eta. \tag{2.3.12}$$

Equations (2.3.10) can be integrated easily by noting that they can be derived from the Hamiltonian

$$H(\Delta,\phi) = -\tfrac{1}{2}\mu\Delta^2 - G(\Delta)\cos\phi. \tag{2.3.13}$$

As a result, ϕ and Δ can be treated as the generalized position and momentum of a fictitious particle, respectively. This analogy permits us to describe the switching dynamics of solitonsin the Δ–ϕ phase plane. The qualitative behavior depends on the parameter μ. To understand soliton switching, we first find the fixed points of Eqs. (2.3.10) by setting the Z derivatives to zero. Two fixed points are given by $\Delta = 0$ with $\phi = 0$ or π. Since both solitons have equal energy when $\Delta = 0$, these fixed points correspond to the even and odd supermodes found earlier in the CW case.

Two other fixed points of Eqs. (2.3.10) correspond to the situation in which the soliton is confined to only one core and are given by

$$\Delta = \pm 1, \qquad \cos\phi = 2\mu/\pi. \tag{2.3.14}$$

They exist only for $\mu < \pi/2$ and are always unstable. For $\mu > \pi/2$, two new fixed points emerge for which $\sin\phi = 0$ and Δ is obtained from the implicit relation $\mu\Delta = -(dG/d\Delta)$. In the limit of small Δ, the integral in Eq. (2.3.13) can be performed analytically with the approximation

$$G(\Delta) \approx (1-\Delta^2)(1-\alpha\Delta^2), \tag{2.3.15}$$

with $\alpha = (\pi^2/6 - 1)/3 \approx 0.215$. These fixed points disappear when $\mu > \mu_c = 2(1+\alpha) \approx 2.43$. In this region, the even-mode fixed point is also unstable.

Figure 2.9 shows trajectories in the Δ–ϕ phase plane in three regimes with different sets of fixed points. The various trajectories correspond to different launch conditions at the input end of the fiber coupler. Consider the case in which a single soliton is launched in one core such that $\eta_2(0) = 0$ or $\Delta(0) = 1$. The parameter μ is then related to the peak power P_0 of the launched soliton as $\mu = P_0/P_c$, where P_c is the CW critical power introduced earlier. As long as $\mu < \pi/2$, the soliton exhibits the same behavior as a low-power CW beam. More specifically, its energy oscillates between the two cores in a periodic manner.

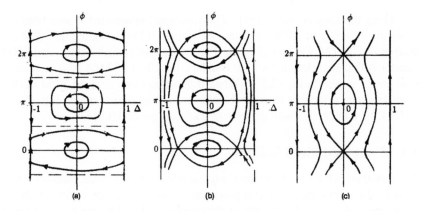

Figure 2.9: Phase-space trajectories in three different regimes corresponding to (a) $\mu < \pi/2$, (b) $\pi/2 < \mu < \mu_c$, and (c) $\mu > \mu_c$. (From Ref. [45]; ©1993 OSA.)

Nonlinear switching occurs in the region $\pi/2 < \mu < \mu_c$ since most of the soliton energy remains in the core in which the pulse is launched initially. This feature is equivalent to the CW switching discussed earlier except that the required peak power P_0 is larger by a factor of $\pi/2$. This increase is not surprising since even the low-power wings of the soliton switch together with its peak as one unit. Finally, when $\mu > \mu_c$, soliton energy oscillates around the stable point located at $\Delta = 0$ and $\phi = \pi$ (the odd supermode). Nonlinear switching is incomplete in this case. The main point to note is that whole-pulse switching of solitons is feasible with a proper control of soliton energy.

The CW-like switching behavior of solitons can be seen even more clearly if the soliton width is assumed to remain constant in spite of amplitude changes; that is, the ansatz (2.3.7) is replaced with [43]

$$u(\xi, \tau) = \eta_1 \mathrm{sech}(\tau) e^{i\phi_1}, \qquad v(\xi, \tau) = \eta_2 \mathrm{sech}(\tau) e^{i\phi_2}. \tag{2.3.16}$$

In this case, the integral in Eq. (2.3.11) can be evaluated analytically. Since $G(\Delta)$ is then known, Eqs. (2.3.10) can be integrated in a closed form in terms of the elliptic functions. Introducing the peak powers P_1 and P_2 using $\eta_j^2 = \gamma L_D P_j$ ($j = 1, 2$), P_1 is found to vary along the coupler length as

$$P_1(z) = \tfrac{1}{2}P_0[1 + \mathrm{cn}(2\kappa z|m)], \tag{2.3.17}$$

where the modulus of the Jacobi elliptic function is given by $m = (2P_0/3P_c)^2$, where P_c is the CW critical power. This solution is identical to the CW case except that the critical power for switching is larger by a factor of 3/2. This value compares reasonably well with the enhancement factor of $\pi/2$ predicted earlier using Eqs. (2.3.10). The variational analysis also predicts a symmetry-breaking bifurcation at $m = 1/2$. At this value, the symmetric solution, with equal peak powers in the two cores, becomes unstable and is replaced with an asymmetric solution [43].

The variational analysis based on Eq. (2.3.7) assumes that solitons remain unchirped. More accurate results are obtained when both pulse width and chirp are allowed to evolve along the coupler length [57]. The variational approach has also been used to study the influence of XPM on soliton switching [46] by including the XPM term appearing in Eqs. (2.1.14) and (2.1.15). For the relatively small values of the XPM parameter σ that are relevant for fiber couplers, the effect of XPM is to increase the critical power as seen in Eq. (2.2.7). When σ becomes close to 1, the XPM modifies the switching characteristics considerably. In the limiting case of $\sigma = 1$, Eqs. (2.1.14) and (2.1.15) are integrable with the inverse scattering method [66].

2.3.3   Coupler-Paired Solitons

A different approach for studying the nonlinear effects in directional couplers focuses on finding the soliton pairs that can propagate through the coupler without changes in their amplitude and width in spite of the coupling induced by the proximity of the two cores. Such soliton pairs are analogous to the XPM-paired solitons discussed in Chapter 7 of Ref. [11], except that the coupling between the two solitons is linear in nature. Several analytic solutions of Eqs. (2.3.1) and (2.3.2) have been obtained under different conditions [48]–[51]. It should be stressed that, strictly speaking, such solutions represent not solitons but solitary waves because Eqs. (2.3.1) and (2.3.2) are not integrable by the inverse scattering method.

The shape-preserving solutions of Eqs. (2.3.1) and (2.3.2) can be found by assuming a solution in the form [49]

$$u(\xi,\tau) = U(\tau)e^{iq\xi}, \qquad v(\xi,\tau) = V(\tau)e^{iq\xi}, \tag{2.3.18}$$

where q is a constant representing change in the wave number (from its value β). The amplitudes U and V are ξ independent and govern the shape of the two pulses representing the soliton pair. By substituting Eq. (2.3.18) in Eqs. (2.3.1) and (2.3.2), U and V are found to satisfy the following set of two coupled ordinary differential equations:

$$\frac{1}{2}\frac{d^2U}{d\tau^2} + U^3 + KV - qU = 0, \tag{2.3.19}$$

$$\frac{1}{2}\frac{d^2V}{d\tau^2} + V^3 + KU - qV = 0, \tag{2.3.20}$$

where the GVD is taken to be anomalous by choosing $s = -1$.

Equations (2.3.19) and (2.3.20) can be solved analytically when $V = \pm U$, as they reduce to a single equation. The resulting solutions are given by

$$U(\tau) = V(\tau) = \sqrt{2(q-K)}\,\text{sech}[\sqrt{2(q-K)}\tau], \tag{2.3.21}$$

$$U(\tau) = -V(\tau) = \sqrt{2(q+K)}\,\text{sech}[\sqrt{2(q+K)}\tau]. \tag{2.3.22}$$

The solution (2.3.21) is called the symmetric state and exists only for $q > K$. The solution (2.3.22) represents an antisymmetric state and exists for all $q > -K$. These two solutions correspond to the even and odd supermodes introduced in Section 2.1.2. In both cases, identical pulses propagate in the two cores with the only difference being that they are in phase ($U = V$) for the even supermode but out of phase ($U = -V$) for

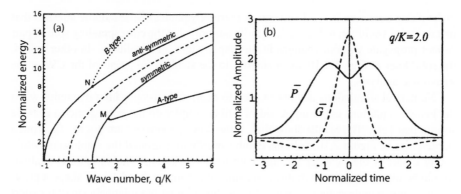

Figure 2.10: (a) Energy $Q/\sqrt{K}$ and wave number q/K of soliton pairs that can propagate along the fiber coupler without change in their shape. (b) Example of B-type asymmetric soliton pair. (From Refs. [49] and [50]; ©1993 APS.)

the odd supermode. The total energy of both solitons can be calculated using Eq. (2.3.12) and is given by $Q(K) = 4\sqrt{2(q \pm K)}$, where the minus sign corresponds to the symmetric state.

The symmetric and antisymmetric states represent soliton pairs with equal pulse energies in two cores of a fiber coupler. Depending on the total energy Q associated with the soliton pair, Eqs. (2.3.19) and (2.3.20) also have asymmetric solutions such that pulse energies are different in the two cores. The pulse shapes for such solutions are found numerically. Figure 2.10(a) shows possible solutions in the q–Q phase space [49]. The point M marks the location ($q/K = 5/3$) where the symmetric state bifurcates and results in soliton pairs with different amplitudes (A-type branch). The point N marks the location ($q/K = 1$) where the antisymmetric state bifurcates toward the B-type branch. The new feature of solitons on this branch is that their shape can be quite complicated with multiple humps. Figure 2.10(b) shows an example of the shapes associated with a soliton pair on the B-type branch.

Stability of soliton pairs can be examined using an extension of the modulation-stability analysis of Section 2.2.4. In this approach, the soliton state is perturbed as

$$u(\xi, \tau) = [U(\tau) + a_1(\xi, \tau)]e^{iq\xi}, \qquad v(\xi, \tau) = [V(\tau) + a_2(\xi, \tau)]e^{iq\xi}, \qquad (2.3.23)$$

where perturbations a_1 and a_2 vary with both ξ and τ. If the perturbations grow exponentially with ξ, the corresponding soliton pair is unstable. The results of such a stability analysis are shown by dashed lines in Figure 2.10 and can be summarized as follows [50]. Symmetric states are stable up to the bifurcation point M in Figure 2.10 and become unstable after that. The antisymmetric states are unstable for $q/K > -0.6$. Asymmetric solutions are always unstable on the B branch and stable on the A branch only if the slope $dQ/dq > 0$. Since the slope is negative in a small range—$5/3 < q/K < 1.85$—the asymmetric solutions on the A branch are stable except for a tiny region near the bifurcation point M. The existence of this tiny unstable region on the A branch implies that the symmetry-breaking bifurcation occurring at the point M is subcritical and leads to hysteresis with respect to pulse energy Q (first-order phase transition in the language of thermodynamics). It should be stressed that instability of a solution in Figure 2.10 indicates only that the corresponding soliton pair cannot propagate without changes in its shape, width, or amplitude. In other words, the solid lines in Figure 2.10 are analogous to the stable fixed points of the CW or the variational analysis.

Numerical simulations have been used to explore the propagation dynamics when the launch conditions at the input end of a fiber coupler do not correspond to a stable soliton pair [51]. The results show that, if the input parameters are not too far from a stable point in Figure 2.10, solitons exhibit oscillations around the stable state while losing a part of their energy through continuum radiation. The variational analysis should be used with caution in this case, since it assumes a fixed "sech" shape a priori and does not include radiative energy losses. Such losses are relatively small for short couplers but must be accounted for when $L \gg L_c$.

In the case of normal dispersion, one should choose $s = 1$ in Eqs. (2.3.1) and (2.3.2). As discussed in Section 5.3 of Ref. [11], the NLS equation supports dark solitons in each core in the absence of coupling. One may therefore ask whether the coupled NLS equations have solutions in the form of dark-soliton pairs. This turns out to be the case. Mathematically, one can follow the same procedure adopted previously and assume the solution of the form given in Eq. (2.3.18). The resulting equations for U and V are identical to Eqs. (2.3.19) and (2.3.20) except for a change in the sign of the second derivative term. These equations have the following symmetric and antisymmetric dark-soliton pairs [56]:

$$U(\tau) = V(\tau) = \sqrt{q-K} \tanh(\sqrt{q-K}\tau), \qquad (2.3.24)$$

$$U(\tau) = -V(\tau) = \sqrt{q+K} \tanh(\sqrt{q+K}\tau). \qquad (2.3.25)$$

Asymmetric dark-soliton pairs also exist after a bifurcation point on the symmetric branch, but their properties are quite different from those associated with the bright-soliton pairs.

2.3.4   Higher-Order Effects

In the case of ultrashort pulses, several higher-order effects must be considered. For example, intermodal dispersion, discussed in Section 2.1.3 and related to the frequency dependence of the coupling coefficient κ, cannot be ignored for femtosecond pulses. Also, the effects of third-order dispersion, self-steepening, and intrapulse Raman scattering should be included for such pulses [11].

Consider first the impact of intermodal dispersion on nonlinear switching [64]. If we include the frequency dependence of κ up to first order through the parameter κ_1, the coupled-mode equations, (2.3.1) and (2.3.2), contain an addition term and become

$$i\frac{\partial u}{\partial \xi} - \frac{s}{2}\frac{\partial^2 u}{\partial \tau^2} + |u|^2 u + Kv + iK_1\frac{\partial u}{\partial \tau} = 0, \qquad (2.3.26)$$

$$i\frac{\partial v}{\partial \xi} - \frac{s}{2}\frac{\partial^2 v}{\partial \tau^2} + |v|^2 v + Ku + iK_1\frac{\partial v}{\partial \tau} = 0, \qquad (2.3.27)$$

where $K_1 = \kappa_1 L_D/T_0$ is a measure of intermodal dispersion.

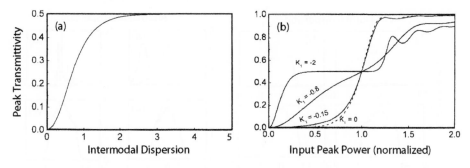

Figure 2.11: Output to input peak power ratio plotted (a) as a function of $|K_1|$ and (b) as a function of input peak power in the anomalous-GVD regime of a coupler of length $L = L_c$. (From Ref. [64]; ©2004 Elsevier.)

Numerical simulations can be used to study how the K_1 term affects the switching process inside a nonlinear coupler. As discussed earlier, when $K_1 = 0$, a fundamental soliton can switch to the cross port of a fiber coupler of length $L_c = (\pi/2)L_D$, while keeping its shape intact. The presence of intermodal dispersion affects this behavior in such a way that only a part of the pulse energy is transferred to the cross port. Figure 2.11(a) shows that only half of the input peak power is transferred to the cross port when $|K_1|$ exceeds a value of 2. The effect of intermodal dispersion on nonlinear switching is shown in Figure 2.11(b) for several values of K_1. For relatively small values of K_1, step-functionlike switching can occur when input peak power is close to P_c, just as predicted in Section 2.3.1 for $K_1 = 0$. However, such sharpness is destroyed when $|K_1|$ exceeds 0.5 and is close to 1. A qualitatively different behavior occurs for $|K_1| > 1.5$, in the sense that the peak powers become equal in the two cores at a relatively low input power level. This behavior is related to the splitting of the input pulse into two supermodes for large values of intermodal dispersion [64].

We now focus on the the effects of third-order dispersion, intrapulse Raman scattering, and self-steepening that become important for femtosecond pulses. In the general case, the two coupled-mode equations, (2.1.31) and (2.1.32), are replaced with

$$\frac{\partial A_1}{\partial z} + \kappa_1 \frac{\partial A_2}{\partial T} + \frac{i\beta_2}{2} \frac{\partial^2 A_1}{\partial T^2} - \frac{\beta_3}{6} \frac{\partial^3 A}{\partial t^3} = i\kappa A_2$$
$$+ i\gamma \left(1 + \frac{i}{\omega_0} \frac{\partial}{\partial T}\right) \left[A_1(z,T) \int_0^\infty R(t')|A_1(z,T-t')|^2 dt'\right], \quad (2.3.28)$$

$$\frac{\partial A_2}{\partial z} + \kappa_1 \frac{\partial A_1}{\partial T} + \frac{i\beta_2}{2} \frac{\partial^2 A_2}{\partial T^2} - \frac{\beta_3}{6} \frac{\partial^3 A}{\partial t^3} = i\kappa A_1$$
$$+ i\gamma \left(1 + \frac{i}{\omega_0} \frac{\partial}{\partial T}\right) \left[A_2(z,T) \int_0^\infty R(t')|A_2(z,T-t')|^2 dt'\right]. \quad (2.3.29)$$

The nonlinear response function has the general form [11]

$$R(t) = (1 - f_R)\delta(t) + f_R h_R(t), \quad (2.3.30)$$

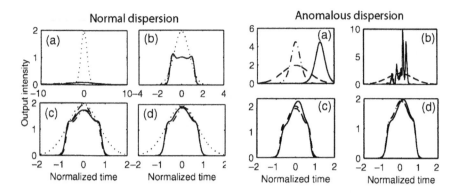

Figure 2.12: Temporal profiles of the switched pulse for a coupler of length $L = L_c = \pi/(2\kappa)$ in the normal (left) and anomalous (right) regimes for κL_D values of (a) 0.5, (b) 20, (c), 200, and (d) 1000. The dashed curves show the intensity profiles expected when the higher-order effects are ignored. The dotted and dot-dashed curves show the input Gaussian pulse launched with a peak power $P_0 \approx 2P_c$. (From Ref. [65]; ©2006 IEEE.)

where the first term represents the electronic contribution, $h_R(t)$ is the Raman response function and $f_R \approx 0.18$ represents its fractional contribution [67]. The Raman effects are often included through a simple model in which $h_R(t)$ has the form [68]

$$h_R(t) = \frac{\tau_1^2 + \tau_2^2}{\tau_1 \tau_2^2} \exp(-t/\tau_2) \sin(t/\tau_1),\qquad(2.3.31)$$

where $\tau_1 = 12.2$ fs and $\tau_2 = 32$ fs are two adjustable parameters chosen to provide a good fit to the actual Raman-gain spectrum.

Equations (2.3.28)–(2.3.31) have been solved numerically to study the impact of higher-order terms on a coupler's performance [65]. The results indicate that their impact depends on the magnitude of the dimensionless parameter κL_D, representing the ratio of the dispersion length L_D to the coupling length. When $\kappa L_D \gg 1$, the dispersive effects play a minor role. In contrast, when $\kappa L_D < 1$, the dispersive effects become quite important, but the coupler exhibits good switching characteristics. In the intermediate region in which $\kappa L_D \sim 10$, switched pulses suffer most from higher-order effects and are distorted considerably. The extent of distortion depends on whether the pulse experiences normal or anomalous GVD. Solid curves in Figure 2.12 show the temporal profiles of the switched pulse for a coupler of length $L = L_c = \pi/(2\kappa)$ in the two GVD regimes for values of κL_D in the range 0.5 to 1000 when a Gaussian pulse is launched with a peak power $P_0 \approx 2P_c$. The dashed curves show, for comparison, the intensity profiles expected when the higher-order effects are ignored. It is evident that such effects play little role when κL_D exceeds 100.

Several features in Figure 2.12 are noteworthy. When $\beta_2 < 0$ and $\kappa L_D = 0.5$, the output pulse is relatively narrow and delayed considerably. These changes are related to the phenomena of soliton formation and intrapulse Raman scattering that shift the pulse spectrum toward the longer wavelengths [11]. For $\kappa L_D = 20$, the output pulse is distorted and exhibits a multipeak structure related to the onset of soliton fission. In the case of normal GVD ($\beta_2 > 0$), the pulse cannot form a soliton and is broadened considerably for $\kappa L_D < 1$. The broadening does not occur for $\kappa L_D \gg 1$, but the pulse is severely distorted.

2.4 Other Types of Couplers

The discussion of nonlinear effects has so far focused on symmetric fiber couplers whose cores are identical in all respects. There are several different ways in which two cores can become dissimilar. For example, the cores may have different shapes or sizes. This case was discussed in Section 2.1 but the nonlinear effects were neglected. Nonlinear phenomena in asymmetric couplers can lead to new effects. An interesting situation occurs when the cores have different dispersive properties (normal versus anomalous). Cores can also be made different by selective doping and pumping. An example is provided by couplers in which one core is doped with erbium ions and pumped externally to provide gain. As another example, a Bragg grating can be integrated in one or both cores; such devices are called *grating-assisted directional couplers*. This section considers several extensions of the basic coupler design and discusses their practical applications.

2.4.1   Asymmetric Couplers

Nonlinear effects in asymmetric couplers with dissimilar cores are also of considerable interest [69]–[75]. Several new effects can occur in couplers with cores of different sizes. Mathematically, we employ Eqs. (2.1.9) and (2.1.10) and write them in terms of the soliton units as

$$i\frac{\partial u}{\partial \xi} + \frac{1}{2}\frac{\partial^2 u}{\partial \tau^2} + |u|^2 u + Kv + d_p u = 0, \qquad (2.4.1)$$

$$i\frac{\partial v}{\partial \xi} + i d_g \frac{\partial u}{\partial \tau} + \frac{d_2}{2}\frac{\partial^2 v}{\partial \tau^2} + d_n |v|^2 v + Ku - d_p u = 0, \qquad (2.4.2)$$

where we used normalized variables as defined in Eq. (2.3.3), assumed that the GVD in the first core is anomalous ($\beta_{21} < 0$), and introduced the following four parameters related to the asymmetric nature of the coupler:

$$d_p = \delta_a L_D, \quad d_g = (\beta_{12} - \beta_{11})L_D/T_0, \quad d_2 = \beta_{22}/\beta_{21}, \quad d_n = \gamma_2/\gamma_1. \qquad (2.4.3)$$

Physically, d_p and d_g represent, respectively, phase- and group-velocity mismatch, while d_2 and d_n account for differences in the dispersive properties and effective core areas, respectively. The parameter d_2 can be negative if the GVD in the second core is normal.

   The presence of four new parameters in the coupled NLS equations makes the analysis of asymmetric couplers quite involved. Differences in the GVD parameters result from the waveguide contribution to GVD that depends on the core size. If the operating wavelength is close to the zero-dispersion wavelength of the fiber, small changes in the core shape and size can induce large enough changes in dispersion that even the nature of GVD (normal versus anomalous) can be different for the two cores. In contrast, if the operating wavelength is far from the zero-dispersion wavelength, the GVD parameters are nearly the same in both cores. We consider the latter case and assume that the two cores are similar enough that we can set $d_g = 0$, $d_2 = 1$, and $d_n = 1$ in Eqs. (2.4.1) and (2.4.2). The asymmetry in such couplers is due only to different phase velocities in the two cores.

We can use the same method used earlier to find the stationary soliton pairs that propagate without change in their shape. By substituting Eq. (2.3.18) in Eqs. (2.4.1) and (2.4.2), U and V are found to satisfy the following set of two ordinary differential equations:

$$\frac{1}{2}\frac{d^2U}{d\tau^2} + U^3 + KV - (q - d_p)U = 0, \tag{2.4.4}$$

$$\frac{1}{2}\frac{d^2V}{d\tau^2} + V^3 + KU - (q + d_p)V = 0. \tag{2.4.5}$$

These equations should be solved numerically to find $U(\tau)$ and $V(\tau)$. A variational technique can be used with a Gaussian-shape ansatz [72]. The phase diagram in the q–Q plane [see Eq. (2.3.12) for the definition of Q] turns out to be quite different to of Figure 2.10 when $d_p \neq 0$. This is not surprising as all solutions for asymmetric couplers must be asymmetric, such that $|U| \neq |V|$. However, one still finds solutions such that U and V have the same sign. Since the relative phase between the two components is zero, such in-phase solitons are analogous to the symmetric state such that $U > V$ when $d_p > 0$. Similarly, one finds out-of-phase soliton pairs that are analogous to the asymmetric state in the sense that U and V have opposite signs. It turns out that $|V| > |U|$ for such solitons when $d_p > 0$. In both cases, more and more energy remains confined to one core as $|d_p|$ becomes larger. This feature can be understood from Eqs. (2.4.4) and (2.4.5) by solving them in the limit of large $|d_p|$. If both the dispersive and nonlinear terms are neglected, q can have two values given by

$$q = \pm\sqrt{d_p^2 + K^2} \approx \pm d_p, \tag{2.4.6}$$

and the solutions corresponding to these values of q satisfy

$$U \approx 2d_p V, \qquad U \approx -V/(2d_p). \tag{2.4.7}$$

Clearly, almost all energy remains in one core of the coupler for large d_p.

A third solution of Eqs. (2.4.4) and (2.4.5) is found when $|d_p|$ exceeds a critical value [72]. The components U and V have opposite signs for this solution, and most of the energy is confined to one of them. The exact range of $|d_p|$ over which the third solution occurs depends on both K and the total energy Q. In fact, depending on the value of Q, only in-phase soliton pairs may exist for some values of $|d_p|$. Bistable behavior can also occur when Q is large enough.

The shapes and energies of the two solitons are quite different for the in-phase and out-of-phase solitons. Numerical solutions of Eqs. (2.4.4) and (2.4.5) show that soliton pairs for which U and V have the same signs are localized in the sense that their amplitude decreases exponentially far away from the center [72]. In contrast, soliton

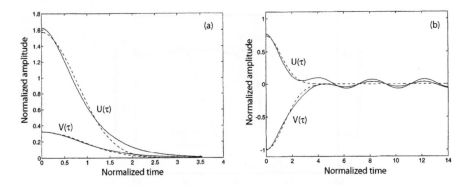

Figure 2.13: Pulse amplitudes for the (a) in-phase and (b) out-of-phase soliton pairs. Dashed curves show the Gaussian approximation based on a variational analysis. (From Ref. [72]; ©1997 OSA.)

pairs for which U and V have opposite signs (out-of-phase solitons) are delocalized such that their amplitude oscillates and does not decrease to zero even far away from the center. Figure 2.13 shows an example of these two types of soliton pairs for $Q = 2$.

The effect of GVD mismatch between the two cores—governed by the parameter d_2 in Eqs. (2.4.1) and (2.4.2)—is even more interesting, especially in the case in which the GVD is normal in the second core [73]. The most striking new feature is related to the existence of gap solitons, similar to those found for Bragg gratings (see Section 1.6), that occur inside a gap region in which light cannot propagate when nonlinear effects are weak. Moreover, such bright solitons carry most of their energy in the core with normal GVD. The shape of the soliton components U and V exhibits oscillatory tails that decay exponentially far away from the pulse center.

2.4.2 Active Couplers

Fiber losses are typically neglected in the context of fiber couplers. This is justified in view of short fiber lengths used in practice (typically $L < 10$ m) and relatively low losses associated with silica fibers. The situation is different when one or both cores of a coupler are doped with a rare-earth element such as erbium. The doped core will absorb considerable light whose wavelength is close to an atomic resonance, or it will amplify the propagating signal if that core is externally pumped to provide gain (see Chapter 4). The pumping level can be different for the two cores, resulting in different gains, or one core may be left unpumped. Because of differences in the amount of gain or loss in the two cores, doped couplers behave asymmetrically even if both cores are identical in shape and size. Such couplers are sometimes called *active directional couplers* and can be useful for a variety of applications [76]–[84].

To understand operation of such devices, we use Eqs. (2.3.1) and (2.3.2), derived for a symmetric coupler, but add an extra gain term [76]:

$$i\frac{\partial u}{\partial \xi} - \frac{s}{2}\frac{\partial^2 u}{\partial \tau^2} + |u|^2 u + Kv = \frac{i}{2}g_1 L_D \left(u + b\frac{\partial^2 u}{\partial \tau^2}\right), \qquad (2.4.8)$$

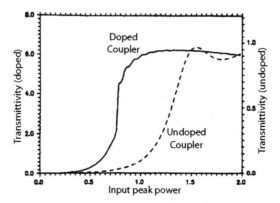

Figure 2.14: Switching characteristics (solid curve) of an active fiber coupler with equal gain in both cores. Dashed curve shows the behavior expected in the absence of gain. (From Ref. [76]; ©1991 OSA.)

$$i\frac{\partial v}{\partial \xi} - \frac{s}{2}\frac{\partial^2 v}{\partial \tau^2} + |v|^2 v + Ku = \frac{i}{2}g_2 L_D\left(v + b\frac{\partial^2 v}{\partial \tau^2}\right), \qquad (2.4.9)$$

where g_1 and g_2 are the gain coefficients whose value depends on the pumping level. The parameter $b = (T_2/T_0)^2$ accounts for the finite gain bandwidth. It originates from the frequency dependence of the gain approximated as $\tilde{g}_j(\omega) = g_j(1 - \omega^2 T_2^2)$, where T_2 is the dipole relaxation time of dopants, related inversely to the gain bandwidth (see Chapter 4). For picosecond pulses (width > 5 ps), the spectrum is narrow enough that all frequency components of the pulse experience nearly the same gain. The parameter b can be set to zero for such pulses. In the absence of pumping, g_j becomes negative and accounts for dopant-induced losses in the jth core.

In the quasi-CW case, the two terms involving time derivatives can be set to zero. The resulting equations can be solved analytically in the low-power case but require a numerical solution to study nonlinear switching [77]. When both cores are pumped to provide equal gains, the switching threshold is reduced at the expense of switch quality. The best performance occurs for active couplers with gain in one core and comparable loss in the other core. With proper choice of device parameters, the switching threshold can be reduced by a factor of more than 10 while maintaining a sharp, step-functionlike response of the switch.

Soliton switching in active fiber couplers has been investigated numerically by choosing $s = -1$ (anomalous GVD), setting $g_1 = g_2 = g_0$ and $b = 0$ in Eqs. (2.4.8) and (2.4.9), and using the input conditions [76]

$$u(0, \tau) = N\mathrm{sech}(\tau), \qquad v(0, \tau) = 0. \qquad (2.4.10)$$

Figure 2.14 shows improvement in switching of picosecond pulses occurring because of amplification for a coupler of length $L = 2\pi L_D$ by choosing $K = 0.25$ and $g_0 L_D = 0.3$. Several features are noteworthy. First, the switching threshold is reduced by about a factor of 2. Second, the switching is much sharper. A relatively small change in peak power of the pulse can switch the soliton from one core to another. Third, the switching contrast is improved because of the amplification provided by the coupler. In fact, the switched pulse is narrower than the input pulse by a factor in the range of 3 to 7, depending on the input peak power. For femtosecond pulses, gain dispersion must be included by choosing $b \neq 0$. Numerical simulations show that the main effect of gain dispersion is to reduce the overall switching efficiency without affecting the pulse quality significantly. It should be stressed that input pulse does not correspond to a fundamental soliton when $N \neq 1$. As a result, the switching is accompanied by dispersive radiation that appears in the other core because of its low power.

Asymmetric active couplers in which two cores have different gains ($g_1 \neq g_2$) can be used as saturable absorbers. Consider the case in which one core is pumped to provide gain while the other core is either undoped or unpumped. Low-energy pulses are then transferred to the second core, while high-energy pulses whose peak power exceeds the switching threshold remain in the core with gain. Such a device acts as a saturable absorber and can be used for many applications. For example, it can be used for passive mode locking of fiber lasers by using the doped core as a gain medium within a cavity [79]. This scheme works even in the normal-GVD regime and can be used to generate picosecond pulse trains in the spectral region below 1.3 μm by using dopants such as neodymium [80]. Such a device can also be used to filter noise associated with solitons, since noise can be transferred selectively to the lossy core because of its low power level [81]. The device acts as an optical amplifier whose gain is power dependent such that low-power signals are attenuated while high-power signals are amplified [82]. It should be stressed that the dopants used to provide gain or loss in fiber couplers can also have their own saturable nonlinearities that can affect the switching behavior significantly [78]. This issue is discussed in Chapter 4 in the context of fiber amplifiers.

2.4.3 Grating-Assisted Couplers

An important class of directional couplers makes use of a Bragg grating to improve the performance of asymmetric couplers. Such couplers are called *grating-assisted couplers* [85]–[96]. They have been studied mostly in the context of planar waveguides, in which grating-induced variations in the thickness of one waveguide lead to periodic modulation of the coupling coefficient. The grating period Λ is chosen such that the mismatch between the modal propagation constants equals the grating wave vector, that is,

$$\bar{\beta}_1 - \bar{\beta}_2 = \beta_g \equiv 2\pi/\Lambda. \qquad (2.4.11)$$

This condition is similar to that of a long-period grating used for coupling the modes in a single-core fiber (see Section 1.7.1). In the case of a grating-assisted coupler, such a long-period grating couples the modes supported by two spatially separated waveguides (or the even and odd modes of the coupled waveguides) and allows complete transfer of a low-power beam between the two waveguides even though little power exchange occurs in the absence of the grating.

In the case of fiber couplers, it is difficult to vary the core diameter in a periodic fashion on a scale of about 10 μm. For this reason, gratings are formed by modulating the refractive index of the core. Since the spacing between cores does not vary

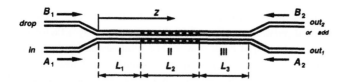

Figure 2.15: A grating-assisted fiber coupler. (From Ref. [102]; ©1997 OSA.)

in this case, the coupling coefficient remains nearly unchanged. Nevertheless, such phase gratings can be quite useful. Indeed, several kinds of grating-assisted fiber couplers have been proposed and analyzed for wavelength-division multiplexing (WDM) applications [97]–[110]. An example of such a coupler is shown in Figure 2.15. Both long- and short-period gratings have been used. In a 1992 experiment, an acoustic wave, excited by a silica horn, coupled the forward-propagating normal modes of an asymmetric dual-core fiber coupler [97]. Such a device can be useful for a variety of applications related to WDM and signal processing. Periodic microbending induced by an acoustic wave or by a fixed mechanical grating has also been used to induce mode coupling [98].

Short-period Bragg gratings have been incorporated into one core of fiber couplers for making add–drop WDM filters [100]. Such gratings produce a backward-propagating wave if the wavelength of the input signal falls within its stop band. When a multichannel WDM signal is injected into the core without the grating and transferred to the second core, a specific channel is selectively reflected back by the grating; it appears at the input end of the second core while the remaining channels appear at the output end of that core. A signal at the same specific wavelength can be added by injecting it from the output port of the core without the grating. The grating period is set by Eq. (2.4.11)—after changing the minus sign to a plus sign because of the backward propagation of the dropped channel—and is a fraction of the wavelength of that channel. Fabrication of a Bragg grating in the coupling region between the two cores allows the same add–drop functionality [101].

Fiber couplers in which both cores contain built-in Bragg gratings can also be used for adding or dropping a channel. In this case, forward- and backward-propagating waves are produced in both cores. Denoting the backward waves by B_1 and B_2 and neglecting the GVD, the operation of such a device is governed by the following four coupled-mode equations [102]:

$$\frac{dA_1}{dz} = i\delta_1 A_1 + i\kappa_{g1} B_1 + i\kappa_{12} A_2 + i\gamma_1(|A_1|^2 + 2|B_1|^2)A_1, \qquad (2.4.12)$$

$$-\frac{dB_1}{dz} = i\delta_1 B_1 + i\kappa_{g1} A_1 + i\kappa_{12} B_2 + i\gamma_1(|B_1|^2 + 2|A_1|^2)B_1, \qquad (2.4.13)$$

$$\frac{dA_2}{dz} = i\delta_2 A_2 + i\kappa_{g2} B_2 + i\kappa_{21} A_1 + i\gamma_2(|A_2|^2 + 2|B_2|^2)A_2, \qquad (2.4.14)$$

$$-\frac{dB_2}{dz} = i\delta_2 B_2 + i\kappa_{g2} A_2 + i\kappa_{21} B_1 + i\gamma_2(|B_2|^2 + 2|A_2|^2)B_2, \qquad (2.4.15)$$

where κ_{g1} and κ_{g2} are the coupling coefficients of the two gratings that can be designed to be different if necessary. The parameters δ_1 and δ_2 represent detuning between the Bragg and the modal propagation constants [107]. These equations can be easily generalized to include fiber dispersion by adding the first and second time-derivative terms as was done in Eqs. (2.1.9) and (2.1.10).

Equations (2.4.12)–(2.4.15) can be solved analytically only in the case of identical gratings and low-power CW beams by setting $\kappa_{g1} = \kappa_{g2}$ and $\gamma_1 = \gamma_2 = 0$; the results confirm the add–drop function offered by such a device [102]. In the general case in which gratings are different, occupy only a fraction of the coupling region, and are allowed to be nonuniform (e.g., apodized gratings), a numerical solution is required to optimize the performance of such add–drop multiplexers [107]. When a broadband WDM signal is launched inside one core of a coupler of length $L = L_c$, the channel whose wavelength falls within the stop band of the grating is reflected back and appears at the unused input port of the second core, while the remaining channels appear at the output end. Such grating-assisted fiber couplers have been fabricated and exhibit large add–drop efficiency ($> 90\%$) with low losses [104].

Nonlinear effects can be studied by solving Eqs. (2.4.12)–(2.4.15) numerically. Similar to the case of grating-assisted codirectional couplers [91], the intensity-dependent shift of the Bragg frequency affects the channel to be dropped. As a result, the device can act as a nonlinear switch, such that the channel is dropped only if its power exceeds a certain value. Propagation of short optical pulses should also lead to interesting nonlinear phenomena since such a grating-assisted fiber coupler can support Bragg solitons in each core but these solitons are coupled by the proximity of two cores.

2.4.4 Birefringent Couplers

Another situation in which one needs to solve a set of four coupled-mode equations occurs when a fiber coupler exhibits large birefringence. In practice, birefringence can be induced either by using elliptical cores or through stress-induced anisotropy, the same techniques used for making polarization-maintaining fibers. Polarization in such fibers is maintained only when light is polarized along the fast or slow axis of the fiber. If incident light is polarized at an angle to these axes, the state of polarization changes along the core length in a periodic fashion.

The mathematical description of birefringent fiber couplers requires four coupled-mode equations that correspond to the two orthogonally polarized components of light in the two cores [111]–[114]. In the general case of asymmetric couplers and arbitrary birefringence, these equations are quite complicated, since all four field components propagate with different group velocities. They can be simplified considerably for symmetric couplers with either very high or very low birefringence.

Consider first the high-birefringence case. Using the notation that A_m and B_m denote the linearly polarized components in the mth core, the coupled-mode equations in this case become

$$\frac{\partial A_1}{\partial z} + \frac{1}{v_{gx}}\frac{\partial A_1}{\partial t} + \frac{i\beta_2}{2}\frac{\partial^2 A_1}{\partial T^2} = i\kappa A_2 + i\gamma(|A_1|^2 + \sigma|B_1|^2)A_1, \quad (2.4.16)$$

$$\frac{\partial B_1}{\partial z} + \frac{1}{v_{gy}}\frac{\partial B_1}{\partial t} + \frac{i\beta_2}{2}\frac{\partial^2 B_1}{\partial T^2} = i\kappa B_2 + i\gamma(|B_1|^2 + \sigma|A_1|^2)B_1, \quad (2.4.17)$$

$$\frac{\partial A_2}{\partial z} + \frac{1}{v_{gx}}\frac{\partial A_2}{\partial t} + \frac{i\beta_2}{2}\frac{\partial^2 A_2}{\partial T^2} = i\kappa A_1 + i\gamma(|A_2|^2 + \sigma|B_2|^2)A_2, \qquad (2.4.18)$$

$$\frac{\partial B_2}{\partial z} + \frac{1}{v_{gy}}\frac{\partial B_2}{\partial t} + \frac{i\beta_2}{2}\frac{\partial^2 B_2}{\partial T^2} = i\kappa B_1 + i\gamma(|B_2|^2 + \sigma|A_2|^2)B_2, \qquad (2.4.19)$$

where v_{gx} and v_{gy} are group velocities for the two polarization components. The XPM parameter takes a value of $\sigma = 2/3$ for linearly polarized components. In a low-birefringence coupler, all components propagate with the same group velocity but one cannot neglect four-wave mixing between the linearly polarized components. It is common to use circularly polarized components in this case. The resulting equations are identical to Eqs. (2.4.16)–(2.4.19), provided we set $v_{gx} = v_{gy}$ and use $\sigma = 2$ for the XPM parameter.

The CW case in which all time-derivative terms in Eqs. (2.4.16)–(2.4.19) are set to zero was analyzed in 1988 using both the Hamiltonian and Stokes-parameter formalisms [111]. The new feature is that the state of polarization of the optical fields in the two cores can be different. This feature can be used to control the behavior of an intense beam launched in one core of the coupler by injecting a weak, orthogonally polarized probe in the other core. It can also be used to perform AND logic operation since the threshold for nonlinear switching is reduced when two orthogonally polarized pulses are launched simultaneously into the same core of the coupler. Another interesting result is that the power-dependent switching exhibits chaotic behavior when light is launched in both cores simultaneously to excite the even or odd supermode of the coupler.

Solutions of Eqs. (2.4.16)–(2.4.19) in the form of coupled soliton pairs have been studied using variational analysis with a Gaussian-shape ansatz [113]. These solutions represent two vector solitons coupled by the proximity of two cores. They can again be classified as being symmetric or antisymmetric with equal energies in the two cores ($|A_1| = |A_2|$ and $|B_1| = |B_2|$) or being asymmetric such that the two cores have pulses with different energies. Stability properties of these soliton pairs are similar to those seen in Figure 2.10, where birefringence effects were ignored [114].

A birefringent coupler can be converted into a dual-core rocking filter if the axis of birefringence is rocked periodically by twisting the preform during the fiber-drawing process [115]. Such periodic rotation of the birefringence axis at the beat length can be included in Eqs. (2.4.16)–(2.4.19) by adding an additional gratinglike term on the right side. More specifically, one should add the term $i\kappa_g B_1 \exp(-4i\pi z/L_B)$ to Eq. (2.4.16), and similar terms to other equations, where L_B is the beat length of the birefringent fiber. Following the approach outlined in Section 1.5.2, the resulting four equations can be reduced to a pair of coupled NLS equations under suitable conditions [52]. These equations support pairs of coupled Bragg solitons that can propagate along the coupler length without changing their shapes.

2.5   Fibers with Multiple Cores

An interesting configuration of fiber couplers consists of fibers with two or more cores. Arrays of planar waveguides (active or passive) were studied extensively during the

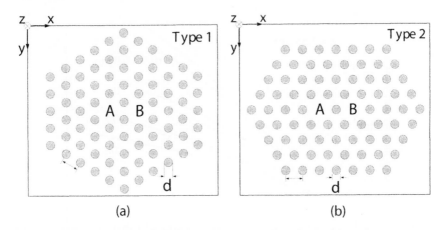

Figure 2.16: Cross sections of two types of PCF couplers with core spacings of $\sqrt{3}\Lambda$ and 2Λ. Shaded regions represent air holes. (From Ref. [125]; ©2003 OSA.)

1980s and were used to make high-power semiconductor lasers [116]–[120]. Multicore fiber couplers were fabricated as early as 1989 [121] and used for making multiplexers and star couplers [122]. The advent of photonic crystal fibers (PCFs) in recent years has provided a new impetus for such couplers. In his section, we first focus on two-core fibers and then consider multicore fiber arrays.

2.5.1 Dual-Core Photonic Crystal Fibers

As discussed in Section 2.2.2, dual-core fibers were used as directional couplers starting in 1985, and several experiments showed evidence of nonlinear switching in such fibers [20]–[24]. Starting around 2000, PCFs containing two closely spaced cores, and thus acting as couplers, attracted considerable attention [123]–[129]. Any PCF contains an array of periodic air holes of fixed diameter d separated by a certain distance Λ within the silica glass (see Section 9.1). If the central hole is missing, the PCF acts as a silica-core fiber surrounded by a cladding whose refractive index is reduced by a multitude of air holes. In a dual-core PCF, two closely spaced silica cores are surrounded by air holes. Figure 2.16 shows the design of two types of PCF couplers with core spacings of $\sqrt{3}\Lambda$ and 2Λ.

Modal properties of a PCF coupler depend strongly on the two parameters, d and Λ, associated with the air holes. Several methods have been developed for calculating mode shapes and the effective mode indices in PCFs couplers [124]–[127]. An important property of such couplers is that their coupling length depends not only on d and Λ but also on the state of polarization of input light because of the hexagonal pattern of air holes used commonly in practice [125]. Figure 2.17 shows how coupling length varies with Λ for the Type 1 coupler with different values of the ratio d/Λ. Solid and dashed curves correspond to light polarized along the x and y directions, respectively, as indicated in Figure 2.16. Depending on the size and spacing of air holes, the cou-

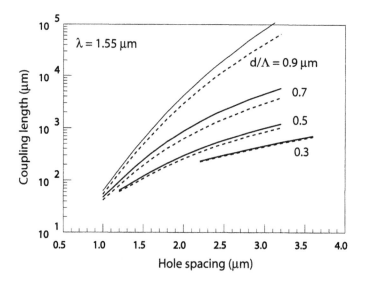

Figure 2.17: Coupling length as a function of Λ for a Type 1 coupler for several values of the ratio d/Λ. Solid and dashed curves correspond to polarization along the x and y axes, respectively. (From Ref. [125]; ©2003 OSA.)

pling length can vary from <0.1 to >10 mm. Its relatively small values indicate that PCF couplers can be quite compact in practice.

The nonlinear supermodes of a dual-core PCF were found numerically in a 2005 study [128]. Whereas a linear coupler supports only one symmetric and one antisymmetric supermode, the Kerr nonlinearity allows for an additional asymmetric supermode in the case of a dual-core PCF coupler. Figure 2.18 displays the amplitude distribution of the three supermodes, together with their spatial profiles, for a PCF designed with $d = 1.5$ μm and $\Lambda = 2$ μm. One can think of these supermodes in terms of spatial solitons supported by the PCF coupler. The nonlinear switching characteristics of such a dual-core PCF were also studied. It was found that the switching behavior is similar to that shown in Figure 2.7. In particular, the soliton launched into one core ceases to transfer its power to the neighboring core when the input power level exceeds a critical value. The nonlinear switching was observed in a 2006 experiment in which a dual-core PCF with $d = 1$ μm and $\Lambda = 2.5$ μm was employed, resulting in a coupling length of about 5 mm [129]. Because of the use of 120 fs pulses with peak intensities of up to 6 TW/cm^2, the generation of new spectral components is expected and was indeed observed. The nonlinear effects are enhanced so much in PCFs that a supercontinuum is easily formed for intense femtosecond pulses [11]. As this process reduces the intensity at the original input wavelength, it interferes with the switching process, and makes it unlikely that such couplers would be useful for nonlinear switching applications.

One can also employ two separate PCFs whose cores are brought close to each other to form a coupler through a conventional technique such as side polishing [130]–[132]. In a 2004 experiment it was not only possible to realize a constant value of

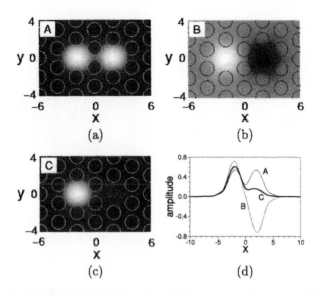

Figure 2.18: Amplitude distribution across the two cores of a PCF for (a) symmetric, (b) anti-symmetric, and (c) asymmetric supermodes. Part (d) shows the corresponding spatial profiles. (From Ref. [128]; ©2005 OSA.)

coupling coefficient over a 400-nm spectral range but also to tune this value over a wide range by adjusting the mating angle between the two side-polished PCFs [131]. In a 2006 study, two double-clad PCFs were used to form a coupler through the fused tapering method [132]. The coupler had a splitting ratio of 97:3 (with an insertion loss of only 1.1 dB) over the entire visible and near-infrared wavelength range and was used to construct a miniaturized microscope that employed two-photon fluorescence for nonlinear imaging.

2.5.2   Multicore Fiber Arrays

It is possible to fabricate fibers, both conventional and PCF types, such that they contain multiple cores sharing the same cladding [133]. Nonlinear effects in such fiber arrays have been analyzed theoretically since the early 1990s using a set of coupled NLS equations [134]–[149]. When all cores are identical, these equations take a simple form and can be written, using soliton units, in the following compact form:

$$i\frac{\partial u_m}{\partial \xi} + \frac{1}{2}\frac{\partial^2 u_m}{\partial \tau^2} + |u_m|^2 u_m + K(u_{m+1} + u_{m-1}) = 0, \qquad (2.5.1)$$

where u_m represents the field amplitude in the mth core and is coupled to the fields in the two neighboring cores. For a linear array of M cores, the cores at the two ends have only one neighbor. The resulting boundary conditions require $u_0 = u_{M+1} = 0$. This asymmetry can be avoided for a circular fiber array in which all cores are spaced equally and their centers lie along a circle, resulting in periodic boundary conditions. Figure 2.19(a) shows such a fiber array schematically.

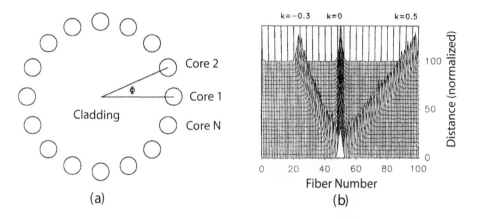

Figure 2.19: (a) A circular fiber array. (b) Optical steering of a CW beam along the array. (From Ref. [144]; ©1994 OSA.)

The specific case of three-core couplers has attracted considerable attention since the resulting three coupled NLS equations permit analytical solution in both the CW and pulsed cases [134]–[140]. The periodic boundary conditions can be used when the core centers form an equilateral triangle. In analogy with the two-core couplers, one can find soliton triplets that propagate through the three-core coupler without changing their shapes [58]. The bifurcation diagram in the q–Q plane is much more complicated in this case [137]. The reason is related to a vast variety of possible solutions that may exist even for couplers with only three cores. At low values of the total energy Q, the symmetric solution for which all three solitons are identical (in-phase solution) is stable. However, an antisymmetric solution also exists. In this case, two solitons are out of phase and have the same energy while the energy in the third core is zero. At a certain value of the energy Q, both the symmetric and antisymmetric solutions become unstable and give rise to partially or totally asymmetric solutions.

The analysis of multicore fiber couplers becomes increasingly more involved as the number of cores increases [141]–[144]. Numerical solutions of Eq. (2.5.1) for a linear array, in which the input CW beam is launched initially at one end of the array, shows that nonlinear switching not only occurs but has a sharper threshold [134]. More specifically, the input beam is transferred to the outermost core at the other end at low powers (if the coupler length is chosen judiciously) but remains in the same core when the input power exceeds a threshold value. However, the threshold power increases and the power-transfer efficiency decreases as the number of intermediate cores is increased [141].

Power also is transferred from core to core in the case of a circular fiber array [144]. Figure 2.19(b) shows this behavior for a 101-core array by solving Eq. (2.5.1) numerically when a CW beam is launched initially with the amplitude

$$u_m = Ka\,\mathrm{sech}[a/\sqrt{2}(m - m_c)]\exp[-ik(m - m_c)], \qquad (2.5.2)$$

where $m_c = 51$ and $a^2 = 1.1$. The integer m is varied from 51 to 55, resulting in the excitation of five cores at $\xi = 0$. The parameter k determines the initial phase

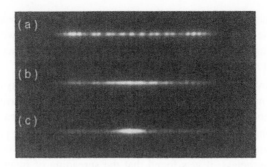

Figure 2.20: Observed output images at peak-power levels of (a) 70 W, (b) 320 W, and (c) 500 W when 100-fs pulses were propagated thorough a 6-mm-long array containing 41 waveguides. (From Ref. [150]; ©2002 OSA.)

difference between the excited cores. The beam remains confined to the same cores when $k = 0$. However, when $k \neq 0$, the power is transferred to successive cores as the CW beam propagates down the array. In all cases, the shape of the envelope governing power distribution among cores is maintained during this process. For this reason, such a structure is referred to as the *discrete soliton* [145]. It should be stressed that the word *soliton* in this context refers to a spatial soliton.

To understand why spatial solitons may form in an array of waveguides, it is useful to take the the quasi-CW limit of the set of equations (2.5.1) in which the effects of GVD are negligible and the second-derivative term can be ignored. By writing Eq. (2.5.1) in the form

$$i\frac{\partial u_m}{\partial \xi} + K(u_{m+1} - 2u_m + u_{m-1}) + 2Ku_m + |u_m|^2 u_m = 0, \qquad (2.5.3)$$

and removing the linear term $2Ku_m$ through the transformation $u'_m = u_m \exp(2iK\xi)$, we obtain the discrete NLS equation. It is easy to see that it is a discrete version of the NLS equation

$$i\frac{\partial u}{\partial \xi} + Kd^2\frac{\partial^2 u}{\partial x^2} + |u|^2 u = 0, \qquad (2.5.4)$$

where $x = md$ represents the position of the mth core along the array and the core spacing d is assumed to be small in the continuum limit. Equation (2.5.4) has spatially localized solutions in the form of spatial solitons. It is therefore likely that Eq. (2.5.3) also possesses localized nonlinear modes such that the power is confined to only few neighboring cores. This indeed turns out to be the case [120].

The discrete NLS equation (2.5.1) has been studied extensively, both analytically and numerically [145]–[151]. Its predictions have also been verified experimentally using an array of planar waveguides [150]–[152]. When the input pulse is launched into a single waveguide, as expected, its energy spreads into neighboring waveguides as the pulse travels down the array. Figure 2.20 shows the observed images at the output facet at three peak-power levels when 100-fs pulses were propagated thorough a 6-mm-long array containing 41 waveguides that were 4 μm wide and spaced apart by 8 μm. At a relatively low peak power of 70 W, pulse energy spread over the entire array. However, the energy distribution narrowed down as peak power was increased, and a discrete soliton formed at 500 W such that the entire pulse energy was confined to a narrow central region.

A similar behavior is expected to occur in multicore fiber arrays. In a 2004 experiment, a 320-μm-diameter fiber containing a two-dimensional array of silica cores separated by air holes was employed for this purpose [153]. Because of manufacturing imperfections, all cores were neither identical nor formed a perfectly uniform array, and the mean core diameter was 6.9 μm with an average core-to-core spacing of 12.1 μm. Nevertheless, it was found that the spreading of input light across the entire array was reduced significantly at high peak-power levels and exhibited features associated with a discrete soliton. A much more uniform array was used in a 2006 experiment in which individual cores were created within bulk silica by using high-energy femtosecond laser pulses [154]. The 5×5 cubic array consisted of 25 waveguides of 7.44-cm length separated 40 μm apart. The formation of a two-dimensional discrete spatial soliton was observed when 100-fs pulses with sufficient peak power were launched into one of the waveguides.

Problems

2.1 Explain in physical terms why the proximity of two cores in a fiber coupler leads to power exchange between the two cores.

2.2 Starting from the wave equation, derive the coupled-mode equations for a fiber coupler in the frequency domain.

2.3 Convert Eqs. (2.1.4) and (2.1.5) into time-domain equations, treating both the propagation constants and the coupling coefficients as frequency dependent. Assume a symmetric coupler to simplify the algebra.

2.4 Simplify the double integral in Eq. (2.1.6) for a symmetric fiber coupler whose core centers are separated by a distance d, assuming a Gaussian shape for the fundamental mode in each waveguide. Use cylindrical coordinates and express your answer as a single integral over the radial coordinate.

2.5 Discuss how κ obtained in previous problem depends on the ratio d/w_0 by plotting it. Find the coupling length when $d/w_0 = 3$, where w_0 is the mode width (FWHM). Compare your results with Eq. (2.1.24) an comment on the resulting poor agreement.

2.6 A CW optical beam with power P_0 is launched into one core of a symmetric fiber coupler. Solve the coupled-mode equations and find the power transferred to the second core. You can neglect the XPM-induced coupling. Discuss what happens when the input power exceeds $4\kappa/\gamma$.

2.7 Show that Eqs. (2.2.11)–(2.2.13) follow from the CW coupled-mode equations when the Stokes vector components are introduced as defined in Eq. (2.2.10).

2.8 Find all solutions of Eqs. (2.2.11)–(2.2.13) that remain invariant with z. Show the location of these fixed points on the Poincaré sphere. What happens when input power exceeds $2\kappa/\gamma$?

2.9 Find the dispersion relation associated with modulation instability when the CW solution of Eqs. (2.2.19) and (2.2.20) corresponds to the odd mode of a symmetric fiber coupler. Discuss the main differences from the even-mode case.

2.10 Show that the coupled NLS equations for a fiber coupler, Eqs. (2.3.1) and (2.3.2), indeed follow from the Lagrangian density in Eq. (2.3.6).

2.11 Evaluate the integral $\mathscr{L} = \int_{-\infty}^{\infty} \mathscr{L}_d d\tau$ using the soliton ansatz given in Eq. (2.3.7) and derive the four equations describing the evolution of the soliton parameters along the coupler length.

2.12 Solve Eq. (2.3.10) numerically for $\mu = 1.5$, 1.6, and 2.5 and plot Δ and ϕ along the coupler length. Interpret your results using phase diagrams of Figure 2.8.

2.13 Repeat the previous problem using the ansatz given in Eq. (2.3.16) and solve the resulting four equations analytically.

2.14 Find the symmetric and antisymmetric shape-preserving soliton pairs by solving Eqs. (2.3.1) and (2.3.2).

2.15 Solve Eqs. (2.4.4) and (2.4.5) numerically and reproduce the pulse shapes shown in Figure 2.10.

References

[1] G. P. Agrawal, *Lightwave Technology: Components and Devices* (Wiley, Hoboken, NJ, 2004).

[2] V. J. Tekippe, *Fiber Integ. Opt.* **9**, 97 (1990).

[3] P. E. Green, Jr., *Fiber-Optic Networks* (Prentice Hall, Upper Saddle River, NJ, 1993), Chap. 3.

[4] J. Hecht, *Understanding Fiber Optics* (Prentice Hall, Upper Saddle River, NJ, 1999), Chap. 15.

[5] A. K. Ghatak and K. Thyagarajan, *Introduction to Fiber Optics* (Cambridge University Press, New York, 1999), Chap. 17.

[6] A. W. Snyder, *J. Opt. Soc. Am.* **62**, 1267 (1972); P. D. McIntyre and A. Snyder, *J. Opt. Soc. Am.* **63**, 1518 (1973).

[7] A. W. Snyder and J. D. Love, *Optical Waveguide Theory* (Chapman and Hall, London, 1983).

[8] D. Marcuse, *Theory of Dielectric Optical Waveguides* (Academic Press, San Diego, CA, 1991), Chap. 6.

[9] H. A. Haus and W. P. Huang, *Proc. IEEE* **79**, 1505 (1991).

[10] W. P. Huang, *J. Opt. Soc. Am. A* **11**, 963 (1994).

[11] G. P. Agrawal, *Nonlinear Fiber Optics*, 4th ed. (Academic Press, Boston, 2007).

[12] R. Tewari and K. Thyagarajan, *J. Lightwave Technol.* **4**, 386 (1986).

[13] K. S. Chiang, *Opt. Lett.* **20**, 997 (1995); *IEEE J. Quantum Electron.* **33**, 950 (1997).

[14] K. S. Chiang, Y. T. Chow, D. J. Richardson, D. Taverner, L. Dong, L. Reekie, and K. M. Lo, *Opt. Commun.* **143**, 189 (1997).

[15] S. M. Jenson, *IEEE J. Quantum Electron.* **QE-18**, 1580 (1982).

[16] A. A. Maier, *Sov. J. Quantum Electron.* **12**, 1490 (1982); *Sov. J. Quantum Electron.* **14**, 101 (1984).

[17] K. Kitayama and S. Wang, *Appl. Phys. Lett.* **43**, 17 (1983).

[18] B. Daino, G. Gregori, and S. Wabnitz, *J. Appl. Phys.* **58**, 4512 (1985).

[19] S. Wabnitz, E. M. Wright, C. T. Seaton, and G. I. Stegeman, *Appl. Phys. Lett.* **49**, 838 (1986).

[20] D. D. Gusovskii, E. M. Dianov, A. A. Maier, V. B. Neustruev, E. I. Shklovskii, and I. A. Shcherbakov, *Sov. J. Quantum Electron.* **15**, 1523 (1985)

[21] D. D. Gusovskii, E. M. Dianov, A. A. Maier, V. B. Neustruev, V. V. Osiko, A. M. Prokhorov, K. Y. Sitarskii, and I. A. Shcherbakov, *Sov. J. Quantum Electron.* **17**, 724 (1987).

[22] S. R. Friberg, Y. Silberberg, M. K. Oliver, M. J. Andrejco, M. A. Saifi, and P. W. Smith, *Appl. Phys. Lett.* **51**, 1135 (1987).

[23] S. R. Friberg, A. M. Weiner, Y. Silberberg, B. G. Sfez, and P. W. Smith, *Opt. Lett.* **13**, 904 (1988).

[24] A. M. Weiner, Y. Silberberg, H. Fouckhardt, D. E. Leaird, M. A. Saifi, M. J. Andrejco, and P. W. Smith, *IEEE J. Quantum Electron.* **25**, 2648 (1989).

[25] S. Trillo, S. Wabnitz, W. C. Banyai, N. Finlayson, C. T. Seaton, G. I. Stegeman, and R. H. Stolen, *IEEE J. Quantum Electron.* **25**, 104 (1989).

[26] G. I. Stegeman and E. M. Wright, *Opt. Quantum Electron.* **22**, 95 (1990).

[27] A. T. Pham and L. N. Binh, *J. Opt. Soc. Am. B* **8**, 1914 (1991).

[28] A. W. Snyder, D. J. Mitchell, L. Poladian, D. R. Rowland, and Y. Chen, *J. Opt. Soc. Am. B* **8**, 2102 (1991).

[29] W. Samir, C. Pask, and S. J. Garth, *J. Opt. Soc. Am. B* **11**, 2193 (1994).

[30] A. A. Maier, *Sov. Phys. Usp.* **165**, 1037 (1995); *Sov. Phys. Usp.* **166**, 1171 (1996).

[31] K. Yasumoto, H. Maeda, and N. Maekawa, *J. Lightwave Technol.* **14**, 628 (1996).

[32] D. Artigas, F. Dios, and F. Canal, *J. Mod. Opt.* **44**, 1207 (1997).

[33] S. R. Vigil, Z. Zhou, B. K. Canfield, J. Tostenrude, and M. G. Kuzyk, *J. Opt. Soc. Am. B* **15**, 895 (1998).

[34] D. Marchese, M. De Sairo, A. Jha, A. K. Kar, and E. C. Smith, *J. Opt. Soc. Am. B* **15**, 2361 (1998).

[35] T. Gabler, A. Brauer, H. H. Horhold, T. Pertsch, and R. Stockmann, *Chem. Phys.* **245**, 507 (1999).

[36] S. Trillo, S. Wabnitz, E. M. Wright, and G. I. Stegeman, *Opt. Lett.* **13**, 672 (1988).

[37] S. Trillo, S. Wabnitz, G. I. Stegeman, and E. M. Wright, *J. Opt. Soc. Am. B* **6**, 899 (1989).

[38] S. Trillo, S. Wabnitz, and G. I. Stegeman, *IEEE J. Quantum Electron.* **25**, 1907 (1989).

[39] S. Trillo and S. Wabnitz, *Opt. Lett.* **16**, 1 (1991).

[40] C. C. Yang, *Opt. Lett.* **16**, 1641 (1991).

[41] S. Wabnitz, S. Trillo, E. M. Wright, and G. I. Stegeman, *J. Opt. Soc. Am. B* **8**, 602 (1991).

[42] E. Caglioti, S. Trillo, S. Wabnitz, B. Crosignani, and P. Di Porto, *J. Opt. Soc. Am. B* **7**, 374 (1990).

[43] C. Paré and M. Florjanczyk, *Phys. Rev. A* **41**, 6287 (1990).

[44] M. Romagnoli, S. Trillo, and S. Wabnitz, *Opt. Quantum Electron.* **24**, S1237 (1992).

[45] Y. S. Kivshar, *Opt. Lett.* **18**, 7 (1993).

[46] Y. S. Kivshar and M. L. Quiroga-Teixerio, *Opt. Lett.* **18**, 980 (1993).

[47] P. L. Chu, G. D. Peng, and B. A. Malomed, *Opt. Lett.* **18**, 328 (1993); *J. Opt. Soc. Am. B* **10**, 1379 (1993).

[48] E. M. Wright, G. I. Stegeman, and S. Wabnitz, *Phys. Rev. A* **40**, 4455 (1989).

[49] N. Akhmediev and A. Ankiewicz, *Phys. Rev. Lett.* **70**, 2395 (1993).

[50] J. M. Soto-Crespo and N. Akhmediev, *Phys. Rev. E* **48**, 4710 (1993).

[51] N. Akhmediev and J. M. Soto-Crespo, *Phys. Rev. E* **47**, 1358 (1993); *Phys. Rev. E* **49**, 4519 (1994).

[52] D. C. Psaila and C. M. de Sterke, *Opt. Lett.* **18**, 1905 (1993).

[53] C. Schmidt-Hattenberger, R. Muschall, F. Lederer, and U. Trutschel, *J. Opt. Soc. Am. B* **10**, 1592 (1993).

[54] W. Samir, S. J. Garth, and C. Pask, *Appl. Opt.* **32**, 4513 (1993).

[55] A. V. Buryak and N. Akhmediev, *Opt. Commun.* **110**, 287 (1994).

[56] A. Ankiewicz, M. Karlsson, and N. Akhmediev, *Opt. Commun.* **111**, 116 (1994).

[57] I. M. Uzunov, R. Muschall, M. Gölles, Y. S. Kivshar, B. A. Malomed, and F. Lederer, *Phys. Rev. E* **51**, 2527 (1995).

[58] N. Akhmediev and A. Ankiewicz, *Solitons: Nonlinear Pulses and Beams* (Chapman and Hall, New York, 1997).

[59] P. M. Ramos and C. R. Paiva, *IEEE J. Sel. Topics Quantum Electron.* **3**, 1224 (1997); *IEEE J. Quantum Electron.* **35**, 983 (1999).

[60] P. Shum, K. S. Chinag, and W. A. Gambling, *IEEE J. Quantum Electron.* **35**, 79 (1999).

[61] T. P. Valkering, J. van Honschoten, and H. J. Hoekstra, *Opt. Commun.* **159**, 215 (1999).

[62] B. A. Umarov, F. K. Abdullaev, and M. R. B. Wahiddin, *Opt. Commun.* **162**, 340 (1999).

[63] S. C. Tsang, K. S. Chiang, and K. W. Chow, *Opt. Commun.* **229**, 431 (2004).

[64] S. Droulias, M. Manousakis, and K. Hizanidis, *Opt. Commun.* **240**, 209 (2004).

[65] Y. Wang and W. Wang, *J. Lightwave Technol.* **24**, 1041 (2006); **24**, 2458 (2006).

[66] P. A. Bélanger and C. Paré, *Phys. Rev. A* **41**, 5254 (1990).

[67] R. H. Stolen, J. P. Gordon, W. J. Tomlinson, and H. A. Haus, *J. Opt. Soc. Am. B* **6**, 1159 (1989).

[68] K. J. Blow and D. Wood, *IEEE J. Quantum Electron.* **25**, 2665 (1989).

[69] P. B. Hansen, A. Kloch, T. Aaker, and T. Rasmussen, *Opt. Commun.* **119**, 178 (1995).

[70] B. A. Malomed, *Phys. Rev. E* **51**, R864 (1995).

[71] B. A. Malomed, I. M. Skinner, P. L. Chu, and G. D. Peng, *Phys. Rev. E* **53**, 4084 (1996).

[72] D. J. Kaup, T. I. Lakoba, and B. A. Malomed, *J. Opt. Soc. Am. B* **14**, 1199 (1997).

[73] D. J. Kaup and B. A. Malomed, *J. Opt. Soc. Am. B* **15**, 2838 (1998).

[74] J. Atai and B. A. Malomed, *Opt. Commun.* **221**, 55 (2003).

[75] W. B. Fraga, J. W. M. Menezes, M. G. da Silva, C.S. Sobrinho, and A. S. B. Sombra, *Opt. Commun.* **262**, 32 (2006).

[76] J. Wilson, G. I. Stegeman, and E. M. Wright, *Opt. Lett.* **16**, 1653 (1991).

[77] Y. Chen, A. W. Snyder, and D. N. Payne, *IEEE J. Quantum Electron.* **28**, 239 (1992).

[78] P. L. Chu and B. Wu, *Opt. Lett.* **17**, 255 (1992).

[79] H. G. Winful and D. T. Walton, *Opt. Lett.* **17**, 1688 (1992).

[80] D. T. Walton and H. G. Winful, *Opt. Lett.* **18**, 720 (1993).

[81] P. L. Chu, B. A. Malomed, H. Hatami-Hanza, and I. M. Skinner, *Opt. Lett.* **20**, 1092 (1995).

[82] B. A. Malomed, G. D. Peng, and P. L. Chu, *Opt. Lett.* **21**, 330 (1996).

[83] M. Liu, K. S. Chiang, and P. Shum, *IEEE J. Quantum Electron.* **40**, 1597 (2004).

[84] R. Ganapathy, B. A. Malomed, and K. Porsezian, *Phys. Lett. A* **354**, 366 (2005).

[85] D. Marcuse, *J. Lightwave Technol.* **5**, 268 (1987); *IEEE J. Quantum Electron.* **26**, 675 (1990).

[86] W. P. Huang and H. A. Haus, *J. Lightwave Technol.* **7**, 920 (1989).

[87] Y. J. Chen and A. W. Snyder, *Opt. Lett.* **16**, 217 (1991).

[88] Y. J. Chen, *J. Mod. Opt.* **38**, 1731 (1991).

[89] W. P. Huang, B. E. Little, and S. K. Chadhuri, *J. Lightwave Technol.* **9**, 721 (1991).

[90] L. P. Yuan, *IEEE J. Quantum Electron.* **29**, 171 (1993).

[91] B. E. Little and W. P. Huang, *J. Lightwave Technol.* **11**, 1990 (1993).

[92] B. E. Little, *J. Lightwave Technol.* **12**, 774 (1994).

[93] S. H. Zhang and T. Tamir, *Opt. Lett.* **20**, 803 (1995); *J. Opt. Soc. Am. A* **13**, 2403 (1997).

[94] N. H. Sun, J. K. Butler, G. A. Evans, L. Pang, and P. Congdon, *J. Lightwave Technol.* **15**, 2301 (1997).

[95] T. Liang and R. W. Ziolkowski, *Microwave Opt. Tech. Lett.* **17**, 17 (1998).

[96] N. Izhaky and A. Hardy, *J. Opt. Soc. Am. A* **16**, 1303 (1999); *Appl. Opt.* **38**, 6987 (1999).

[97] H. Sabert, L. Dong, and P. S. J. Russell, *Int. J. Optoelectron.* **7**, 189 (1992).

[98] L. Dong, T. A. Birks, M. H. Ober, and P. S. J. Russell, *J. Lightwave Technol.* **12**, 24 (1994).

[99] J. L. Archambault, P. S. J. Russell, S. Bacelos, P. Hua, and L. Reekie, *Opt. Lett.* **19**, 180 (1994).

[100] L. Dong, P. Hua, T. A. Birks, L. Reekie, and P. S. J. Russell, *IEEE Photon. Technol. Lett.* **8**, 1656 (1996).

[101] F. Bakhti, P. Sansonetti, C. Sinet, L. Gasca, L. Martineau, S. Lacroix, X. Daxhelet, and F. Gonthier, *Electron. Lett.* **33**, 803 (1997).

[102] S. S. Orlov, A. Yariv, and S. van Essen, *Opt. Lett.* **22**, 688 (1997).

[103] A. Ankiewicz and G. D. Peng, *Electron. Lett.* **33**, 2151 (1997).

[104] A. S. Kewitsch, G. A. Rakuljic, P. A. Willems, and A. Yariv, *Opt. Lett.* **23**, 106 (1998).

[105] T. Erdogan, *Opt. Commun.* **157**, 249 (1998).

[106] B. Ortega, L. Dong, and L. Reekie, *Appl. Opt.* **37**, 7712 (1998).

[107] J. Capmany, P. Muñoz, and D. Pastor, *IEEE J. Sel. Topics Quantum Electron.* **5**, 1392 (1999).

[108] H. An, B. Ashton, and S. Fleming, *Opt. Lett.* **29**, 343 (2004).

[109] M. Kulishov and J. Azaña, *Opt. Express* **12**, 2699 (2004).

[110] F. Y. U. Chan and K. S. Chiang, *J. Lightwave Technol.* **24**, 1008 (2006).

[111] S. Trillo and S. Wabnitz, *J. Opt. Soc. Am. B* **5**, 483 (1988).

[112] D. C. Psaila and C. M. de Sterke, *Opt. Lett.* **18**, 1905 (1993).

[113] T. I. Lakoba, D. J. Kaup, and B. A. Malomed, *Phys. Rev. E* **55**, 6107 (1997).

[114] T. I. Lakoba and D. J. Kaup, *Phys. Rev. E* **56**, 4791 (1997).

[115] R. H. Stolen, A. Ashkin, W. Pleibel, and J. M. Dziedzic, *Opt. Lett.* **9**, 200 (1984).

[116] H. A. Haus and L. Molter-Orr, *IEEE J. Quantum Electron.* **19**, 840 (1983).

[117] E. Kapon, J. Katz, and A. Yariv, *Opt. Lett.* **9**, 125 (1984).

[118] G. P. Agrawal, *J. Appl. Phys.* **58**, 2922 (1985).

[119] M. Kuznetsov, *IEEE J. Quantum Electron.* **21**, 1893 (1985).

[120] D. N. Christodoulides and R. I. Joseph, *Opt. Lett.* **13**, 794 (1988).

[121] J. W. Arkwright and D. B. Mortimore, *Electron. Lett.* **26**, 1534 (1989).

[122] D. B. Mortimore and J. W. Arkwright, *Appl. Opt.* **29**, 1814 (1990).

[123] B. J. Mangan, J. C. Knight, T. A. Birks, P. St. J. Russell, and A. H. Greenaway, *Electron. Lett.* **36**, 1358 (2000).

[124] F. Fogli, L. Saccomandi, P. Bassi, G. Bellanca, and S. Trillo, *Opt. Express* **10**, 54 (2002).

[125] K. Saitoh, Y. Sato, and M. Koshiba, *Opt. Express* **11**, 3188 (2003).

[126] F. Cuesta-Soto, A. Martínez, J. García, F. Ramos, P. Sanchis, J. Blasco, and J. Martí, *Opt. Express* **12**, 161 (2004).

[127] L. Zhang and C. Yang, *J. Lightwave Technol.* **22**, 1367 (2004).

[128] J. R. Salgueiro and Y. S. Kivshar, *Opt. Lett.* **30**, 1858 (2005).

[129] A. Betlej, S. Suntsov, K. G. Makris, L. Jankovic, D. N. Christodoulides, G. I. Stegeman, J. Fini, R. T. Bise, andD. J. DiGiovanni, *Opt. Lett.* **31**, 1480 (2006).

[130] B. H. Lee, J. B. Eom, J. Kim, D. S. Moon, U. C. Paek, and G. H. Yang, *Opt. Lett.* **27**, 812 (2002).

[131] H. Kim, J. Kim, U. C. Paek, B. H. Lee, and K. T. Kim, *Opt. Lett.* **29**, 1194 (2004).

[132] L. Fu and M. Gu, *Opt. Lett.* **31**, 1471 (2006).

[133] U. Röpke, H. Bartelt, S. Unger, K. Schuster and J. Kobelke, *Opt. Express* **15**, 6894 (2007).

[134] Y. Chen, A. W. Snyder, and D. J. Mitchell, *Electron. Lett.* **26**, 77 (1990).

[135] N. Finlayson and G. I. Stegeman, *Appl. Phys. Lett.* **56**, 2276 (1990).

[136] C. Schmidt-Hattenberger, R. Muschall, U. Trutschel, and F. Lederer, *Opt. Quantum Electron.* **24**, 691 (1992).

[137] N. Akhmediev and A. V. Buryak, *J. Opt. Soc. Am. B* **11**, 804 (1994).

[138] D. Artigas, J. Olivas, F. Dios, and F. Canal, *Opt. Commun.* **131**, 53 (1996).

[139] M. G. da Silva, A. F. Teles, and A. S. B. Sombra, *J. Appl. Phys.* **84**, 1834 (1998).

[140] K. G. Kalonakis and E. Paspalakis, *J. Mod. Opt.* **52**, 1885 (2005).

[141] C. Schmidt-Hattenberger, U. Trutschel, and F. Lederer, *Opt. Lett.* **16**, 294 (1991).

[142] P. E. Langridge and W. J. Firth, *Opt. Quantum Electron.* **24**, 1315 (1992).

[143] Y. S. Kivshar, *Opt. Lett.* **18**, 1147 (1993); *Phys. Lett. A* **173**, 172 (1993).

[144] W. Królikowski, U. Trutschel, M. Cronin-Golomb, and C. Schmidt-Hattenberger, *Opt. Lett.* **19**, 320 (1994).

[145] Y. S. Kivshar and G. P. Agrawal, *Optical Solitons: From Fibers to Photonic Crystals* (Academic Press, Boston, 2003).

[146] A. B. Aceves, C. De Angelis, A. M. Rubenchik, and S. K. Turitsyn, *Opt. Lett.* **19**, 329 (1994).

[147] A. B. Aceves, G. G. Luther, C. De Angelis, A. M. Rubenchik, and S. K. Turitsyn, *Phys. Rev. Lett.* **75**, 73 (1995); *Opt. Fiber Technol.* **1**, 244 (1995).

[148] A. B. Aceves, M. Santagiustina, and C. De Angelis, *J. Opt. Soc. Am. B* **14**, 1807 (1997).

[149] P. M. Ramos and C. R. Paiva, *J. Opt. Soc. Am. B* **17**, 1125 (2000).

[150] H. S. Eisenberg, R. Morandotti, Y. Silberberg, J. M. Arnold, G. Pennelli, and J. S. Aitchison, *J. Opt. Soc. Am. B* **19**, 2938 (2002).

[151] U. Peschel, R. Morandotti, J. M. Arnold, J. S. Aitchison, H. S. Eisenberg, Y. Silberberg, T. Pertsch, and F. Lederer, *J. Opt. Soc. Am. B* **19**, 2637 (2002).

[152] D. Cheskis, S. Bar-Ad, R. Morandotti, J. S. Aitchison, H. S. Eisenberg, Y. Silberberg, and D. Ross, *Phys. Rev. Lett.* **91**, 223901 (2003).

[153] T. Pertsch, U. Peschel, J. Kobelke, K. Schuster, and H. Bartelt, S. Nolte, A. Tünnermann, and F. Lederer, *Phys. Rev. Lett.* **93**, 053901 (2004).

[154] A. Szameit, J. Burghoff, T. Pertsch, S. Nolte, A. Tünnermann, and F. Lederer, *Opt. Express* **14**, 6055 (2006).

Chapter 3

Fiber Interferometers

The two fiber components covered in Chapters 1 and 2 can be combined to form a variety of fiber-based optical devices. Four common ones among them are the fiber version of the well-known Fabry–Perot, Mach–Zehnder, Sagnac, and Michelson interferometers [1]. They exhibit interesting nonlinear effects that are useful for optical-switching applications, when power levels are large enough for the nonlinear phenomenon of self-phase modulation (SPM) or cross-phase modulation (XPM) to become important [2]. This chapter is devoted to the nonlinear effects occurring in such four types fiber interferometers. Section 3.1 focuses on the Fabry–Perot and ring resonators and considers several nonlinear effects, such as optical bistability, chaos, and and modulation instability, that may occur because of the optical feedback. Nonlinear fiber-loop mirrors, whose operation is based on Sagnac interferometers, are covered in Section 3.2. Such devices are commonly employed for switching an optical beam nonlinearly to a different output port by changing the incident power. Both the SPM- and XPM-based switching schemes are discussed in this section together with their potential applications. Nonlinear switching in Mach–Zehnder interferometers is described in Section 3.3. Finally, Section 3.4 is devoted to Michelson interferometers.

3.1 Fabry–Perot and Ring Resonators

Fabry–Perot and ring resonators are well-known devices used commonly for making lasers. A fiber-based Fabry–Perot resonator can be constructed by simply making two ends of an optical fiber partially reflecting. This can be realized in practice by using external mirrors or by depositing high-reflectivity coatings at the two ends. An alternative approach, shown schematically in Figure 3.1, splices a fiber grating at each end of the fiber. The construction of a fiber-ring resonator is even simpler. It can be made by connecting the two ends of a piece of fiber to an input and an output port of a fiber coupler, as shown schematically in Figure 3.1. This section is devoted to the nonlinear effects occurring in such resonators. The continuous-wave (CW) case is considered first with focus on optical bistability and chaos. It is followed by a discussion of modulation instability and other temporal phenomena.

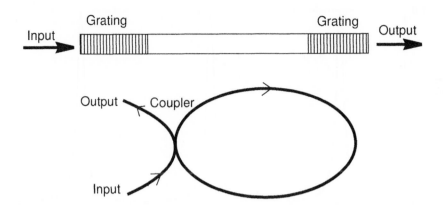

Figure 3.1: Fiber-based Fabry–Perot and ring resonators made by using Bragg gratings and directional couplers.

3.1.1   Transmission Resonances

Several types of fiber-based Fabry–Perot interferometers have been developed for wavelength-division multiplexing (WDM) applications [3]–[5]. Some of them function in a linear fashion if air is used as an intracavity medium [6]. Others use a piece of fiber between two Bragg gratings [7] and thus are capable of exhibiting the nonlinear effects. Fiber-ring resonators were made as early as 1982 using a directional coupler [8], and a finesse of 1260 was realized by 1988 [9].

Transmittivity of a Fabry–Perot resonator, formed by using two identical mirrors (or Bragg gratings) of reflectivity R_m, can be calculated by adding coherently the optical fields transmitted on successive round trips [1]. Consider a CW optical beam at the frequency ω incident at the left mirror located at $z = 0$. The optical field inside the resonator at a distance z consists of forward- and backward-propagating waves and can be expressed as

$$E(\mathbf{r},t) = \tfrac{1}{2}F(x,y)\{A(z)\exp[i(\tilde{\beta}z - \omega t)] + B(z)\exp[-i(\tilde{\beta}z + \omega t)] + \text{c.c.}\}, \quad (3.1.1)$$

where $F(x,y)$ is the spatial distribution and $\tilde{\beta}$ is the propagation constant associated with the fundamental mode supported by the fiber. The transmitted field is obtained by adding contributions of an infinite number of round trips and is given by

$$A(L) = \frac{(1 - R_m)A(0)}{1 - R_m\exp(i\tilde{\beta}L_R)}, \quad (3.1.2)$$

where $L_R \equiv 2L$ is the round-trip distance for a fiber of length L.

Transmittivity of the resonator is obtained from Eq. (3.1.2) and is given by the well-known Airy formula [1]

$$T_R = \frac{P_t}{P_i} = \left|\frac{A(L)}{A(0)}\right|^2 = \frac{(1 - R_m)^2}{(1 - R_m)^2 + 4R_m\sin^2(\phi_R/2)}, \quad (3.1.3)$$

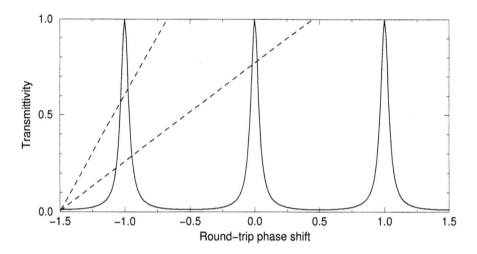

Figure 3.2: Transmittivity of a Fabry–Perot resonator as a function of $\phi_0/2\pi$ for $R_m = 0.8$. Dashed lines show changes in ϕ_R because of SPM at two power levels.

where $P_i = |A(0)|^2$ is the input power, P_t is the transmitted power, and $\phi_R = \tilde{\beta}L_R$ is the phase shift occurring over one round trip inside the resonator. The nonlinear and dispersive effects enter through this phase shift, which can be separated into two parts

$$\phi_R(\omega) \equiv \phi_0(\omega) + \phi_{NL} = [\beta(\omega) + \Delta\beta_{NL}]L_R. \tag{3.1.4}$$

The nonlinear part ϕ_{NL} represents the contribution of SPM and can be related to the nonlinear parameter γ as [2]

$$\phi_{NL} = \gamma \int_0^L [|A(z)|^2 + |B(z)|^2]\,dz = \gamma P_{av}L_R, \tag{3.1.5}$$

where P_{av} is the average power level inside the resonator.

At low power levels such that $\phi_{NL} \ll 1$, the nonlinear effects can be neglected. In that case, 100% of the incident light is transmitted ($T_R = 1$) whenever $\phi_0 = 2m\pi$, where m is an integer. Frequencies that satisfy this condition correspond to the longitudinal modes of the resonator. Transmission drops as the frequency of incident light is detuned from the resonance. The solid curve in Figure 3.2 shows the transmittivity of a fiber resonator as a function of ϕ_0 for $R_m = 0.8$. The frequency spacing Δv_L between the successive transmission peaks is known as the *free-spectral range*; Δv_L is also called the *longitudinal-mode spacing* in laser literature. It is obtained using the phase-matching condition

$$[\beta(\omega + 2\pi\Delta v_L) - \beta(\omega)]L_R = 2\pi \tag{3.1.6}$$

and is approximately given by $\Delta v_L = v_g/L_R \equiv 1/T_R$, where $v_g \equiv 1/\beta_1$ is the group velocity and T_R is the round-trip time within the resonator. Because of group-velocity dispersion (GVD), the free spectral range of a fiber resonator becomes frequency dependent. It can vary considerably in a Fabry–Perot resonator made by using Bragg gratings because of the large GVD associated with them [10]. The sharpness of the resonance peaks in Figure 3.2 is quantified through the resonator finesse F_R defined as

$$F_R = \frac{\Delta \nu_L}{\Delta \nu_R} = \frac{\pi \sqrt{R_m}}{1 - R_m}, \tag{3.1.7}$$

where $\Delta \nu_R$ is the width of each resonance peak (at half maximum).

Equation (3.1.3) changes somewhat for a fiber-ring resonator [8]. The changes are related to the constant phase shift of $\pi/2$ occurring when light crosses over from one core to another inside a fiber coupler (see Section 2.1). Note also that $B(z) = 0$ in Eq. (3.1.1) for a ring resonator since a backward-propagating wave is not generated in this case. This feature simplifies the mathematical description and has considerable implications for nonlinear phenomena since the XPM-induced coupling between the forward- and backward-propagating waves cannot occur in unidirectional ring resonators.

3.1.2 Optical Bistability

The nonlinear phenomenon of optical bistability has been studied in nonfiber resonators since 1976 by placing the nonlinear medium inside a cavity formed by using multiple mirrors [11]–[15]. The single-mode fiber was used in 1983 as the nonlinear medium inside a ring cavity [16]. Since then, the study of nonlinear phenomena in fiber resonators has remained a topic of considerable interest [17]–[31].

The origin of the nonlinear effects in fiber resonators is evident from Eq. (3.1.3), where the round-trip phase shift ϕ_R depends on input power because of the SPM-induced phase shift ϕ_{NL}. For high-finesse resonators, $P_t \approx (1 - R_m)P_{av}$. Using this relation in Eq. (3.1.5), the transmitted power from Eq. (3.1.3) is found to satisfy the transcendental equation

$$P_t \left\{ 1 + \frac{4R_m}{(1 - R_m)^2} \sin^2 \left[\frac{\phi_0}{2} + \frac{\gamma P_t L_R}{2(1 - R_m)} \right] \right\} = P_i. \tag{3.1.8}$$

It is clear from this equation that multiple values of P_t are possible at a fixed value of the incident power P_i because of SPM. Dashed lines in Figure 3.2 show ϕ_R as a function of ϕ_0 for two values of P_i using Eq. (3.1.4). The intersection points of the dashed lines with the solid curve correspond to the multiple solutions of Eq. (3.1.8). At low powers, the dashed lines become nearly vertical, and only one solution is possible. With increasing input power, the dashed lines tilt, and the number of solutions increases from one to three, then to five and beyond. We focus on the case of three solutions since it requires the least input power.

Multiple solutions of Eq. (3.1.8) lead to dispersive optical bistability, a nonlinear phenomenon that has been observed using several different nonlinear media [15]. It occurs in fiber resonators when the linear phase shift ϕ_0 does not correspond to a resonance of the resonator so that little light is transmitted at low power levels. For a given detuning $\delta \equiv 2\pi M - \phi_0$ of the input signal from the nearest Mth resonance, the SPM-induced phase shift reduces the net detuning toward zero, resulting in higher transmission. However, the transmitted power P_t does not increase linearly with P_0,

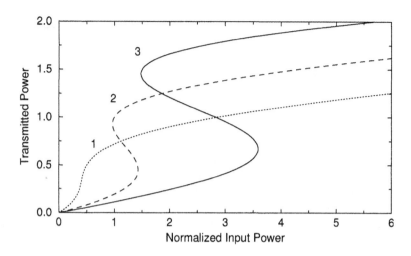

Figure 3.3: Bistable response of a fiber resonator with $R_m = 0.5$ for three values of detuning δ. Powers are normalized using $P_n = (\gamma L_R)^{-1}$.

as is evident from the nonlinear nature of Eq. (3.1.8). Figure 3.3 shows the expected behavior for three values of δ. Over a certain range of δ, three solutions of Eq. (3.1.8) produce the well-known S-shaped curve associated with optical bistability. The middle branch with a negative slope is always unstable [15]. As a result, the transmitted power jumps up and down at specific values of P_i in such a way that it exhibits hysteresis. The switching power is on the order of $(\gamma L_R)^{-1}$. Its numerical values are ~ 1 W for $L_R \sim 500$ m and $\gamma = 2$ W^{-1}/km but can be reduced by a factor of 10 or more by employing highly nonlinear fibers.

Experimental observation of optical bistability using CW optical beams is hampered by a relatively low threshold of stimulated Brillouin scattering (SBS) in fiber resonators [32]. The evidence of bistability in a ring cavity was first seen in a 1983 experiment in which the onset of SBS was avoided using picosecond pulses [16]. In a later experiment, SBS was suppressed by placing an optical isolator inside the ring cavity that was formed using 13 m of low-birefringence fiber [18]. Bistable behavior was observed in this experiment at CW power levels below 10 mW. The nonlinear phase shift ϕ_{NL} at this power level is relatively small in magnitude (below 0.01 rad) but still large enough to induce bistability.

In all experiments on optical bistability, it is important to stabilize the cavity length to subwavelength accuracy. An improved stabilization scheme was used in a 1998 experiment [28]. Figure 3.4 shows the observed behavior at four values of the detuning δ. The experiment used mode-locked pulses (width ~ 1 ps) emitted from a Ti:sapphire laser. The length of ring resonator (about 7.4 m) was adjusted precisely so that an entering laser pulse overlapped in time with another pulse already circulating inside the cavity (synchronous pumping). The observed bistable behavior was in qualitative agreement with the CW theory in spite of the use of short optical pulses since the GVD played a relatively minor role [19].

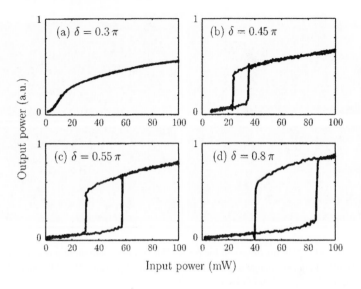

Figure 3.4: Hysteresis cycles observed in a fiber-ring resonator at four values of detuning δ. (From Ref. [28]; ©1998 OSA.)

3.1.3 Nonlinear Dynamics and Chaos

It was discovered in 1979 that the nonlinear response of a ring resonator can initiate a period-doubling route to optical chaos [12]. The basic idea consists of recognizing that the dynamics in a ring cavity correspond to that of a nonlinear map in the sense that the intracavity field is mapped to a different function on each round trip inside the cavity [33]–[35]. Mathematically, the map can be written as

$$A^{(n+1)}(0,t) = \sqrt{\rho} A^{(n)}(L_R,t) \exp(i\phi_0) + i\sqrt{(1-\rho)P_i}, \qquad (3.1.9)$$

where the superscript denotes the number of round trips inside the resonator and ρ represents the fraction of the power remaining in the resonator after the coupler (see Figure 3.1). Evolution of the intracavity field $A(z,t)$ during each round trip is governed by the usual nonlinear Schrödinger (NLS) equation:

$$i\frac{\partial A}{\partial z} - \frac{\beta_2}{2}\frac{\partial^2 A}{\partial T^2} + \gamma|A|^2 A = 0, \qquad (3.1.10)$$

where $T = t - z/v_g$ is the reduced time and β_2 is the GVD parameter. If the effect of GVD can be neglected in a CW or quasi-CW situation, this equation can be solved analytically to obtain the simple result

$$A(L_R,t) = A(0,t) \exp[i\gamma|A(0,t)|^2 L_R]. \qquad (3.1.11)$$

Using Eq. (3.1.11) in Eq. (3.1.9), the nonlinear map can be iterated for a given value of the input power P_i. The results show that the output of the ring resonator can become time dependent even for a CW input. Moreover, the output becomes

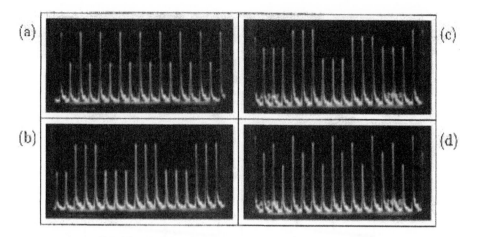

Figure 3.5: Period-2 patterns observed for $\delta = 0.35\pi$ at an average input power of (a) 200 mW and (b) 250 mW. Patterns change to period 4 for $\delta = 0.65\pi$ [(c) and (d)] at the same two power levels. (From Ref. [28]; ©1998 OSA.)

chaotic following a period-doubling route in a certain range of input parameters [12]. This behavior was observed experimentally in a 1983 experiment by launching 140-ps pulses (obtained from a Q-switched, mode-locked Nd:YAG laser) into a fiber-ring cavity [16]. The cavity length was selected to precisely match the round-trip time to the 7.6-ns interval between the neighboring pulses (synchronous pumping).

In a 1998 experiment, a mode-locked Ti:sapphire laser was used to launch short pulses (width $\sim$1 ps) into a well-stabilized fiber-ring resonator [28]. Figure 3.5 shows the period-2 and period-4 patterns observed using two different values of detuning at two different power levels. At higher power levels, the output became chaotic over a wide range of values for the detuning parameter δ, with period-3 windows embedded within the chaos. These features are consistent with the general theory of nonlinear dynamical systems [33]–[35].

3.1.4   Modulation Instability

Even in the absence of feedback, the combination of GVD and SPM can lead to modulation instability when a CW beam propagates inside optical fibers [2]. An interesting question is how the presence of feedback modifies this nonlinear phenomenon. Modulation instability in fiber resonators is of considerable interest as it can be used to convert a CW beam into a train of ultrashort pulses [36]–[41].

The theory of modulation instability has been extended to include the effects of feedback occurring inside a fiber resonator. The analysis is quite involved in the case of a Fabry–Perot cavity, since one must use the coupled NLS equations describing the evolution of the forward- and backward-propagating waves [41]. It simplifies considerably for a ring resonator [39]. In fact, one can use Eqs. (3.1.9) and (3.1.10). The approach is similar to that used in Section 5.1 of Ref. [2]. It is useful to normalize Eq.

(3.1.10) in the usual way and write it as

$$i\frac{\partial u}{\partial \xi} - \frac{s}{2}\frac{\partial^2 u}{\partial \tau^2} + |u|^2 u = 0, \tag{3.1.12}$$

where $s = \mathrm{sgn}(\beta_2) = \pm 1$ and we have introduced

$$\xi = z/L_R, \quad \tau = T/\sqrt{|\beta_2|L_R}, \quad u = (\gamma L_R)^{1/2}A. \tag{3.1.13}$$

Note that the resonator length L_R is used to define the time scale.

The CW solution of Eq. (3.1.12) is given by $u = u_0 \exp(iu_0^2\xi)$. To examine its stability, we perturb it at a frequency Ω such that

$$u(\xi, \tau) = [u_0 + a_1 \exp(-i\Omega\tau) + a_2 \exp(i\Omega\tau)] \exp(iu_0^2\xi), \tag{3.1.14}$$

where a_1 and a_2 represent weak perturbations whose growth results in the two sidebands associated with modulation instability. When the NLS equation is linearized in terms of a_1 and a_2, we obtain the coupled linear differential equations:

$$da_1/d\xi = i(\tfrac{1}{2}s\Omega^2 + u_0^2)a_1 + iu_0^2 a_2, \tag{3.1.15}$$

$$da_2/d\xi = i(\tfrac{1}{2}s\Omega^2 + u_0^2)a_1 + iu_0^2 a_1 \tag{3.1.16}$$

These equations should be solved subject to the boundary conditions imposed by the ring cavity:

$$a_j^{(n+1)}(0) = \sqrt{\rho}a_j^{(n)}(1)\exp[i(\phi_0 + u_0^2)], \quad (j = 1, 2), \tag{3.1.17}$$

where the superscript denotes the round-trip number.

Equations (3.1.15)–(3.1.17) relate the perturbation amplitudes a_1 and a_2 on two successive round trips. Modulation instability occurs if they grow after each round trip for a given set of parameters. The growth rate depends not only on the frequency Ω and the input power P_i but also on the fiber-ring parameters β_2, γ, ρ, and ϕ_0. The interesting new feature is that modulation instability can occur even in the normal-GVD region of the fiber [40]. Moreover, the instability occurs either close to a cavity resonance, $\phi_0 \approx 2m\pi$, or close to the antiresonance condition $\phi_0 \approx (2m+1)\pi$. Modulation instability in the latter case is called period-2 type since the phase of perturbation is restored after two round trips inside the cavity. Figure 3.6 shows the gain spectra in the normal-GVD region of a ring cavity using $\rho = 0.95$ and $u_0 = 1$. Different peaks correspond to detuning of the CW beam such that ϕ_0 deviates from the resonance (thin line) or the antiresonance (thick line) condition by 0.1π. In real units, $\Omega = 1$ corresponds to a frequency of about 0.3 THz when $\beta_2 = 30$ ps^2/km and $L_R = 10$ m.

Evidence of modulation instability in a fiber-ring resonator has been seen experimentally [40] with the same setup used for Figure 3.5. The 7.38-m ring cavity was driven synchronously using 1.25 ps from a 980-nm, mode-locked Ti:sapphire laser. When the peak power of input pulses exceeded a threshold value (about 500 W), the pulse spectrum developed peaks at the location corresponding to antiresonances of the fiber resonator. The spectrum exhibited peaks at cavity resonances also. However, such peaks appear even below the modulation-instability threshold. In contrast, the antiresonance spectral peaks appear only above the instability threshold, and their presence constitutes a clear evidence of the cavity-induced modulation instability.

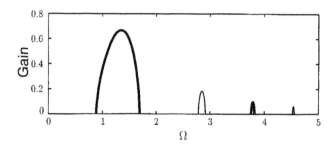

Figure 3.6: Gain spectrum of modulation instability in the normal-GVD region of a fiber-ring resonator. Thin and thick lines correspond to the resonance and antiresonance conditions, respectively. (From Ref. [40]; ©1997 APS.)

3.1.5 Ultrafast Nonlinear Effects

When short optical pulses are fed into a fiber resonator whose length is much larger than the dispersion and nonlinear length scales associated with the pulse, considerable pulse shaping is likely to occur over a single round trip. The combined effects of GVD and SPM on pulses circulating in a fiber resonator can lead to quite interesting nonlinear dynamics [42]–[47]. Depending on the input and fiber parameters, a steady-state pattern (along the fiber length) may or may not evolve (in the sense that it remains unchanged from one round trip to next). Moreover, evolution of pulses within the fiber ring depends on whether pulses experience normal or anomalous GVD.

Consider the situation in which ultrashort optical pulses are injected into a fiber-ring cavity synchronously using a mode-locked laser. Evolution of pulses over multiple round trips is governed by a generalized NLS equation [2]. Including the terms related to third-order dispersion and intrapulse Raman scattering [2], Eq. (3.1.10) becomes

$$i\frac{\partial A}{\partial z} - \frac{\beta_2}{2}\frac{\partial^2 A}{\partial T^2} - \frac{i\beta_3}{6}\frac{\partial^3 A}{\partial T^3} + \gamma|A|^2 A - \gamma T_R A\frac{\partial |A|^2}{\partial T} = 0, \qquad (3.1.18)$$

where T_R is the Raman parameter with a value of about 3 fs for silica fibers. For a fiber ring of length L_R, this equation should be solved with the following boundary condition at the coupler after each round trip:

$$A^{(n+1)}(0,T) = \sqrt{\rho}A^{(n)}(L_R,T)\exp(i\phi_0) + i\sqrt{1-\rho}A_i(T), \qquad (3.1.19)$$

where a superscript denotes the round-trip number. The input amplitude A_i for "sech" pulses can be written as

$$A_i(T) = \sqrt{P_0}\,\text{sech}(T/T_0), \qquad (3.1.20)$$

where and P_0 is the peak power for a pulse of width T_0. The full width at half maximum (FWHM) of the pulse is related to T_0 as $T_p = 2\ln(1+\sqrt{2})T_0 \approx 1.763T_0$.

Numerical simulations for 10-ps pulses propagating inside a 100-m fiber ring show that each input pulse develops an internal substructure consisting of many subpulses of widths ~1 ps. Moreover, a steady state is reached only if the input peak power is below a certain value. In the steady state, subpulses have a uniform spacing that does

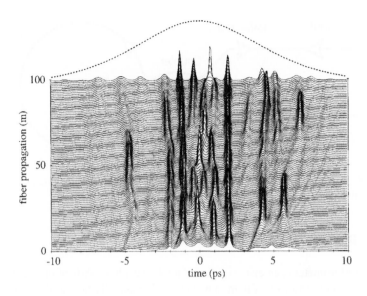

Figure 3.7: Evolution of substructure in a 100-m ring resonator when 10-ps pulses (dotted curve) are injected into it synchronously. (From Ref. [46]; ©1997 APS.)

not change from one round trip to next. Each subpulse corresponds to a fundamental soliton whose width and peak power are related such that the soliton order $N \approx 1$. When the input peak power exceeds the critical value, a phase-transition-like behavior occurs such that the position and width of subpulses change continuously in an apparently random manner. Since most subpulses retain their soliton character (in the sense that $N \approx 1$), such an ensemble of subpulses is referred to as the *soliton gas*. Figure 3.7 shows the evolution of substructure along the ring over one round trip.

Direct experimental observation of such a pattern is difficult because of the ultrashort time scale involved. However, the autocorrelation and spectral measurements agree with the theoretical predictions based on the NLS equation. In a 1997 experiment, 2-ps mode-locked pulses, obtained from a color-center laser operating at 1.57 μm, were injected into a ring resonator made with 6 m of polarization-maintaining fiber [46]. Both the autocorrelation trace and the spectrum changed qualitatively as the peak power increased beyond a certain value, resulting in the transition from a regular to irregular pattern of subpulses.

In the case of normal GVD, nonlinear dynamics becomes even more complex [44]. Numerical simulations show that each pulse still develops an internal substructure but the resulting pattern is not governed by soliton shaping. Depending on the linear detuning ϕ_0 of the ring resonator, the substructure varies from pulse to pulse and exhibits period-doubling bifurcations and chaos. Experiments performed using 12-ps pulses (obtained from a Nd:YAG laser operating near 1.32 μm) show that the pulse energy also varies from pulse to pulse and exhibits a period-doubling route toward chaos as ϕ_0 is varied in the vicinity of a cavity resonance.

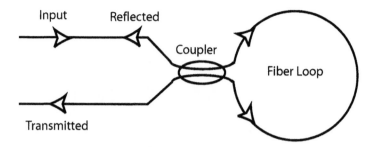

Figure 3.8: Schematic illustration of an all-fiber Sagnac interferometer acting as a nonlinear fiber-loop mirror.

3.2 Sagnac Interferometers

Sagnac interferometers can also exploit the nonlinear phase shift of optical fibers for optical switching [48]–[50]. Figure 3.8 shows schematically how a fiber coupler can be used to make a Sagnac interferometer. It is made by connecting a piece of long fiber to the two output ports of a fiber coupler to form a loop. It appears similar to a fiber-ring resonator but behaves quite differently because of two crucial differences. First, there is no feedback mechanism since all light entering from the input port exits from the resonator after a single round trip. Second, the entering optical field is split into two counterpropagating parts that share the same optical path and interfere at the coupler coherently.

The relative phase difference between the counterpropagating beams determines whether an input beam is reflected or transmitted by the Sagnac interferometer. In fact, if a 3-dB fiber coupler is used, any input is totally reflected, and the Sagnac loop acts as a perfect mirror. Such a device can be designed to transmit a high-power signal while reflecting it at low power levels, thus acting as an all-optical switch. For this reason, it is referred to as the *nonlinear fiber-loop mirror* and has attracted considerable attention, not only for optical switching but also for mode locking and wavelength demultiplexing.

3.2.1 Nonlinear Transmission

The physical mechanism behind nonlinear switching can be easily understood by considering a CW or a quasi-CW input beam. When such an optical signal is incident at one port of the fiber coupler, the transmittivity of a Sagnac interferometer depends on the power-splitting ratio of the coupler. If a fraction ρ of the input power P_0 travels in the clockwise direction, the transmittivity for a loop of length L is obtained by calculating the phase shifts acquired during a round trip by the counterpropagating optical waves, then recombining them interferometrically at the coupler. It is important to include any relative phase shift introduced by the coupler. If we use the transfer matrix of a fiber coupler given in Eq. (2.1.22) together with $A_2(0) = 0$, the amplitudes of the forward- (clockwise) and backward- (counterclockwise) propagating fields are given by

$$A_f = \sqrt{\rho}\, A_0, \qquad A_b = i\sqrt{1-\rho}\, A_0, \qquad (3.2.1)$$

where $\rho = \cos^2(\kappa l_c)$ for a coupler of length l_c. Notice the $\pi/2$ phase shift for A_b introduced by the coupler. After one round trip, both fields acquire a linear phase shift as well as the SPM- and XPM-induced nonlinear phase shifts. As a result, the two fields reaching at the coupler take the following form:

$$A'_f = A_f \exp[i\phi_0 + i\gamma(|A_f|^2 + 2|A_b|^2)L], \qquad (3.2.2)$$

$$A'_b = A_b \exp(i\phi_0 + i\gamma(|A_b|^2 + 2|A_f|^2)L], \qquad (3.2.3)$$

where $\phi_0 \equiv \beta L$ is the linear phase shift for a loop of length L and β is the propagation constant within the loop.

The reflected and transmitted fields can now be obtained by using the transfer matrix of the fiber coupler and are given by

$$\begin{pmatrix} A_t \\ A_r \end{pmatrix} = \begin{pmatrix} \sqrt{\rho} & i(1-\rho)^{1/2} \\ i(1-\rho)^{1/2} & \sqrt{\rho} \end{pmatrix} \begin{pmatrix} A'_f \\ A'_b \end{pmatrix}. \qquad (3.2.4)$$

Using Eqs. (3.2.1)–3.2.4), the transmittivity $T_S \equiv |A_t|^2/|A_0|^2$ of the Sagnac loop is given by [50]

$$T_S = 1 - 2\rho(1-\rho)\{1 + \cos[(1-2\rho)\gamma P_0 L]\}, \qquad (3.2.5)$$

where $P_0 = |A_0|^2$ is the input power. The linear phase shift does not appear in this equation because of its exact cancellation. For $\rho = 0.5$, T_S equals zero, and the loop reflectivity is 100% at all power levels (hence the name *fiber-loop mirror*). Physically, if the power is equally divided between the counterpropagating waves, the nonlinear phase shift is equal for both waves, resulting in no relative phase difference between the counterpropagating waves. However, if the power-splitting factor ρ is different than 0.5, the fiber-loop mirror exhibits different behavior at low and high powers and can act as an optical switch.

Figure 3.9 shows the transmitted power as a function of P_0 for two values of ρ. At low powers, little light is transmitted if ρ is close to 0.5 since $T_S \approx 1 - 4\rho(1-\rho)$. At high powers, the SPM-induced phase shift leads to 100% transmission of the input signal whenever

$$|1 - 2\rho|\gamma P_0 L = (2m-1)\pi, \qquad (3.2.6)$$

where m is an integer. As seen in Figure 3.9, the device switches from low to high transmission periodically as input power increases. In practice, only the first transmission peak ($m = 1$) is likely to be used for switching because it requires the least power. The switching power for $m = 1$ can be estimated from Eq. (3.2.6) and is 31 W for a 100-m-long fiber loop when $\rho = 0.45$ and $\gamma = 10$ W^{-1}/km. It can be reduced by increasing the loop length, but one should then consider the effects of fiber loss and GVD that were neglected in deriving Eq. (3.2.5).

3.2.2 Nonlinear Switching

Nonlinear switching in all-fiber Sagnac interferometers was observed beginning in 1989 in several experiments [51]–[57]. Most experiments used short optical pulses

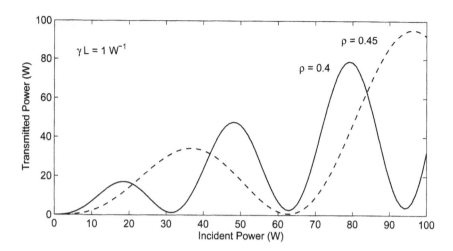

Figure 3.9: Transmitted power as a function of incident power for two values of ρ, showing the nonlinear response of an all-fiber Sagnac interferometer.

with high peak powers. In this case, the power dependence of loop transmittivity in Eq. (3.2.5) can lead to considerable pulse distortion, since only the central part of a pulse is intense enough to undergo switching. In a 1989 experiment, 180-ps pulses obtained from a Q-switched, mode-locked Nd:YAG laser were injected into a 25-m Sagnac loop [51]. Transmission increased from a few percent to 60% when peak power was increased beyond 30 W. Transmitted pulses were narrower than input pulses, as expected, because only the central part of the pulse was switched. As discussed in Section 2.3, the shape-induced deformation of optical pulses can be avoided in practice by using soliton effects, since solitons have a uniform nonlinear phase across the entire pulse. Their use requires ultrashort pulses (width < 10 ps) propagating in the anomalous-GVD regime of the fiber. The XPM-induced coupling between the counter-propagating solitons can be ignored for optical pulses short enough that they overlap for a relatively short time compared with the round-trip time. As a result, one can use two uncoupled NLS equations in the form of Eq. (3.1.12) for counterpropagating solitons inside the fiber loop.

SPM-Induced Switching

Soliton switching in Sagnac interferometers was observed in 1989 by launching ultra-short pulses at a wavelength in the anomalous-GVD regime of the fiber loop. In one experiment, mode-locked pulses (width about 0.4 ps), obtained from a color-center laser operating near 1.5 μm, were launched into a 100-m Sagnac loop formed using a 58:42 fiber coupler [52]. In another experiment, a 25-m-long loop was formed using a polarization-maintaining fiber (having its zero-dispersion wavelength near 1.58 μm), and 0.3-ps input pulses were obtained from a color-center laser operating near 1.69 μm [53]. Figure 3.10 shows the switching characteristics observed in this experiment. Energies of the transmitted and reflected pulses (E_{OUT} and E_{REFL}, respectively) vary with

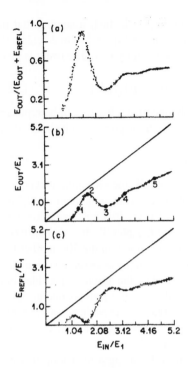
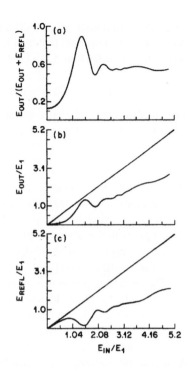

Figure 3.10: Measured (left) and simulated (right) switching characteristics of a nonlinear fiber-loop mirror. The energy level for forming a fundamental soliton was $E_1 = 33.2$ pJ for 0.3-ps input pulses used in the experiment. (From Ref. [53]; ©1989 OSA.)

the input pulse energy E_{IN}, showing clear evidence of nonlinear switching. The energy of transmitted pulses increases from a few percent to 90% as the input pulse energy is ramped up close to 55 pJ (peak power $\sim$ 100 W).

The experimental results obtained with 0.3-ps pulses cannot be explained using the simple CW theory given earlier since soliton effects play an important role. Good agreement was obtained by solving the generalized NLS equation given in Eq. (3.1.18) numerically with the appropriate boundary conditions [53]. The inclusion of intrapulse Raman scattering, a higher-order nonlinear effect that shifts the spectrum of solitons (see Section 5.5 of Ref. [2]), was found to be important for such short pulses and limited the peak transmission from reaching 100%. It also led to pulse breakup at high powers.

The switching threshold of a Sagnac interferometer can be reduced by incorporating a fiber amplifier within the loop [56]. If the amplifier is located close to the fiber coupler, its presence introduces an asymmetry as the counterpropagating pulses are not amplified simultaneously. Since the Sagnac interferometer is unbalanced by the amplifier, even a 50:50 coupler ($\rho = 0.5$) can be used. The switching behavior in this case can be understood by noting that one wave is amplified at the entrance to the loop while the counterpropagating wave experiences amplification just before exiting the loop. Since the intensities of the two waves differ by a large amount throughout the loop, the differential phase shift can be quite large. In fact, assuming that the clockwise wave is amplified first by a factor G, we can use Eq. (3.2.4) to calculate the transmittivity, provided that A_f in Eq. (3.2.2) is multiplied by $\sqrt{G}$. The result is given by

$$T_S = 1 - 2\rho(1-\rho)\{1 + \cos[(1-\rho-G\rho)\gamma P_0 L]\}. \qquad (3.2.7)$$

The condition for complete transmission is obtained from Eq. (3.2.6) by replacing $(1-2\rho)$ with $(1-\rho-G\rho)$. For $\rho = 0.5$, the switching power is given by (using $m = 1$)

$$P_0 = 2\pi/[(G-1)\gamma L]. \qquad (3.2.8)$$

Since the amplification factor G can be as large as 30 dB, the switching power is reduced by a factor of up to 1000. Such a device, referred to as the *nonlinear amplifying-loop mirror*, can switch at peak power levels below 1 mW. Its implementation is relatively simple with the advent of fiber amplifiers (see Chapter 4). In a demonstration of the basic concept, 4.5 m of Nd-doped fiber was spliced within the 306-m fiber loop formed using a 3-dB coupler [56]. Quasi-CW-like switching was observed using 10-ns pulses. The switching power was about 0.9 W even when the amplifier provided only a 6-dB gain (a factor of 4). In a later experiment, the use of a semiconductor optical amplifier, providing different gains for counterpropagating waves, inside a 17-m fiber loop resulted in switching powers of less than 250 μW when 10-ns pulses obtained from a semiconductor laser were injected into the loop [57].

A Sagnac interferometer can also be unbalanced by using a fiber loop in which GVD is not constant but varies along the loop [58]–[65]. The GVD can vary continuously as in a dispersion-decreasing fiber, or in a steplike fashion (using fibers with different dispersive properties connected in series). The simplest situation corresponds to the case in which the Sagnac loop is made with two types of fibers and is similar to a dispersion-management scheme used in lightwave systems for GVD compensation. Dispersion-varying fiber loops unbalance a Sagnac interferometer since the counterpropagating waves experience different GVD as they complete a round trip. The most noteworthy feature of such Sagnac loops is that they remain balanced for CW beams of any power levels since GVD does not affect them. However, evolution of optical pulses is affected by both GVD and SPM, resulting in a net relative phase shift between the counterpropagating waves. As a result, optical pulses can be switched to the output port while any CW background noise is reflected by dispersion-imbalanced Sagnac loops. An extinction ratio of 22 dB for the CW background was observed in an experiment [60] in which the 20-m loop was made using equal lengths of standard telecommunication fiber ($\beta_2 = -23$ ps^2/km) and dispersion-shifted fiber ($\beta_2 = -2.3$ ps^2/km).

XPM-Induced Switching

An important class of applications is based on the XPM effects occurring when a control or pump signal is injected into the Sagnac loop such that it propagates in only one direction and induces a nonlinear phase shift on one of the counterpropagating waves through XPM while the other is not affected by it. In essence, the control signal is used to unbalance the Sagnac interferometer in a way similar to how an optical amplifier can be used to produce different SPM-induced phase shifts. As a result, the loop can be made using a 50:50 coupler so that a low-power CW beam is reflected in the absence of the control but transmitted when a control pulse is applied. Many experiments have shown the potential of XPM-induced switching [66]–[75]. As early as 1989, transmittivity of a 632-nm CW signal (obtained from a He–Ne laser) was switched from zero to close to 100% by using intense 532-nm picosecond pump pulses with peak powers of about 25 W [66].

When the signal and control wavelengths are far apart, one should consider the walk-off effects induced by the group-velocity mismatch. In the absence of GVD effects, the XPM-induced relative phase shift at the coupler is given by [2]

$$\phi_{XPM} = 2\gamma \int_0^L |A_p(T - d_w z)|^2 dz, \tag{3.2.9}$$

where A_p is the pump-pulse amplitude, $T = t - z/v_{gs}$ is the reduced time in the frame moving with the signal pulse, and $d_w = v_{gp}^{-1} - v_{gs}^{-1}$ represents the group-velocity mismatch between the pump and signal pulses. The integral can be evaluated analytically for certain shapes of the pump pulse. For example, for a "sech" pump pulse with $A_p(T) = \sqrt{P_p}\text{sech}(T/T_0)$, the phase shift becomes [67]

$$\phi_{XPM}(\tau) = (\gamma P_p/\delta_w)[\tanh(\tau) - \tanh(\tau - \delta_w)], \tag{3.2.10}$$

where $\tau = T/T_0$ and $\delta_w = d_w L/T_0$. The relative phase is not only time dependent but its shape is also affected considerably by the group-velocity mismatch. Since loop transmittivity remains high as long as the phase shift is close to an odd multiple of π, the transmitted signal shape changes considerably with the shape and peak power of pump pulses.

The problem of pulse walk-off can be solved by using a fiber whose zero-dispersion wavelength lies between the pump and signal wavelengths such that the two waves have the same group velocity ($d_w = 0$). Indeed, such a 200-m-long Sagnac loop was built in 1990 using polarization-maintaining fiber [68]. It was employed to switch the 1.54-μm signal using 120-ps pump pulses with 1.8-W peak power at 1.32 μm. In a later experiment, 14-ps pump pulses, obtained from a gain-switched 1.55-μm DFB laser and amplified using a fiber amplifier, were able to switch a CW signal in the wavelength region near 1.32 μm.

The pulse walk-off occurring because of wavelength difference between the pump and signal can also be avoided by using an orthogonally polarized pump at the same wavelength as that of the signal [69]. There is still a group-velocity mismatch because of polarization-mode dispersion, but it is relatively small. Moreover, it can be used to advantage by constructing a Sagnac loop in which the slow and fast axes of polarization-maintaining fibers are interchanged in a periodic fashion. In one implementation of this idea [70], a 10.2-m loop consisted of 11 such sections. Two orthogonally polarized pump and signal pulses (width about 230 fs) were injected into the loop and propagated as solitons. The pump pulse was polarized along the fast axis and delayed initially such that it overtook the signal pulse in the first section. In the second section, the signal pulse traveled faster because of the reversing of slow and fast axes and overtook the pump pulse. This process repeated in each section. As a result, two solitons collided multiple times inside the Sagnac loop, and the XPM-induced phase shift was enhanced considerably.

3.2.3   Applications

By exploiting different nonlinear effects such as XPM, SPM, and four-wave mixing (FWM) occurring inside the fiber used to make the Sagnac loop, one can employ a nonlinear fiber-loop mirror for many practical applications. The main advantage of using the nonlinearity of fibers is its ultrafast nature that permits all-optical signal processing at femtosecond time scales. The advent of highly nonlinear fibers [2], in which the nonlinear parameter γ is enhanced by as much as a factor of 1000, has made the use of a Sagnac interferometer much more practical by reducing the required length of the nonlinear fiber loop.

Pulse Shaping and Noise Reduction

A nonlinear Sagnac interferometer acts as a high-pass intensity filter in the sense that it reflects low-intensity signals but transmits high-intensity radiation without affecting it. This feature is similar to that of saturable absorbers, which absorb weak signals but become transparent at high intensities, with one crucial difference. The speed of saturable absorbers is limited in practice to time scales longer than 10 ps while the nonlinear response of silica fibers is almost instantaneous (< 10 fs).

A simple application of Sagnac interferometers consists of using them for pulse shaping and pulse cleanup. For example, if a short optical pulse contains a broad low-intensity pedestal, the pedestal can be removed by passing it through such a device [76]. Similarly, a soliton pulse train, corrupted by continuum radiation or amplified spontaneous emission, can be cleaned by passing it through an all-fiber Sagnac loop. Since solitons can be switched as one unit, they are transmitted by the loop while the low-energy dispersive radiation or noise is reflected back. The Sagnac loop can also be used for pulse compression (see Chapter 6) and for generating a train of short optical pulses at a high repetition rate by injecting a dual-wavelength signal [77].

Saturable absorbers are routinely used for passive mode locking of lasers to generate picosecond pulses. However, their use is limited by their sluggish nonlinear response. Since a nonlinear fiber-loop mirror responds on femtosecond timescales, its passive use for mode-locked lasers was suggested as early as 1990 [78]. Indeed, this approach led to a new class of fiber lasers known as figure-8 lasers [79]. Such lasers can generate femtosecond pulses and are covered in Chapter 5.

Another approach makes use of XPM-induced switching in a Sagnac loop for wavelength conversion. The basic idea is to launch a CW beam together with control pulses at a different wavelength. In the absence of control signal, the CW light is reflected from a balanced Sagnac interferometer since it acts as a perfect mirror. However, each control pulse shifts the optical phase through XPM and directs a time slice of the CW beam to the output end, producing a pulse train at the CW-laser wavelength. In effect, the Sagnac loop acts as an all-optical gate that is open for the duration of each control pulse. Clearly, such a device acts as a wavelength converter, and this mode of operation should be useful for WDM networks. An added benefit is that the wavelength-converted pulse train can be of higher quality than the control pulses themselves. In a 1992 experiment, control pulses from a gain-switched DFB laser operating near 1533 nm were used to convert the 1554-nm CW radiation into a pulse train [80]. Even though 60-ps control pulses were highly chirped, the pulses produced by the Sagnac loop were nearly transform limited. The pulse quality was high enough that pulses could be propagated over 2400 km using a recirculating fiber loop [81]. By 2000, a nonlinear optical loop mirror was used for wavelength conversion at a bit rate of 40 Gb/s [82]. Chapter 9 discusses this application in more detail.

Sagnac interferometers are also useful for all-optical signal regeneration in lightwave systems, as they can reshape pulses while reducing the noise level [83]. The pulse-shaping capability of such interferometers can be improved significantly by concatenating several Sagnac loops in series [84]. The loop length can be reduced by using fibers with a relatively high value of the nonlinear parameter γ. Such fibers were used in a 2000 experiment to form two concatenated Sagnac loops [85]. The XPM-induced switching was used in the first loop to convert the wavelength of the 10-Gb/s data channel by using it as a control signal. The output of the first loop became the control signal for the second loop where the wavelength was switched back to the original wavelength. The net result was regeneration (noise reduction and pulse shaping) of the data without change in its wavelength. Chapter 9 provides more details on all-optical regeneration.

All-Optical Logic Operations

An important category of applications is related to format conversion and logic operation on digital bit streams used in lightwave systems. Nonlinear Sagnac loops can be used as analog-to-digital and digital-to-analog converters [86]. They can also be used for converting frequency modulation into amplitude modulation [87]. The possibility of using a nonlinear Sagnac interferometer for all-optical logic operations was pointed out as early as 1983 [48]. A polarization-maintaining Sagnac loop was used in 1991 to demonstrate the elementary logic operations in the form of AND, XOR, and $\overline{\text{XOR}}$ gates [88]. Two control signals in counterpropagating directions were used to realize this functionality.

To understand how a Sagnac loop performs digital logic, consider the situation in which a regular pulse train (an optical clock) is launched into the loop through a 3-dB coupler. In the absence of control signals, all 1 bits are reflected. If two data streams (random sequences of 1 and 0 bits) are launched inside the loop as control signals, the clock pulse will be reflected if both controls have the same bit in that time slot but get transmitted otherwise. The reason is that a net XPM-induced phase shift is produced when the two controls have different types of bits in a given time slot. The transmitted and reflected signals thus correspond to the XOR and $\overline{\text{XOR}}$ gates, respectively. The AND gate requires only one control signal since a pulse is transmitted only when both the control and signal bits are present simultaneously. Inversion operation can also be carried out using only one control.

In a 1991 experiment, logic operations were performed by injecting optical pulses, obtained from a 1.54-μm DFB laser, into a polarization-maintaining Sagnac loop together with 100-ps control pulses from a 1.32-μm Nd:YAG laser [88]. The loop was 200-m long and required power levels of about 1 W to realize the π phase shift. The system-level applications of Sagnac logic gates have also been studied [89]. The use

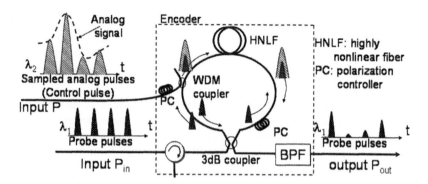

Figure 3.11: Schematic of the Sagnac-loop configuration for one of the encoder used for analog-to-digital conversion. (From Ref. [91]; ©2006 IEEE.)

of a Sagnac loop for such applications requires consideration of factors such as signal-clock walk-off, GVD-induced pulse broadening, and timing jitter [90].

The use of Sagnac fiber loops for analog-to-digital conversion has attracted considerable attention in recent years [91]. The analog signal is first sampled to generate relatively broad pulses and then split into N parts, where N is the number of bits used to represent each sample. A parallel sequence of N encoders and N thresholders is then used to digitize the analog signal. Figure 3.11 shows the configuration employed for the encoder. Sampled pulses, propagating in the clockwise direction, act as a control signal inside a Sagnac loop in which much shorter probe pulses (at a different wavelength) circulate in both directions. Each control pulse shifts the phase of a clockwise-traveling probe pulse through XPM in direct proportion to its peak power. The XPM-induced phase shift for the counterpropagating probe pulse is small as it is proportional to the average power of the analog signal. The net result is that the transmitted probe-pulse peak power is related to the peak power of the control pulse that overlaps with it in the clockwise direction. By suitably adjusting the transfer function of the encoders and the thresholds levels, a digitized version of the analog signal can be produced. All-optical digital-to-analog conversion has also been performed with nonlinear optical loop mirrors [92].

Parametric Loop Mirror

Another class of applications is based on four-wave mixing occurring inside a nonlinear Sagnac interferometer [93]–[103]. As discussed in Chapter 10 of Ref. [2], simultaneous propagation of the pump and signal waves of different wavelengths inside an optical fiber generates an idler wave through the FWM process. Both the signal and idler waves experience gain through parametric amplification. Moreover, the phase of the idler wave is related to that of the signal wave through the phase-matching condition. For this reason, such a FWM process is also known as *phase conjugation*.

The FWM inside a Sagnac loop is considerably modified by the counterpropagating nature of the pump and signal fields and the nonlinear phase shifts induced by SPM and XPM. Such a device is referred to as the *optical parametric loop mirror* to

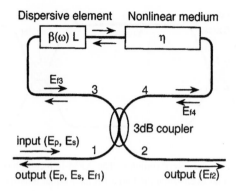

Figure 3.12: A Sagnac interferometer acting as a parametric-loop mirror through the FWM process. (From Ref. [93]; ©1995 OSA.)

emphasize the importance of the parametric gain [93]. Figure 3.12 shows the device configuration schematically. The pump and signal fields (E_p and E_s) are launched into the loop from the same port of the coupler. If the Sagnac interferometer is balanced by using a 3-dB coupler so that both pump and signal fields are split equally, they will be reflected by the loop mirror. On the other hand, the FWM component E_f (idler wave) generated inside the loop behaves asymmetrically if a piece of dispersive fiber is placed near the coupler to unbalance the interferometer.

To understand the operation of a parametric loop mirror, we need to consider the relative phase difference between the counterpropagating components of the idler wave (E_{f3} and E_{f4}). Since the propagation constant β inside a dispersive fiber is different for the pump, signal, and conjugate fields because of their frequencies (ω_p, ω_s, and ω_c, respectively), a net relative phase shift ϕ_d is introduced by a dispersive fiber of length L_f. As a result, the FWM power coming out from the output port 2 of the Sagnac loop depends on this phase shift and is given by [93]

$$P_{\text{out}} = P_c \sin^2(\phi_d/2), \qquad \phi_d = [2\beta(\omega_p) - \beta(\omega_s) - \beta(\omega_c)]L_f, \qquad (3.2.11)$$

where P_c is the total power generated through FWM. The remaining power exits from the input port. Therefore, when ϕ_d is an odd multiple of π, the FWM signal exits from the output port. In contrast, when ϕ_d is an even multiple of π, the loop acts as a phase-conjugate mirror since all FWM power appears to be reflected. From a practical standpoint, the FWM power at the frequency ω_c can be separated from both the pump and signal fields by choosing $\phi_d = \pi$ without requiring an optical filter. At the same time, low-power noise associated with the signal (e.g., amplified spontaneous emission) is filtered by the Sagnac loop since it gets reflected.

FWM in a Sagnac loop has been used for many applications. The phase-sensitive nature of parametric amplification can be used for all-optical storage of data packets consisting of a random string of 1 and 0 bits in the form of picosecond pulses [102]. It can also be used to produce amplitude-squeezed solitons using an asymmetric Sagnac loop [95]. FWM in a nonlinear Sagnac interferometer has been used to make parametric oscillators. Pulses shorter than 1 ps can be generated through synchronous pumping

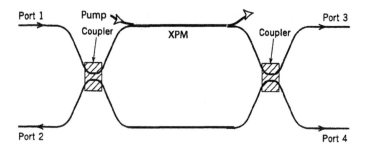

Figure 3.13: A Mach–Zehnder interferometer. The XPM-induced phase shift can be used to unbalance the interferometer even when the two arms have equal lengths.

of a Sagnac loop [96]. Moreover, such parametric oscillators are tunable over a range as wide as 40 nm [97]. Parametric amplification is also useful for reducing the noise figure of amplifiers below the 3-dB quantum limit (see Chapter 4). In a 2000 experiment, 16-dB amplification was realized with a noise figure of only 2 dB [98].

Another application consists of using the nonlinear Sagnac interferometer for phase conjugation. In one experiment, two orthogonally polarized pump waves were fed into different ports of the Sagnac interferometer to realize a phase conjugator that was not only polarization insensitive but also wavelength-shift free [99]. In another experiment, a semiconductor optical amplifier was used for phase conjugation within a Sagnac loop [100]. Such a device was capable of compensating dispersion over 106 km of standard fiber at a bit rate of 80 Gb/s when the phase conjugator was placed nearly in the middle of the fiber span. This technique of dispersion compensation is also known as *midway spectral inversion* as the spectrum of the FWM signal is a mirror image of the signal spectrum because of phase conjugation [104].

3.3   Mach–Zehnder Interferometers

An all-fiber Mach–Zehnder interferometer (MZI) is constructed by connecting two fiber couplers in series, as shown schematically in Figure 3.13. The first coupler splits the input signal into two parts, which acquire different phase shifts if arm lengths are different, before they interfere at the second coupler. Such a device has the same functionality as a Sagnac loop but has an added advantage that nothing is reflected back toward the input port. Moreover, a MZI can be unbalanced by simply using different lengths for its two arms since the two optical fields inside it take physically separated paths. However, the same feature also makes the interferometer susceptible to environmental fluctuations. Nonlinear effects in MZIs were considered starting in 1987 and have continued to be of interest [105]–[118].

3.3.1   Nonlinear Characteristics

The theory of nonlinear switching in a MZI is similar to that of Sagnac interferometers. The main difference is that the two fields produced at the output of the first fiber coupler take different physical paths, and thus acquire only SPM-induced phase shifts. In general, two couplers need not be identical and can have different power-splitting fractions, ρ_1 and ρ_2. Two arms of the interferometer can also have different lengths and propagation constants. We consider such an asymmetric MZI and find the powers transmitted from the two output ports when a single CW beam with power P_0 is incident at one input port. Using Eq. (3.2.1) at the first coupler and taking into account both the linear and nonlinear phase shifts, the optical fields at the second coupler are given by

$$A_1 = \sqrt{\rho_1} A_0 \exp(i\beta_1 L_1 + i\rho_1 \gamma |A_0|^2 L_1), \qquad (3.3.1)$$

$$A_2 = i\sqrt{1-\rho_1} A_0 \exp[i\beta_2 L_2 + i(1-\rho_1)\gamma|A_0|^2 L_2], \qquad (3.3.2)$$

where L_1 and L_2 are the lengths and β_1 and β_2 are the propagation constants for the two arms of the MZI.

The optical fields exiting from the output ports of a MZI are obtained by using the transfer matrix of the second fiber coupler:

$$\begin{pmatrix} A_3 \\ A_4 \end{pmatrix} = \begin{pmatrix} \sqrt{\rho_2} & i(1-\rho_2)^{1/2} \\ i(1-\rho_2)^{1/2} & \sqrt{\rho_2} \end{pmatrix} \begin{pmatrix} A_1 \\ A_2 \end{pmatrix}. \qquad (3.3.3)$$

The fraction of power transmitted from the bar port of the MZI is obtained using $T_b = |A_3|^2/|A_0|^2$ and is given by

$$T_b = \rho_1\rho_2 + (1-\rho_1)(1-\rho_2) - 2[\rho_1\rho_2(1-\rho_1)(1-\rho_2)]^{1/2}\cos(\phi_L + \phi_{NL}), \quad (3.3.4)$$

where the linear and nonlinear parts of the relative phase shift are given by

$$\phi_L = \beta_1 L_1 - \beta_2 L_2, \qquad \phi_{NL} = \gamma P_0[\rho_1 L_1 - (1-\rho_1)L_2]. \qquad (3.3.5)$$

This equation simplifies considerably for a symmetric MZI, made using two 3-dB couplers so that $\rho_1 = \rho_2 = \frac{1}{2}$. The nonlinear phase shift vanishes for such a coupler when $L_1 = L_2$, and the transmittivity of the bar port is given as $T_b = \sin^2(\phi_L/2)$. Since the linear phase shift ϕ_L is frequency dependent, the output depends on the wavelength of light. Thus, an MZI acts as an optical filter. The spectral response can be improved by using a cascaded chain of such interferometers with relative path lengths adjusted suitably.

The nonlinear response of an MZI is similar to that of a Sagnac loop in the sense that the output from one of the ports can be switched from low to high (or vice versa) by changing the input peak power of the incident signal. Figure 3.14 shows the experimentally observed transmittance from the bar port (circles) and the cross port (crosses) as input peak power is varied over a range of 0 to 25 W for two values of ϕ_L [111]. Predictions of Eq. (3.3.4) are also shown for comparison using $\rho_1 = 0.34$ and $\rho_2 = 0.23$ for the power-splitting ratios of the two couplers. The arm lengths were identical in this experiment ($L_1 = L_2$) as the MZI was made using a dual-core fiber whose two identical cores were connected on each side to a fiber coupler. This configuration avoids temporal fluctuations occurring on a millisecond timescale. Such fluctuations occur invariably when two separate fiber pieces are used in each arm of the MZI and require an active stabilization scheme for controlling them [106].

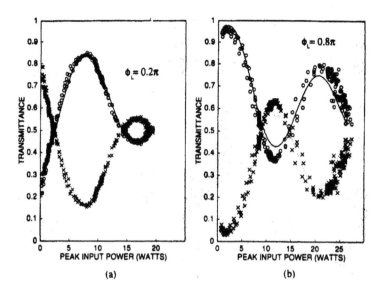

Figure 3.14: Nonlinear switching in a Mach–Zehnder interferometer for two values of ϕ_L. Data from the bar and cross ports are shown by circles and crosses. Theoretical predictions are shown as solid and dotted curves. (From Ref. [111]; ©1991 OSA.)

Similar to the case of Sagnac interferometers, switching can also be accomplished through a XPM-induced phase shift by injecting pump light in an arm of the MZI as shown schematically in Figure 3.13. In one experiment, one arm of the MZI incorporated 1.6 m of Yb-doped fiber while the fiber in the other arm was undoped [114]. Doping did not affect the signal launched in one of the input ports of the MZI using 1.31-μm and 1.55-μm semiconductor lasers, and most of the power appeared at the cross port. However, when a 980-nm pump was injected in the arm with doped fiber, the signal switched to the bar port at pump power levels of less than 5 mW. The physical mechanism behind switching is the phase shift induced at the signal wavelength resulting from saturation of absorption near 980 nm. Remarkably, phase shifts of π or more can be induced with only a few milliwatts of the pump power. This mechanism should be distinguished from the XPM-induced phase shift, discussed earlier in the context of Sagnac interferometers, since the phase shift is induced by the dopants rather than fiber nonlinearity.

In the case of ultrashort optical pulses, dispersive effects must be included for an accurate description of the nonlinear switching process . If XPM-induced switching is realized by launching a pump pulse at a different wavelength, the walk-off effects may reduce the XPM-induced phase shift considerably. They can be avoided by choosing the pump pulse at the same wavelength as the signal, such that it is polarized orthogonal to the signal pulses. This configuration was studied numerically in a 1995 study [112]. The switching contrast was found to be reduced when the peak power of pump pulses was increased beyond an optimum value because of GVD. The situation is worse for copolarized pump and signal pulses because of the walk-off effects.

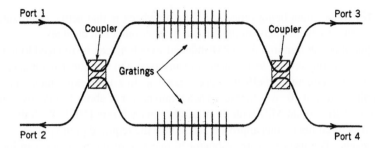

Figure 3.15: An add–drop multiplexer based on a Mach–Zehnder interferometer. Fiber gratings are used for adding or dropping a specific channel whose wavelength coincides with the Bragg wavelength.

3.3.2 Applications

MZIs are used for a variety of applications. Most of them are based on the ability of an MZI to produce large changes in its output with small changes in the refractive index in one of its arms. MZIs made by using $LiNbO_3$ or semiconductor waveguides are used routinely as high-speed modulators since such electro-optic materials permit voltage-induced changes in the refractive index. Silica fibers do not have this property, but their refractive index can be changed either optically (through SPM and XPM) or through changes in the environment (such as temperature or pressure). The latter property is useful for making fiber sensors [119]–[121]. Such applications are not discussed here since they do not make use of fiber nonlinearity.

Another class of applications uses MZIs as optical filters. Several kinds of add–drop filters have been developed using MZIs [122]–[126]. The simplest scheme uses a series of interconnected fiber couplers, forming a chain of MZIs. Such a device is sometimes referred to as a *resonant coupler* since it resonantly couples out a specific-wavelength channel from a WDM signal to one output port while the remaining channels appear at the other output port. Its performance can be optimized by controlling the power-splitting ratios of various directional couplers [122]. The wavelength selectivity of Bragg gratings can also be used to make add–drop filters [7]. In one scheme, two identical Bragg gratings are formed in the center of each arm of an MZI [123]. Operation of such a device can be understood from Figure 3.15. A single channel, whose wavelength λ_g falls within the stop band of the Bragg grating, is totally reflected and appears at port 2. The remaining channels are not affected by the gratings and appear at port 4. The same device can add a channel at the wavelength λ_g if the signal at that wavelength is injected from port 3. Stability of the MZI is of primary concern in these devices and requires active phase control in practice [124]. Although Such devices are important for WDM networks [126], their operation does not require fiber nonlinearity.

Nonlinear applications of MZIs make use of the SPM- or XPM-induced phase shifts. In fact, nearly all applications discussed in Section 3.2.3 in the context of Sagnac interferometers can use an MZI in place of the Sagnac loop. As an example, demultiplexers based on MZIs are of considerable interest [117]. The main advantage is that multiple MZIs can be cascaded as remaining channels appear at the output end of the MZI (rather than being reflected). The drawback is that active stabilization is often necessary to avoid fluctuations induced by environmental changes.

The pump power required for XPM-induced switching can be reduced to manageable levels by using several techniques. For instance, the nonlinear parameter γ can be increased by reducing the effective core area A_{eff}. In a 1988 experiment, an XPM-induced phase shift of $10°$ was measured at a pump power of about 15 mW by reducing A_{eff} to only 2 μm^2 in an MZI with 38 m of fiber in each arm [107]. With the advent of microstructured fibers, this approach can reduce the required pump powers to manageable levels. The use of a ring resonator in one arm of the resonator can increase the XPM-induced phase shift by several orders of magnitude [115]. The pump power required for the π phase shift is reduced to under 10 mW for a 10-m-long fiber ring, although the switching speed is also reduced to below 1 GHz for such devices [118].

3.4   Michelson Interferometers

A Michelson interferometer is made by connecting two separate pieces of fiber to the output ports of a fiber coupler and attaching 100% reflecting mirrors or Bragg gratings at the other end of the fibers [7]. Bragg gratings reflect completely the light whose wavelength falls within the stop band of the grating (see Figure 3.12). A Michelson interferometer functions much like an MZI with the crucial difference that the light propagating in its two arms is forced to interfere at the same coupler where it was split. Because of this feature, a Michelson interferometer acts as a nonlinear mirror, similar to a Sagnac interferometer, with the important difference that the interfering optical fields do not share the same physical path. Nonlinear Michelson interferometers can also be made using bulk optics (beam splitters and mirrors) with a long piece of fiber in one arm acting as a nonlinear medium. Nonlinear effects in Michelson interferometers have been mostly studied in the context of passive mode locking [127]–[131].

We can apply the analysis of Section 3.3.1 developed for an MZI to the case of a Michelson interferometer because of the similarity between the two. In both cases, an optical field is split into two parts at a fiber coupler, each part acquires a phase shift, and the two parts recombine interferometrically at the coupler. Since the same coupler is used for splitting and combining the optical fields in the case of a Michelson interferometer, we should set $\rho_1 = \rho_2 \equiv \rho$ in Eq. (3.3.4). For the same reason, transmission from the bar port of the coupler turns into reflection from the input port, and the reflectivity is given by

$$R_M = \rho^2 + (1-\rho)^2 - 2\rho(1-\rho)\cos(\phi_L + \phi_{\text{NL}}). \tag{3.4.1}$$

The lengths L_1 and L_2 appearing in Eq. (3.3.5) should be interpreted as round-trip lengths in each arm of the Michelson interferometer. The transmittivity is, of course, given by $T_M = 1 - R_M$. The reflection and transmission characteristics of a Michelson interferometer are similar to those of a Sagnac loop with two major differences. First, the round-trip path lengths L_1 and L_2 can be different for a Michelson interferometer. Second, the reflectivity and transmittivity are reversed for the Sagnac loop. Indeed, Eq. (3.4.1) reduces to Eq. (3.2.5) if $\phi_L = 0$.

Because of the SPM-induced nonlinear phase shift, the reflectivity of a Michelson interferometer is power dependent. As a result, such an interferometer tends to shorten an optical pulse and acts effectively as a fast-responding saturable absorber [127]. The pulse-shortening mechanism can be understood as follows. When the relative linear phases are set appropriately, the nonlinear phase shift may lead to constructive interference near the peak of the pulse, while the wings of the pulse experience destructive interference. The pulse-shortening capability of Michelson interferometers can be exploited for passive mode locking of lasers. This technique is commonly referred to as *additive-pulse mode locking* since the interferometric addition of an optical pulse at the coupler is responsible for mode locking [132]. The discovery of additive-pulse mode locking led to a revolution in the field of lasers and resulted in mode-locked lasers capable of generating pulses shorter than 10 fs. SPM in optical fibers played an important role in this revolution. This topic is discussed further in Chapter 5 in the context of mode-locked fiber lasers.

The use of Michelson interferometer for quantum applications has attracted considerable attention in recent years [133]–[136]. For example, a fiber-based source of polarization-entangled photon pairs employed FWM inside a dispersion-shifted fiber in combination with a Michelson interferometer and Faraday mirrors to realize practically useful features such as stable performance and turnkey operation [135]. The quantum applications of fiber nonlinearities are discussed in much more detail in Chapter 10.

Problems

3.1 Derive Eq. (3.1.3) by considering multiple round trips inside a Fabry–Perot resonator.

3.2 Derive an expression for the transmittivity of a fiber-ring resonator of length L formed using a fiber coupler with bar-state transmission of p.

3.3 Prove that the free spectral range of a ring resonator of length L is given by v_g/L, where v_g is the group velocity. How much does it change for a 10-m ring when the input wavelength is changed by 10 nm in the wavelength region near 1.55 μm? Assume that GVD of the fiber near this wavelength is -20 ps^2/km and $n_g = 1.46$.

3.4 Reproduce the bistability curves shown in Figure 3.3 using Eq. (3.1.8). Explore the impact of resonator finesse on bistability by varying R_m in the range from 0.4 to 0.8. Explain your results qualitatively.

3.5 Iterate the nonlinear map given in Eq. (3.1.9) numerically assuming that the phase changes during each round trip inside the ring resonator as indicated in Eq. (3.1.11). Plot the transmittivity as a function of round-trip number for values of $\gamma P_i L_R = 1, 5$, and 10. Assume $p = 0.75$.

3.6 Derive Eq. (3.2.5) by considering the phase shifts experienced by the counter-propagating waves inside a Sagnac loop. Use it to estimate the minimum switching power required when $p = 0.4$ and $\gamma L = 0.1$ W^{-1}.

3.7 Use Eq. (3.2.5) for a Gaussian pulse for which $P_i(t) = P_0 \exp[-(t/T_0)^2]$. Plot the shape of the transmitted pulse using $T_0 = 1$ ps, $\rho = 0.45$, and $\gamma P_0 L = 1, 2$, and 4. Estimate the compression factor in each case.

3.8 Derive an expression for the transmittivity of a Sagnac loop containing an optical amplifier next to the fiber coupler. Assume G is the amplifier gain, ρ is the bar-state transmission of the coupler, and a CW beam with power P_0 is injected into the loop.

3.9 Use the expression derived in the previous problem to find the switching power when a 3-dB coupler is used ($\rho = 0.5$) to make the Sagnac loop. Estimate its numerical value for a 100-m loop when $G = 30$ dB. Use $\gamma = 2$ W^{-1}/km.

3.10 Show that the XPM-induced phase shift for a "sech" pump pulse is given by Eq. (3.2.10).

3.11 Explain how a Sagnac loop can be used for demultiplexing a single channel from an optical time-division multiplexed (OTDM) bit stream.

3.12 Derive an expression for the bar-state transmittivity of a Mach–Zehnder interferometer. Allow for different path lengths of the two arms and different power-splitting ratios of the two couplers.

References

[1] M. Born and E. Wolf, *Principles of Optics*, 7th ed. (Cambridge University Press, New York, 1999), Chap. 7.

[2] G. P. Agrawal, *Nonlinear Fiber Optics*, 4th ed. (Academic Press, Boston, 2007).

[3] I. P. Kaminow, *IEEE J. Sel. Areas Commun.* **8**, 1005 (1990).

[4] P. E. Green, Jr., *Fiber-Optic Networks* (Prentice-Hall, Englewood Cliffs, NJ, 1993), Chap. 4.

[5] G. P. Agrawal, *Lightwave Technnology: Components and Devices* (Wiley, New York, 2004).

[6] J. Stone and L. W. Stulz, *Electron. Lett.* **23**, 781 (1987); *Electron. Lett.* **26**, 1290 (1990).

[7] R. Kashyap, *Fiber Bragg Gratings* (Academic Press, San Diego, CA, 1999).

[8] L. F. Stokes, M. Chodorow, and H. J. Shaw, *Opt. Lett.* **7**, 288 (1982).

[9] C. Y. Yue, J. D. Peng, Y. B. Liao, and B. K. Zhou, *Electron. Lett.* **24**, 622 (1988).

[10] S. Legoubin, M. Douay, P. Bernage, and P. Niay, *J. Opt. Soc. Am. A* **12**, 1687 (1995).

[11] H. M. Gibbs, S. L. McCall, and T. N. C. Venkatesan, *Phys. Rev. Lett.* **36**, 1135 (1976).

[12] K. Ikeda, *Opt. Commun.* **30**, 257 (1979).

[13] G. P. Agrawal and H. J. Carmichael, *Phys. Rev. A* **19**, 2074 (1979).

[14] L. A. Lugiato, in *Progess in Optics*, Vol. 21, E. Wolf, Ed. (North-Holland, Amsterdam, 1984).

[15] H. M. Gibbs, *Optical Bistability: Controlling Light with Light* (Academic Press, Boston, 1985).

[16] H. Nakatsuka, S. Asaka, H. Itoh, K. Ikeda, and M. Matsuoka, *Phys. Rev. Lett.* **50**, 109 (1983).

[17] B. Crosignani, B. Daino, P. Di Porto, and S. Wabnitz, *Opt. Commun.* **59**, 309 (1986).

[18] R. M. Shelby, M. D. Levenson, and S. H. Perlmutter, *J. Opt. Soc. Am. B* **5**, 347 (1988).

[19] R. Vallée, *Opt. Commun.* **81**, 419 (1991).

[20] K. Ogusu and S. Yamamoto, *J. Lightwave Technol.* **11**, 1774 (1993).

[21] Y. H. Ja, *Appl. Opt.* **32**, 5310 (1993); *IEEE J. Quantum Electron.* **30**, 329 (1994).

[22] J. Capmany, F. J. Fraile-Pelaez, and M. A. Muriel, *IEEE J. Quantum Electron.* **30**, 2578 (1994).

[23] M. Haelterman, S. Trillo, and S. Wabnitz, *J. Opt. Soc. Am. B* **11**, 446 (1994).

[24] G. Steinmeyer, D. Jasper, and F. Mitschke, *Opt. Commun.* **104**, 379 (1994).

[25] M. Haelterman and M. D. Tolley, *Opt. Commun.* **108**, 165 (1994).

[26] T. Fukushima and T. Sakamoto, *Opt. Lett.* **20**, 1119 (1995).

[27] K. Ogusu, *IEEE J. Quantum Electron.* **32**, 1537 (1996).

[28] S. Coen, M. Haelterman, P. Emplit, L. Delage, L. M. Simohamed, and F. Reynaud, *J. Opt. Soc. Am. B* **15**, 2283 (1998).

[29] S. Coen and M. Haelterman, *Opt. Lett.* **24**, 80 (1999).

[30] A. L. Steele, *Opt. Commun.* **236**, 109 (2004).

[31] G. Genty, M. Lehtonen, and H. Ludvigsen, *Appl. Phys. B* **81**, 357 (2005).

[32] L. F. Stokes, M. Chodorow, and H. J. Shaw, *Opt. Lett.* **7**, 509 (1982).

[33] E. Ott, *Chaos in Dynamical Systems* (Cambridge University Press, New York, 1993).

[34] O. Williams, *Nonlinear Dynamics and Chaos* (World Scientific, Boston, 1997).

[35] E. Infeld and G. Rowlands, *Nonlinear Waves, Solitons, and Chaos*, 2nd ed. (Cambridge University Press, New York, 2000).

[36] M. Nakazawa, K. Suzuki, and H. A. Haus, *Phys. Rev. A* **38**, 5193 (1988); *Phys. Rev. A* **39**, 5788 (1989); *IEEE J. Quantum Electron.* **25**, 2036 (1989).

[37] M. Nakazawa, K. Suzuki, H. Kubota, and H. A. Haus, *IEEE J. Quantum Electron.* **25**, 2045 (1989).

[38] E. J. Greer, D. M. Patrick, and P. G. J. Wigley, *Electron. Lett.* **25**, 1246 (1989).

[39] M. Haelterman, S. Trillo, and S. Wabnitz, *Opt. Lett.* **17**, 745 (1992).

[40] S. Coen and M. Haelterman, *Phys. Rev. Lett.* **79**, 4139 (1997).

[41] M. Yu, C. J. McKinstrie, and G. P. Agrawal, *J. Opt. Soc. Am. B* **15**, 607 (1998); *J. Opt. Soc. Am. B* **15**, 617 (1998).

[42] R. Vallée, *Opt. Commun.* **93**, 389 (1992).

[43] M. B. van der Mark, J. M. Schins, and A. Lagendijk, *Opt. Commun.* **98**, 120 (1993).

[44] G. Steinmeyer, A. Buchholz, M. Hänsel, M. Heuer, A. Schwache, and F. Mitschke, *Phys. Rev. A* **52**, 830 (1995).

[45] F. Mitschke, G. Steinmeyer, and A. Schwache, *Physica D* **96**, 251 (1996).

[46] A. Schwache and F. Mitschke, *Phys. Rev. E* **55**, 7720 (1997).

[47] F. Mitschke, I. Halama, and A. Schwache, *Chaos, Solitons and Fractals* **10**, 913 (1999).

[48] K. Otsuka, *Opt. Lett.* **8**, 471 (1983).

[49] D. B. Mortimore, *J. Lightwave Technol.* **6**, 1217 (1988).

[50] N. J. Doran and D. Wood, *Opt. Lett.* **13**, 56 (1988).

[51] N. J. Doran, D. S. Forrester, and B. K. Nayar, *Electron. Lett.* **25**, 267 (1989).

[52] K. J. Blow, N. J. Doran, and B. K. Nayar, *Opt. Lett.* **14**, 754 (1989).

[53] M. N. Islam, E. R. Sunderman, R. H. Stolen, W. Pleibel, and J. R. Simpson, *Opt. Lett.* **14**, 811 (1989).

[54] N. Takato, T. Kaminato, A. Sugita, K. Jinguji, H. Toba, and M. Kawachi, *IEEE J. Sel. Areas Commun.* **8**, 1120 (1990).

[55] K. J. Blow, N. J. Doran, and B. P. Nelson, *Electron. Lett.* **26**, 962 (1990).

[56] M. E. Fermann, F. Haberl, M. Hofer, and H. Hochstrasser, *Opt. Lett.* **15**, 752 (1990).

[57] A. W. O'Neill and R. P. Webb, *Electron. Lett.* **26**, 2008 (1990).

[58] A. L. Steele, *Electron. Lett.* **29**, 1972 (1993).

[59] A. L. Steele and J. P. Hemingway, *Opt. Commun.* **123**, 487 (1996).

[60] W. S. Wong, S. Namiki, M. Margalit, H. A. Haus, and I. P. Ippen, *Opt. Lett.* **22**, 1150 (1997).

[61] M. G. da Silva and A. S. B. Sombra, *Opt. Commun.* **145**, 281 (1998).

[62] I. Y. Khrushchev, R. V. Penty, and I. H. White, *Electron. Lett.* **34**, 1009 (1998)

[63] I. Y. Khrushchev, I. D. Philips, A. D. Ellis, R. J. Manning, D. Nesset, D. G. Moodie, R. V. Penty, and I. H. White, *Electron. Lett.* **35**, 1183 (1999).

[64] J. L. S. Lima and A. S. B. Sombra, *Opt. Commun.* **163**, 292 (1999).

[65] Y. J. Chai, I. Y. Khrushchev, and I. H. White, *Electron. Lett.* **36**, 1565 (2000).

[66] M. C. Farries and D. N. Payne, *Appl. Phys. Lett.* **55**, 25 (1989).

[67] K. J. Blow, N. J. Doran, B. K. Nayar, and B. P. Nelson, *Opt. Lett.* **15**, 248 (1990).

[68] M. Jinno and T. Matsumoto, *IEEE Photon. Technol. Lett.* **2**, 349 (1990); *Electron. Lett.* **27**, 75 (1991).

[69] H. Avramopoulos, P. M. W. French, M. C. Gabriel, H. H. Houh, N. A. Whitaker, and T. Morse, *IEEE Photon. Technol. Lett.* **3**, 235 (1991).

[70] J. D. Moores, K. Bergman, H. A. Haus, and E. P. Ippen, *Opt. Lett.* **16**, 138 (1991); *J. Opt. Soc. Am. B* **8**, 594 (1991).

[71] A. D. Ellis and D. A. Cleland, *Electron. Lett.* **28**, 405 (1992).

[72] M. Jinno and T. Matsumoto, *IEEE J. Quantum Electron.* **28**, 875 (1992).

[73] M. Jinno, *J. Lightwave Technol.* **10**, 1167 (1992); *Opt. Lett.* **18**, 726 (1993); *Opt. Lett.* **18**, 1409 (1993).

[74] N. A. Whitaker, P. M. W. French, M. C. Gabriel, and H. Avramopoulos, *IEEE Photon. Technol. Lett.* **4**, 260 (1992).

[75] H. Bülow and G. Veith, *Electron. Lett.* **29**, 588 (1993).

[76] K. Smith, N. J. Doran, and P. G. J. Wigley, *Opt. Lett.* **15**, 1294 (1990).

[77] S. V. Chernikov and J. R. Taylor, *Electron. Lett.* **29**, 658 (1993).

[78] A. G. Bulushev, E. M. Dianov, and O. G. Okhotnikov, *IEEE Photon. Technol. Lett.* **2**, 699 (1990); *Opt. Lett.* **26**, 968 (1990).

[79] I. N. Duling III, *Electron. Lett.* **27**, 544 (1991); *Opt. Lett.* **16**, 5394 (1991).

[80] R. A. Betts, J. W. Lear, S. J. Frisken, and P. S. Atherton, *Electron. Lett.* **28**, 1035 (1992).

[81] R. A. Betts, J. W. Lear, N. T. Dang, R. D. Shaw, and P. S. Atherton, *IEEE Photon. Technol. Lett.* **4**, 1290 (1992).

[82] J. Yu, X. Zheng, C. Peucheret, A. T. Clausen, H. N. Poulsen, and P. Jeppesen, *J. Lightwave Technol.* **18**, 1001 (2000).

[83] J. K. Lucek and K. Smith, *Opt. Lett.* **15**, 1226 (1993).

[84] B. K. Nayar, N. Finlayson, and N. J. Doran, *J. Mod. Opt.* **40**, 2327 (1993).

[85] S. Watanabe and S. Takeda, *Electron. Lett.* **36**, 52 (2000).

[86] J. M. Jeong and M. E. Marhic, *Opt. Commun.* **91**, 115 (1992).

[87] F. Mogensen, B. Pedersen, and B. Nielsen, *Electron. Lett.* **29**, 1469 (1993).

[88] M. Jinno and T. Matsumoto, *Opt. Lett.* **16**, 220 (1991).

[89] A. Huang, N. Whitaker, H. Avramopoulos, P. French, H. Houh, and I. Chuang, *Appl. Opt.* **33**, 6254 (1994).

[90] M. Jinno, *J. Lightwave Technol.* **12**, 1648 (1994).

[91] K. Ikeda, J. M. Abdul, H. Tobioka, T. Inoue, S. Namiki, and K. Kitayama, *J. Lightwave Technol.* **24**, 2618 (2006).

[92] S. Oda and A. Murata, *IEEE Photon. Technol. Lett.* **18**, 703 (2006).

[93] K. Mori, T. Morioka, and M. Saruwatari, *Opt. Lett.* **20**, 1424 (1995).

[94] S. Schmitt, J. Ficker, M. Wolff, F. Konig, A. Sizmann, and G. Leuchs, *Phys. Rev. Lett.* **81**, 2446 (1998).

[95] D. Krylov and K. Bergman, *Opt. Lett.* **23**, 1390 (1998).

[96] D. K. Serkland, G. D. Bartiolini, A. Agarwal, P. Kumar, and W. L. Kath, *Opt. Lett.* **23**, 795 (1998).

[97] D. K. Serkland and P. Kumar, *Opt. Lett.* **24**, 92 (1999).

[98] W. Imajuku, A. Takada, and Y. Yamabayashi, *Electron. Lett.* **36**, 63 (2000).

[99] H. C. Lim, F. Futami, and K. Kikuchi, *IEEE Photon. Technol. Lett.* **11**, 578 (1999).

[100] U. Feiste, R. Ludwig, C. Schmidt, E. Dietrich, S. Diez, H. J. Ehrke, E. H. Patzak, G. Weber, and T. Merker, *IEEE Photon. Technol. Lett.* **11**, 1063 (1999).

[101] H. C. Lim, F. Futami, K. Taira, and K. Kikuchi, *IEEE Photon. Technol. Lett.* **11**, 1405 (1999).

[102] A. Agarwal, L. Wang, Y. Su, and P. Kumar, *J. Lightwave Technol.* **23**, 2229 (2005).

[103] T. Torounidis, B. E. Olsson, H. Sunnerud, M. Karlsson, and P. A. Andrekson, *IEEE Photon. Technol. Lett.* **17**, 321 (2005).

[104] G. P. Agrawal, *Lightwave Technnology: Telecommunication Systems* (Wiley, New York, 2005).

[105] N. J. Doran and D. Wood, *J. Opt. Soc. Am. B* **4**, 1843 (1987).

[106] N. Imoto, S. Watkins, and Y. Sasaki, *Opt. Commun.* **61**, 159 (1987).

[107] I. H. White, R. V. Penty, and R. E. Epworth, *Electron. Lett.* **24**, 340 (1988).

[108] M. N. Islam, S. P. Dijaili, and J. P. Gordon, *Opt. Lett.* **13**, 518 (1988).

[109] D. V. Khaidatov, *Sov. J. Quantum Electron.* **20**, 379 (1990).

[110] T. V. Babkina, F. G. Bass, S. A. Bulgakov, V. V. Grogoryants, and V. V. Konotop, *Opt. Commun.* **78**, 398 (1990).

[111] B. K. Nayar, N. Finlayson, N. J. Doran, S. T. Davey, W. L. Williams, and J. W. Arkwright, *Opt. Lett.* **16**, 408 (1991).

[112] M. Asobe, *J. Opt. Soc. Am. B* **12**, 1287 (1995).

[113] K. I. Kang, T. G. Chang, I. Glesk, and P. R. Prucnal, *Appl. Opt.* **35**, 1485 (1996).

[114] P. Elango, J. W. Arkwright, P. L. Chu, and G. R. Atkins, *IEEE Photon. Technol. Lett.* **8**, 1032 (1996).

[115] J. E. Heebner and R. W. Boyd, *Opt. Lett.* **24**, 847 (1999).

[116] M. Fiorentino, J. E. Sharping, P. Kumar, and A. Porzio, *Opt. Express* **10**, 128 (2003).

[117] J. L. S. Lima, C. S. N. Rios, M. G. da Silva, C. S. Sobrinho, E. F. de Almeida, and A. S. B. Sombra, *Opt. Fiber Technol.* **11**, 167 (2005).

[118] J. Li, L. Li, L. Jin. and C. Li, *Opt. Commun.* **260**, 318 (2006).

[119] F. T. S. Yu and S. Yin, Eds., *Fiber Optic Sensors* (Marcel Dekker, New York, 2002).

[120] M. A. Marcus, *Fiber Optic Sensor Technology and Applications*, Vol. 4 (SPIE, Bellingham, WA, 2005).

[121] E. Udd, *Fiber Optic Sensors: An Introduction for Engineers and Scientists* (Wiley, Hoboken, NJ, 2006).

[122] M. Kuznetsov, *J. Lightwave Technol.* **12**, 226 (1994).

[123] T. J. Cullen, H. N. Rourke, C. P. Chew, S. R. Baker, T. Bircheno, K. Byron, and A. Fielding, *Electron. Lett.* **30**, 2160 (1994).

[124] G. Nykolak, M. R. X. de Barros, T. N. Nielsen, and L. Eskildsen, *IEEE Photon. Technol. Lett.* **9**, 605 (1997).

[125] K. N. Park, T. T. Lee, M. H. Kim, K. S. Lee, and Y. H. Won, *IEEE Photon. Technol. Lett.* **10**, 555 (1998).

[126] T. Mizuochi, T. Kitayama, K. Shimizu, and K. Ito, *J. Lightwave Technol.* **16**, 265 (1998).

[127] F. Ouellette and M. Piché, *Opt. Commun.* **60**, 99 (1986); *Canadian J. Phys.* **66**, 903 (1988).

[128] E. M. Dianov and O. G. Okhotnikov, *IEEE Photon. Technol. Lett.* **3**, 499 (1991).

[129] C. Spielmann, F. Krausz, T. Brabec, E. Wintner, and A. J. Schmidt, *Appl. Phys. Lett.* **58**, 2470 (1991).

[130] P. Heinz, A. Reuther, and A. Laubereau, *Opt. Commun.* **97**, 35 (1993).

[131] C. X. Shi, *Opt. Lett.* **18**, 1195 (1993).

[132] H. A. Haus, J. G. Fujimoto, and E. P. Ippen, *J. Opt. Soc. Am. B* **7**, 2068 (1991).

[133] X. F. Mo, B. Zhu, Z. F. Han, Y. Z. Gui, and G. C. Guo, *Opt. Lett.* **30**, 2632 (2005).

[134] S. Odate, H. B. Wang, and T. Kobayashi, *Phys. Rev. A* **72**, 068312 (2005).

[135] C. Liang, K. F. Lee, T. Levin, J. Chen, and P. Kumar, *Opt. Express* **14**, 6936 (2005).

[136] H. Rehbein, J. Harms, R. Schnabel, and K. Danzmann, *Phys. Rev. Lett.* **95**, 193001 (2005).

Chapter 4

Fiber Amplifiers

Optical fibers attenuate light like any other material. In the case of silica fibers, losses are relatively small, especially in the wavelength region near 1.55 μm ($\alpha \approx$ 0.2 dB/km). For this reason, losses can simply be ignored if fiber length is 1 km or less. In the case of long-haul fiber-optic communication systems, transmission distances may exceed thousands of kilometers. Several types of fiber amplifiers can be used to overcome transmission losses and restore the optical signal in such systems. Section 4.1 discusses general concepts, such as gain spectrum and amplifier bandwidth, that are common to all amplifiers. Section 4.2 describes the operating characteristics of erbium-doped fiber amplifiers (EDFAs). The nonlinear and dispersive effects are included in Section 4.3 using the Maxwell–Bloch formalism. The resulting Ginzburg–Landau equation is used in Sections 4.4 to 4.6 to discuss a variety of nonlinear effects in EDFAs, including modulation instability, pulse splitting, the formation of autosolitons, and the self-similar evolution toward parabolic pulses in the case of normal dispersion. Sections 4.7 is devoted to fiber-optic Raman amplifiers in which an optical fiber itself provides amplification, without any dopants, when it is pumped suitably.

4.1 Basic Concepts

Although fiber amplifiers were made as early as 1964 [1], their use became practical only after 1986 when the techniques for fabrication and characterization of low-loss, rare-earth-doped fibers were perfected [2]. The rare earths (or lanthanides) form a group of 14 similar elements with atomic numbers in the range from 58 to 71. When these elements are doped in silica or other glass fibers, they become triply ionized. Many different rare-earth ions, such as erbium, holmium, neodymium, samarium, thulium, and ytterbium, can be used to make fiber amplifiers that operate at wavelengths covering a wide range from visible to infrared. Amplifier characteristics, such as the operating wavelength and the gain bandwidth, are determined by dopants rather than by the fiber, which plays the role of a host medium. EDFAs have attracted the

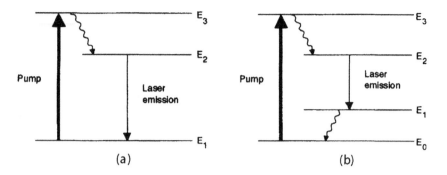

Figure 4.1: (a) Three-level and (b) four-level pumping schemes. Wavy arrows indicate fast relaxation of the level population through nonradiative processes.

most attention as they operate near 1.55 μm and are useful for modern fiber-optic communication systems [3]–[6].

4.1.1 Pumping and Gain Coefficient

Fiber amplifiers amplify incident light through stimulated emission, the same mechanism used by lasers. Indeed, an optical amplifier is just a laser without feedback. Its main ingredient is the optical gain, occurring when the amplifier is pumped optically to realize population inversion. Depending on the energy levels of the dopant, pumping schemes can be classified as a three- or four-level scheme [7]–[9]. Figure 4.1 shows the two kinds of pumping schemes. In both cases, dopants absorb pump photons to reach a higher energy state and then relax rapidly to a lower-energy excited state (level 2). The stored energy is used to amplify the incident signal through stimulated emission. The main difference between the three- and four-level pumping schemes is related to the energy state occupied by the dopant after each stimulated-emission event. In the case of a three-level scheme, the ion ends up in the ground state, whereas it remains in an excited state in the case of a four-level pumping scheme. It will be seen later that this difference affects the amplifier characteristics significantly. EDFAs make use of a three-level pumping scheme.

For understanding the physics behind signal amplification, details of pumping are not important. Optical pumping creates the necessary population inversion between the two energy states, which in turn provides the optical gain $g = \sigma(N_1 - N_2)$, where σ is the transition cross section and N_1 and N_2 are atomic densities in the two energy states. The gain coefficient g can be calculated for both the three- and four-level pumping schemes by using the appropriate rate equations [7]–[9].

The gain coefficient of a homogeneously broadened gain medium can be written as [8]

$$g(\omega) = \frac{g_0}{1 + (\omega - \omega_a)^2 T_2^2 + P/P_s}, \qquad (4.1.1)$$

where g_0 is the peak value, ω is the frequency of the incident signal, ω_a is the atomic transition frequency, and P is the optical power of the continuous-wave (CW) signal

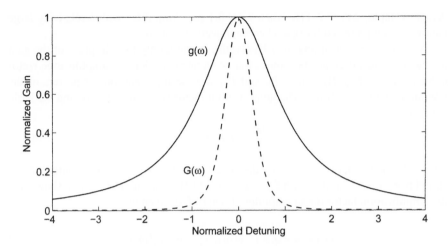

Figure 4.2: Lorentzian spectrum $g(\omega)$ and the corresponding amplifier-gain spectrum $G(\omega)$ for a fiber amplifier.

being amplified. The saturation power P_s depends on dopant parameters such as the fluorescence time T_1 and the transition cross section σ. The parameter T_2 in Eq. (4.1.1) is known as the dipole relaxation time and is typically quite small (~ 0.1 ps) for fiber amplifiers. The fluorescence time T_1 varies in the range from 0.1 μs to 10 ms, depending on the dopant. Equation (4.1.1) can be used to discuss the important characteristics of optical amplifiers such as gain bandwidth, amplification factor, and output saturation power. We begin by considering the case in which $P/P_s \ll 1$ throughout the amplifier. This is referred to as the *unsaturated regime* as the gain remains unsaturated during amplification.

4.1.2 Amplifier Gain and Bandwidth

By neglecting the term P/P_s in Eq. (4.1.1), the gain coefficient becomes

$$g(\omega) = \frac{g_0}{1 + (\omega - \omega_a)^2 T_2^2}. \tag{4.1.2}$$

This equation shows that the gain is maximum when the signal frequency ω coincides with the atomic transition frequency ω_a. The gain reduction for $\omega \neq \omega_a$ is governed by a Lorentzian profile (see Figure 4.2) that is characteristic of homogeneously broadened systems [7]–[9]. As discussed later, the actual gain spectrum of fiber amplifiers can deviate considerably from the Lorentzian profile. The gain bandwidth is defined as the full width at half maximum (FWHM) of the gain spectrum $g(\omega)$. For the Lorentzian spectrum, the gain bandwidth is given by

$$\Delta v_g = \frac{\Delta \omega_g}{2\pi} = \frac{1}{\pi T_2}. \tag{4.1.3}$$

As an example, $\Delta v_g \approx 3$ THz when $T_2 = 0.1$ ps. Amplifiers with a relatively large bandwidth are preferred for optical communication systems.

A related concept of amplifier bandwidth is commonly used in place of the gain bandwidth. The difference becomes clear when one considers the amplification factor defined as $G = P_{out}/P_{in}$, where P_{in} and P_{out} are the input and output powers of the CW signal being amplified. The amplification factor is obtained by solving the basic equation,

$$\frac{dP}{dz} = g(\omega)P(z), \tag{4.1.4}$$

where $P(z)$ is the optical power at a distance z from the input end of the amplifier. A straightforward integration with the conditions $P(0) = P_{in}$ and $P(L) = P_{out}$ shows that the amplification factor for an amplifier of length L is given by

$$G(\omega) = \exp\left[\int_0^L g(\omega)\,dz\right] = \exp[g(\omega)L], \tag{4.1.5}$$

where g is assumed to be constant along the amplifier length.

Both $G(\omega)$ and $g(\omega)$ are maximum at $\omega = \omega_a$ and decrease when $\omega \neq \omega_a$. However, $G(\omega)$ decreases much faster than $g(\omega)$ because of the exponential dependence seen in Eq. (4.1.5). The amplifier bandwidth Δv_A is defined as the FWHM of $G(\omega)$ and is related to the gain bandwidth Δv_g as

$$\Delta v_A = \Delta v_g \left(\frac{\ln 2}{\ln G_0 - \ln 2}\right)^{1/2}, \tag{4.1.6}$$

where G_0 is the peak value of the amplifier gain. Figure 4.2 shows the gain profile $g(\omega)$ and the amplification factor $G(\omega)$ by plotting both g/g_0 and G/G_0 as a function of $(\omega - \omega_a)T_2$. As expected, the amplifier bandwidth is smaller than the gain bandwidth, and the difference depends on the amplifier gain itself.

The origin of gain saturation lies in the power dependence of the gain coefficient in Eq. (4.1.1). Since g is reduced when P becomes comparable to P_s, the amplification factor G is also expected to decrease. To simplify the discussion, let us consider the case in which the signal frequency is exactly tuned to the atomic transition frequency ω_a. By substituting g from Eq. (4.1.1) in Eq. (4.1.4), we obtain

$$\frac{dP}{dz} = \frac{g_0 P}{1 + P/P_s}. \tag{4.1.7}$$

This equation can be easily integrated over the amplifier length. By using the initial condition $P(0) = P_{in}$ together with $P(L) = P_{out} = GP_{in}$, the amplifier gain is given by the implicit relation

$$G = G_0 \exp\left(-\frac{G-1}{G}\frac{P_{out}}{P_s}\right). \tag{4.1.8}$$

Figure 4.3 shows the saturation characteristics by plotting G as a function of P_{out}/P_s for several values of G_0. A quantity of practical interest is the output saturation power

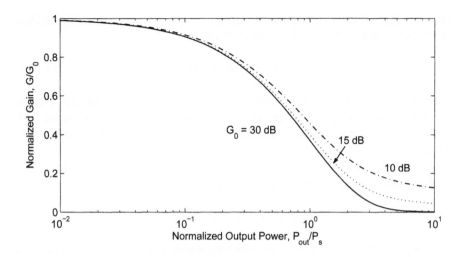

Figure 4.3: Saturated amplifier gain as a function of output power (normalized to the saturation power) for several values of the unsaturated amplifier gain G_0.

P_{out}^s, defined as the output power at which the amplifier gain G is reduced by a factor of 2 from its unsaturated value G_0. By using $G = G_0/2$ in Eq. (4.1.8), P_{out}^s is given by

$$P_{out}^s = \frac{G_0 \ln 2}{G_0 - 2} P_s. \tag{4.1.9}$$

By noting that $G_0 \gg 2$ in practice, $P_{out}^s \approx (\ln 2)P_s \approx 0.69 P_s$. As seen in Figure 4.3, P_{out}^s becomes nearly independent of G_0 for $G_0 > 20$ dB.

4.1.3 Amplifier Noise

All amplifiers degrade the signal-to-noise ratio (SNR) of the amplified signal because of spontaneous emission that is added to the signal during its amplification. The SNR degradation is quantified through the noise figure F_n defined as

$$F_n = (\text{SNR})_{in}/(\text{SNR})_{out}, \tag{4.1.10}$$

where both SNRs refer to the electrical power generated when an optical signal is converted to electric current by using a photodetector. In general, F_n depends on several parameters governing the shot and thermal noises associated with the detector. We can obtain a simple expression for F_n by considering an ideal detector whose performance is limited by shot noise only.

In the shot-noise limit, the SNR of the input signal is given by [10]

$$(\text{SNR})_{in} = \frac{I^2}{\sigma_s^2} = \frac{(R_d P_{in})^2}{2q(R_d P_{in})\Delta f} = \frac{P_{in}}{2h\nu\Delta f}, \tag{4.1.11}$$

where $I = R_d P_{\text{in}}$ is the average photocurrent, $R_d = q/h\nu$ is the responsivity of an ideal photodetector with 100% quantum efficiency, and

$$\sigma_s^2 = 2q(R_d P_{\text{in}})\Delta f \qquad (4.1.12)$$

represents the contribution of shot noise. Here Δf is the detector bandwidth, ν is the optical frequency, and q is the magnitude of the electron's charge. To evaluate the SNR of the amplified signal, we should add the contribution of spontaneous emission to the detector noise.

The spectral density of spontaneous-emission noise is nearly constant for broad-band amplifiers (white noise) and is given by [11]

$$S_{\text{sp}}(\nu) = (G-1)n_{\text{sp}}h\nu, \qquad (4.1.13)$$

where n_{sp}, called the *spontaneous-emission* (or *population-inversion*) *factor*, is defined as

$$n_{\text{sp}} = N_2/(N_2 - N_1). \qquad (4.1.14)$$

The effect of spontaneous emission is to add fluctuations to the amplified signal, which are converted to current fluctuations during the detection process.

The dominant contribution to the noise current comes from the beating of spontaneous emission with the signal. This beating phenomenon is similar to heterodyne detection: Spontaneously emitted radiation mixes coherently with the amplified signal at the photodetector and produces a heterodyne component of the photocurrent. The variance of the photocurrent can be written as [10]

$$\sigma^2 = 2q(R_d G P_{\text{in}})\Delta f + 4(R_d G P_{\text{in}})(R S_{\text{sp}})\Delta f, \qquad (4.1.15)$$

where the first term is due to shot noise and the second term results from signal–spontaneous emission beating. Since $I = R_d G P_{\text{in}}$ is the average current, the SNR of the amplified signal is given by

$$(\text{SNR})_{\text{out}} = \frac{(R G P_{\text{in}})^2}{\sigma^2} \approx \frac{G P_{\text{in}}}{(4S_{\text{sp}} + 2h\nu)\Delta f}. \qquad (4.1.16)$$

The amplifier noise figure is obtained by substituting Eqs. (4.1.11) and (4.1.16) in Eq. (4.1.10) and is given by

$$F_n = 2n_{\text{sp}}\left(1 - \frac{1}{G}\right) + \frac{1}{G} \approx 2n_{\text{sp}}, \qquad (4.1.17)$$

where the last approximation is valid for $G \gg 1$. This equation shows that the SNR of the amplified signal is degraded by a factor of 2 (or 3 dB) even for an ideal amplifier for which $n_{\text{sp}} = 1$. In practice, F_n exceeds 3 dB. For its application in optical communication systems, an optical amplifier should have F_n as low as possible.

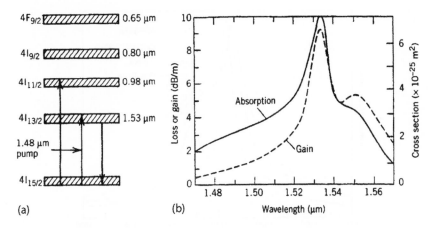

Figure 4.4: (a) Energy levels of erbium ions in silica fibers. (b) Absorption and gain spectra of an erbium-doped fiber.

4.2 Erbium-Doped Fiber Amplifiers

In this section we focus on EDFAs because of their importance for lightwave systems. Figure 4.4(a) shows the relevant energy levels of Er^{3+} in silica glasses. The amorphous nature of silica broadens each energy level into bands. Many transitions can be used for pumping. Initial experiments used visible pump wavelengths even though their use is relatively inefficient [12], [13]. Efficient pumping is possible using semiconductor lasers operating near 0.98-μm and 1.48-μm wavelengths [14]–[17]. High gains in the range of 30 to 40 dB can be obtained with pump powers ~10 mW. The transition $^4I_{15/2} \rightarrow ^4 I_{9/2}$ allows the use of GaAs pump lasers operating near 0.8 μm, but the pumping efficiency is relatively poor [18]. It can be improved by codoping the fiber with aluminum and phosphorus [19]. EDFAs can also be pumped in the wavelength region near 650 nm. In one experiment, 33-dB gain was realized with 27 mW of pump power at 670 nm [20].

The pump and signal beams inside an EDFA may propagate in the same or opposite directions. The performance is nearly the same in the two pumping configurations when the signal power is small enough for the amplifier to remain unsaturated. In the saturation regime, power-conversion efficiency is better in the backward-pumping configuration because of lower amplified spontaneous emission [21]. In the bidirectional pumping configuration, the amplifier is pumped in both directions simultaneously using two semiconductor lasers located at the two fiber ends. This configuration requires two pump lasers but has the advantage that the small-signal gain remains relatively constant along the entire amplifier length.

4.2.1 Gain Spectrum

The gain spectrum of an EDFA is affected considerably by the amorphous nature of silica and by the presence of other codopants such as germania and alumina within the

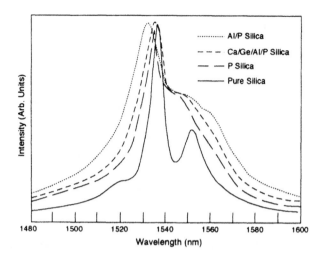

Figure 4.5: Gain spectra of four EDFAs with different core compositions. Doping of silica core with aluminum or phosphorus (in addition of erbium) broadens the emission spectrum considerably. (From Ref. [22];, ©1991 IEEE.)

fiber core [22]. The gain spectrum of isolated erbium ions is homogeneously broadened, and its bandwidth is determined by the dipole relaxation time T_2. However, it is considerably broadened by the silica host. Structural disorders lead to inhomogeneous broadening of the gain profile, whereas Stark splitting of various energy levels is responsible for additional homogeneous broadening [4]. Mathematically, the gain is obtained by averaging over the distribution of transition frequencies ω_a so that

$$g(\omega) = \int_{-\infty}^{\infty} g_h(\omega, \omega_a) f_a(\omega_a)\, d\omega_a, \qquad (4.2.1)$$

where $g_h(\omega, \omega_a)$ is the homogeneously broadened gain profile and $f_a(\omega_a)$ is the distribution function whose form depends on the glass composition within the fiber core.

Figure 4.4(b) shows the gain and absorption spectra of an EDFA whose core was doped with germania. The gain spectrum is quite broad with a double-peak structure. The shape and the width of the gain spectrum is sensitive to core composition. Figure 4.5 shows the emission spectra for four core compositions. The gain spectrum is narrowest in the case of pure silica but can be broadened considerably by codoping the core with alumina. Attempts have been made to isolate the relative contributions of homogeneous and inhomogeneous broadening. For silica-based EDFAs, the contribution of homogeneous broadening, as deduced from spectral hole-burning measurements, is in the range of 4 to 10 nm, depending on the signal wavelength [4]. With a proper choice of dopants and a host fiber, the spectral bandwidth over which an EDFA is able to amplify optical signals can exceed 30 nm. However, the gain is not uniform over the entire bandwidth.

With the advent of the WDM technique, a single EDFA is used to amplify a large number of channels simultaneously. Moreover, the WDM signal is often propagated through a chain of such cascaded EDFAs covering long distances. If the gain spectrum of EDFAs is not flat over the entire bandwidth of the WDM signal and the gain varies as little as a few percent from channel to channel, large variations (>10 dB) among channel powers occur when the signal arrives at the receiver; such variations can degrade the system performance considerably. For this reason, many techniques have been developed for flattening the gain spectrum of EDFAs and extending the usable bandwidth to a range as large as 80 nm [4]. One solution consists of using an optical filter whose spectral response is tailored such that the filter transmits more light at wavelengths where gain is lower (and vice versa).

Optical filters based on Mach–Zehnder interferometers were used as early as 1991 [23]. More recently, long-period fiber gratings have been used for this purpose with considerable success [24]–[26]. Other approaches makes use of an acoustooptic filter [27], a fiber-loop mirror [28], or a fiber coupler [29]. The last technique employs a dual-core fiber, with only one core doped with erbium. The coupling efficiency from the doped core to the undoped core is maximized for the wavelength where the EDFA gain is maximum. This feature introduces wavelength-dependent losses whose magnitude can be controlled to provide a flatter gain spectrum.

With a proper design, the use of optical filters can provide flat gain over a bandwidth as large as 35 nm. However, WDM systems, designed to transmit more than 50 channels, may require uniform gain over a bandwidth exceeding 50 nm. It is difficult to achieve such large gain bandwidths with a single amplifier. A hybrid two-stage approach is commonly used in practice. In one design, two amplifiers are cascaded to produce flat gain (to within 0.5 dB) over the wavelength range of 1544 to 1561 nm [30]. The second EDFA is codoped with ytterbium and phosphorus and is optimized such that it acts as a power amplifier. In a variation of this idea, the second EDFA uses fluoride fiber as a host and is pumped at 1480 nm [31]. Another approach combines Raman amplification with one or two EDFAs to realize uniform gain over a 65-nm bandwidth extending from 1549 to 1614 nm [32].

A two-arm design has also been developed to solve the gain-flattening problem for dense WDM systems [33]. In this approach, the WDM signal is divided into two bands, the conventional or C band (1530 to 1565 nm) and the long-wavelength or L band (1565 to 1625 nm). The incoming WDM signal is split into two branches containing optimized EDFAs for C and L bands. The L-band EDFA requires long fiber lengths (>100 m) since the inversion level is kept relatively low. By 1999, the two-arm design produced a relatively uniform gain of 24 dB over a bandwidth as large as 80 nm when pumped with 980-nm semiconductor lasers while maintaining a noise figure of about 6 dB [4]. By 2004, the bandwidth could be increased to 105 nm using dual-core erbium-doped fibers [34]. EDFAs can also provide amplification in the short-wavelength band (the so-called S band). Designing such amplifiers is not easy, but by 2003 they were able to provide more than 21 dB gain in the wavelength range extending from 1490 to 1520 nm [35].

4.2.2 Amplifier Gain

The gain of EDFAs depends on a large number of parameters such as erbium-ion concentration, amplifier length, core radius, and pump power [36]–[40]. A three-level rate-equation model, used commonly for lasers, can be adapted for EDFAs [4]. It is sometimes necessary to add a fourth level to include the effects of excited-state absorption. Another complication stems from the nonuniform nature of inversion along the amplifier length. Since a fiber amplifier is pumped from one end, the pump power decreases along the fiber length. As a result, it is necessary to include axial variations of the pump, the signal, and the atomic-level populations. In general, the resulting set of coupled equations must be solved numerically.

Much insight can be gained by using a simple model that neglects amplified spontaneous emission and excited-state absorption. The model assumes that the pump level of the three-level system remains nearly unpopulated because of a rapid transfer of the pumped population to the excited state 2 (see Figure 4.1). It also neglects differences between the emission and absorption cross sections. With these simplifications, the excited-state density $N_2(z,t)$ is obtained by solving the following rate equation [9]

$$\frac{\partial N_2}{\partial t} = W_p N_1 - W_s(N_2 - N_1) - \frac{N_2}{T_1}, \tag{4.2.2}$$

where $N_1 = N_t - N_2$, N_t is the total ion density, and W_p and W_s are the transition rates for the pump and signal, respectively. These rates are given by

$$W_p = \frac{\Gamma_p \sigma_p P_p}{a_p h \nu_p}, \qquad W_s = \frac{\Gamma_s \sigma_s P_s}{a_s h \nu_s}, \tag{4.2.3}$$

where Γ_p is the overlap factor representing the fraction of pump power P_p within the doped region of the fiber, σ_p is the transition cross section at the pump frequency ν_p, and a_p is the mode area of the pump inside the fiber. The quantities Γ_s, σ_s, P_s, a_s, and ν_s are defined similarly for the signal. The steady-state solution of Eq. (4.2.2) is given by

$$N_2 = \frac{(P_p' + P_s')N_t}{1 + 2P_s' + P_p'}, \tag{4.2.4}$$

where $P_p' = P_p/P_p^{\text{sat}}$, $P_s' = P_s/P_s^{\text{sat}}$, and the saturation powers are defined as

$$P_p^{\text{sat}} = \frac{a_p h \nu_p}{\Gamma_p \sigma_p T_1}, \qquad P_s^{\text{sat}} = \frac{a_s h \nu_s}{\Gamma_s \sigma_s T_1}. \tag{4.2.5}$$

The pump and signal powers vary along the amplifier length because of absorption, stimulated emission, and spontaneous emission. Their variation also depends on whether the signal and pump waves propagate in the same or opposite directions. If the contribution of spontaneous emission is neglected and forward pumping is assumed, P_p and P_s satisfy

$$\frac{dP_p}{dz} = -\Gamma_p \sigma_p N_1 - \alpha' P_p, \qquad \frac{dP_s}{dz} = \Gamma_s \sigma_s (N_2 - N_1) - \alpha P_s, \tag{4.2.6}$$

where α and α' take into account fiber losses at the signal and pump wavelengths, respectively. By substituting N_2 from Eq. (4.2.4) together with $N_1 = N_t - N_2$, we obtain a set of two coupled equations,

$$\frac{dP_p'}{dz} = -\frac{(P_s' + 1)\alpha_p P_p'}{1 + 2P_s' + P_p'} - \alpha' P_p', \tag{4.2.7}$$

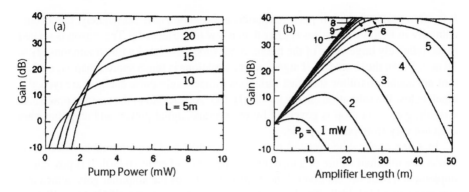

Figure 4.6: Small-signal gain at 1.55 μm as a function of (a) pump power and (b) amplifier length for an EDFA pumped at 1.48 μm. (From Ref. [37]; ©1991 IEEE.)

$$\frac{dP'_s}{dz} = \frac{(P'_p - 1)\alpha_s P'_s}{1 + 2P'_s + P'_p} - \alpha P'_s, \qquad (4.2.8)$$

where $\alpha_p \equiv \Gamma_p \sigma_p N_t$ and $\alpha_s \equiv \Gamma_s \sigma_s N_t$ are the absorption coefficients at the pump and signal wavelengths, respectively. These equations govern the evolution of signal and pump powers inside an EDFA. Their predictions are in good agreement with experiments as long as the amplified spontaneous emission (ASE) remains negligible [39]. The inclusion of fiber losses is essential for distributed-gain amplifiers, which amplify signals over long fiber lengths. For lumped amplifiers with fiber lengths under 1 km, α and α' can be set to zero.

A drawback of this model is that the absorption and emission cross sections are taken to be the same for both the pump and signal beams. As was seen in Figure 4.4(b), these cross sections are generally different. It is easy to extend the model to include such differences [37]. An analytic solution can still be obtained [36]. Figure 4.6 shows the small-signal gain at 1.55 μm as a function of the pump power and the amplifier length by using typical parameter values. For a given amplifier length L, the gain increases exponentially with pump power initially, but at a much reduced rate when pump power exceeds a certain value [corresponding to the "knee" in Figure 4.6(a)]. For a given pump power, amplifier gain becomes maximum at an optimum value of L and drops sharply when L exceeds this optimum value. The reason is that the end portion of the amplifier remains unpumped and absorbs the amplified signal.

Since the optimum value of L depends on the pump power P_p, it is necessary to choose both L and P_p appropriately. Figure 4.6(b) shows that, for 1.48-μm pumping, 35-dB gain can be realized at a pump power of 5 mW for $L = 30$ m. It is possible to design high-gain amplifiers using fiber lengths as short as a few meters. The qualitative features shown in Figure 4.6 are observed in all EDFAs; the agreement between theory and experiment is generally quite good [39].

The preceding analysis assumes that both pump and signal waves are in the form of CW beams. In practice, EDFAs are pumped by using CW semiconductor lasers, but the signal is generally not a CW beam. For example, in lightwave system applications the signal is in the form of a pulse train (containing a random sequence of 1 and 0 bits).

It is often required that all pulses experience the same gain. Fortunately, this occurs naturally in EDFAs for pulses shorter than a few microseconds. The reason is related to the relatively large value of the fluorescence time associated with erbium ions ($T_1 \approx$ 10 ms). When the timescale of signal power variations is much shorter than T_1, erbium ions are unable to follow such fast variations. Since single-pulse energies are typically much below the saturation energy (~ 10 μJ), EDFAs respond to the average power. As a result, gain saturation is governed by the average signal power, and amplifier gain does not vary from pulse to pulse.

In some applications related to packet-switched and reconfigurable WDM networks, the transient nature of gain dynamics becomes of concern [40]. It is possible to implement a built-in gain-control mechanism that keeps the amplifier gain pinned at a constant value [41]–[48]. The basic idea consists of forcing the EDFA to oscillate at a controlled wavelength outside the range of interest (typically below 1.5 μm). Since the gain remains clamped at the threshold value for a laser, the signal is amplified by the same factor in spite of variations in the signal power. A simple scheme uses an all-optical feedback loop at a specific wavelength to initiate lasing [41]. In another implementation, an EDFA is forced to oscillate at 1.48 μm by fabricating fiber Bragg gratings at the two ends of the amplifier [42]. One of the gratings can also be replaced by a fiber-loop mirror [45]. With this change, the signal wavelength can be close to the lasing wavelength without affecting the amplifier performance.

4.2.3 Amplifier Noise

Since amplifier noise is the ultimate limiting factor for system applications, it has been studied extensively [49]–[59]. As discussed earlier, amplifier noise is quantified through the noise figure $F_n = 2n_{sp}$, where the spontaneous-emission factor n_{sp} depends on the relative populations N_1 and N_2 of the two energy states, as indicated in Eq. (4.1.14). Since EDFAs operate on the basis of a three-level pumping scheme, N_1 is not negligible and n_{sp} exceeds 1. Therefore, the noise figure of EDFAs is expected to be larger than the ideal value of 3 dB.

The spontaneous-emission factor for EDFAs can be calculated by using the three-level rate-equation model discussed earlier. However, one should take into account the fact that both N_1 and N_2 vary along the fiber length because of their dependence on the pump and signal powers [see Eq. (4.2.4)], and n_{sp} should be averaged along the amplifier length. As a result, the noise figure depends both on the amplifier length L and the pump power P_p, just as the amplifier gain does. Figure 4.7(a) shows the variation of F_n with the amplifier length for several values of P_p/P_p^{sat} when a 1.53-μm signal is amplified with an input power of 1 mW [53]. The amplifier gain under the same conditions is shown in Figure 4.7(b). The results show that a noise figure close to 3 dB can be obtained for high-gain amplifiers.

The experimental results confirm that F_n close to 3 dB can be realized in EDFAs. A noise figure of 3.2 dB was measured in a 30-m long EDFA, pumped at 0.98 μm with 11 mW of power [51]. A similar value was measured in another experiment with only 5.8 mW of pump power [52]. In general, it is difficult to achieve high gain, low noise, and high pumping efficiency simultaneously. The main limitation is imposed by the ASE traveling backward toward the pump and depleting the pump power. An

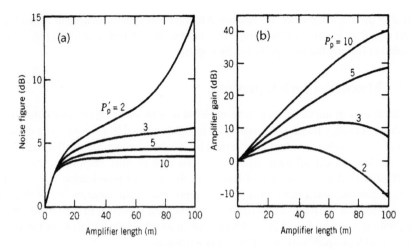

Figure 4.7: (a) Noise figure and (b) amplifier gain at several pumping levels as a function of fiber length. (From Ref. [53]; ©1990 IEE.)

internal isolator alleviates this problem to a large extent. In one implementation, a 51-dB gain was realized with a 3.1-dB noise figure at a pump power of only 48 mW [56]. The relatively low noise levels of EDFAs make them an ideal choice for WDM light-wave systems. In spite of low noise, the performance of long-haul systems employing multiple EDFAs is often limited by the ASE.

The effects of amplifier noise are most severe when a low-power signal is amplified by a large factor. In practice, the noise in a chain of cascaded EDFAs can be reduced by decreasing the amplifier spacing. For this reason, considerable attention has focused on *distributed* fiber amplifiers in which the gain is distributed over long lengths (~50 km) of lightly doped silica fibers such that fiber losses are nearly compensated all along the fiber length [60]–[67]. Such fibers are referred to as being *transparent*, although they become nearly transparent only when pumped at a suitable wavelength. The optimal pumping wavelength is 1.48 μm since fiber losses for the pump at this wavelength are minimal compared to other pumping wavelengths (such as 0.98 μm). In general, one should consider the effect of stimulated Raman scattering (SRS) in distributed EDFAs pumped at 1.48 μm since the signal wavelength lies within the Raman gain bandwidth [63]. As a result, the signal experiences not only the gain provided by the dopants but also the gain provided by SRS. In practice, SRS increases the net gain and reduces the noise figure for a given amount of pump power. Nonlinear and dispersive effects associated with the silica host play an important role in distributed fiber amplifiers. We turn to them in the following section.

4.3   Dispersive and Nonlinear Effects

Because of their large bandwidths, fiber amplifiers can amplify short optical pulses without introducing distortion. Indeed, EDFAs were used to amplify ultrashort pulses soon after their development. We discuss in this section how the nonlinear Schrödinger (NLS) equation, useful for describing pulse propagation in undoped fibers [68], can be extended to include the gain provided by dopants.

4.3.1 Maxwell–Bloch Equations

Rare-earth ions in doped fibers can be modeled as a two-level system by considering only the two energy levels that participate in light-induced transitions. The dynamic response of a two-level system is governed by the well-known Maxwell–Bloch equations [8]. We can extend these equations to the case of fiber amplifiers. In the formalism presented in Section 2.1 of Ref. [68], the induced polarization $\mathbf{P}(\mathbf{r},t)$ in Eq. (2.1.8) should include a third term, $\mathbf{P}_d(\mathbf{r},t)$ representing the contribution of dopants. This contribution is calculated by using a semiclassical approach in which dopants interact with the optical field $\mathbf{E}(\mathbf{r},t)$ through the induced dipole moment. In the slowly varying envelope approximation, $\mathbf{P}_d(\mathbf{r},t)$ is written as

$$\mathbf{P}_d(\mathbf{r},t) = \tfrac{1}{2}\hat{x}[P(\mathbf{r},t)\exp(-i\omega_0 t)+\text{c.c.}], \tag{4.3.1}$$

where $\hat{x}$ is the polarization unit vector associated with the optical field $\mathbf{E}(\mathbf{r},t)$. The slowly varying part $P(\mathbf{r},t)$ is obtained by solving the Bloch equations, which can be written as [8]

$$\frac{\partial P}{\partial t} = -\frac{P}{T_2} - i(\omega_a - \omega_0)P - \frac{i\mu^2}{\hbar}EW, \tag{4.3.2}$$

$$\frac{\partial W}{\partial t} = \frac{W_0 - W}{T_1} + \frac{1}{\hbar}\text{Im}(E^*P), \tag{4.3.3}$$

where μ is the dipole moment, ω_a is the atomic transition frequency, $W = N_2 - N_1$ is the population-inversion density with the initial value W_0, T_1 and T_2 are the population and dipole relaxation times, and $E(\mathbf{r},t)$ is the slowly varying amplitude associated with the optical field. Following the details presented in Section 2.3 of Ref. [68], the net effect of dopants is to modify the NLS equation as

$$\frac{\partial A}{\partial z} + \beta_1 \frac{\partial A}{\partial t} + \frac{i\beta_2}{2}\frac{\partial^2 A}{\partial t^2} + \frac{\alpha}{2}A = i\gamma|A|^2 A + \frac{i\omega_0}{\varepsilon_0 c}\langle P\exp(-i\beta_0 z)\rangle, \tag{4.3.4}$$

where angle brackets denote spatial averaging over the mode profile $|F(x,y)|^2$. An average over the atomic transition frequencies should also be performed if one wants to include the effects of inhomogeneous broadening.

The set of Maxwell–Bloch equations (4.3.2)–(4.3.4) must be solved for pulses whose width is shorter or comparable to the dipole relaxation time ($T_2 < 0.1$ ps). However, the analysis is simplified considerably for broader optical pulses, since one can make the rate-equation approximation in which the dopants respond so fast that the induced polarization follows the optical field adiabatically [8].

Dispersive effects associated with dopants can be included by working in the Fourier domain and defining the dopant susceptibility through the standard relation

$$\tilde{P}(\mathbf{r},\omega) = \varepsilon_0 \chi_d(\mathbf{r},\omega)\tilde{E}(\mathbf{r},\omega), \tag{4.3.5}$$

where ε_0 is the vacuum permittivity and the tilde represents the Fourier transform. The susceptibility is found to be given by

$$\chi_d(\mathbf{r}, \omega) = \frac{\sigma_s W(\mathbf{r}) n_0 c / \omega_0}{(\omega - \omega_a) T_2 + i},$$ (4.3.6)

where the transition cross section σ_s is related to the dipole moment μ through the realtion $\sigma_s = \mu^2 \omega_0 T_2 / (\varepsilon_0 n_0 \hbar c)$ and n_0 is the background linear refractive index of the host medium at the frequency ω_0.

4.3.2 Ginzburg–Landau Equation

The propagation equation for optical pulses is obtained from Eqs. (4.3.2) through (4.3.6) by following the analysis of Section 2.3 in Ref. [68]. The major change is that $\Delta\beta$ becomes frequency dependent because of the frequency dependence of χ_d. When the optical field is transformed back to the time domain, we must expand both β and $\Delta\beta$ in a Taylor series to include the dispersive effects associated with the dopants. Writing $\omega - \omega_a = (\omega - \omega_0) + (\omega_0 - \omega_a)$ and using the Taylor-series expansion, the resulting equation is given by [69]

$$\frac{\partial A}{\partial z} + \beta_1^{\text{eff}} \frac{\partial A}{\partial t} + \frac{i}{2} \beta_2^{\text{eff}} \frac{\partial^2 A}{\partial t^2} + \frac{1}{2}(\alpha + \alpha_2 |A|^2) A = i\gamma |A|^2 A + \frac{g_0}{2} \frac{1 + i\delta}{1 + \delta^2} A,$$ (4.3.7)

where

$$\beta_1^{\text{eff}} = \beta_1 + \frac{g_0 T_2}{2} \left[\frac{1 - \delta^2 + 2i\delta}{(1 + \delta^2)^2} \right],$$ (4.3.8)

$$\beta_2^{\text{eff}} = \beta_2 + g_0 T_2^2 \left[\frac{\delta(\delta^2 - 3) + i(1 - 3\delta^2)}{(1 + \delta^2)^3} \right],$$ (4.3.9)

and the detuning parameter $\delta = (\omega_0 - \omega_a) T_2$. The gain g_0 is defined as

$$g_0(z,t) = \frac{\sigma_s \iint_{-\infty}^{\infty} W(\mathbf{r}, t) |F(x, y)|^2 \, dx \, dy}{\iint_{-\infty}^{\infty} |F(x, y)|^2 \, dx \, dy},$$ (4.3.10)

where integration is over the entire range of x and y. Equation (4.3.7) includes the effects of two-photon absorption through the parameter α_2. Even though two-photon absorption is negligible for silica fibers, it may become important for fibers made using materials with high nonlinearities [70].

Equation (4.3.7) shows how the dispersion parameters of the host fiber change because of the dopant contribution. Since v_g equals β_1^{-1}, changes in β_1 indicate that the group velocity of the pulse is affected by the dopants. However, the dopant-induced change in the group velocity is negligible in practice because the second term in Eq. (4.3.8) is smaller by more than a factor of 10^4 under typical operating conditions. In contrast, changes in β_2 cannot be neglected, as the two terms in Eq. (4.3.9) can become comparable, especially near the zero-dispersion wavelength of the amplifier. Even in the special case $\delta = 0$, β_2^{eff} does not reduce to β_2. In fact, Eq. (4.3.9) shows that for $\delta = 0$,

$$\beta_2^{\text{eff}} = \beta_2 + i g_0 T_2^2$$ (4.3.11)

is a complex parameter whose imaginary part results from the gain provided by dopants. The physical origin of this contribution is related to the finite gain bandwidth of fiber amplifiers and is referred to as *gain dispersion*, since it originates from the frequency dependence of the gain. Equation (4.3.11) is a consequence of the parabolic-gain approximation in which the gain spectrum of fiber amplifiers is approximated by a parabola over the spectral bandwidth of the pulse.

It is difficult to perform integration in Eq. (4.3.10) because the inversion W depends not only on the spatial coordinates x, y, and z but also on the mode profile $|F(x,y)|^2$ owing to gain saturation. In practice, only a small portion of the fiber core is doped with rare-earth ions. If the mode intensity and the dopant density are nearly uniform over the doped portion, W can be assumed to be a constant in the doped region and zero outside it. The integration is then readily performed to yield the simple relation

$$g_0(z,t) = \Gamma_s \sigma_s W(z,t), \tag{4.3.12}$$

where Γ_s represents the fraction of mode power within the doped portion of the fiber. Using Eqs. (4.3.3) and (4.3.12), the gain dynamics is governed by

$$\frac{\partial g_0}{\partial t} = \frac{g_{ss} - g_0}{T_1} - \frac{g_0 |A|^2}{T_1 P_s^{\text{sat}}}, \tag{4.3.13}$$

where $g_{ss} = \Gamma_s \sigma_s W_0$ is the small-signal gain and the saturation power P_s^{sat} is defined in Eq. (4.2.5). Note that g_0 is not constant along the fiber length because of pump power variations. The z dependence of g_0 depends on the pumping configuration and requires the use of Eq. (4.2.6).

In general, one must solve Eqs. (4.3.7) and (4.3.13) together in a self-consistent manner. However, for most fiber amplifiers, the fluorescence time T_1 is so long (0.1–10 ms) compared with typical pulse widths that we can assume that spontaneous emission and pumping do not occur over the pulse duration. Equation (4.3.13) is then readily integrated to obtain the result

$$g_0(z,t) \approx g_{ss} \exp\left(-\frac{1}{E_s} \int_{-\infty}^{t} |A(z,t)|^2 dt\right), \tag{4.3.14}$$

where the saturation energy is defined as $E_s = \hbar\omega_0(a_s/\sigma_s)$. Typical values of E_s for fiber amplifiers are close to 1 μJ. However, pulse energies used in practice are much smaller than the saturation energy E_s. As a result, gain saturation is negligible over the duration of a single pulse. However, it cannot be neglected for a long pulse train because the amplifier gain will saturate over timescales longer than T_1. The average power within the amplifier then determines the saturated gain as $g_0 = g_{ss}(1 + P_{\text{av}}/P_s^{\text{sat}})^{-1}$.

Pulse propagation in fiber amplifiers is thus governed by a generalized NLS equation, with coefficients β_1^{eff} and β_2^{eff} that are not only complex but also vary with z along the fiber length. In the specific case in which $\delta = 0$, Eq. (4.3.7) simplifies considerably and can be written as

$$\frac{\partial A}{\partial z} + \frac{i}{2}(\beta_2 + ig_0 T_2^2)\frac{\partial^2 A}{\partial T^2} = i\left(\gamma + \frac{i}{2}\alpha_2\right)|A|^2 A + \frac{1}{2}(g_0 - \alpha)A, \tag{4.3.15}$$

where $T = t - \beta_1^{\text{eff}} z$ is the reduced time. This equation governs amplification of optical pulses in fiber amplifiers. The T_2 term accounts for decrease in gain for spectral components of an optical pulse located far from the gain peak. Equation (4.3.15) is a generalized NLS equation with complex coefficients. It can be reduced to a Ginzburg–Landau equation, which has been studied extensively in the context of fluid dynamics. We discuss in the next section the stability of its steady-state solutions.

4.4 Modulation Instability

Modulation instability, discussed in Section 5.1 of Ref. [68], should play an important role if a CW beam propagates in a distributed fiber amplifier in which amplification occurs along long fiber lengths. Also, a new type of modulation instability can occur if signals are periodically amplified in a chain of short-length amplifiers, a situation that occurs in soliton communication systems (see Section 7.3). In this section we discuss the two cases separately.

4.4.1 Distributed Amplification

Consider the propagation of CW or quasi-CW signals inside a long fiber amplifier. The steady-state solution can be obtained by neglecting the time-derivative term in Eq. (4.3.15). Assuming for simplicity that g_0 is z independent, the solution is given by

$$\bar{A}(z) = \sqrt{P_0} \exp[b(z)], \tag{4.4.1}$$

where P_0 is the incident power and

$$b(z) = \frac{1}{2}(g_0 - \alpha)z + i\gamma P_0 \int_0^z \exp[(g_0 - \alpha)z]dz. \tag{4.4.2}$$

Equation (4.4.1) shows that the CW signal is amplified exponentially and acquires a nonlinear phase shift induced by self-phase modulation (SPM).

Following a standard procedure, we perturb the steady state slightly such that

$$A(z,T) = [\sqrt{P_0} + a(z,T)] \exp[b(z)] \tag{4.4.3}$$

and examine the evolution of the weak perturbation $a(z,T)$ using a linear stability analysis. By substituting Eq. (4.4.3) in Eq. (4.3.15) and linearizing in a, we obtain a linear equation that can be solved approximately and has a solution in the form

$$a(z,T) = a_1 \exp[i(\int_0^z K(z)dz - \Omega T)] + a_2 \exp[-i(\int_0^z K(z)dz - \Omega T)], \tag{4.4.4}$$

where Ω is the frequency of perturbation. The wave number K is z dependent because of the gain provided by the amplifier and is found to satisfy the following dispersion relation [71]:

$$K(\Omega, z) = \frac{1}{2} i g_0 T_2^2 \Omega^2 \pm \frac{1}{2}|\beta_2 \Omega|[\Omega^2 + (4\gamma P_0 / \beta_2)e^{(g_0 - \alpha)z}]^{1/2}. \tag{4.4.5}$$

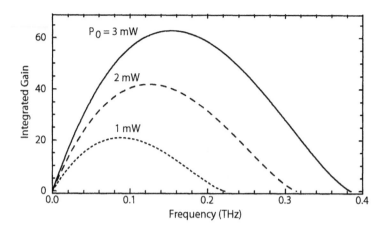

Figure 4.8: Gain spectra of modulation instability at three power levels for a distributed fiber amplifier with 30-dB gain over 10 km. Values of other parameters are $\beta_2 = -20$ ps^2/km and $\gamma = 10$ W^{-1}/km.

The dispersion relation (4.4.5) reduces to that obtained for undoped fibers in Section 5.1 of Ref. [68] when gain and loss are neglected. Modulation instability occurs when $K(\Omega, z)$ has a negative imaginary part over a large length of the fiber amplifier. It is useful to define the total integrated gain at a frequency Ω as

$$h(\Omega) = -2 \int_0^L \text{Im}[K(\Omega, z)]dz, \qquad (4.4.6)$$

where L is the amplifier length. Stability of the steady state depends critically on whether light experiences normal or anomalous GVD inside the amplifier. In the case of normal GVD, $h(\Omega)$ is negative for all values of Ω, and the steady state is stable against small perturbations.

The situation is quite different in the case of anomalous GVD ($\beta_2 < 0$). Similar to the case of undoped fibers, $h(\Omega)$ becomes positive in a certain range of Ω. Figure 4.8 shows the gain spectrum of modulation instability by plotting $h(\Omega)$ at three input power levels for a fiber amplifier with 30-dB gain distributed over a length of 10 km. Modulation instability occurs for input power levels of about 1 mW. It can transform a CW beam into a pulse train at a repetition rate around 100 GHz.

4.4.2 Periodic Lumped Amplification

Most long-haul fiber-optic communication systems use optical amplifiers in which the doped fiber is only a few meters long. The length of such amplifiers is much shorter than both the dispersion and nonlinear length scales. In essence, the role of a fiber amplifier is to amplify the signal without introducing any temporal or spectral changes. Such amplifiers are called *lumped amplifiers* since they amplify the signal by a factor of 20 dB or so over a length of about 10 m and compensate for fiber losses acquired over a distance as large as 100 km. With this scheme, optical signals can be transmitted over distances ~ 1000 km by simply placing multiple amplifiers periodically along the fiber link.

Modulation instability affects the performance of such periodically amplified fiber-optic communication systems in several ways. As early as 1990, computer simulations showed that modulation instability can be a limiting factor for lightwave systems employing the nonreturn-to-zero (NRZ) format for data transmission [72]. Since then, the impact of modulation instability has been studied, both analytically and experimentally, for single-channel as well as WDM systems [73]–[83].

The use of optical amplifiers can induce modulation instability in both the normal and anomalous GVD regimes of optical fibers because of the periodic nature of amplification [74]. The new instability mechanism has its origin in the periodic sawtooth variation of the optical power along the link length. To understand the physics more clearly, note that a periodic variation of power in z is equivalent to formation of an index grating, since the nonlinear part of the refractive index depends on the local power level. The period of this grating is equal to the amplifier spacing and is typically in the range of 40 to 80 km. Such a long-period grating provides a new coupling mechanism between the modulation-instability sidebands and allows them to grow when the perturbation frequency Ω satisfies the Bragg condition.

Mathematically, the evolution of the optical field outside fiber amplifiers is governed by the standard NLS equation

$$i\frac{\partial A}{\partial z} - \frac{\beta_2}{2}\frac{\partial^2 A}{\partial T^2} + \gamma|A|^2 A = -\frac{i\alpha}{2}A, \qquad (4.4.7)$$

where α accounts for fiber losses. Within each amplifier, $-\alpha$ is replaced by the net gain g_0, and the dispersive and nonlinear effects are negligible. By introducing a new variable B through $A = B\exp(-\alpha z/2)$, Eq. (4.4.7) can be written as

$$i\frac{\partial B}{\partial z} - \frac{\beta_2}{2}\frac{\partial^2 B}{\partial T^2} + \gamma f(z)|B|^2 B = 0, \qquad (4.4.8)$$

where $f(z)$ is a periodic function such that it decreases exponentially as $f(z) = \exp(-\alpha z)$ in each fiber section between amplifiers and jumps to 1 at the location of each amplifier.

The preceding analysis can be extended to include periodic variations of $f(z)$. If we expand $f(z)$ in a Fourier series as

$$f(z) = \sum_{n=-\infty}^{\infty} c_n \exp(2\pi i n z/L_A), \qquad (4.4.9)$$

the frequency at which the gain of modulation instability peaks is given by [74]

$$\Omega_m = \pm\left(\frac{2\pi m}{\beta_2 L_A} - \frac{2\gamma P_0 c_0}{\beta_2}\right)^{1/2}, \qquad (4.4.10)$$

where the integer m represents the order of Bragg diffraction, L_A is the spacing between amplifiers (grating period), and the Fourier coefficient c_0 is related to the fiber loss α, or the amplifier gain $G \equiv \exp(\alpha L_A)$, as

$$c_0 = \frac{1 - \exp(-\alpha L_A)}{\alpha L_A} = \frac{G-1}{G \ln G}. \qquad (4.4.11)$$

In the absence of periodic gain–loss variations or when $m = 0$, Ω_0 exists only when the CW signal experiences anomalous GVD. However, when $m \neq 0$, modulation sidebands can occur even for normal GVD ($\beta_2 > 0$). For this reason, this instability is referred to as *sideband instability*. Physically, the creation of sidebands can be understood by noting that the nonlinear index grating helps to satisfy the phase-matching condition necessary for four-wave mixing when $m \neq 0$. This phenomenon can be avoided in practice by ensuring that the amplifier spacing is not uniform along the fiber link.

4.4.3 Noise Amplification

Modulation instability can degrade the system performance considerably in the presence of noise produced by optical amplifiers. Physically, spontaneous emission within fiber amplifiers adds broadband noise to the amplified signal. This noise can seed the growth of modulation-instability sidebands and is thus amplified through *induced modulation instability* [77]–[83]. Such noise amplification affects system performance in two ways. First, it degrades the SNR at the receiver. Second, it broadens the signal spectrum. Since GVD-induced broadening of optical signals depends on their spectral bandwidth, system performance is compromised.

We can study the noise amplification process in each section of the fiber between two optical amplifiers by adding noise to the CW solution of Eq. (4.4.8) so that

$$B(z,T) = [\sqrt{P_0} + a(z)e^{i\Omega T}] \exp(i\phi_{NL}), \qquad (4.4.12)$$

where $\phi_{NL} = \gamma P_0 \int_0^z f(z) dz$ is the SPM-induced nonlinear phase shift and $a(z)$ is the noise amplitude at the frequency Ω. Substituting Eq. (4.4.12) in Eq. (4.4.8), we obtain

$$\frac{da}{dz} = \frac{i}{2}\beta_2\Omega^2 + i\gamma P_0 f(z)(a + a^*). \qquad (4.4.13)$$

This equation can be solved easily in the lossless case in which $\alpha = 0$ and $f(z) \equiv 1$ is z independent [77]. It can also be solved when $\alpha \neq 0$, but the solution is quite complicated as it involves the Hankel functions [79]. An approximate solution is obtained when $f(z)$ is replaced by its average value c_0 and is given by [83]

$$\begin{pmatrix} a_1(z) \\ a_2(z) \end{pmatrix} = \begin{pmatrix} \cos(Kz) & -r_0^{-1}\sin(Kz) \\ r_0\sin(Kz) & \cos(Kz) \end{pmatrix} \begin{pmatrix} a_1(0) \\ a_2(0) \end{pmatrix}, \qquad (4.4.14)$$

where a_1 and a_2 are the real and imaginary parts of the noise amplitude ($a = a_1 + ia_2$) and K and r_0 are defined as

$$K = \tfrac{1}{2}\beta_2\Omega^2 r_0, \qquad r_0 = [1 + 4\gamma P_0 c_0/(\beta_2\Omega^2)]^{1/2}. \qquad (4.4.15)$$

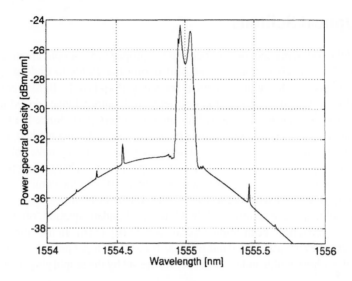

Figure 4.9: Optical spectrum for a 2500-km fiber link with 50 amplifiers showing effects of modulation instability. Values of fiber parameters are $\beta_2 = -1$ ps^2/km, $\gamma = 2$ W^{-1}/km, and $\alpha = 0.22$ dB/km. (From Ref. [83]; ©1999 IEEE.)

Fiber amplifiers generate noise over their entire gain bandwidth (typically > 30 nm). Frequency components of the noise that fall within the gain spectrum of modulation instability are amplified exponentially since r_0, and hence the propagation constant K becomes purely imaginary for them. In practice, optical filters are placed just after each amplifier to reduce the noise bandwidth. Figure 4.9 shows an example of a numerically simulated optical spectrum at the end of a 2500-km fiber link with 50 amplifiers placed 50 km apart [83]. A 1-mW signal at the 1.55-μm wavelength is transmitted through the amplifier chain. Optical filters with a 8-nm passband (Lorentzian shape) are placed after every amplifier. The broad pedestal represents the contribution of ASE to the signal spectrum located at 1.55 μm. The double-peak structure near this wavelength is due to the standard modulation instability occurring even in the absence of amplifiers. The weak satellite peaks result from the nonlinear index grating formed by periodic power variations. Their location is in agreement with the prediction of Eq. (4.4.10).

The enhancement of amplifier noise degrades the SNR of the signal at the receiver. Experimental results for a lightwave system operating at 10 Gb/s showed considerable degradation after a transmission distance of only 455 km [80]. The system performance improved considerably when the GVD was partially compensated using a dispersion-compensating fiber. In the case of WDM systems, a signal in one channel can act as a seed for induced modulation instability, resulting in interchannel crosstalk if channel spacing falls within the gain bandwidth of the instability. This phenomenon, called *resonant FWM* can occur because of SPM-mediated phase matching, in spite of large GVD [82]. In general, the impact of modulation instability on lightwave systems can be reduced by reducing amplifier spacing and by using the normal-GVD regime of the fiber link for signal transmission.

4.5   Optical Solitons

In this and the following section we focus on the propagation of optical pulses in fiber amplifiers. Considerable attention was paid during the 1990s to understanding the amplification process through theoretical modeling [84]–[108]. Before considering amplification of ultrashort pulses, it is instructive to inquire whether doped fibers can support solitons similar to those found for undoped fibers [68]. Since the Ginzburg–Landau equation (4.3.15) is not integrable by the inverse scattering method, it does not support solitons in a strict mathematical sense. However, it may have solitary-wave solutions that represent optical pulses whose shape does not change on propagation. Such a solution of Eq. (4.3.15) was found as early as 1977 in the context of fluid dynamics [109]; it was rediscovered in 1989 in the context of nonlinear fiber optics [84]. Since then, solitary-wave solutions of the Ginzburg–Landau equation, often called the *dissipative solitons*, have been studied extensively [110]–[116]. In the context of optical amplifiers, they are also called *autosolitons* because all input pulses evolve toward a specific pulse whose width and other properties are set by the amplifying medium [87].

4.5.1   Properties of Autosolitons

Similar to the case of conventional solitons in undoped fibers, it is useful to introduce the dimensionless variables (soliton units)

$$\xi = z/L_D, \qquad \tau = T/T_0, \qquad u = \sqrt{\gamma L_D}A, \tag{4.5.1}$$

where $L_D = T_0^2/|\beta_2|$ is the dispersion length. Equation (4.3.15) then takes the normalized form [92]

$$i\frac{\partial u}{\partial \xi} - \frac{1}{2}(s + id)\frac{\partial^2 u}{\partial \tau^2} + (1 + i\mu_2)|u|^2 u = \frac{i}{2}\mu u, \tag{4.5.2}$$

where $s = \mathrm{sgn}(\beta_2) = \pm 1$ and the other parameters are defined as

$$d = g_0 L_D (T_2/T_0)^2, \qquad \mu = (g_0 - \alpha)L_D, \qquad \mu_2 = \alpha_2/2\gamma. \tag{4.5.3}$$

Equation (4.5.2) reduces to the standard NLS equation when the three parameters d, μ, and μ_2 are set to zero. Physically, d is related to the amplifier bandwidth (through the parameter T_2), μ is related to the amplifier gain, and μ_2 governs the effect of two-photon absorption. Numerical values of these parameters for most EDFAs are such that $\mu \sim 1$, $d \sim 10^{-3}$, and $\mu_2 \sim 10^{-4}$ when $T_0 \sim 1$ ps.

   An extended version of Eq. (4.5.2), known as the *cubic–quintic Ginzburg–Landau equation*, has also attracted considerable attention [111]–[116]. It adds a fifth-order term $\varepsilon|u|^4 u$ to Eq. (4.5.2), where ε is a constant parameter that may be complex in general. Physically, the quintic term results from saturation of the fiber nonlinearity and is negligible for silica fibers at practical power levels. For this reason, its effects are not considered in this chapter.

   Since the inverse scattering method is not applicable, the solitary-wave solutions of Eq. (4.5.2) are found by trial and error. In this method, an analytic form of the solution is guessed, and the constants are adjusted to satisfy Eq. (4.5.2). An appropriate functional form of the solitary-wave solution of this equation is [109]

$$u(\xi, \tau) = N_s [\text{sech}(p\tau)]^{1+iq} \exp(iK_s\xi). \tag{4.5.4}$$

The constants N_s, p, q, and K_s are determined by substituting this solution in Eq. (4.5.2) and are

$$N_s^2 = \tfrac{1}{2}p^2[s(q^2 - 2) + 3qd], \tag{4.5.5}$$

$$p^2 = \mu[d(q^2 - 1) - 2sq]^{-1}, \tag{4.5.6}$$

$$K_s = \tfrac{1}{2}p^2[s(q^2 - 1) + 2qd], \tag{4.5.7}$$

where q is a solution of the quadratic equation

$$(d - \mu_2 s)q^2 - 3(s + \mu_2 d)q - 2(d - \mu_2 s) = 0. \tag{4.5.8}$$

The general solution in Eq. (4.5.4) exists for both the positive and negative values of β_2. It is easy to verify that, when $s = -1$ (anomalous GVD) and d, μ, and μ_2 are set to zero, this solution reduces to the standard soliton of the NLS equation [68]. The parameter p remains undetermined in that limit because the NLS equation supports a whole family of fundamental solitons such that $N_s = p$. By contrast, both p and N_s are fixed for the Ginzburg–Landau equation by the amplifier parameters μ and d. This is a fundamental difference introduced by the dopants: Fiber amplifiers select a single soliton from the entire family of solitons supported by the undoped fiber. The width and the peak power of this soliton are uniquely determined by the amplifier parameters (such as its gain and bandwidth), justifying the name *autosoliton* given to such pulses [87].

An important property of autosolitons is that, unlike conventional NLS solitons, they represent chirped pulses. This feature is seen clearly from Eq. (4.5.4) by noting that the phase of the soliton becomes time dependent when $q \neq 0$. In fact, Eq. (4.5.4) can be written as

$$u(\xi, \tau) = N_s \text{sech}(p\tau) \exp[iK_s\xi - iq\ln(\cosh p\tau)]. \tag{4.5.9}$$

By defining the frequency chirp as $\delta\omega = -\partial\phi/\partial\tau$, the chirp is given by

$$\delta\omega(\tau) = qp\tanh(p\tau). \tag{4.5.10}$$

The parameter q governs the magnitude of chirp. As seen from Eq. (4.5.8), $q \neq 0$ only when d or μ_2 are nonzero. For silica fiber amplifiers, μ_2 is small enough that it can be set to zero. The parameter q is then given by

$$q = [3s \pm (9 + 8d^2)^{1/2}]/2d, \tag{4.5.11}$$

where the sign is chosen such that both p and N_s are real.

The existence of solitons in a fiber amplifier is somewhat surprising. For the soliton to preserve its shape and energy in spite of the gain provided by the amplifier, a loss mechanism must exist. Both gain dispersion and two-photon absorption provide such

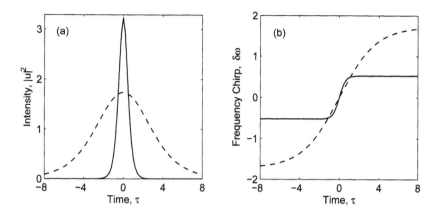

Figure 4.10: (a) Intensity and (b) chirp profile of an autosoliton when $d = 0.5$. Solid and dashed curves correspond to the cases of normal and anomalous GVD, respectively.

a loss mechanism. Although the role of two-photon absorption as a loss mechanism is easily understood, it is not obvious how gain dispersion leads to loss. Because of a finite gain bandwidth, the frequency dependence of the gain is such that spectral wings of an optical pulse experience less gain—and may even experience loss—if the pulse spectrum becomes wider than the gain bandwidth. Thus, gain dispersion can act as a loss mechanism for pulses with a wide spectrum. The frequency chirp imposed on the soliton of Eq. (4.5.9) helps to maintain balance between gain and loss since it can tailor the pulse spectrum through the chirp parameter q. This is why autosolitons are chirped. This mechanism also explains why amplifier solitons can exist even in the normal-GVD region of a doped fiber.

Figure 4.10 compares the intensity and chirp profiles of an amplifier soliton in the cases of normal (dashed curve) and anomalous (solid curve) GVD using $d = 0.5$, $\mu = 0.5$, and $\mu_2 = 0$. In both cases, the chirp is nearly linear over most of the intensity profile, but the soliton is considerably broader in the case of normal GVD. Dependence of soliton parameters on the gain-dispersion parameter d is shown in Figure 4.11, where the width parameter p^{-1} and the chirp parameter q are plotted as a function of d using $\mu = d$ and $\mu_2 = 0$. Solid and dashed curves correspond to the cases of normal ($s = 1$) and anomalous ($s = -1$) GVD, respectively. For large values of d, the difference between normal and anomalous GVD disappears since the soliton behavior is determined by gain dispersion (rather than index dispersion of the silica host). In contrast, both the width and chirp parameters are much larger in the case of normal GVD when $d < 1$. Indeed, both of these parameters tend to infinity as $d \to 0$ since undoped fibers do not support bright solitons in the case of normal GVD. In the presence of two-photon absorption, the soliton amplitude decreases and its width increases. For most fiber amplifiers μ_2 is so small that its effects can be ignored.

Since gain dispersion and two-photon absorption permit the existence of bright solitons in the normal-GVD region, one is justified in asking whether the Ginzburg–Landau equation has solutions in the form of dark solitary waves that exist in both the

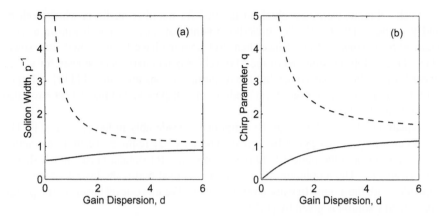

Figure 4.11: (a) Soliton width p^{-1} and (b) chirp parameter q plotted as a function of d. Solid and dashed curves correspond to the cases of normal and anomalous GVD, respectively.

normal- and anomalous-GVD regions. This turns out to be the case. Recalling that $\text{sech}(\tau)$ is replaced by $\tanh(\tau)$ for dark solitons in undoped fibers [68], a simple guess replaces Eq. (4.5.9) with

$$u(\xi, \tau) = N_s \tanh(p\tau) \exp[iK_s \xi - iq\ln(\cosh p\tau)].\qquad(4.5.12)$$

Equation (4.5.12) is indeed a solution of the Ginzburg–Landau equation [84]. The parameters N_s, p, q, and K_s are determined by a set of equations similar to Eqs. (4.5.5)–(4.5.8). The qualitative behavior of dark solitons is also similar to that of bright solitons governed by Eq. (4.5.9). In particular, gain dispersion determines the frequency chirp imposed on the dark soliton.

Just as modulation instability can destabilize the CW solution, a solitary-wave solution of the Ginzburg–Landau equation can become unstable under some conditions. For this reason, the stability of such solutions has been studied extensively [112]. It is evident from Eqs. (4.5.4)–(4.5.8) that the parameters N_s and p can have real positive values only in a certain range of the three parameters μ, d, and μ_2 for $s = \pm 1$. A stable autosoliton exists only when $\mu > 0$ and amplifier parameters are such that both N_s and p are positive numbers. However, when $\mu > 0$, the background is not stable because any small fluctuation can be amplified by the fiber gain. This instability of background noise has important implications for fiber amplifiers and lasers.

4.5.2 Maxwell–Bloch Solitons

The soliton solution in Eq. (4.5.9) shows that the width of autosolitons can become comparable to the dipole relaxation time T_2 (see Figure 4.11). The validity of the Ginzburg–Landau equation then becomes questionable, because the rate-equation approximation used in its derivation [see Eq. (4.3.5)] becomes invalid under such conditions. In its place, we should look for the solitary-wave solutions of the Maxwell–Bloch equations themselves by solving Eqs. (4.3.2)–(4.3.4).

Such solutions were first obtained in 1967 for nonfiber media in which both β_2 and γ are negligible [117]. The underlying nonlinear phenomenon is known as *self-induced transparency* (SIT). Since then, SIT solitons have been studied extensively [118]–[122]. Soliton solutions of the Maxwell–Bloch equations exist even for a nonlinear host (without dispersion), but the resulting solitons are chirped [119]. Chirped solitons for an amplifying two-level medium in a dispersive nonlinear host have also been obtained [121].

Equations (4.3.2)–(4.3.4) can be simplified considerably in the case of SIT. The terms containing T_1 and T_2 can be neglected because SIT requires coherent interaction between atoms and the optical field that occurs only for optical pulses much shorter than T_1 and T_2. The amplitude $A(z,t)$ can be assumed real if the laser frequency ω_0 coincides with the atomic transition frequency ω_a. For a two-level absorber (no pumping), the SIT soliton is given by [117]

$$A(z,t) = N_0 \text{sech}\left(\frac{t - z/V}{\tau_p}\right),\tag{4.5.13}$$

where the pulse velocity V and the pulse width τ_p are related as

$$\frac{1}{V} = \frac{1}{c} + \frac{\rho\mu^2\omega_0}{2\varepsilon_0 hc}\int_{-\infty}^{\infty}\frac{\tau_p^2 h(\Delta)d\Delta}{1 + (\Delta\tau_p)^2}.\tag{4.5.14}$$

In this equation, ρ is the atomic density, $\Delta = \omega - \omega_0$, and $h(\Delta)$ is the distribution function for an inhomogeneously broadened two-level system.

Equation (4.5.13) shows that a "sech" pulse can propagate without change in its shape, width, or amplitude—even in an absorbing medium—provided that its input amplitude N_0 is related to its width to form a 2π pulse [122]. The effect of absorption is to slow down the optical pulse. Indeed, the soliton velocity V may be reduced by several orders of magnitude ($V/c \sim 10^{-3}$). Physically, the pulse slows down because of continuous absorption and emission of radiation occurring inside the medium. Qualitatively speaking, energy is absorbed from the leading edge of the pulse and is emitted back near the trailing edge. For the pulse amplitude given in (4.5.13), the two processes can occur coherently in such a manner that the pulse shape remains unchanged on propagation. In essence, the role of dispersion is played by absorption for SIT solitons. Similar to the case of fiber solitons, SIT solitons describe an entire family of solitons whose width and group velocities are related by Eq. (4.5.14).

The situation becomes much more interesting in the case of fiber amplifiers in which the two kinds of solitons can exist simultaneously. The silica host supports NLS solitons, whereas the dopants can support SIT solitons. The question thus arises whether both types of solitons can exist simultaneously inside a fiber amplifier. To answer this question, we should look for soliton solutions of Eqs. (4.3.2)–(4.3.4) in the coherent limit in which the terms containing T_1 and T_2 can be neglected [93]–[101]. Detuning effects can be ignored by setting $\omega_a = \omega_0$. It turns out that the solution given in Eq. (4.5.13) remains valid but group velocity of the soliton is given by [94]

$$\frac{1}{V} = \left(\frac{1}{v_g^2} + \frac{2n_2 n_0 \omega_0^2 h^2}{\mu^2 c^2}\right)^{1/2},\tag{4.5.15}$$

where v_g is the group velocity in the undoped fiber.

A remarkable feature of Eq. (4.5.15) is that the soliton velocity depends on the nonlinear parameter n_2, but it is independent of the dopant density and the soliton width. Another noteworthy feature is that both the width and the peak power of the soliton are uniquely determined by the amplifier. More specifically, the peak power and the width of the soliton must satisfy not only the fundamental soliton condition $N = 1$ but also the SIT condition that the pulse area equals 2π [95]. Such an SIT soliton exists for both normal and anomalous GVD. The situation is similar to that occurring for autosolitons [see Eq. (4.5.4)] in the sense that a single soliton is selected from the entire family of SIT solitons. The surprising feature is that an SIT soliton can be chirp free, in contrast to the solitary-wave solution of the Ginzburg–Landau equation.

Experimental realization of SIT solitons is difficult in practice because of a relatively small value of the dipole relaxation time ($T_2 \sim 100$ fs) for the dopants inside an optical fiber. In the coherent regime, the soliton width should be smaller than T_2. The required peak power for such a 2π pulse is prohibitively large ($P_0 > 1$ GW). Nevertheless, coherent effects associated with the SIT solitons were observed in a 1992 experiment in which an EDFA was cooled down to 4.2 K [123]. Cooling of a doped fiber to such low temperatures increases T_2 by orders of magnitude ($T_2 \sim 1$ ns) because of reduced phonon-related effects. As a result, SIT solitons can be observed by using pulse widths of about 100 ps and peak powers of about 10 W. Indeed, 400-ps pulses with peak power levels of about 50 W were used in the 1991 experiment. When such pulses were propagated inside a 1.5-m-long cooled fiber, they formed the SIT soliton when their peak power was large enough to form the 2π pulse. Another coherent effect, known as the *photon echo*, has also been observed in EDFAs cooled to liquid-helium temperatures [124].

The study of SIT solitons in optical fibers has been extended in several directions [125]–[131]. As early as 1998, SIT solitons were predicted to form inside a photonic bandgap medium (such as a fiber grating) doped with two-level atoms to provide optical gain [126]. Such solitons propagated close to but outside of the photonic bandgap. It was found later that even a pulse whose spectrum lies inside the stop band can be transmitted through the medium if it coexists simultaneously as a SIT soliton and a Bragg soliton [128]. In a 2006 study, the core of a photonic crystal fiber was assumed to be filled with a Raman-active medium that generated a Stokes wave through stimulated Raman scattering when a pump pulse was launched inside the fiber [131]. It was found that both the pump and Stokes pulses could propagate as SIT solitons under suitable conditions. Such solitons are distinct from the NLS-type solitons because their formation does not require dispersion. In general, the use of highly nonlinear fibers is advantageous for observing the nonlinear effects inside fiber amplifiers.

4.6 Pulse Amplification

Amplification of short optical pulses can be studied by solving the Ginzburg–Landau equation numerically. Since that equation is valid only for pulses of duration $T_0 \gg T_2$, picosecond pulses are considered first; femtosecond pulses require the use of full

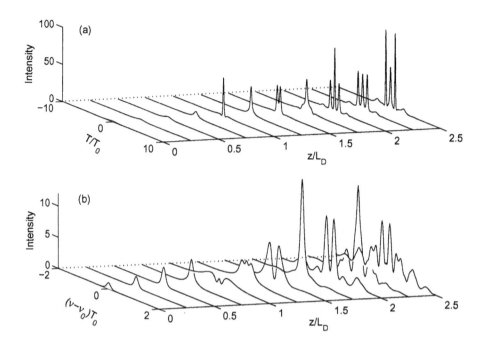

Figure 4.12: Evolution of a fundamental soliton ($N = 1$) in a fiber amplifier with parameters $d = 0.092$, $\mu = 2.3$, and $\mu_2 = 0$.

Maxwell–Bloch equations. The split-step Fourier method (see Section 2.4 of Ref. [68]) can be easily extended for solving these equations.

4.6.1   Anomalous-Dispersion Regime

We first focus on the case in which the input pulse propagates in the anomalous-GVD regime of the fiber amplifier. Since soliton effects govern the amplification process in this regime, we take the input amplitude in the form $A(0,t) = \sqrt{P_0}\,\mathrm{sech}(t/T_0)$, and assume that the pulse width T_0 is related to the peak power P_0 such that the pulse will propagate as a fundamental soliton in the absence of dopants and fiber losses. The evolution of such a pulse inside a fiber amplifier is studied by solving Eq. (4.5.2) numerically with the input $u(0,\tau) = \mathrm{sech}(\tau)$. The amplification process depends strongly on the value of the gain parameter μ. One can distinguish two regions depending on whether $\mu \ll 1$ or becomes comparable to or larger than 1. In the former case, the soliton is amplified adiabatically. In fact, one can treat Eq. (4.5.2) as a perturbed NLS equation (since all three parameters d, μ, and μ_2 are much less than 1) and apply soliton perturbation theory [98]. In essence, the soliton adjusts its parameters adiabatically and evolves toward the chirped amplifier soliton given in Eq. (4.5.9).

Practical fiber amplifiers can provide gains of 30 dB or more over a length of only a few meters. For such high-gain amplifiers, the parameter μ can easily exceed 1 and the amplification process is not adiabatic. Figure 4.12 shows the evolution of pulse shape and spectrum for a fundamental soliton over a distance of $2.5L_D$ for an EDFA pumped to provide 10-dB gain over each dispersion length $[\exp(\mu) = 10$ or $\mu \approx 2.3]$. The width T_0 is chosen such that $T_2/T_0 = 0.2$ $(d = 0.092)$. Two-photon absorption is neglected by setting $\mu_2 = 0$. The input pulse is compressed by a large factor at $z = L_D$, a feature that can be used to simultaneously amplify and compress ultrashort optical pulses by passing them through an EDFA. The soliton develops additional structure in the form of subpulses as it propagates beyond $z = 1.5L_D$ [69].

The compression stage seen in Figure 4.12 is similar to that occurring for higher-order solitons and can be understood by noting that the initial stage of amplification raises the peak power such that N exceeds 1. As discussed in Section 5.2 of Ref. [68], the pulse tries to maintain $N = 1$ by reducing its width. As long as the amplification process remains adiabatic, this process continues and the pulse width keeps decreasing, as seen in Figure 4.12 for distances up to $\xi = z/L_D = 1$. However, by that time, the pulse has become so short and its spectrum has become so broad (comparable to the gain bandwidth) that the effects of gain dispersion take over. Gain dispersion reduces the spectral bandwidth and broadens the pulse in the propagation region beyond $\xi >$ 1.5. Both spectral broadening and narrowing are clearly evident in Figure 4.12 in the range between $\xi = 1$ and 2. The pulse spectrum develops multiple peaks for $\xi > 2$.

The amplified pulse is also considerably chirped because of SPM and gain dispersion. The chirp profile is similar to that expected from SPM alone for $\xi < 1$, but it develops rapid oscillations beyond that distance because of the formation of subpulses seen in Figure 4.12. The number of such subpulses grows with further propagation. Figure 4.13 displays pulse shapes at $\xi = 3$, 3.5, 4, and 4.5 and shows how new subpulses are created continuously. Each subpulse, once it has stabilized, has nearly the same width and about the same amplitude. Spacing among subpulses changes during the formation of new subpulses, but it eventually becomes nearly uniform, as seen by the temporal trace at a distance $\xi = 4$. These features can be understood qualitatively in terms of chirped autosolitons. The width and the peak power (parameters p and N_s) of such solitons are fixed by the amplifier parameters [μ and d in Eq. (4.5.2)]. Thus, the input pulse evolves toward such a soliton by reducing its width and increasing its peak power (see Figure 4.12). However, during this process it sheds a part of its energy as dispersive waves. Because of the gain provided by the amplifier and instability of the background, parts of the dispersive wave can grow and evolve toward another chirped soliton. This mechanism explains continuous generation of subpulses during the amplification process.

In the absence of soliton interactions, each subpulse will correspond to the solitary-wave solution given in Eq. (4.5.9). However, soliton interactions cannot be ignored, especially for chirped solitons, for which pulses and chirp profiles overlap considerably. It is this interaction of chirped solitons that leads to the oscillatory structure in the pulse spectra of Figure 4.12. This effect has been studied for the Ginzburg–Landau equation by using perturbation theory [132]. The results show that the origin of multiple-pulse solitons, similar to those seen in Figure 4.13, lies in the frequency chirp associated with such solitons.

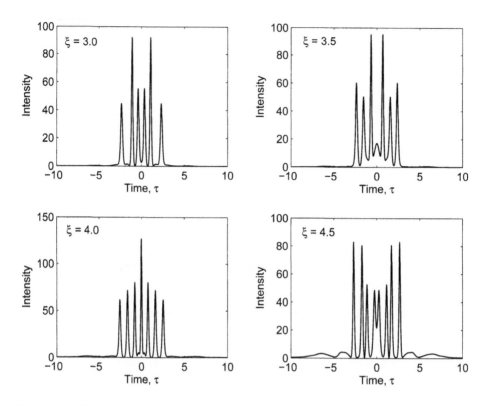

Figure 4.13: Temporal structure of the amplified pulse at $\xi = 3$, 3.5, 4, and 4.5 for the same value of parameters used in Figure 4.12.

4.6.2 Normal-Dispersion Regime

This subsection focuses on the case in which optical pulses propagate in the normal-dispersion region of a fiber amplifier. Solitary-wave solutions of the Ginzburg–Landau equation exist in the form of a chirped soliton even in the case of normal GVD ($\beta_2 > 0$). Thus, one should expect an input pulse to evolve toward this soliton, similar to the case of anomalous GVD. Numerical simulations confirm this expectation in a limited parameter regime in which the parameter d plays an important role. The surprising feature turns out to be the absence of pulse splitting. Moreover, because of normal GVD, the pulse broadens as it is amplified. For relatively small values of d, no autosoliton is formed because the pulse width tends toward infinity.

So far we have focused on solitons that maintain not only their shape but also their width (and other parameters) during propagation. A more general class of shape-preserving waves includes self-similar pulses that maintain their general shape, but not their width or amplitude, during propagation [133]. In the context of optical fibers, parabolic-shape pulses were first discovered in 1993 in the context of optical wave breaking [134]. In the context of optical amplifiers, they were discovered in 1996 though numerical simulations performed for an EDFA designed to provide normal dispersion [135]. It was found later analytically that optical amplifiers support self-similar waves in the form of pulses whose shape is parabolic [136]–[143]. Even when the pulse shape is not parabolic initially, the pulse evolves asymptotically to become nearly parabolic, while maintaining a linear chirp across it. Pulses that evolve in a self-similar fashion are sometimes called *similaritons*.

To obtain the self-similar solution for optical amplifiers, we use the NLS equation (4.3.15) with $T_2 = 0$ and $\alpha_2 = 0$ and write it as

$$\frac{\partial A}{\partial z} + \frac{i\beta_2}{2}\frac{\partial^2 A}{\partial T^2} = i\gamma|A|^2A + \frac{g}{2}A, \tag{4.6.1}$$

where $g = g_0 - \alpha$ is the net gain. We then look for a self-similar solution of the form [137]

$$A(z,T) = a_p(z)F(\tau)\exp[ic_p(z)T^2 + \phi_p(z)], \tag{4.6.2}$$

where the three pulse parameters, $a_p(z)$, $c_p(z)$, and $\phi_p(z)$, are allowed to evolve with z. The function $F(\tau)$, with $\tau = T(a_p^2 e^{-gz})$ acting as the self-similarity variable, governs the pulse shape. The use of Eq. (4.6.2) in Eq. (4.6.1) results in the following equation:

$$\left(\frac{dc_p}{dz} - 2\beta_2 c_p^2\right)\frac{\tau^2}{a_p^6}e^{2gz} + \frac{1}{a_p^2}\frac{d\phi_p}{dz} = \gamma F^2(\tau) - \frac{\beta_2}{2}\frac{a_p^2}{F}\frac{d^2F}{d\tau^2}e^{-2gz}. \tag{4.6.3}$$

together with an ordinary differential equation for $a_p(z)$:

$$\frac{da_p}{dz} = \beta_2 c_p a_p + \frac{g}{2}a_p. \tag{4.6.4}$$

In the asymptotic limit $z \to \infty$, the last term in Eq. (4.6.3) can be neglected. Moreover, since its right side is then a function of only τ, this equation can be satisfied if and only if the chirp and phase parameters evolve with z as [138]

$$\frac{dc_p}{dz} = 2\beta_2 c_p^2 - \frac{\gamma}{\tau_0^2}a_p^6 e^{-2gz}, \qquad \frac{d\phi_p}{dz} = \gamma a_p^2, \tag{4.6.5}$$

where τ_0 is an arbitrary constant. The shape function depends on this parameter through the relation $F(\tau) = (1 - \tau^2/\tau_0^2)^{1/2}$ for $|\tau| \leq \tau_0$; outside of this range, $F(\tau)$ must vanish for physical regions. We have thus found a pulse whose intensity profile $F^2(\tau)$ becomes a perfect parabola for large z and whose parameters evolve with z in a self-similar fashion.

It turns out that Eqs. (4.6.4) and (4.6.5) can be integrated analytically to find how the three pulse parameters evolve with z [138]. The final result is given by [136]

$$a_p(z) = \tfrac{1}{2}(gE_0)^{1/3}(\gamma\beta_2/2)^{-1/6}\exp(gz/3), \tag{4.6.6}$$

$$T_p(z) = 6g^{-1}(\gamma\beta_2/2)^{1/2}a_p(z), \tag{4.6.7}$$

$$c_p(z) = g/(6\beta_2), \qquad \phi_p(z) = \phi_0 + (3\gamma/2g)a_p^2(z), \tag{4.6.8}$$

where E_0 is the input energy and ϕ_0 is an arbitrary constant phase of a pulse whose shape evolves as $F(\tau)$ with $\tau = T/T_p(z)$. Thus, $T_p(z)$ can be interpreted as a pulse

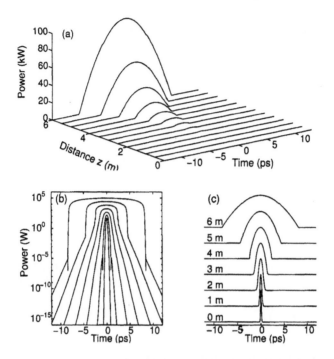

Figure 4.14: (a) Evolution of a 200-fs Gaussian input pulse over a 6-m-long Yb-doped fiber amplifier providing 50-dB gain. Intensity profiles of the amplified pulse plotted on (b) a logarithmic and (c) a linear scale. (From Ref. [138]; ©2002 OSA.)

width that scales linearly with the amplitude $a_p(z)$, as evident from Eq. (4.6.7). Because of self-similarity, the pulse maintains its parabolic shape even though its width and amplitude increase with z in an exponential fashion. Notice that this solution makes sense for fiber amplifiers (for which $\gamma > 0$) only when dispersion is normal ($\beta_2 > 0$).

The most noteworthy feature of the preceding self-similar solution is that the amplified pulses are linearly chirped across their entire temporal profile. A purely linear chirp is possible through SPM only when pulse shape is parabolic [134]. Optical amplifiers facilitate the production of a parabolic shape since new frequency components are generated continuously through SPM in such a way that the pulse maintains a linear chirp as it is amplified. Another important property of this self-similar solution is that the amplified pulse depends on the energy of the input pulse, but not on its other properties, such as its shape and width. Thus, even a Gaussian input pulse should evolve toward a parabolic pulse, provided it propagates in the normal-GVD regime of a fiber amplifier. Ytterbium-doped fiber amplifiers, operating near 1050 nm, often exhibit normal dispersion and can be used to observe the self-similar behavior.

The evolution toward a parabolic shape has been verified numerically by solving Eq. (4.6.1) with the split-step Fourier method [138]. Figure 4.14 shows the simulated evolution of a 200-fs Gaussian input pulse of 12-pJ energy, launched inside a 6-m-long fiber amplifier doped with ytterbium and pumped to provide a net amplification

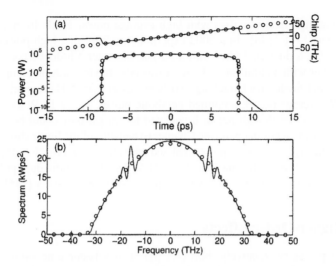

Figure 4.15: (a) Pulse shape and chirp profile and (b) pulse spectrum at the output of a 6-m-long Yb-doped fiber amplifier providing 50-dB gain. The predictions of the analytic solution are marked with open circles. (From Ref. [138]; ©2002 OSA.)

of 50 dB at its output ($g = 1.92$ m^{-1}). Realistic values of $\beta_2 = 25$ ps^2/km and $\gamma = 5.8$ W^{-1}/km were used in numerical simulations. It is clear from Figure 4.14 that the pulse shape indeed becomes parabolic near the output end of the amplifier. Also, the pulse width increases as $\sqrt{P_0}$ as the peak power P_0 grows with amplification, as predicted by Eq. (4.6.7). Indeed, as shown in Figure 4.15, the shape, the chirp profile, and the spectrum of the output pulse at the output end of the amplifier (solid lines) agree well with the predictions of the analytic self-similar solution (circles). The frequency chirp should vary linearly across the entire pulse for a quadratic phase profile. Numerical simulations indeed show a linear variation over most of the pulse, except near the pulse edges, where pulse intensity deviates from a perfect parabola. Oscillations seen in the spectrum near pulse edges correspond to such chirp deviations. Their origin is related to the fact that the analytical solution is valid in the asymptotic limit $z \to \infty$ that is not reached for a 6-m-long fiber. At what distance this limit is approached depends on the input pulse parameters. For example, if the input pulse width is chosen to nearly match the value $T_p(0)$ obtained from Eq. (4.6.7), this limit can be approached at shorter distances.

Parabolic pulses has been observed in several experiments in which picosecond or femtosecond pulses were amplified in the normal-dispersion regime of a fiber amplifier. In a 2000 experiment, 200-fs pulse with 12-pJ energy was launched into a 3.6-m-long Yb-doped fiber amplifier pumped to provide 30-dB gain [136]. The intensity and phase profiles of the amplified output pulses were deduced using the technique of frequency-resolved optical gating (FROG). The experimental results agreed well with both the numerical results obtained by solving the NLS equation and the asymptotic parabolic-pulse solution.

From a practical standpoint, the discovery of self-similar parabolic pulses is useful for producing high-quality pulses with high average power levels from a Yb-doped fiber amplifier [144]–[148]. In a 2002 experiment, a 9-m-long high-power amplifier was made using a dual-clad fiber with a 30-μm-diameter Yb-doped core [144]. By coiling the fiber with a radius of <10 cm, it was possible to guide and amplify mostly the fundamental mode. By launching 180-fs pulses at a 75-MHz repetition rate with only 10-mW average power, amplified parabolic pulses with a 17 W average power were obtained. Even though pulses broadened to 5.6 ps, they could be compressed down to 80 fs using a grating pair (see Chapter 6) because of the linear chirp associated with them. The highest average power realized with such an approach [145] was 76 W in 2003 and could be further increased to 131 W by 2005. Individual pulses in this later experiment had an energy of 1.8 μJ with a 8.2-MW peak power [146].

4.6.3 Higher-Order Effects

When input pulses are relatively short (<1 ps), it may become necessary to include the higher-order nonlinear and dispersive effects. The parabolic-gain approximation made in deriving the Ginzburg–Landau equation should also be relaxed for such short pulses. This can be done by keeping the denominator in Eq. (4.3.6) intact while using Eqs. (4.3.4) and (4.3.5). Following the method outlined in Section 2.3 of Ref. [68], the generalized Ginzburg–Landau equation can be written in terms of soliton units as

$$i\frac{\partial u}{\partial \xi} \pm \frac{1}{2}\frac{\partial^2 u}{\partial \tau^2} + |u|^2 u - i\delta_3 \frac{\partial^3 u}{\partial \tau^3} + is_0 \frac{\partial |u|^2 u}{\partial \tau} - \tau_R u \frac{\partial |u|^2}{\partial \tau}$$
$$= \frac{i}{2}g_0 L_d \int_{-\infty}^{\infty} \frac{\tilde{u}(\xi,f)\exp(-if\tau)df}{1 - i(f - f_0)(T_2/T_0)} - \frac{i}{2}\alpha L_D u, \qquad (4.6.9)$$

where $\tilde{u}(\xi,f)$ is the Fourier transform of $u(\xi,\tau)$, $f_0 = \omega_0 T_0$, and δ_3, s_0, and τ_R are the same three parameters introduced in Section 5.5 of Ref. [68]. The self-steepening parameter s_0 is negligible except for extremely short pulses ($\sim$10 fs) for which Eq. (4.6.9) itself is likely to break down. Third-order dispersive effects are also negligible unless a fiber amplifier operates very close to the zero-dispersion wavelength. In contrast, the parameter τ_R governs the frequency shift induced by intrapulse Raman scattering, and its effects should be included for pulse widths below 5 ps.

As early as 1988, it was found that the Raman-induced spectral shift of solitons may be suppressed in fiber amplifiers because of gain-dispersion effects [149]. Indeed, in an early experiment in which gain was provided by SRS (rather than dopants), little frequency shift was observed even for 100-fs pulses [150]. Figure 4.16 shows the results of numerical simulations when a 50-fs fundamental soliton is amplified by using $\tau_R = 0.1$, $\delta_3 = 0.01$, and $s_0 = 0$ in Eq. (4.6.9) with $g_0 L_D = 2.3$ (10-dB gain over each dispersion length) and $T_2/T_0 = 0.2$. The spectrum shifts slightly toward the red side, but this shift is less than that expected for such short pulses in the absence of amplification. Physically, this behavior can be understood by noting that a shift of the pulse spectrum from the gain peak reduces the gain experienced by the center frequency of the pulse. At the same time, spectral components located near the gain peak are amplified more. Thus, the amplifier has a built-in mechanism that tries to pull

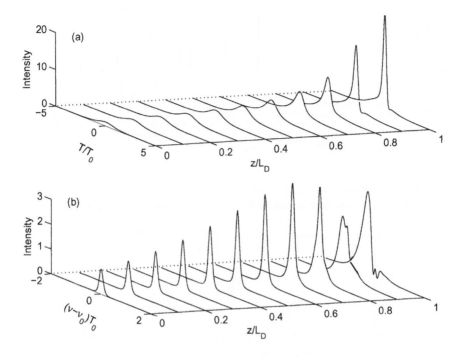

Figure 4.16: Evolution of (a) pulse shape and (b) spectrum when a relatively narrow fundamental soliton is amplified by 10 dB over one dispersion length and the higher-order effects are included.

the pulse spectrum toward the gain peak, resulting in a decrease in the Raman-induced frequency shift of solitons.

Pulse splitting, seen earlier in Figure 4.12, occurs even when higher-order effects are included. As an example, Figure 4.17 shows the pulse evolution over 2.5 dispersion lengths under conditions identical to those of Figure 4.16. Two subpulses form soon after $\xi = 1$ and evolve toward four pulses a a distance of $\xi = 2.5$. Each pulse shifts toward the right side as it slows down because of the Raman-induced spectral shift. The widths and amplitudes of these subpulses are still governed by the autosoliton solution given earlier.

Several experiments have focused on amplification of ultrashort pulses inside an amplifier operating near 1.55 μm [151]–[160]. Pulse shortening for femtosecond input pulses was observed in several of these experiments because of the soliton effects occurring in the anomalous-GVD regime. In one experiment, the dependence of the pulse width and the spectrum on amplifier gain was studied by using 240-fs input pulses [155]. Pulses as short as 60-fs were observed at the output end of a 3-m-long fiber amplifier. This experiment also showed that the Raman-induced frequency shift was nearly absent at low pump powers—an effect referred to as *soliton trapping*—but became dominant when the amplifier gain was large enough. Figure 4.18 shows the experimental pulse spectra for three values of pump powers and compares them to the input spectrum (dashed curve). The pulse spectrum did not shift significantly for

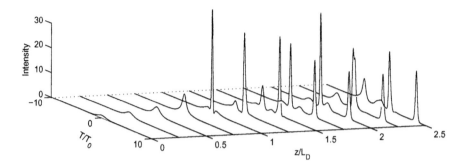

Figure 4.17: Evolution of a fundamental soliton over 2.5 dispersion lengths under conditions of Figure 4.16 showing pulse splitting and the Raman-induced temporal delay of subpulses.

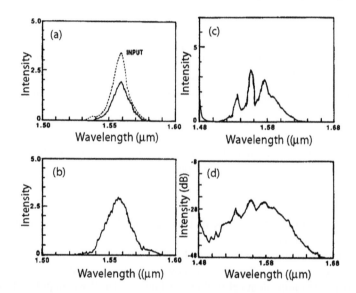

Figure 4.18: Experimental pulse spectra at three pump powers showing effects of Raman-induced frequency shift; (a)–(c) correspond to pump powers of 7, 13, and 25 mW, respectively; (d) shows the 25-mW spectrum on a logarithmic scale. (From Ref. [155]; ©1990 AIP.)

pump powers of 7 and 13 mW but exhibited a shift of more than 20 nm for 25 mW of pump power. At high power levels, pulse spectrum becomes quite broad (>200 nm) and the pulse shape develops a narrow spike on top of a broad pedestal [158]. These features are related to the compression of a high-order soliton and the generation of a supercontinuum initiated by the higher-order effects [68].

In recent years the use photonic crystal and other microstructured fibers has become common. The core of such fibers can be doped with erbium or ytterbium to realize fiber amplifiers. In one experiment, the doped core had an effective diameter of 26 μm so that it could support high peak intensities [160]. Such a 9-m-long EDFA was used to amplify 700-fs pulses at 1557 nm by pumping it through a double-clad con-

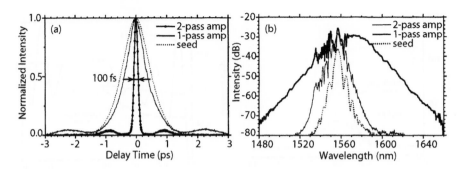

Figure 4.19: Experimental (a) autocorrelation traces and (b) pulse spectra after single (thin line) and double passes (thick line) through a large-mode area EDFA. The dotted curves show the shape and spectrum of input pulses. (From Ref. [160]; ©2005 OSA.)

figuration. The fiber was coiled with 32-cm-diameter to suppress higher-order modes, and a double-pass configuration was used with orthogonal polarization during the second pass. The amplified pulse was compressed by a factor of about 7 and its energy increased by 10-fold to more than 7 nJ. Figure 4.19 shows changes in the pulse shape and spectrum after single and double passes through the amplifier. Pulse compression is related to the nearly adiabatic amplification of a fundamental soliton. The shift of the spectral peak toward the longer wavelength is due to the Raman-induced frequency shift [68].

For optical pulses shorter than T_2, in principle, one should use the complete set of Maxwell–Bloch equations in place of the Ginzburg–Landau equation. These equations have been solved numerically by using the split-step Fourier method [97]. The results show significant deviations between the exact and approximate solutions. For example, whereas the soliton amplitude increases exponentially and its width decreases exponentially in the case of $T_0 \gg T_2$ (see Figure 4.12), in the coherent regime in which $T_0 \ll T_2$, the changes are linear in ξ rather than exponential [93]. Even when $T_0 > T_2$ initially, the coherent effects should be included whenever the pulse width becomes comparable to T_2 during propagation. Both qualitative and quantitative differences were found to occur in a numerical study in which $T_0 = 3T_2$ initially but the amplifier gain of 10 dB per dispersion length was large enough to lead to considerable pulse narrowing during the amplification process [97].

One may ask how the solitary-wave solution of the Ginzburg–Landau equation, obtained in the parabolic-gain approximation and given by Eq. (4.5.4), changes when the Lorentzian shape of the gain spectrum is taken into account through Eq. (4.6.9). Numerical solutions show that autosolitons still exist, in the sense that any input pulse evolves toward a unique solitary pulse whose shape, amplitude, width, and chirp are determined by the amplifier parameters [106]. However, the pulse characteristics are quite different than those of the solitary-wave solution (4.5.4). A new feature is that the parameter α/g_0 plays an important role in determining the properties of the autosoliton. As an example, Figure 4.20 shows the intensity and chirp profiles obtained numerically for $\alpha/g_0 = 0.6$ and 0.8 using $T_2 = 0.2$ ps. The corresponding profiles for the parabolic-gain soliton obtained from Eq. (4.5.4) are also shown for comparison.

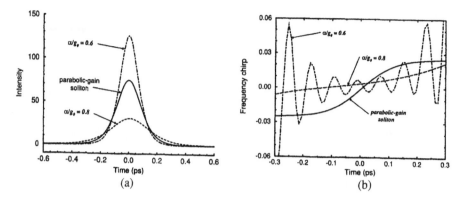

Figure 4.20: (a) Intensity and (b) frequency chirp across the autosoliton for two values α/g_0 using $T_2 = T_0 = 0.2$ ps, $\delta_3 = s_0 = \tau_R = 0$, and $\exp(g_0 L_D) = 2$. The solid line shows the chirp in the parabolic-gain approximation. (From Ref. [106]; ©1996 Elsevier.)

In general, the autosoliton becomes narrower and more intense as α/g_0 decreases. The reason can be understood by noting that the pulse spectrum can expand as long as the gain in the spectral wings exceeds the loss level, resulting in shorter pulses for smaller values of α/g_0. However, chirp variations along the pulse then become large and nonuniform with a periodic structure. It should be stressed that such autosolitons are not absolutely stable since background noise is amplified by the amplifier.

From a practical standpoint, fiber amplifiers can be used to simultaneously amplify and compress picosecond optical pulses, but the amplification process becomes less useful for femtosecond pulses because of temporal and spectral distortions occurring as a result of the higher-order nonlinear effects. One can use the technique of chirped-pulse amplification to advantage in this situation (see Section 6.5). Numerical results show that highly efficient and practically distortionless amplification of femtosecond pulses can be realized by this method [161]. The technique of chirped-pulse amplification is often used to generate ultrashort pulses with high energy levels and is discussed in Chapter 6.

4.7 Fiber-Optic Raman Amplifiers

It is not always necessary to make use of dopants because several nonlinear effects occurring within optical fibers can provide significant amount of gain [68]. Two such processes are SRS and four-wave mixing (FWM), which are used to make Raman and parametric amplifiers, respectively. Both require a pump laser whose energy is transferred to the signal being amplified. Fiber-optic parametric amplifiers are covered in Chapter 10 of Ref. [68]. This section focuses on the Raman amplifiers in which SRS is used to provide the optical gain.

Raman amplifiers have attracted considerable attention recently as they are useful for optical communication systems [162]–[164]. Their properties are also discussed in Chapter 8 of Ref. [68]. Because the nonlinear effects play a relatively minor role in the

CW case, this section focuses on the amplification of short optical pulses. Although the anomalous-dispersion regime has been studied since the early 1980s, Raman amplification in the normal-dispersion regime is of current interest from the standpoint of the self-similar evolution of optical pulses [165]–[170].

4.7.1 Pulse Amplification through Raman Gain

Before discussing the self-similar evolution, we review briefly the main features of Raman amplifiers [68]. When an optical pulse is launched inside a silica fiber together with a copolarized CW pump beam whose frequency is upshifted by 13.2 THz (the so-called Raman shift for which the Raman gain becomes maximum), the pulse is amplified with an effective local gain coefficient $g(z) = g_R I_p(z)$, where $I_p(z)$ is the pump intensity at a distance z from the input end and g_R is the Raman-gain parameter whose value is about 6×10^{-14} m/W at a wavelength near 1.5 μm and scales inversely with the wavelength. The z dependence of the gain is a consequence of fiber losses at the pump wavelength governed by α_p that reduce the pump intensity as $I_p(z) = I_0 \exp(-\alpha_p z)$, where I_0 is the input pump intensity. If a backward pumping configuration is employed in which the pump beam is launched at the output end of a Raman amplifier of length L, the gain varies with z as $g(z) = g_R I_0 \exp[-\alpha_p(L-z)]$. Bidirectional pumping is possible by launching two pump beams at two ends of the fiber; it results in a different dependence of the gain on z.

If gain saturation were negligible, the evolution of a Raman pulse would be governed by Eq. (4.6.1), the same equation used earlier for doped fibers, with the only difference that g varies with z. The NLS equation with gain is found to have self-similar solutions even when the parameters g, β_2, and γ vary along the fiber length, provided they satisfy a certain relation [171]–[174]. These solutions exists in the cases of both normal and anomalous GVD and can take a variety of shapes depending on the parameter values.

In the case of a Raman amplifier designed to amplify pulse energy by a large factor, gain saturation cannot be neglected. In this case, one should employ a generalized NLS equation, derived in Section 2.3 of Ref. [68] and having the form

$$\frac{\partial A}{\partial z} + \frac{1}{2}\left(\alpha(\omega_0) + i\alpha_1 \frac{\partial}{\partial t}\right)A + \frac{i\beta_2}{2}\frac{\partial^2 A}{\partial T^2} - \frac{\beta_3}{6}\frac{\partial^3 A}{\partial T^3}$$

$$= i\left(\gamma(\omega_0) + i\gamma_1 \frac{\partial}{\partial t}\right)\left(A(z,t)\int_0^\infty R(t')|A(z,t-t')|^2 dt'\right), \quad (4.7.1)$$

where the frequency dependence of fiber loss and of the nonlinear parameter is included through the derivatives $\alpha_1 = d\alpha/d\omega$ and $\gamma_1 = (d\gamma/d\omega)$, both evaluated at the reference frequency ω_0. In practice, γ_1 can be approximated by γ/ω_0 if the effective mode area of the fiber is nearly the same at the pump and signal wavelengths.

If ω_0 is taken to be the carrier frequency ω_s of the signal pulse, A in Eq. (4.7.1) has the form $A = A_s + A_p e^{-i\Omega t}$, where $\Omega = \omega_p - \omega_s$ and ω_p is the pump frequency. The Raman gain is included through the nonlinear response function

$$R(t) = (1 - f_R)\delta(t) + f_R h_R(t), \quad (4.7.2)$$

where the first term is due to the Kerr effect (of electronic origin) and f_R represents the fractional contribution of the delayed Raman response related to nuclear motion within silica molecules. The Raman response function $h_R(t)$ satisfies $\int_0^\infty h_R(t)dt = 1$. The imaginary part of its Fourier transform is related to the Raman gain spectrum.

Equation (4.7.1) was used to study Raman amplification in the normal-GVD regime of a fiber starting in 2003 [165]. Its use is time consuming numerically because a large number of points must be used for the temporal mesh. The numerical effort can be reduced considerably by splitting this equations into two separate equations for A_s and A_p. The evolution of the signal pulse is then governed by [167]

$$\frac{\partial A_s}{\partial z} + \frac{\alpha_s}{2}A_s + \frac{i\beta_2}{2}\frac{\partial^2 A_s}{\partial T^2} - \frac{\beta_3}{6}\frac{\partial^3 A_s}{\partial T^3} = i\gamma\left(1 + \frac{i}{\omega_s}\frac{\partial}{\partial t}\right)\left[(1-f_R)(|A_s|^2 + 2|A_p|^2)A_s\right.$$

$$+ f_R A_s \int_0^\infty h_R(t')[|A_s(z,t-t')|^2 + |A_p(z,t-t')|^2]dt'$$

$$\left. + f_R A_p \int_0^\infty h_R(t')A_s(z,t-t')A_p^*(z,t-t')e^{-i\Omega t'}dt'\right], \quad (4.7.3)$$

where $\alpha_s \equiv \alpha(\omega_s)$ accounts for fiber losses. The pump equation has a similar form except that it includes an additional term related to different group velocities at the pump and signal wavelengths. It can be simplified for a CW pump, especially when the backward-pumping configuration is employed [168].

4.7.2   Self-Similar Evolution and Similariton Formation

The form of Eq. (4.7.3) is so different from Eq. (4.6.1) that it is not apparent that self-similar parabolic pulses should form in the normal-dispersion region of Raman amplifiers. However, numerical simulations as well as experiments have shown that such similaritons do form under suitable conditions [165]–[168]. In a 2004 experiment, a 5.3-km-long fiber was pumped in the forward direction at 1455 nm to amplify picosecond pulses at a wavelength 1550 nm [167]. The pulse shape and chirp of the amplified pulse were deduced from FROG traces obtained through sum-frequency generation inside a nonlinear crystal. Figure 4.21(a) shows the experimental intensity and chirp profiles (circles), together with the numerical predictions (crosses), for 7-ps input pulses launched with 2.16-pJ energy. The solid, dashed, and dotted curves show the intensity profiles for parabolic, Gaussian, and "sech" pulses. It is evident that the experimental pulse shapes are nearly parabolic, deviations occurring only in the pulse wings where intensity is below 0.1% of the peak value.

Figure 4.21(b) shows the evolution of the pulse along fiber length by solving Eq. (4.7.3) numerically together with the corresponding pump equation in the case of forward pumping [167]. The solid curves show the experimental pulse shapes at the input and output ends of the 5-3-km-long Raman amplifier. The 7-ps input pulse has a "sech" shape, but it transforms into a parabolic pulse within 1 km or so, which then evolves in a self-similar fashion over the entire fiber length. The agreement between the experimental and numerical pulse shapes at the amplifier output is quite reasonable, especially if we note that the amplifier model does not include polarization-mode dispersion effects that are likely to play some role in a 5-3-km-long fiber.

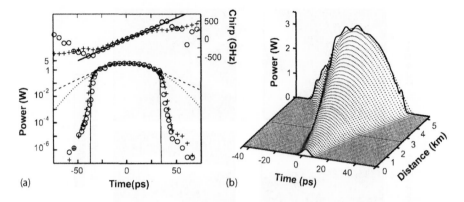

Figure 4.21: (a) Experimental intensity and chirp profiles (circles) and numerical predictions (crosses) for 7-ps input pulses launched with 2.16-pJ energy inside a 5-3-km-long Raman amplifier. Solid, dashed, and dotted curves show the intensity profile for parabolic, Gaussian, and "sech" pulses. (b) Simulated evolution of the pulse along amplifier length. Solid curves show the experimental pulse shapes at the input and output end. (From Ref. [167]; ©2004 IEEE.)

The formation of parabolic pulses in a Raman amplifier with a linear chirp across them indicates that such amplifiers behave in the same fashion as a doped-fiber amplifier, and the amplified pulse follows the self-similar evolution predicted by Eq. (4.6.1). Indeed, as expected, the output pulse characteristics did not change much in the 2004 experiment when the width of input pulses was doubled to 14 ps [167]. Changes in input pulse energy affected the width and energy of output pulses, as indicated in Eqs. (4.6.6)–(4.6.8), but the pulse shape remained parabolic. To understand why the predictions of Eq. (4.6.1) apply to a Raman amplifier, one should realize that for picosecond pulses, Eq. (4.7.3) reduces to that form approximately [68]. The reason is that A_p and A_s vary little over the short time scale over which the Raman response function $h_R(t)$ changes. Treating A_p and A_s as constants, the two integrals in Eq. (4.7.3) are related to the Fourier transform $\tilde{h}_R(\Omega)$ of $h_R(t)$. Introducing the Raman gain as $g_R = 2\gamma f_R \text{Im}[\tilde{h}_R(\Omega)]$ and neglecting the higher-order dispersive and nonlinear effects, Eq. (4.7.3) reduces to

$$\frac{\partial A_s}{\partial z} + \frac{i\beta_2}{2}\frac{\partial^2 A_s}{\partial T^2} = i\gamma\left[(|A_s|^2 + (2 - f_R)|A_p|^2)A_s + \frac{1}{2}(g_R|A_p|^2 - \alpha_s)A_s\right]. \quad (4.7.4)$$

Except for the presence of the cross-phase modulation (XPM) term, this equation has the form of Eq. (4.6.1) with $g = g_R|A_p|^2 - \alpha_s$.

It is well known that two solitons of different wavelengths, moving at different speeds and crossing each other, remain intact after they separate [68]. An interesting question is how two similaritons behave under such conditions. It was found in a 2005 numerical study that similaritons remain stable after they collide and separate inside a Raman amplifier [169]. Figure 4.22 shows the temporal and spectral evolution of two pulses (initially Gaussian with a 6-ps width) separated in their carrier frequencies by 1.25 THz (left column) or 0.75 THz (right column). Each pulse evolves to become a parabolic similariton whose width continues to increase. The two similaritons interact

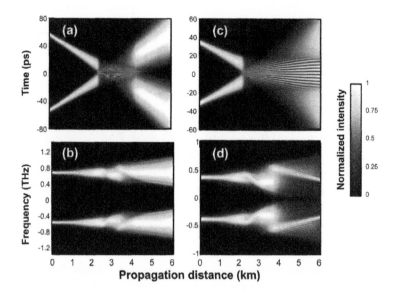

Figure 4.22: Temporal (top row) and spectral (bottom row) evolution of two pulses (initially Gaussian with a 6-ps width), separated in their carrier frequencies by 1.25 THz (left columm) or 0.75 THz (right column), showing collision of two similaritons inside a Raman amplifier. (From Ref. [169]; ©2005 OSA.)

only during their overlap and regain their parabolic shape after they separate from each other. The fringe pattern forming during their overlap indicates a temporal modulation within the overlapping region. Under certain conditions, this pattern can be used to generate a train of dark solitons at ultrahigh repetition rates [170]. Collisions of two similaritons have been observed experimentally. The results indicate that, similar to the case of soliton collisions, similaritons interact with each other through XPM.

The interaction of two similaritons inside a fiber amplifier can lead to features that are quite distinct from those occurring in the case of solitons. It was found in a recent study [175] that character of similariton interaction depends crucially on the sign and magnitude of the of the linear chirp. In particular, two similaritons colliding inside a dispersion-decreasing fiber can form a moleculelike bound state under certain conditions. As the emerging field of similaritons in still in its infancy, further progress is likely to occur in the near future.

Problems

4.1 Assuming that the gain spectrum of a fiber amplifier can be approximated by a Lorentzian profile of 30-nm bandwidth (FWHM), calculate the amplifier bandwidths when it is operated to provide 20- and 30-dB gain. Neglect gain saturation.

4.2 A fiber amplifier can amplify a 1-μW signal to the 1-mW level. What is the output power when a 1-mW signal is incident on the same amplifier? Assume a saturation power of 10 mW.

4.3 Explain the concept of noise figure for a fiber amplifier. Why does the SNR of the amplified signal degrade by 3 dB even for an ideal amplifier?

4.4 Derive an expression for the small-signal gain of an EDFA by using rate equations for the three-level pumping scheme. Assume a rapid transfer of the pumped population to the excited state.

4.5 Solve Eqs. (4.2.7) and (4.2.8) analytically, or numerically if an analytic solution is not possible. Plot the saturated amplifier gain as a function of the pump power for $\alpha_p L = 5$ and $\alpha_s L = 2$, where L is the amplifier length. Neglect fiber losses and use $P'_s = 0.1$.

4.6 Derive the Ginzburg–Landau equation (4.3.7) by following the method of Section 2.3 of Ref. [68] and adding the contribution of dopants to the nonlinear polarization.

4.7 Show by direct substitution that the chirped soliton given by Eq. (4.5.4) is indeed a solution of the Ginzburg–Landau equation (4.5.2) when the soliton parameters are given by Eqs. (4.5.5)–(4.5.8).

4.8 Use the chirped soliton solution given by Eq. (4.5.4) to obtain an expression for the frequency chirp. How would you operate a fiber amplifier to minimize the chirp?

4.9 Solve the Ginzburg–Landau equation of the form in Eq. (4.5.2) numerically by using the split-step Fourier method. Use this code to reproduce the results shown in Figure 4.12 when a fundamental soliton is amplified in a fiber amplifier.

4.10 Modify the numerical scheme used in the preceding problem and solve Eq. (4.6.9) numerically. Use your computer code to reproduce Figure 4.17.

4.11 Integrate Eqs. (4.6.4) and (4.6.5) and verify the expressions given in Eqs. (4.6.6) through (4.6.8) for parabolic pulses forming in the case of normal dispersion.

4.12 Derive Eq. (4.7.3) using Eq. (4.7.1) with $A = A_s + A_p e^{-i\Omega t}$.

References

[1] C. J. Koester and E. Snitzer, *Appl. Opt.* **3**, 1182 (1964).

[2] S. B. Poole, D. N. Payne, R. J. Mears, M. E. Fermann, and R. E. Laming, *J. Lightwave Technol.* **4**, 870 (1986).

[3] E. Desuvire, *Erbium-Doped Fiber Amplifiers: Principles and Applications* (Wiley, Hoboken, NJ, 1994).

[4] P. C. Becker, N. A. Olsson, and J. R. Simpson, *Erbium-Doped Fiber Amplifiers: Fundamentals and Technology* (Academic Press, San Diego, CA, 1999).

[5] M. J. F. Digonnet, Ed., *Rare-Earth Doped Fiber Lasers and Amplifiers* (Marcel Dekker, New York, 2001).

[6] E. Desuvire, D. Bayart, B. Desthieux, and S. Bigo, *Erbium-Doped Fiber Amplifiers: Device and System Development* (Wiley, Hoboken, NJ, 2002).

[7] A. E. Siegman, *Lasers* (University Science Books, Mill Valley, CA, 1986).

[8] P. W. Milonni and J. H. Eberly, *Lasers* (Wiley, Hoboken, NJ, 1988).

[9] O. Svelto, *Principals of Lasers*, 4th ed. (Plenum, New York, 1998).

[10] G. P. Agrawal, *Lightwave Technnology: Telecommunication Systems* (Wiley, Hoboken, NJ, 2005).

[11] H. Kogelnik and A. Yariv, *Proc. IEEE* **52**, 165 (1964).

[12] R. J. Mears, L. Reekie, I. M. Jauncey, and D. N. Payne, *Electron. Lett.* **23**, 1026 (1987).

[13] E. Desurvire, J. R. Simpson, and P. C. Becker, *Opt. Lett.* **12**, 888 (1987).

[14] M. Nakazawa, Y. Kimura, and K. Suzuki, *Appl. Phys. Lett.* **54**, 295 (1989).

[15] P. C. Becker, J. R. Simpson, N. A. Olsson, and N. K. Dutta, *IEEE Photon. Technol. Lett.* **1**, 267 (1989).

[16] M. Yamada, M. Shimizu, T. Takeshita, M. Okayasu, M. Horiguchi, S. Uehara, and E. Sugita, *IEEE Photon. Technol. Lett.* **1**, 422 (1989)

[17] M. Shimizu, M. Yamada, M. Horiguchi, T. Takeshita, and M. Okayasu, *Electron. Lett.* **26**, 1641 (1990).

[18] M. Nakazawa, Y. Kimura, E. Yoshida, and K. Suzuki, *Electron. Lett.* **26**, 1936 (1990).

[19] B. Pederson, A. Bjarklev, H. Vendeltorp-Pommer, and J. H. Povlsen, *Opt. Commun.* **81**, 23 (1991).

[20] M. Horiguchi, K. Yoshino, M. Shimizu, and M. Yamada, *Electron. Lett.* **29**, 593 (1993).

[21] R. I. Laming, J. E. Townsend, D. N. Payne, F. Meli, G. Grasso, and E. J. Tarbox, *IEEE Photon. Technol. Lett.* **3**, 253 (1991).

[22] W. J. Miniscalco, *J. Lightwave Technol.* **9**, 234 (1991).

[23] K. Inoue, T. Korninaro, and H. Toba, *IEEE Photon. Technol. Lett.* **3**, 718 (1991).

[24] R. Kashyap, *Fiber Bragg Gratings* (Academic Press, San Diego, CA, 1999).

[25] J. Bae, J, Chun, and S. B. Lee, *J. Lightwave Technol.* **22**, 1976 (2004).

[26] A. P. Zhang, X. W. Chen, J. H. Yan, Z. G. Guan, S. L. He, and H. Y. Tam, *IEEE Photon. Technol. Lett.* **17**, 2559 (2005).

[27] R. Feced, C. Algeria, M. N. Zervas, and R. I. Laming, *IEEE J. Sel. Topics Quantum Electron.* **5**, 1278 (1999).

[28] N. Kumar, M. R. Shenoy, and B. P. Pal, *IEEE Photon. Technol. Lett.* **17**, 2056 (2005).

[29] Y. B. Lu and P. L. Chu, *IEEE Photon. Technol. Lett.* **12**, 1616 (2000).

[30] P. F. Wysocki, N. Park, and D. DiGiovanni, *Opt. Lett.* **21**, 1744 (1996).

[31] H. Ono, M. Yamada, T. Kanamori, and Y. Ohishi, *Electron. Lett.* **33**, 1477 (1997).

[32] M. Masuda, K. I. Suzuki, S. Kawai, and K. Aida, *Electron. Lett.* **33**, 753 (1997).

[33] M. Yamada, H. Ono, T. Kanamori, S. Sudo, and Y. Ohishi, *Electron. Lett.* **33**, 710 (1997).

[34] Y. B. Lu, P. L. Chu, A. Alphones, and P. Shum, *IEEE Photon. Technol. Lett.* **16**, 1640 (2004).

[35] H. Ono, M. Yamada, and M. Shimizu, *J. Lightwave Technol.* **21**, 2240 (2003).

[36] A. A. M. Saleh, R. M. Jopson, J. D. Evankow, and J. Aspell, *IEEE Photon. Technol. Lett.* **2**, 714 (1990).

[37] C. R. Giles and E. Desurvire, *J. Lightwave Technol.* **9**, 271 (1991).

[38] B. Pedersen, A. Bjarklev, J. H. Povlsen, K. Dybdal, and C. C. Larsen, *J. Lightwave Technol.* **9**, 1105 (1991).

[39] K. Nakagawa, S. Nishi, K. Aida, and E. Yoneda, *J. Lightwave Technol.* **9**, 198 (1991).

[40] Y. Sun, J. L. Zyskind, and A. K. Srivastava, *IEEE J. Sel. Topics Quantum Electron.* **3**, 991 (1997).

[41] M. Zirngibl, *Electron. Lett.* **27**, 560 (1991).

[42] E. Delevaque, T. Georges, J. F. Bayon, M. Monerie, P. Niay, and P. Benarge, *Electron. Lett.* **29**, 1112 (1993).

[43] A. Yu and M. J. O'Mahony, *IEEE J. Sel. Topics Quantum Electron.* **3**, 1013 (1997).

[44] N. Takahashi, T. Hirono, H. Akashi, S. Takahashi, and T. Sasaki, *IEEE J. Sel. Topics Quantum Electron.* **3**, 1019 (1997).

[45] K. Inoue, *IEEE Photon. Technol. Lett.* **11**, 533 (1999).

[46] J. T. Ahn and K. H. Kim, *IEEE Photon. Technol. Lett.* **16**, 84 (2004).

[47] S. W. Harun, N. Tamchek, P. Poopalan, and H. Ahmad, *IEEE Photon. Technol. Lett.* **16**, 422 (2004).

[48] J. H. Ji, L. Zhan, L. L. Yi, C. C. Tang, Q. H. Ye, and Y. X. Xia, *J. Lightwave Technol.* **23**, 1375 (2005).

[49] R. Olshansky, *Electron. Lett.* **24**, 1363 (1988).

[50] C. R. Giles, E. Desurvire, J. L. Zyskind, and J. R. Simpson, *IEEE Photon. Technol. Lett.* **1**, 367 (1989).

[51] M. Yamada, M. Shimizu, M. Okayasu, T. Takeshita, M. Horiguchi, Y. Tachikawa, and E. Sugita, *IEEE Photon. Technol. Lett.* **2**, 205 (1990).

[52] R. I. Laming and D. N. Payne, *IEEE Photon. Technol. Lett.* **2**, 418 (1990).

[53] K. Kikuchi, *Electron. Lett.* **26**, 1851 (1990).

[54] B. Pedersen, J. Chirravuri, and W. J. Miniscalco, *IEEE Photon. Technol. Lett.* **4**, 351 (1992).

[55] R. G. Smart, J. L. Zyskind, J. W. Sulhoff, and D. J. DiGiovanni, *IEEE Photon. Technol. Lett.* **4**, 1261 (1992).

[56] R. I. Laming, M. N. Zervas, and D. N. Payne, *IEEE Photon. Technol. Lett.* **4**, 1345 (1992).

[57] M. N. Zervas, R. I. Laming, and D. N. Payne, *IEEE J. Quantum Electron.* **31**, 472 (1995).

[58] R. Lebref, B. Landousies, T. Georges, and E. Delevaque, *J. Lightwave Technol.* **15**, 766 (1997).

[59] M. Yamada, H. Ono, and Y. Ohishi, *Electron. Lett.* **34**, 1747 (1998).

[60] G. R. Walker, D. M. Spirit, D. L. Williams, and S. T. Davey, *Electron. Lett.* **27**, 1390 (1991).

[61] D. N. Chen and E. Desurvire, *IEEE Photon. Technol. Lett.* **4**, 52 (1992).

[62] K. Rottwitt, J. H. Povlsen, A. Bjarklev, O. Lumholt, B. Pedersen, and T. Rasmussen, *Electron. Lett.* **28**, 287 (1992); *IEEE Photon. Technol. Lett.* **5**, 218 (1993).

[63] S. Wen and S. Chi, *J. Lightwave Technol.* **10**, 1869 (1992).

[64] K. Rottwitt, J. H. Povlsen, and A. Bjarklev, *J. Lightwave Technol.* **11**, 2105 (1993).

[65] C. Lester, K. Rottwitt, J. H. Povlsen, P. Varming, M. A. Newhouse, and A. J. Antos, *Opt. Lett.* **20**, 1250 (1995).

[66] A. Altuncu, L. Noel, W. A. Pender, A. S. Siddiqui, T. Widdowson, A. D. Ellis, M. A. Newhouse, A. J. Antos, G. Kar, and P. W. Chu, *Electron. Lett.* **32**, 233 (1996).

[67] M. Nissov, H. N. Poulsen, R. J. Pedersen, B. F. Jørgensen, M. A. Newhouse, and A. J. Antos, *Electron. Lett.* **32**, 1905 (1996).

[68] G. P. Agrawal, *Nonlinear Fiber Optics*, 4th ed. (Academic Press, San Diego, CA, 2007).

[69] G. P. Agrawal, *Phys. Rev. A* **44**, 7493 (1991).

[70] M. Asobe, T. Kanamori, and K. Kubodera, *IEEE Photon. Technol. Lett.* **4**, 362 (1992); *IEEE J. Quantum Electron.* **2**, 2325 (1993).

[71] G. P. Agrawal, *IEEE Photon. Technol. Lett.* **4**, 562 (1992).

[72] J. P. Hamide, P. Emplit, and J. M. Gabriagues, *Electron. Lett.* **26**, 1452 (1990).

[73] K. Kikuchi, *IEEE Photon. Technol. Lett.* **5**, 221 (1993).

[74] F. Matera, A. Mecozzi, M. Romagnoli, and M. Settembre, *Opt. Lett.* **18**, 1499 (1993).

[75] A. Mecozzi, *J. Opt. Soc. Am. B* **11**, 462 (1994).

[76] S. B. Cavalcanti, G. P. Agrawal, and M. Yu, *Phys. Rev. A* **51**, 4086 (1995).

[77] M. Yu, G. P. Agrawal, and C. J. McKinstrie, *J. Opt. Soc. Am. B* **12**, 1126 (1995).

[78] C. Lorattanasane and K. Kikuchi, *IEEE J. Quantum Electron.* **33**, 1084 (1997).

[79] A. Carena, V. Curri, R. Gaudino, P. Poggiolini, and S. Benedetto, *IEEE Photon. Technol. Lett.* **9**, 535 (1997).

[80] R. A. Saunders, B. A. Patel, and D. Garthe, *IEEE Photon. Technol. Lett.* **9**, 699 (1997).

[81] R. Q. Hui, M. O'Sullivan, A. Robinson, and M. Taylor, *J. Lightwave Technol.* **15**, 1071 (1997).

[82] D. F. Grosz, C. Mazzali, S. Celaschi, A. Paradisi, and H. L. Fragnito, *IEEE Photon. Technol. Lett.* **11**, 379 (1999).

[83] M. Norgia, G. Giuliani, and S. Donati, *J. Lightwave Technol.* **17**, 1750 (1999).

[84] P. A. Bélanger, L. Gagnon, and C. Paré, *Opt. Lett.* **14**, 943 (1989).

[85] G. P. Agrawal, *IEEE Photon. Technol. Lett.* **2**, 875 (1990); *Opt. Lett.* **16**, 226 (1991).

[86] L. Gagnon and P. A. Bélanger, *Phys. Rev. A* **43**, 6187 (1991).

[87] V. S. Grigoryan and T. S. Muradyan, *J. Opt. Soc. Am. B* **8**, 1757 (1991).

[88] I. R. Gabitov, M. Romagnoli, and S. Wabnitz, *Appl. Phys. Lett.* **59**, 1811 (1991).

[89] E. M. Dianov, K. K. Konstantinov, A. N. Pilipetskii, and A. N. Starodumov, *Sov. Lightwave Commun.* **1**, 169 (1991).

[90] W. Hodel, J. Schülz, and H. P. Weber, *Opt. Commun.* **88**, 173 (1992).

[91] M. Romagnoli, F. S. Locati, F. Matera, M. Settembre, M. Tamurrini, and S. Wabnitz, *Opt. Lett.* **17**, 1456 (1992).

[92] G. P. Agrawal, *Phys. Rev. E* **48**, 2316 (1993).

[93] V. Petrov and W. Rudolph, *Opt. Commun.* **76**, 53 (1990).

[94] I. V. Melnikov, R. F. Nabiev, and A. V. Nazarkin, *Opt. Lett.* **15**, 1348 (1990).

[95] T. Y. Wang and S. Chi, *Opt. Lett.* **16**, 1575 (1991).

[96] M. Nakazawa, E. Yamada, and H. Kubota, *Phys. Rev. Lett.* **66**, 2625 (1991); *Phys. Rev. A* **44**, 5973 (1991).

[97] B. Gross and J. T. Manassah, *Opt. Lett.* **17**, 340 (1992).

[98] V. V. Afanasjev, V. N. Serkin, and V. A. Vysloukh, *Sov. Lightwave Commun.* **2**, 35 (1992).

[99] A. I. Maimistov, *Sov. J. Quantum Electron.* **22**, 271 (1992); *Opt. Commun.* **94**, 33 (1992).

[100] S. Chi, T. Y. Wang, and S. Wen, *Phys. Rev. A* **47**, 3371 (1993).

[101] S. Chi, C. W. Chang, and S. Wen, *Opt. Commun.* **106**, 183 (1994); *Opt. Commun.* **111**, 132 (1994).

[102] M. N. Islam, L. Rahman, and J. R. Simpson, *J. Lightwave Technol.* **12**, 1952 (1994).

[103] K. Porsezian and K. Nakkeeran, *Phys. Rev. Lett.* **74**, 2941 (1995); *J. Mod. Opt.* **42**, 1953 (1995).

[104] H. Ammamm, W. Hodel, and H. P. Weber, *Opt. Commun.* **115**, 347 (1995).

[105] J. T. Manassah and B. Gross, *Opt. Commun.* **122**, 71 (1995).

[106] L. W. Liou and G. P. Agrawal, *Opt. Commun.* **124**, 500 (1996).

[107] K. Nakkeeran and K. Porsezian, *Opt. Commun.* **123**, 169 (1996); *J. Mod. Opt.* **43**, 693 (1996).

[108] G. H. M. van Tartwijk and G. P. Agrawal, *J. Opt. Soc. Am. B* **14**, 2618 (1997).

[109] N. N. Pereira and L. Stenflo, *Phys. Fluids* **20**, 1733 (1977).

[110] W. van Saarloos and P. Hohenberg, *Physica D* **56**, 303 (1992).

[111] N. N. Akhmediev, V. V. Afanasjev, and J. M. Soto-Crespo, *Phys. Rev. E* **53**, 1190 (1996).

[112] N. N. Akhmediev and A. Ankiewicz, *Solitons: Nonlinear Pulses and Beams* (Chapman and Hall, London, 1997), Chap. 13.

[113] J. M. Soto-Crespo, N. Akhmediev, and A. Ankiewicz, *Phys. Rev. Lett.* **85**, 2937 (2000).

[114] S. C. V. Latas, M. F. S. Ferreira, and A. S. Rodrigues, *Opt. Fiber Technol.* **11**, 292 (2005).

[115] N. N. Akhmediev and A. Ankiewicz, Eds., *Dissipative Solitons* (Springer, Berlin, 2005).

[116] V. Skarka and N. B. Aleksić, *Phys. Rev. Lett.* **96**, 013903 (2006).

[117] S. L. McCall and E. L. Hahn, *Phys. Rev. Lett.* **18**, 908 (1967); *Phys. Rev.* **183**, 457 (1969).

[118] J. E. Armstrong and E. Courtens, *IEEE J. Quantum Electron.* **4**, 411 (1968); *IEEE J. Quantum Electron.* **5**, 249 (1969).

[119] J. H. Eberly and L. Matulic, *Opt. Commun.* **1**, 241 (1969).

[120] L. Matulic and J. H. Eberly, *Phys. Rev. A* **6**, 822 (1972).

[121] S. B. Barone and S. Chi, *Opt. Commun.* **3**, 343 (1973).

[122] A. A. Maimistov, A. M. Basharov, S. O. Elyutin, and Yu. M. Sklyarov, *Phys. Rep.* **191**, 1 (1990).

[123] M. Nakazawa, K. Suzuki, Y. Kimura, and H. Kubota, *Phys. Rev. A* **45**, 2682 (1992).

[124] V. L. da Silva, Y. Siberberg, J. P. Heritage, E. W. Chase, M. A. Saifi, and M. J. Andrejco, *Opt. Lett.* **16**, 1340 (1991).

[125] V. V. Kozlov, *J. Opt. Soc. Am. B* **14**, 1765 (1998).

[126] N. Aközbek and S. John, *Phys. Rev. E* **58**, 3876 (1998).

[127] I. V. Mel'nikov, D. Mihalache, N. C. Panoiu, F. Ginovart, and A. Z. Lara, *Opt. Commun.* **191**, 133 (2001).

[128] H. Y. Tseng and S. Chi, *Phys. Rev. E* **66**, 056606 (2002).

[129] B. Luo, H. Y. Tseng, and S. Chi, *J. Opt. Soc. Am. B* **20**, 1866 (2003).

[130] K. Porsezian, P. Seenuvasakumaran, and R. Ganapathy, *Phys. Lett. A* **348**, 233 (2006).

[131] D. V. Skryabin, A. V. Yulin, and F. Biancalana, *Phys. Rev. E* **73**, 045603 (2006).

[132] B. A. Malomed, *Phys. Rev. A* **44**, 6954 (1991); *Phys. Rev. E* **47**, 2874 (1993).

[133] G. I. Barenblatt, *Scaling, Self-similarity, and Intermediate Asymptotics* (Cambridge University Press, New York, 1996).

[134] D. Anderson, M. Desaix, M. Karlsson, M. Lisak, and M. L. Quiroga-Teixeiro, *J. Opt. Soc. Am. B* **10**, 1185 (1993).

[135] K. Tamura and M. Nakazawa, *Opt. Lett.* **21**, 680 (1996).

[136] M. E. Fermann, V. I. Kruglov, B. C. Thomsen, J. M. Dudley, and J. D. Harvey, *Phys. Rev. Lett.* **84**, 6010 (2000).

[137] V. I. Kruglov, A. C. Peacock, J. M. Dudley, and J. D. Harvey, *Opt. Lett.* **25**, 1753 (2000).

[138] V. I. Kruglov, A. C. Peacock, J. D. Harvey, and J. M. Dudley, *J. Opt. Soc. Am. B* **19**, 461 (2002).

[139] S. Boscolo, S. K. Turitsyn, V. Y. Novokshenov, and J. H. B. Nijhof, *Theor. Math. Phys.* **133**, 1647 (2002).

[140] C. Billet, J. M. Dudley, N. Joly, and J. C. Knight, *Opt. Express* **13**, 323 (2005).

[141] C. Finot, F. Parmigiani, P. Petropoulos, and D. Richardson, *Opt. Express* **14**, 3161 (2006).

[142] V. I. Kruglov and J. D. Harvey, *J. Opt. Soc. Am. B* **23**, 2541 (2006).

[143] S. Wabnitz, *IEEE Photon. Technol. Lett.* **19** 507 (2007).

[144] J. Limpert, T. Schreiber, T. Clausnitzer, K. Zöllner, H.-J. Fuchs, E.-B. Kley, H. Zellmer, and A. Tünnermann, *Opt. Express* **10** 628 (2002).

[145] J. Limpert, T. Clausnitzer, A. Liem, T. Schreiber, H.-J. Fuchs, H. Zellmer, E.-B. Kley, and A. Tünnermann, *Opt. Lett.* **28**, 1984 (2003).

[146] F. Röser, J. Rothhard, B. Ortac, A. Liem, O. Schmidt, T. Schreiber, J. Limpert, and A. Tünnermann, *Opt. Lett.* **30**, 2754 (2005).

[147] T. Schreiber, C. K. Nielsen, B. Ortac, J. Limpert, and A. Tünnermann, *Opt. Lett.* **31**, 574 (2006).

[148] P. Dupriez, C. Finot, A. Malinowski, J. K. Sahu, J. Nilsson, D. J. Richardson, K. G. Wilcox, H. D. Foreman, and A. C. Tropper, *Opt. Express* **14** 9611 (2006).

[149] K. J. Blow, N. J. Doran, and D. Wood, *J. Opt. Soc. Am. B* **5**, 1301 (1988).

[150] A. S. Gouveia-Neto, A. S. L. Gomes, and J. R. Taylor, *Opt. Lett.* **12**, 1035 (1987).

[151] K. Suzuki, Y. Kimura, and M. Nakazawa, *Opt. Lett.* **14**, 865 (1989).

[152] B. J. Ainslie, K. J. Blow, A. S. Gouveia-Neto, P. G. J. Wigley, A. S. B. Sombra, and J. R. Taylor, *Electron. Lett.* **26**, 186 (1990).

[153] I. Y. Khrushchev, A. B. Grudinin, E. M. Dianov, D. V. Korobkin, V. A. Semenov, and A. M. Prokhorov, *Electron. Lett.* **26**, 186 (1990).

[154] A. B. Grudinin, E. M. Dianov, D. V. Korobkin, A. Y. Makarenko, A. M. Prokhorov, and I. Y. Khrushchev, *JETP Lett.* **51**, 135 (1990).

[155] M. Nakazawa, K. Kurokawa, H. Kubota, K. Suzuki, and Y. Kimura, *Appl. Phys. Lett.* **57**, 653 (1990).

[156] M. Nakazawa, K. Kurokawa, H. Kubota, and E. Yamada, *Phys. Rev. Lett.* **65**, 1881 (1990).

[157] K. Kurokawa and M. Nakazawa, *IEEE J. Quantum Electron.* **28**, 1992 (1992).

[158] K. Tamura, E. Yoshida, T. Sugawa, and M. Nakazawa, *Opt. Lett.* **20**, 1631 (1995).

[159] D. S. Peter, W. Hodel, and H. P. Weber, *Opt. Commun.* **130**, 75 (1996).

[160] A. Shirakawa, J. Ota, M. Musha, K. Nakagawa, K. Ueda, J. R. Folkenberg, and J. Broeng, *Opt. Express* **13** 1211 (2005).

[161] W. Hodel, D. S. Peter, and H. P. Weber, *Opt. Commun.* **97**, 233 (1993).

[162] M. N. Islam, Ed., *Raman Amplifiers for Telecommunications* (Springer, New York, 2004).

[163] J. Bromage, *J. Lightwave Technol.* **22**, 79 (2004).

[164] C. Headley and G. P. Agrawal, Eds., *Raman Amplification in Fiber Optical Communication Systems* (Academic Press, San Diego, CA, 2005).

[165] A. C. Peacock, N. G. R. Broderick, and T. M. Monro, *Opt. Commun.* **218**, 167 (2003).

[166] C. Finot, G. Millot, C. Billet, and J. M. Dudley, *Opt. Express* **11**, 1547 (2003).

[167] C. Finot, G. Millot, S. Pitois, C. Billet, and J. M. Dudley, *IEEE J. Sel. Topics Quantum Electron.* **10**, 1211 (2004).

[168] C. Finot, *Opt. Commun.* **249**, 553 (2005).

[169] C. Finot and G. Millot, *Opt. Express* **13**, 5825 (2005); *Opt. Express* **13**, 7653 (2005).

[170] C. Finot, J. M. Dudley, and G. Millot, *Opt. Fiber Technol.* **12**, 217 (2006).

[171] J. D. Moores, *Opt. Lett.* **21**, 555 (1996).

[172] V. N. Serkin and A. Hasegawa, *Phys. Rev. Lett.* **85**, 4502 (2000); *IEEE J. Sel. Topics Quantum Electron.* **8**, 418 (2002).

[173] V. I. Kruglov, D. Méchin, and J. D. Harvey, *Opt. Express* **12**, 6198 (2004).

[174] V. I. Kruglov, A. C. Peacock, and J. D. Harvey, *Phys. Rev. Lett.* **90**, 113902 (2003); *Phys. Rev. E* **71**, 056619 (2005).

[175] S. A. Ponomarenko and G. P. Agrawal, *Opt. Express* **15** 2963 (2007).

Chapter 5

Fiber Lasers

A fiber amplifier can be converted into a laser by placing it inside a cavity designed to provide optical feedback. Such lasers are called *fiber lasers*, and this chapter is devoted to them. Section 5.1 covers general concepts such as pumping, cavity design, and laser threshold. The characteristics of continuous-wave (CW) fiber lasers are covered in Section 5.2. It includes a discussion of high-power Yb-doped fiber lasers and the nonlinear phenomena that limit the power level. Sections 5.3 and 5.4 focus on the Q-switching and mode-locking techniques used to generate short optical pulses from fiber lasers. Section 5.3 covers Q-switching as well as active mode-locking schemes, whereas Section 5.4 is devoted to passive mode-locking techniques for which the nonlinear effects play a dominant role. The effects of fiber dispersion and fiber nonlinearities on the mode-locking process are considered in more detail in Section 5.5. The Ginzburg–Landau equation that was the focus of Chapter 4 is used for this purpose and applied to the cases of both the anomalous and normal types of dispersion.

5.1   Basic Concepts

Many rare-earth ions, such as erbium, neodymium, and ytterbium, can be used to make fiber lasers capable of operating over a wide wavelength range extending from 0.4 to 4 μm. The first fiber laser, demonstrated in 1961, used a Nd-doped fiber with the 300-μm core diameter [1]. Low-loss silica fibers were used to make diode-laser-pumped fiber lasers in 1973, soon after such fibers became available [2]. Although there was some research activity in between [3], it was not until the late 1980s that fiber lasers were fully developed. The initial emphasis was on Nd- and Er-doped fibers [4]–[14], but other dopants, such as holmium, samarium, thulium, and ytterbium, were also used [15]–[18]. Starting in 1989, erbium-doped fiber lasers (EDFLs) became the object of intense focus [19]–[23]. Such lasers are capable of producing short optical pulses in the 1.55-μm spectral region and are useful for a variety of applications [24]–[26]. Ytterbium-doped fiber lasers attracted renewed attention after 2000 because of their potential for producing high power levels [27].

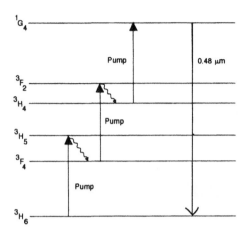

Figure 5.1: Pumping scheme for an up-conversion laser. Energy levels of Tm^{3+} ions are shown. Wavy arrows indicate rapid relaxation between the states. Three 1.06-μm pump photons are absorbed for each photon emitted at 0.48 μm.

5.1.1   Pumping and Optical Gain

Pumping schemes for lasers can be classified as three- or four-level schemes [28]–[30]; both are shown in Figure 4.1. A third kind of pumping scheme is also possible in lasers known as *up-conversion lasers* [31]–[35]. As an example, Figure 5.1 shows pumping of an up-conversion Tm-doped laser. In this pumping scheme, two or more photons from the same pump laser (or from different lasers) are absorbed by the dopant such that it is raised to an excited state whose transition energy exceeds the energy of individual pump photons. As a result, the laser operates at a frequency higher than that of the pump laser, a phenomenon known as *up-conversion* in nonlinear optics. This kind of pumping has attracted attention since it can be used to make "blue" fiber lasers that are pumped with semiconductor lasers operating in the infrared region. In the example shown in Figure 5.1, three 1.06-μm pump photons raise the Tm^{3+} ion to the excited state $^1G_4$. Blue light near 475 nm is emitted though the $^1G_4 \rightarrow {}^3H_5$ transition. Each level in Figure 5.1 is actually an energy band because of host-induced broadening of the atomic transition.

Three- and four-level pumping schemes were discussed in Section 4.2 in the context of fiber amplifiers. EDFLs use a three-level pumping scheme and can be pumped efficiently using semiconductor lasers operating at 0.98 or 1.48 μm. To illustrate the case of a four-level fiber laser, Figure 5.2 shows the energy levels involved in the operation of Nd-doped fiber lasers. Such lasers can be pumped efficiently through the $^4I_{9/2} \rightarrow {}^4F_{5/2}$ transition by using 0.8-μm GaAs semiconductor lasers. They can be designed to operate in the spectral regions near 0.92, 1.06, and 1.35 μm. Pumping is most efficient for the 1.06-μm transition. Although the 1.35-μm transition also corresponds to a four-level pumping scheme, it suffers from the problem of excited-state absorption through the transition $^4F_{3/2} \rightarrow {}^4G_{7/2}$. The first fiber laser in 1961 used the 1.06-μm transition [1]. Fiber lasers pumped using semiconductor lasers were built in

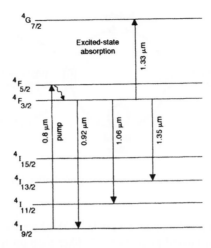

Figure 5.2: Energy levels of a Nd-doped fiber laser pumped at wavelengths near 0.8-μm. Such a laser can emit coherent light in three wavelength regions near 0.92, 1.06, and 1.35 μm.

1973 [2]. Modern Yb-doped fiber lasers can generate CW power levels in excess of 1 kW using arrays of semiconductor lasers for pumping.

5.1.2 Cavity Design

Fiber lasers can be designed with a variety of choices for the laser cavity [24]. The most common type of laser cavity is known as the *Fabry–Perot cavity*, which is made by placing the gain medium between two high-reflecting mirrors. In the case of fiber lasers, mirrors are often butt-coupled to the fiber ends to avoid diffraction losses. This approach was adopted in 1985 for a Nd-doped fiber [4]. The dielectric mirrors were highly reflective at the 1.088-μm laser wavelength but, at the same time, highly transmissive at the pump wavelength of 0.82 μm. Cavity losses were small enough that the laser reached threshold at a remarkably low pump power of 100 μW. Alignment of such a cavity is not easy, since cavity losses increase rapidly with a tilt of the fiber end or the mirror, where tolerable tilts are less than 1°. This problem can be solved by depositing dielectric mirrors directly onto the polished ends of a doped fiber [8]. However, end-coated mirrors are quite sensitive to imperfections at the fiber tip. Furthermore, since pump light passes through the same mirrors, dielectric coatings can be easily damaged when high-power pump light is coupled into the fiber.

Several alternatives exist to avoid passing the pump light through dielectric mirrors. For example, one can take advantage of fiber couplers. It is possible to design a fiber coupler such that most of the pump power comes out of the port that is a part of the laser cavity. Such couplers are called *wavelength-division multiplexing* (WDM) couplers. Another solution is to use fiber gratings as mirrors [36]. As discussed in Chapter 1, a fiber Bragg grating can act as a high-reflectivity mirror for the laser wavelength while being transparent to pump radiation. The use of two such gratings results in an all-fiber Fabry–Perot cavity [37]. An added advantage of Bragg gratings is that the laser can be

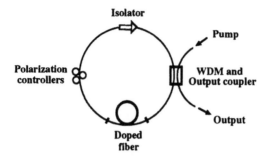

Figure 5.3: A unidirectional ring cavity used often for fiber lasers.

forced to operate in a single longitudinal mode. A third approach makes use of fiber-loop mirrors [38]. As discussed in Section 3.2, fiber-loop mirrors can be designed to reflect the laser light but transmit pump radiation.

Ring cavities are often used to realize unidirectional operation of a laser. In the case of fiber lasers, an additional advantage is that a ring cavity can be made without using mirrors, resulting in an all-fiber cavity. In the simplest design, two ports of a WDM coupler are connected together to form a ring cavity containing the doped fiber, as shown in Figure 5.3. An isolator is inserted within the loop for unidirectional operation. A polarization controller is also needed for conventional doped fiber that does not preserve polarization.

A ring cavity was used as early as 1985 for making an Nd-doped fiber laser [4]. Since then, several new designs have emerged. Figure 5.4 shows a specific design used for mode-locked fiber lasers. This configuration is referred to as the *figure-8 cavity* because of its appearance. The ring cavity on the right acts as a nonlinear amplifying-loop mirror, whose switching characteristics were discussed in Section 3.2. Indeed, the nonlinear effects play an important role in the operation of figure-8 lasers. At low powers, loop transmissivity is relatively small, resulting in relatively large cavity losses for CW operation. The Sagnac loop becomes fully transmissive for pulses whose peak power attains a critical value [see Eq. (3.2.8)]. For this reason, a figure-8 cavity favors mode locking. An isolator in the left cavity ensures unidirectional operation. The laser

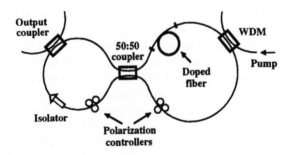

Figure 5.4: A figure-8 cavity used sometimes for mode locking a fiber laser.

output is taken through a fiber coupler with low transmission ($<10\%$) to minimize cavity losses. An interesting property of the figure-8 cavity is that it permits passive mode locking without a saturable absorber.

Many other cavity designs are possible. For example, one can use two coupled Fabry–Perot cavities. In the simplest scheme, one mirror is separated from the fiber end by a controlled amount. The 4% reflectivity of the fiber–air interface acts as a low-reflectivity mirror that couples the fiber cavity with the empty air-filled cavity. Such a compound resonator has been used to reduce the line width of a fiber laser [19]. Three fiber gratings in series also produce two coupled Fabry–Perot cavities. Still another design makes use of a Fox–Smith resonator [39].

5.1.3 Laser Threshold and Output Power

The two most important parameters characterizing a laser are the threshold pump power and the efficiency with which the laser converts the pump power into laser power once it has reached threshold. Laser threshold is determined by requiring that the gain compensate total cavity losses during each round trip [28]–[30]. If we consider a Fabry–Perot cavity, formed by placing two mirrors of reflectivities R_1 and R_2 at the two ends of a fiber of length L, the threshold condition becomes

$$G^2 R_1 R_2 \exp(-2\alpha_{int} L) = 1, \tag{5.1.1}$$

where G is the single-pass amplification factor and α_{int} accounts for internal losses within the cavity.

The single-pass amplification factor should include the nonuniform nature of the gain coefficient and is obtained using

$$G = \exp\left[\int_0^L g(z)dz\right], \qquad g(z) = \sigma_s[N_2(z) - N_1(z)], \tag{5.1.2}$$

where σ_s is the transition cross section and N_1 and N_2 are the dopant densities in the two energy states participating in the stimulated-emission process. By substituting Eq. (5.1.2) in Eq. (5.1.1), the threshold condition becomes

$$\frac{1}{L}\int_0^L g(z)dz = \alpha_{mir} + \alpha_{int} \equiv \alpha_{cav}, \tag{5.1.3}$$

where $\alpha_{mir} = -\ln(R_1 R_2)/2L$ is the effective mirror loss and α_{cav} is the total cavity loss.

The population inversion $N_2 - N_1$ depends on the pumping strength. In general, it is obtained by using a set of three or four rate equations for the energy levels involved in the pumping process. It was calculated in Section 4.2.3 for a three-level laser, and a similar procedure can be followed for a four-level laser. In fact, the calculation is even simpler since $N_1 \approx 0$ and $N_2 \ll N_t$ for a four-level laser, where N_t is the total ion density. For this reason, Eq. (4.2.2) can be replaced with

$$\frac{\partial N_2}{\partial t} = W_p N_t - W_s N_2 - \frac{N_2}{T_1}, \tag{5.1.4}$$

where the transition rates W_p and W_s are given in Eq. (4.2.3). The steady-state solution of Eq. (5.1.4) is given by

$$N_2 = \frac{(P_p/P_p^{\text{sat}})N_t}{1 + P_s/P_s^{\text{sat}}},\qquad(5.1.5)$$

where the saturation powers P_p^{sat} and P_s^{sat} are defined as in Eq. (4.2.5).

The z dependence of N_2 stems from variations in the pump and signal powers along the cavity length. Below or near the laser threshold, gain saturation can be neglected since $P_s/P_s^{\text{sat}} \ll 1$. Using the exponential decrease in the pump power through $P_p(z) = P_p(0)\exp(-\alpha_p z)$, where α_p accounts for pump losses, the integration in Eq. (5.1.3) is easily performed. The pump power needed to reach threshold is found to be given by

$$P_p(0) = \frac{\alpha_{\text{cav}}L}{1 - \exp(-\alpha_p L)}\left(\frac{\alpha_p}{\alpha_s}\right)P_p^{\text{sat}},\qquad(5.1.6)$$

where $\alpha_p = \sigma_p N_t$ and $\alpha_s = \sigma_s N_t$ are the absorption coefficients at the pump and signal wavelengths, respectively. This expression shows how the laser threshold depends on the cavity length. It is common to write the threshold power in terms of the absorbed pump power using

$$P_{\text{abs}} = P_p(0)[1 - \exp(-\alpha_p L)].\qquad(5.1.7)$$

From Eqs. (5.1.6) and (5.1.7), the threshold power P_{th} is given by

$$P_{\text{th}} = P_{\text{abs}} = \alpha_{\text{cav}}L(\alpha_p/\alpha_s)P_p^{\text{sat}} \equiv \alpha_{\text{cav}}L(a_p h\nu_p/\Gamma_s \sigma_s T_1),\qquad(5.1.8)$$

where P_p^{sat} was obtained from Eq. (4.2.5). This equation shows how laser threshold depends on parameters associated with the gain medium (dopants) and the laser cavity.

The output power can also be obtained from the threshold condition (5.1.3), since the saturated gain remains clamped to its threshold value once the pump power exceeds the threshold. By using Eqs. (5.1.2) and (5.1.5) in Eq. (5.1.3), we obtain

$$\frac{\alpha_s}{L}\int_0^L \frac{P_p/P_p^{\text{sat}}}{1 + P_s/P_s^{\text{sat}}}dz = \alpha_{\text{cav}}.\qquad(5.1.9)$$

The integral is difficult to evaluate analytically since the intracavity laser power P_s varies with z along the fiber. However, in most cases of practical interest, mirror reflectivities are large enough that P_s can be treated approximately as constant. The integral then reduces to that evaluated earlier, and P_s is given by the remarkably simple expression

$$P_s = P_s^{\text{sat}}(P_{\text{abs}}/P_{\text{th}} - 1),\qquad(5.1.10)$$

where P_{abs} is the absorbed pump power.

A fraction of the intracavity power P_s is transmitted from each mirror as the output power. The output from the mirror of reflectivity R_1 (or from a port of the output fiber coupler in the case of a ring cavity) can be written as

$$P_{\text{out}} = (1 - R_1)P_s = \eta_s(P_{\text{abs}} - P_{\text{th}}).\qquad(5.1.11)$$

This equation shows that the laser power increases linearly with the absorbed pump power. The slope efficiency, defined as the ratio dP_{out}/dP_{abs}, is given by

$$\eta_s = \left(\frac{1 - R_1}{\alpha_{cav}L} \right) \left(\frac{a_s h\nu_s}{a_p h\nu_p} \right). \tag{5.1.12}$$

The slope efficiency is a measure of the efficiency with which the laser converts pump power into output power once it has reached threshold. It can be maximized by reducing cavity losses as much as possible. Typical values of η_s are around 30% although values as high as 85% are possible in some fiber lasers.

5.2   CW Fiber Lasers

Fiber lasers can be used to generate CW radiation as well as ultrashort optical pulses. This section focuses on the CW operation. The nonlinear effects associated with the host fiber play a relatively minor role for such lasers until power levels exceed several watts. They become important for Yb-doped fiber lasers that are capable of reaching power levels >1 kW.

5.2.1   Nd-Doped Fiber Lasers

Nd-doped fiber lasers were the first to attract attention because they can be pumped with GaAs semiconductor lasers operating near 800 nm. Indeed, such a laser was made as early as 1973 using a silica fiber whose core was codoped with alumina [2]. The graded-index fiber had a core diameter of 35 μm. A Fabry–Perot cavity was made by coating polished ends of an 1-cm-long fiber with dielectric mirrors with high reflectivity (>99.5% at the laser wavelength of 1060 nm). The laser reached threshold its 0.6 mW of pump power; the absorbed pump power was estimated to be 0.2 mW.

Single-mode silica fibers were first used in 1985 for making Nd-doped fiber lasers [4]. The 2-m-long fiber had Nd^{3+} concentration of about 300 ppm (parts per million). The laser cavity was made by butt-coupling the cleaved fiber ends against dielectric mirrors having >99.5% reflectivity at the laser wavelength of 1088 nm and >80% transmission at the 820-nm pump wavelength. The laser threshold was reached at an absorbed pump power of only 0.1 mW. The output power was relatively low. In a later experiment, an Nd-doped fiber laser, pumped with a dye laser, was tunable from 790 to 850 nm [11]. The laser spectrum changed with the pump wavelength λ_p and became 20-nm wide for $\lambda_p > 815$ nm. Figure 5.5 shows the output power as a function of the absorbed pump power at a pump wavelength of 822 nm. The laser reached threshold at a pump power of 1.3 mW, and its output power varied linearly with a further increase in pump power, in agreement with Eq. (5.1.10). Ring cavities have also been used for Nd-doped fiber lasers. As early as 1985, a 2.2-m ring-cavity laser produced 2 mW of output power in each direction at an absorbed pump power of 20 mW when pumped using a 595-nm dye laser [4].

Fiber lasers normally operate in many longitudinal modes because of a large gain bandwidth (>30 nm) and a relatively small longitudinal-mode spacing (<100 MHz). The spectral bandwidth of laser output can exceed 10 nm under CW operation [11].

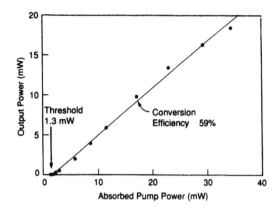

Figure 5.5: Output power as a function of absorbed pump power for an Nd-doped fiber laser. (From Ref. [11]; ©1988 IEE.)

Although a large gain bandwidth is a boon for generating ultrashort pulses, many applications of CW lasers require operation in a narrow-linewidth single mode whose wavelength is tunable over a large part of the gain bandwidth. Several methods have been used to realize narrow-linewidth fiber lasers [40]. An intracavity etalon, commonly used for solid-state lasers, can be used for fiber lasers as well. However, fiber Bragg gratings are preferred for this purpose, since they can be fabricated with a reflectivity spectrum less than 0.1 nm wide. A 1986 experiment used such a grating to realize narrow-band operation (about 16 GHz bandwidth) of a Nd-doped fiber laser [6]. The laser provided output powers in excess of 1.5 mW at 10 mW of input pump power. This laser did not operate in a single longitudinal mode because of a relatively small mode spacing. In a later experiment, operation in a single longitudinal mode was achieved by using a fiber length of only 5 cm [10]. The Nd^{3+} concentration was relatively high to ensure pump absorption over such a short length. Figure 5.6 shows the output power as a function of absorbed pump power together with the observed spectral line shape measured with a self-heterodyne technique. The spectral line width was only 1.3 MHz for this laser.

The large gain bandwidth of fiber lasers is useful for tuning them over a wavelength range exceeding 50 nm [40]. The simplest scheme for tuning replaces one mirror of the Fabry–Perot cavity by a dispersive grating. An Nd-doped fiber laser was tuned in 1986 over the range of 1.07 to 1.14 μm by this technique [5]. An intracavity birefringent filter can also be used for tuning [7]. However, both of these techniques make use of bulky optical components. A remarkably simple technique employs a ring cavity with a fiber coupler, whose coupling efficiency can be varied mechanically (see Figure 5.3). Such lasers are tuned by varying the wavelength for which reflectivity of the fiber coupler becomes maximum and the cavity loss becomes minimum. In a 1989 experiment, an Nd-doped fiber laser could be tuned over 60 nm with this technique [41].

Nd-doped fiber lasers can also operate in the wavelength regions near 0.92 and 1.35 μm. Operation at 0.92 μm requires higher pump powers because of the three-level nature of the laser transition involved. It is necessary to use cavity mirrors that

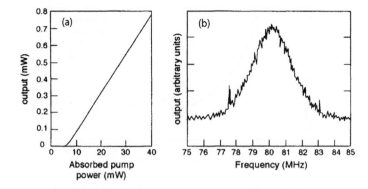

Figure 5.6: (a) Output power versus absorbed pump power and (b) measured spectral line shape for an Nd-doped fiber laser forced to oscillate in a single longitudinal mode through an internal Bragg grating. (From Ref. [10]; ©1988 IEE.)

exhibit a high reflectivity near 0.92 μm with a low reflectivity near 1.06 μm so that cavity losses are high for the latter transition. In a 1986 experiment, such an Nd-doped fiber laser was tuned over 45 nm by using a birefringent filter as a tuning device [7]. It is difficult to operate Nd-doped silica fiber lasers at the 1.35-μm transition because of excited-state absorption (see Figure 5.2). With a proper design, such lasers can be made to operate at wavelengths >1.36 μm because excited-state absorption is less important for such wavelengths [42]. Shorter wavelengths (<1.35 μm) can be obtained by Nd-doping of fluorozirconate fibers. In a 1988 experiment, the laser operated near 1.33 μm when its cavity mirrors were designed to defavor the 1.06-μm transition [10].

A natural question is whether fiber lasers can provide the high power levels (>1 W) that are typical of solid-state lasers. Until 1990, power levels did not exceed 50 mW in most experiments. The situation changed around 1993 with the advent of the double-clad fibers [43]–[45]. In such fibers, the doped core is surrounded by an inner cladding that is used to guide the pump light. The shape and size of this cladding are tailored to match the pump source. A large size and a large numerical aperture of the inner cladding permit efficient coupling of the pump power using GaAs laser-diode bars capable of emitting high powers (>10 W) near 800 nm [44]. Since the pump light is guided by the inner cladding, the laser is pumped all along the length of the doped fiber (side pumping in place of the commonly used end pumping). As early as 1993, 5 W of power was obtained from such a Nd-doped fiber laser [43]. In a 1995 experiment, a double-clad Nd-doped fiber laser emitted 9.2 W of CW power in the form of a high-quality beam when 35 W of pump power was launched into the inner cladding of 400-μm diameter [45]. The 12-μm-diameter core of the double-clad fiber was doped with 1300 ppm of Nd ions.

5.2.2 Yb-Doped Fiber Lasers

Both the Nd and Yb dopants emit light near 1060 nm when the fiber doped with them is pumped at a suitable wavelength. The use of Yb has several advantages, such as

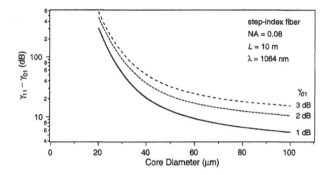

Figure 5.7: Loss difference for the two lowest-order modes as a function of core diameter for a 10-m-long fiber coiled to induce a loss of 1, 2, or 3 dB for the fundamental mode. (From Ref. [56]; ⓒ2000 OSA.)

the absence of excited-state absorption, the wide gain and absorption spectra, and the possibility of obtaining high power levels. Yb-doped fiber lasers tunable over 60 nm were made in 1988 [17], and the tuning range was soon extended to 152 nm covering the range from 1010 to 1162 nm. Output powers of more than 400 mW were obtained by 1995 by pumping a Yb-doped fiber at 850 nm, with only end facets providing 4% reflection to form a cavity [46].

In recent years, the quest for high power has led to a rapid development of Yb-doped fiber lasers and amplifiers [47]–[54]. Already in 1999, a double-clad fiber laser emitted up to 110 W of power under continuous operation at a wavelength near 1.12 μm, while providing a 58.3% optical conversion efficiency [47]. The inner cladding of the Yb-doped fiber had a rectangular cross section. Four semiconductor-laser bars, each emitting 45 W of power near the 915-nm wavelength, were used for pumping the fiber laser. The laser operated in a single spatial mode with 9-μm diameter supported by the narrow fiber core. The use of a single-mode core does not allow increasing the output power much more, because a high power density (approaching 1 GW/cm^2) leads to the onset of the nonlinear effects [55], such as stimulated Raman and Brillouin scattering (SRS and SBS). This problem can be solved only by increasing the core size, but the core would then support multiple spatial modes, unless the core-cladding index difference is also reduced to keep the V parameter of the fiber below 2.4. This is possible, but it reduces numerical aperture of the fiber to below 0.1 even for a core diameter of only 15 μm.

A solution is offered by the fiber-coiling technique [56]. If a multimode fiber is coiled with a suitable radius, continuous bending of the fiber enhances losses for higher-order modes drastically, while the loss of fundamental mode increases by an acceptable amount. Figure 5.7 shows the loss difference for the two lowest-order modes as a function of core diameter for a 10-m-long fiber with a numerical aperture of 0.08. For each value of core diameter, fiber is coiled to induce a loss of 1, 2, or 3 dB for the fundamental mode (bending radii from 1.5 to 4 cm). As is evident, bending losses for all higher-order modes can exceed 20 dB for a fiber with 40-μm core diameter, while loss of the fundamental mode increases by only 1 dB. When such a fiber is used as a

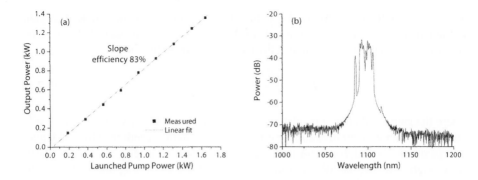

Figure 5.8: (a) Output power as a function of launched pump power for a high-power Yb-doped fiber laser. (b) Optical spectrum at the 1.36-kW power level. (From Ref. [53]; ©2004 OSA.)

fiber amplifier or placed inside a laser cavity, most of the output power appears in the fundamental mode. With such an approach, 500 W of output power was generated in a 2003 experiment using a 35-m-long fiber designed with a 30-μm core diameter and a 0.06 numerical aperture [49]. The power level could be increased to 1.36 kW by 2004 by employing a coiled, 12-m-long, large-mode-area fiber with 40-μm core diameter [53]. Figure 5.8 shows the output power as a function of the launched pump power, together with the measured optical spectrum at the maximum power of 1.36 kW. This laser operated near 1.1 μm with a relatively high slope efficiency of 83% and exhibited nearly diffraction-limited beam quality. Such a high efficiency and kilowatt power levels point to potential industrial applications of fiber lasers (such as welding and metal cutting).

The use of a coiled large-core fiber with a low numerical aperture also suffers from a practical issue. High power levels approaching 10 kW or more would require core diameters approaching 100 μm, and a correspondingly large inner-cladding diameter (up to 1 mm) to accommodate higher pump powers. It would become difficult to coil such fibers to a radius small enough to suppress higher-order modes. An interesting solution consists of developing a helical-core fiber whose core follows a helical trajectory within the inner cladding [57]–[59]. Such a fiber can be made by fabricating a preform with an off-center core and rotating it during the fiber-drawing stage. The core offset also makes pump absorption more efficient. Calculations show that continuous bending of the helical core introduces much larger loss for higher-order modes compared with that for the fundamental mode [59], thus ensuring single-mode output from such a fiber laser. In a preliminary experiment, a Yb-doped fiber laser, containing a fiber with 30-μm-diameter helical core, provided 60.4 W of power at 1043 nm when it was pumped with 92.6 W of power at 976 nm [58]. With further improvements, it should be possible to generate much high power levels.

What limits the maximum power that can be attained with Yb-doped fiber lasers? The answer is related to the onset of the SRS and SBS, two nonlinear phenomena discussed in Chapters 8 and 9 of Ref. [55]. Both of them can transfer power to a new Stokes wave at a different wavelength, resulting in nonlinear losses for the laser beam. Consider first the SRS process. The threshold power level for the onset of SRS is given by $P_{cr} = 16A_{eff}/(g_R L_{eff})$, where A_{eff} is the effective mode area, and the Raman gain coefficient $g_R \approx 1 \times 10^{-13}$ m/W for silica fibers at wavelengths near 1060 nm. The effective fiber length L_{eff} depends on the gain of the doped fiber as

$$L_{eff} = \frac{e^{gL} - 1}{g} = \frac{(G-1)L}{\ln G}, \tag{5.2.1}$$

where the gain coefficient g is related to the unsaturated single-pass gain as $G = e^{gL}$. For an amplifier with 10-dB gain, the effective fiber length exceeds the physical length by a factor of 3.9, and this factor increases to 21.5 when the single-pass gain is 20 dB. As a rough estimate, the SRS threshold is 4.1 kW for a 10-m-long fiber with $G = 10$ and $A_{eff} = 1000$ μm^2. It can be increased to 41 kW by doubling the effective mode area and halving the fiber length, provided the dopant density is increased to maintain the same gain.

In the case of SBS, g_R is replaced with the Brillouin gain coefficient whose value is $g_B \approx 5 \times 10^{-11}$ m/W for silica fibers. At the same time, the factor of 16 is replaced with 21 to account for the backward propagation of the Stokes wave [55]. Clearly, the onset of SBS can occur at much lower power levels as g_B is larger by a factor of about 500 compared with g_R. However, the narrowband nature of the Brillouin gain spectrum (bandwidth $\Delta \nu_B \sim 50$ MHz) helps considerably. If the spectral bandwidth $\Delta \nu_L$ of the laser is larger than $\Delta \nu_B$, only a fraction of its power can fall with the SBS gain bandwidth. Although a complete analysis of this situation is complicated [60], when $\Delta \nu_L \gg \Delta \nu_B$, the SBS threshold power level can be approximated by

$$P_{cr} = \frac{21A_{eff}}{g_B L_{eff}} \left(\frac{\Delta \nu_L}{\Delta \nu_B} \right). \tag{5.2.2}$$

As the threshold power increases by a large factor, the ratio $\Delta \nu_L / \Delta \nu_B$ can offset the larger value of g_B compared with g_R when the laser bandwidth exceeds 25 GHz. However, the onset of SBS becomes a serious problem if a narrowband signal with $\Delta \nu_L \ll \Delta \nu_B$ is amplified inside a Yb-doped fiber amplifier.

5.2.3   Erbium-Doped Fiber Lasers

EDFLs can operate in several wavelength regions, ranging from visible to far infrared. The 1.55-μm region has attracted the most attention because it coincides with the low-loss region of silica fibers used for optical communications. At first sight, 1.55-μm EDFLs do not appear very promising because the transition $^4I_{13/2} \rightarrow {}^4I_{15/2}$ terminates in the ground state of the Er^{3+} ion. Since a three-level laser requires that at least half of the ion population be raised to the excited state, it has a high threshold. Indeed, early attempts to make EDFLs used high-power argon-ion lasers as a pump source [5]. The threshold pump powers were ~ 100 mW with slope efficiencies $\sim 1\%$. In one experiment, the slope efficiency was improved to 10%, but the laser reached threshold at 44 mW [19].

EDFLs pumped near 0.8-μm using GaAs semiconductor lasers suffer from the problem of excited-state absorption. The situation improves in silica fibers sensitized with ytterbium [13]. The core of such fibers is codoped with Yb_2O_3 such that the ratio of Yb^{3+} to Er^{3+} concentrations is more than 20. Such EDFLs can be pumped using 0.8-μm semiconductor lasers or miniature 1.06-μm Nd:YAG lasers. The improved performance of Yb-sensitized EDFLs is due to the near coincidence of the $^2F_{5/2}$ state of Yb^{3+} ions with the $^4I_{11/2}$ state of Er^{3+} ions. The excited state of Yb^{3+} is broad enough that it can be pumped in the range from 0.85 to 1.06 μm. In one Er:Yb fiber laser, pumped with a 820-nm semiconductor laser, the threshold pump power was 5 mW with a slope efficiency of 8.5% [12].

The performance of EDFLs improves considerably when they are pumped at wavelengths near 0.98 or 1.48 μm because of the absence of excited-state absorption. Indeed, semiconductor lasers operating at these wavelengths were initially developed solely for the purpose of pumping Er-doped fibers. Their use by 1995 resulted in commercial 1.55-μm fiber lasers. As early as 1989, a 980-nm-pumped EDFL exhibited a slope efficiency of 58% against absorbed pump power [20], a figure that is close to the quantum limit of 63% obtained by taking the ratio of signal to pump photon energies. EDFLs pumped at 1.48 μm also exhibit good performance. In fact, the choice between 0.98 and 1.48 μm is not always clear, since each pumping wavelength has its own merits. Both pumping wavelengths have been used for developing practical EDFLs with excellent performance characteristics [61]–[68].

An important property of continuously operating EDFLs from a practical standpoint is their ability to provide output that is tunable over a wide range. Similar to the case of Nd-doped fiber lasers, many techniques can be used to reduce the spectral bandwidth of tunable EDFLs [40]. In a 1989 experiment [19], an intracavity etalon formed between a bare fiber end and the output mirror led to a 620-MHz line width even though the fiber was 13 m long. The laser wavelength can also be tuned by using an external grating in combination with an etalon. Figure 5.9 shows the experimental setup together with the tuning curves obtained for two different fiber lengths. This laser was tunable over a 70-nm range [23]. The output power was more than 250 mW in the wavelength range from 1.52 to 1.57 μm.

Ring cavities can also be used to make tunable EDFLs [62]–[67]. A common technique uses a fiber intracavity etalon that can be tuned electrically. By 1991, such EDFLs exhibited a low threshold (absorbed pump power of 2.9 mW) with 15% slope efficiency and they could be tuned over 60 nm [64]. EDFLs can also be designed to provide a line width as small as 1.4 kHz [63]. In an optimized EDFL, 15.6-mW of output power was obtained with 48% slope efficiency (68% with respect to absorbed pump power) while the tuning range (at the 3-dB point) was 42 nm [67].

Many other tuning techniques have been used for fiber lasers. In one experiment, a fiber laser was tuned over 33 nm through strain-induced birefringence [66]. In another, a fiber laser could be tuned over 39 nm by using a reflection Mach–Zehnder interferometer that acts as a wavelength-selective loss element within the ring-laser cavity [68]. The wavelength for which cavity losses are minimum is changed by controlling the optical path length in one of the interferometer arms either electro-optically or by applying stress.

Fiber gratings can also be used to improve the performance of EDFLs [69]. As early as 1990, a Bragg grating was used to realize a line width of about 1 GHz [36]. Since then, fiber gratings have been used for a variety of reasons [70]–[77]. The sim-

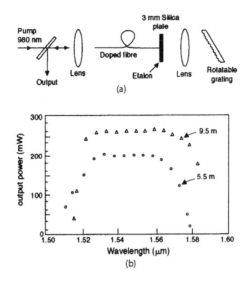

Figure 5.9: (a) Experimental setup for a broadly tunable EDFL. (b) Tuning curves for two fiber lengths at 540 mW of launched power. (From Ref. [23]; ©1989 IEE.)

plest configuration splices a Bragg grating at each end of an erbium-doped fiber, forming a Fabry–Perot cavity. Such devices are called *distributed Bragg reflector* (DBR) lasers, following the terminology used for semiconductor lasers [78]. DBR fiber lasers can be tuned continuously while exhibiting a narrow line width [70]. They can also be made to oscillate in a single longitudinal mode by decreasing the fiber length. In a novel scheme, an EDFL was made to oscillate at two distinct wavelengths, with a narrow line width at each wavelength, by fabricating two different gratings or by using a single grating with dual-peak reflectivity [71].

Multiple fiber gratings can be used to make coupled-cavity fiber lasers. Such lasers have operated at two wavelengths (0.5 nm apart) simultaneously such that each spectral line was stable to within 3 MHz and had a line width of only 16 kHz [72]. Fiber gratings have been used to make efficient, low-noise EDFLs. In one such laser, up to 7.6 mW of output power was obtained without self-pulsation while the relative intensity-noise level was below −145 dB/Hz at frequencies above 10 MHz [73]. Even higher powers can be obtained by using the master oscillator–power amplifier (MOPA) configuration, in which a fiber laser acting as a master oscillator is coupled to a fiber amplifier through an intracore Bragg grating. Output powers of up to 62 mW have been obtained by using such a configuration through active feedback while maintaining intensity-noise levels below −110 dB/Hz at all frequencies [74].

5.2.4   DFB Fiber Lasers

An interesting approach consists of making a distributed feedback (DFB) fiber laser [79]–[86]. In analogy with DFB semiconductor lasers [78], a Bragg grating is formed directly into the erbium-doped fiber that provides gain. As discussed in Chapter 1,

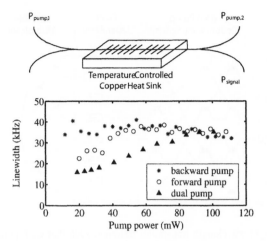

Figure 5.10: Laser configuration and the measured line width as a function of pump power for three pumping configurations. (From Ref. [85]; ©2006 IEEE.)

several variations of the basic design, such as a phase-shifted grating, a chirped grating, or a sampled grating, can be employed to improve the performance of DFB fiber lasers. Multiple gratings with slightly different Bragg wavelengths can also be formed into the same doped fiber, resulting in several DFB lasers cascaded together.

The chief advantage of a DFB laser is that it can operate in a single longitudinal mode whose line width can be made relatively narrow ($\sim$10 kHz). Figure 5.10 shows the laser configuration together with the measured line widths as a function of pump power for three pumping configurations [85]. The DFB laser consists of a 5-cm-long fiber Bragg grating with a phase shift offset from the center by 3 mm. The laser is placed on a copper heat sink and is pumped at 977-nm using two semiconductor lasers. The measured laser line widths vary in the range of 20–40 kHz depending on the pump powers and the pumping scheme. A surprising aspect of the data is that they do not follow the inverse dependence on laser power that is expected from theory [78]. In fact, the line width appears to increase with output power almost linearly in the case of bidirectional pumping. It turns out that pump-noise-induced temperature fluctuations are mainly responsible for this anomalous behavior.

It is useful from a practical standpoint if the laser wavelength can be tuned, while maintaining a narrow line width. Since this wavelength is set by the Bragg wavelength λ_B related to the grating period Λ as $\lambda_B = 2\bar{n}\Lambda$, it is possible to tune the laser wavelength by changing the effective mode index $\bar{n}$ or the physical length Λ. The effective index $\bar{n}$ can be varied to some extent by changing temperature or by applying stress but the tuning range is often limited to 6 nm or so. The grating period Λ can be changed by stretching or compressing the fiber. A compression technique was employed in 1994 to tune a DBR fiber laser over 32 nm [74]. In a 2002 experiment, a tuning range of 27 nm was realized by embedding the laser into a bendable material [83].

Multiwavelength optical sources, capable of emitting light at several well-defined wavelengths simultaneously, are useful for WDM lightwave systems. Fiber lasers can

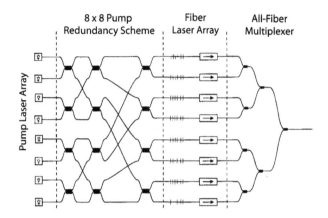

Figure 5.11: An eight-wavelength transmitter designed such that each pump laser pumps all DFB lasers. (From Ref. [82]; ©1999 IEEE.)

be used for this purpose, and several schemes have been developed [87]–[91]. A dual-frequency fiber laser was demonstrated in 1993 by using a coupled-cavity configuration [71]. Later, simultaneous operation of a fiber laser at up to 29 wavelengths was realized by cooling the doped fiber to 77 K to reduce the homogeneous broadening of the gain spectrum to below 0.5 nm [89]. Long cavities with several meters of doped fibers can also be used. Wavelength selection is then made using an intracavity comb filter [87]. In a dual-filter approach, a tunable comb filter in combination with a set of fiber gratings provides a multiwavelength source that is switchable on a microsecond timescale to precise preselected wavelengths [90].

In a different approach, the outputs of multiple DFB fiber lasers are combined to produce a multiwavelength source. In a 1997 experiment, five gratings with slightly different Bragg wavelengths were cascaded serially and pumped with a single semiconductor laser [81]. In another design, multiple DFB lasers operating at wavelengths 0.4-nm apart were combined in a parallel configuration to realize a 16-wavelength transmitter suitable for WDM applications [82]. Fiber couplers were used to combine the output of different lasers as well as to use pump lasers such that each pump provided input to all lasers. Figure 5.11 shows the design of such an eight-wavelength transmitter schematically. A single high-power laser can also be used to pump all DFB lasers simultaneously. In a 2003 experiment, a Yb-doped fiber laser with 1-W output fiber was used for this purpose [84].

Some applications require high power levels with a narrow spectral bandwidth and low intensity noise. For this reason, there is considerable interest in fiber DFB lasers capable of operating at high powers [92]–[95]. The MOPA configuration, in which the output of a DFB master oscillator is amplified inside a high-gain power amplifier, is most suitable for this purpose. As early as 1994, the MOPA configuration was used to provide 60-mW output [74]. The output power could be increased to 166 mW in a 1999 design while maintaining a relative intensity noise (RIN) of less than −160 dB/Hz. By 2003, MOPAs were able to provide close to 100-W output power [94].

Figure 5.12 shows the design of a MOPA emitting 83 W at 1552 nm [95]. The

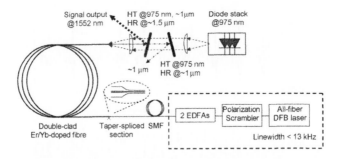

Figure 5.12: A high-power MOPA. The low-power stage (dashed box) consist of a DFB fiber laser and two EDFAs. SMF, HR, and HT stand for single-mode fiber, high reflection, and high transmission, respectively. (From Ref. [95]; ©2004 IEEE.)

low-power stage (dashed box) consist of a DFB fiber laser (master oscillator) emitting 10 mW of power with a 13-KHz line width, a polarization scrambler, and two EDFAs that boost the power to close to 2 W. The high-power amplifier is made with 3.5-m-long double-clad fiber (core diameter 30 μm) pumped backward at 975 nm. Two dichroic mirrors ensure that most of the pump power is injected into the fiber without any feedback. The line width of the amplified output was close to its input value. The maximum power that could be extracted was limited by the onset of SBS. For a narrowband signal, the SBS threshold power is obtained from Eq. (5.2.2) after setting $\Delta v_L / \Delta v_B = 1$ (estimated to be 84 W). As mentioned earlier, SBS is the most limiting nonlinear phenomenon when a narrowband signal is amplified inside a fiber amplifier. Nevertheless, with a suitable design, power levels close to 500 W have been realized in Yb-doped fiber amplifiers [96]. It should also be noted that amplification of optical pulses does not suffer from the SBS problem if the pulse is much shorter than the lifetime of acoustic phonons [55].

5.2.5 Self-Pulsing and Chaos

Some fiber lasers emit a train of optical pulses even when pumped continuously. This phenomenon is referred to as *self-pulsing* and is a specific example of laser instabilities that occur in many kinds of lasers [97]. Its occurrence requires a nonlinear mechanism within the laser cavity. Self-pulsing in EDFLs has been observed, and its origin is attributed to two nonlinear mechanisms [98]–[102]. In one study, ion–ion interactions in erbium clusters were found to produce self-pulsing [98]. Another model shows that self-pulsing can result from destabilization of relaxation oscillations [100], the same mechanism that leads to self-pulsing in semiconductor lasers [97]. This origin of self-pulsing was confirmed in an experiment in which the Er-doped fiber was codoped with alumina to minimize production of erbium-ion clusters within the silica core [102]. In fact, the repetition rate of pulses agreed quite well with relaxation–oscillation frequency. A rate-equation model, generalized to include the excited-state absorption of pump radiation, reproduced most of the features of self-pulsing observed experimentally.

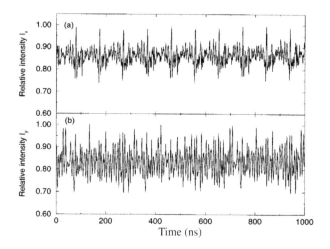

Figure 5.13: Chaotic power fluctuations for the two polarization components observed experimentally at the output of an EDFL. (From Ref. [106]; ©1997 APS.)

It is well known that self-pulsing often leads to optical chaos in the laser output, following a period-doubling or a quasi-periodic route [97]. Chaotic behavior in fiber lasers is attributed to several different nonlinear mechanisms [103]–[108]. Since fiber lasers constitute an example of class-B lasers, the single-mode rate equations do not predict chaos [97]. However, chaos can be induced through pump modulation, optical feedback, or external injection [99]. In the case of fiber lasers, chaos can also occur because of nonlinear coupling between the orthogonally polarized components of the optical field. In an interesting set of experiments, chaos in an EDFL originated from the nonlinear polarization dynamics occurring on a timescale shorter than the round-trip time inside the laser cavity [106]. The two polarization components inside the laser cavity were coupled nonlinearly through cross saturation and gain sharing. A polarization controller inside the cavity acted as a half-wave plate and introduced additional coupling. Figure 5.13 shows an example of chaotic power fluctuations occurring for the two polarization components. Depending on the pumping and loss levels, a variety of chaotic patterns were observed experimentally. The experimental data can be modeled quite well using a stochastic delay difference model.

In general, the outputs of two chaotic fiber lasers are not synchronized. This feature can be used for transmitting data in a secure manner if the signal is generated by modulating the output of a chaotic laser. Several experiments have used chaotic fiber lasers to demonstrate the possibility of secure optical communications [109]–[111]. In one set of experiments, the signal was imposed on the chaotic waveform by injecting it into the EDFA within the laser cavity [109]. The laser output was then transmitted through a fiber link (as long as 35 km). At the receiver end, a part of the chaotic signal was injected into another EDFL, designed to be nearly identical to the one at the transmitter, for chaos synchronization. The data could be recovered from the remaining received signal because of this synchronization. Further details are available in a 2005 review of this topic [111].

5.3 Short-Pulse Fiber Lasers

Two techniques used for generating short optical pulses from lasers are known as *Q-switching* and *mode locking* [28]–[30]. Both of them can be initiated actively or passively. Q-switching generates nanosecond pulses while the width of mode-locked pulses can be <100 fs. With a suitable design, fiber lasers can produce pulses whose energy exceeds 1 mJ and peak power exceeds 100 kW. This section focuses on Q-switching and active mode locking. The techniques employed for passive mode locking are discussed in Section 5.4.

5.3.1 Q-Switched Fiber Lasers

Active Q-switching requires a device that increases cavity losses but acts like a gate that opens periodically for a short duration over which losses are reduced for allowing the buildup of a Q-switched pulse. An intracavity acousto-optic modulator was used as early as 1986 to generate Q-switched optical pulses from a fiber laser [5]. Passive techniques for Q-switching were used soon after. In a 1999 experiment, passively Q-switched pulses with 0.1 mJ energy were generated using a MOPA configuration in which two erbium-doped fibers (lengths 60 and 79 cm) with a relatively large mode area of 300 μm^2 were employed [112]. A semiconductor Bragg mirror was used as a saturable absorber for initiating the passive Q-switching.

Q-switching remains a useful technique for fiber lasers as it can generate high peak-power (>1 kW) pulses whose wavelengths are tunable over a wide range covering the entire gain spectrum [113]–[119]. In a 1999 experiment, Q-switched pulses from a double-clad Yb-doped fiber laser could be tuned from 1060 to 1100 nm, while maintaining peak powers as large as 2 kW [113]. Soon after, pulse energies as high as 2.3 mJ at a repetition rate of 500 Hz and average powers >5 W at higher repetition rates were obtained from cladding-pumped Yb-doped fiber lasers [114]. By 2004, 37-ns wide pulses with energies of 1.2 mJ were generated at a 10-kHz repetition rate [118]. The laser operated in a single spatial mode with an output beam of high quality in spite of the 40-μm diameter of the fiber. Most of these experiments employed an active Q-switching technique and produced pulses whose widths depended on the fiber length and cavity losses (through the cavity lifetime) as well as on the pump power and varied from 40 ns to ~10 μs. In general, pulses become shorter at higher pump powers and for shorter fiber lengths.

The SRS and SBS nonlinear processes affect Q-switched fiber lasers in several ways. Similar to the case of CW lasers, both of them can limit the maximum peak power that can be realized in practice. For this reason, relatively short fibers (<10 m) with a large effective mode area (>400 μm^2) are often used for generating high peak powers. It turns out, however, that SRS and SBS can also be used to advantage to enhance the Q-switching process [120]–[125]. In a 1997 experiment, self-Q-switching of a Yb-doped fiber laser was initiated by distributed Rayleigh and Brillouin backscattering and followed with the onset of SBS [120]. Moreover, the spectrum of the pulse train was broadened by cascaded SRS and formed almost a supercontinuum.

In a 1998 experiment, peak powers of Q-switched pulses were observed to be enhanced by more than a factor of 10 by the onset of SBS inside a Nd-doped fiber laser

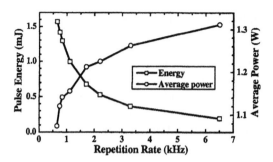

Figure 5.14: Measured energy and average power of Q-switched pulses as a function of the repetition rate for a Yb-doped fiber laser in which SBS helped the Q-switching process. (From Ref. [123]; ©2004 OSA.)

[121]. The laser also emitted much shorter pulses (as short as 2 ns) compared with those forming with active Q-switching. The role of SBS became clear when a 10-m-long piece of single-mode fiber was added to the Nb-doped fiber. The backward propagating Stokes wave generated inside the passive fiber enters the laser cavity in the form of short pulses whose duration is governed by the SBS dynamics. This SBS-induced feedback increases the Q factor of the laser cavity by a large factor over the short duration of each pulse. Because of the stochastic nature of the SBS process, the repetition rate of these short pulses is not stable. The use of an acousto-optic modulator stabilizes the pulse repetition rate. With this SBS-enhanced Q-switching, a peak power of 175-kW for Q-switched pulses was realized by 2004 by using a 20-m-long Yb-doped fiber (with a 10.6-μm core) pumped with up to 8.3 W of power at 975 nm [123]. Figure 5.14 shows how the energy and average power of pulses varied with the repetition rate set by the modulator. Emitted Q-switched pulses were about 5 ns wide with energies $\sim$1 mJ. Moreover, their center wavelength could be tuned over more than 60 nm with the help of a diffraction grating.

The interesting question from a physics standpoint is why the onset of SBS leads to self-Q-switching with relatively short pulses. It is well known that SBS can lead to self-pulsing even in undoped fibers [55]. It is thus not surprising that it can occur in lasers containing doped fibers that provide gain and make it easier to reach the SBS threshold. Indeed, self-pulsing was observed in a 2000 experiment in which a 4-m-long Yb-doped fiber was placed inside a Fabry–Perot cavity and pumped continuously [122]. Pulses were close to 100-ns wide and their width and peak power varied from pulse to pulse. This regime is distinct from the one in which much narrower Q-switch pulses with high peak powers are formed. To understand the formation of such pulses, it is necessary to employ the coupled amplitude equations governing SBS dynamics and given in Section 9.4 of Ref. [55]. Moreover, one should consider multiple Stokes wave generated through cascaded SBS and include noise sources responsible for spontaneous Rayleigh and Brillouin scattering [124].

SRS can also initiate passive Q-switching if the Yb-doped fiber is followed with a relatively long Ge-doped fiber in which Raman scattering occurs [125]. The second passive fiber acts initially as a high-loss element but provides Raman gain after the

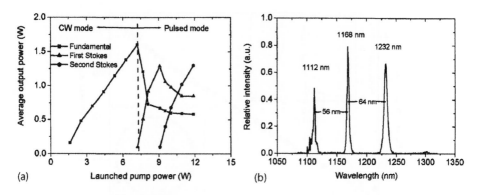

Figure 5.15: (a) Average output power as a function of launched pump power and (b) the three-peak spectrum observed at a high power for a Yb-doped fiber laser in which SRS helped the Q-switching process. (From Ref. [125]; ©2006 OSA.)

onset of SRS, the two features similar to that of a saturable absorber. Such a laser operates continuously at low pump powers but emits Q-switched pulses when pump power is large enough to initiate SRS within the passive fiber. Figure 5.15 shows the average output power of such a laser as a function of launched pump power together with the three-peak spectrum observed at high powers. The 1232-nm peak is formed when the Stokes power at 1168 nm becomes large enough to reach the SRS threshold. The temporal duration of pulses emitted at the three wavelengths was 150, 70, and 60 ns. Their repetition rate depended on the cavity length and was <1 MHz under the experimental conditions.

In a different approach, high-energy pulses are produced by employing a MOPA design in which optical pulses from a nonfiber master oscillator are amplified inside a Yb-doped fiber amplifier [126]–[128]. By 2005, this approach provided pulses with the highest peak powers (>1 MW). In one experiment, a semiconductor laser operating at 1064 nm acted as a maser oscillator. When 500-ns pulses from this laser were amplified through a chain of three Yb-doped fiber amplifiers, output pulses with energies of up to 82 mJ could be produced [126]. The last amplifier employed a 3.5-m-long double-clad fiber containing a 200-μm-diameter core with a 600-μm-diameter inner cladding. In a 2006 experiment, the three-stage fiber amplifier was seeded by a passively Q-switched microchip laser emitting 1-ns-wide pulses with 6 μJ energy at a 9.6-kHz repetition rate [128]. The last amplifier was made with a 90-cm-long, Yb-doped, rodlike photonic crystal fiber with 100-μm-diameter core, 290-μm-diameter inner cladding, and 1500-μm-diameter outer cladding. Figure 5.16 shows energy of the amplified pulse as a function of incident pump power, with two insets displaying the pulse spectrum at 4.3 μJ and the cross section of the fiber used for the last stage.

It is clear from the preceding discussion that Q-switching is a useful technique for fiber lasers designed to produce intense nanosecond pulses. If shorter pulses are required, one must employ the mode-locking technique capable of generating pulses shorter than 100 fs. Historically, Nd-doped fiber lasers produced 120-ps mode-locked pulses in 1988 using a laser-diode array for pumping [14]. Starting in 1989, attention

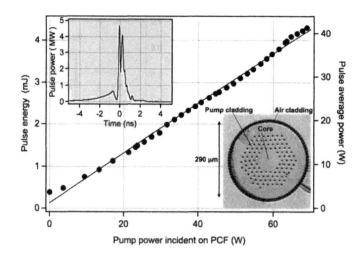

Figure 5.16: Energy of amplified pulses (circles) and their average power (solid line) as a function of incident pump power for a MOPA designed with three stages of Yb fiber amplifiers. Two insets display the pulse spectrum for 4.3-μJ pulses and the cross section of the fiber used for the last stage. (From Ref. [128]; ©2006 AIP.)

focused on the development of mode-locked EDFLs because of their potential applications in lightwave systems [26]. We turn next to the physics behind mode locking.

5.3.2   Physics of Mode Locking

Fiber lasers operate simultaneously in a large number of longitudinal modes falling within the gain bandwidth. The frequency spacing among the modes is given by $\Delta v = c/L_{\text{opt}}$, where L_{opt} is the optical length during one round trip inside the cavity. Multimode operation is due to a wide gain bandwidth compared with a relatively small mode spacing of fiber lasers ($\Delta v \sim 10$ MHz). The total optical field can be written as $E(t) = \sum_{m=-M}^{M} E_m \exp(i\phi_m - i\omega_m t)$, where E_m, ϕ_m, and ω_m are the amplitude, phase, and frequency of a specific mode among $2M + 1$ modes. If all modes operate independently of each other with no definite phase relationship among them, the interference terms in the total intensity $|E(t)|^2$ averages out to zero. This is the situation in multimode CW lasers.

Mode locking occurs when phases of various longitudinal modes are synchronized such that the phase difference between any two neighboring modes is locked to a constant value ϕ such that $\phi_m - \phi_{m-1} = \phi$. Such a phase relationship implies that $\phi_m = m\phi + \phi_0$. The mode frequency ω_m can be written as $\omega_m = \omega_0 + 2m\pi\Delta v$. If we use these relations and assume for simplicity that all modes have the same amplitude E_0, the sum over modes can be performed analytically. The result is given by [30]

$$|E(t)|^2 = \frac{\sin^2[(2M+1)\pi\Delta vt + \phi/2]}{\sin^2(\pi\Delta vt + \phi/2)} E_0^2. \tag{5.3.1}$$

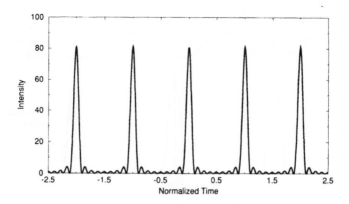

Figure 5.17: Pulse train formed when nine modes of equal amplitudes are mode locked.

The total intensity $|E(t)|^2$ is shown in Figure 5.17 for nine coupled modes ($M = 4$). It is a periodic function of time with period $\tau_r = 1/\Delta v$, which is just the round-trip time inside the laser cavity. The laser output is in the form of a pulse train whose individual pulses are spaced by τ_r. A simple way to interpret this result is that a single pulse circulates inside the laser cavity and a fraction of its energy is emitted by the laser each time the pulse arrives at the output coupler.

The pulse width is estimated from Eq. (5.3.1) to be $\tau_p = [(2M+1)\Delta v]^{-1}$. Since $(2M+1)\Delta v$ represents the total bandwidth of all phase-locked modes, the pulse width is inversely related to the spectral bandwidth over which phases of various longitudinal modes can be synchronized. The exact relationship between the pulse width and the gain bandwidth Δv_g depends on the nature of gain broadening (homogeneous versus inhomogeneous).

5.3.3   Active Mode Locking

Active mode locking requires modulation of either the amplitude or the phase of the intracavity optical field at a frequency f_m equal to (or a multiple of) the mode spacing Δv. It is referred to as AM (amplitude modulation) or FM (frequency modulation) mode locking depending on whether amplitude or phase is modulated. One can understand the locking process as follows. Both the AM and FM techniques generate modulation sidebands, spaced apart by the modulation frequency f_m. These sidebands overlap with the neighboring modes when $f_m \approx \Delta v$. Such an overlap leads to phase synchronization. The mode-locking process can be modeled by using a set of multimode rate equations in which the amplitude of each mode is coupled to its nearest neighbors [28].

One can also understand the process of pulse formation in the time domain. Figure 5.18 shows the case of AM mode locking in which cavity losses are modulated at the frequency Δv. Since the laser generates more light at the loss minima, the intracavity field is modulated at the same frequency. This slight intensity difference builds up on successive round trips, and the laser emits a train of mode-locked pulses in the steady

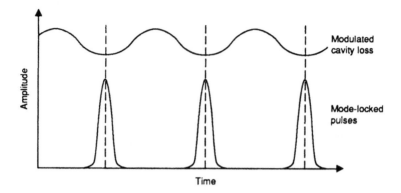

Figure 5.18: Active AM mode locking through modulation of cavity losses. Pulses form at locations where cavity loss is minimum.

state. Stated another way, the laser threshold is lower for pulsed operation. As a result, the laser emits a train of pulses in place of the CW output.

The time-domain theory of mode locking considers the evolution of a mode-locked pulse inside the laser cavity over one round trip [129]–[131]. As usual, even though the pulse amplitude $A(z,t)$ is modified by the gain medium and the modulator, it should recover its original value after one round trip under steady-state conditions. We can adapt the Ginzburg–Landau equation, derived in Section 4.3 for fiber amplifiers, to the case of fiber lasers by adding the losses introduced by the amplitude modulator and cavity mirrors. This requires replacing the loss parameter α in Eq. (4.3.15) with

$$\alpha = \alpha_c + \alpha_M[1 - \cos(\omega_M t)], \qquad (5.3.2)$$

where α_c accounts for all cavity losses and α_M is the additional loss, periodic at the frequency ω_M, introduced by the modulator. In the context of fiber lasers, Eq. (4.3.15) becomes

$$\frac{\partial A}{\partial z} + \frac{i}{2}(\beta_2 + ig_c T_2^2)\frac{\partial^2 A}{\partial t^2} = i\left(\gamma + \frac{i}{2}\alpha_2\right)|A|^2 A + \frac{1}{2}(g_c - \alpha)A, \qquad (5.3.3)$$

where g_c is the saturated gain averaged over the cavity length L. The parameter T_2 is related inversely to the gain bandwidth as $T_2 = 1/\Omega_g$. This equation is sometimes called the *master equation of mode locking* [131].

Consider first the case in which the effects of group-velocity dispersion (GVD) and self-phase modulation (SPM) can be ignored by setting $\beta_2 = 0$ and $\gamma = 0$ in Eq. (5.3.3). Two-photon absorption can also be neglected ($\alpha_2 = 0$). If we use $\cos(\omega_M t) \approx 1 - \frac{1}{2}(\omega_M t)^2$ in Eq. (5.3.2), assuming that the pulse width is much shorter than a modulation cycle, Eq. (5.3.3) takes the form

$$\frac{\partial A}{\partial z} = \frac{g_c}{2}\left(A + \frac{1}{\Omega_g^2}\frac{\partial^2 A}{\partial t^2}\right) - \frac{1}{2}\left(\alpha_c + \frac{1}{2}\alpha_M \omega_M^2 t^2\right)A. \qquad (5.3.4)$$

In the steady state, we look for solutions of the form $A(z,t) = B(t)\exp(iKz)$, where $B(t)$ governs the pulse shape that does not change from one round trip to next. The resulting ordinary differential equation for $B(t)$ is identical to that of a harmonic oscillator and has the following solution in terms of the Hermite–Gauss functions [131]:

$$B_n(t) = C_n H_n(t/T_0) \exp[-\tfrac{1}{2}(t/T_0)^2], \tag{5.3.5}$$

where $n = 0, 1, 2, \ldots$, H_n is the Hermite polynomial of order n,

$$T_0 = [2g_c/(\alpha_M \Omega_g^2 \omega_M^2)]^{1/4} \tag{5.3.6}$$

is a measure of the width of mode-locked pulses, and C_n is a normalization constant related to the pulse energy.

The propagation constant K should be real in the steady state. This requirement leads to the condition

$$g_c = \alpha_c + \alpha_M \omega_M^2 T_0^2 (n + \tfrac{1}{2}), \tag{5.3.7}$$

and it provides the saturated gain needed for various Hermite–Gauss temporal modes supported by the laser. Since the lowest gain occurs for $n = 0$, an actively mode-locked laser emits a Gaussian pulse with a full width at half maximum (FWHM) $T_p \approx 1.665 T_0$. The pulse width depends on both the gain bandwidth Ω_g and the modulator frequency $\Delta \nu = \omega_M / 2\pi$, where $\Delta \nu$ is the longitudinal-mode spacing of the laser cavity.

The situation changes considerably when the effects of GVD and SPM are included in Eq. (5.3.3). In the absence of gain and losses, this equation reduces to the nonlinear Schrödinger (NLS) equation and has the soliton solutions discussed in Chapter 5 of Ref. [55]. The solution corresponding to the fundamental soliton is given by

$$A(z,t) = \sqrt{P_0} \operatorname{sech}(t/T_s) \exp(iz/2L_D), \tag{5.3.8}$$

where the peak power P_0 and the width T_0 are related by the usual soliton condition

$$N = \gamma P_0 T_0^2 / |\beta_2| = 1 \tag{5.3.9}$$

for a soliton of order N.

The important question is how the NLS solitons are affected by the gain and losses inside the laser cavity. If the pulse formation is dominated by the GVD and SPM effects, one should expect the mode-locked pulse to behave as a fundamental soliton, and it should have a "sech" shape in place of the Gaussian shape predicted in the absence of GVD and SPM. Soliton perturbation theory has been used to find the width of the steady-state soliton pulse. The results show that the mode-locked pulse is shorter than that predicted by Eq. (5.3.5) when soliton effects are significant. The maximum possible reduction factor is limited by [132]

$$(T_0/T_s)^2 < \tfrac{1}{2}\left(3d_g + \sqrt{9d_g^2 - \pi^2}\right), \qquad d_g = \operatorname{Re}\left(\sqrt{1 + i|\beta_2|\Omega_g^2/g_c}\right). \tag{5.3.10}$$

In general, pulses become shorter as the amount of anomalous GVD increases; reduction by a factor of 2 occurs when $|\beta_2| = 5g_c/\Omega_g^2$. Note also that a minimum amount of GVD is required for solitons to form.

In the case of FM mode locking, a phase modulator affects the phase of light passing through it. This case can be treated by replacing Eq. (5.3.2) with

$$\alpha = \alpha_c + i\Delta_M \cos(\omega_M t)], \tag{5.3.11}$$

where Δ_M is the modulation depth. In the absence of dispersion and nonlinearity, the master equation (5.3.3) predicts that mode-locked pulses remain Gaussian in shape (similar to the AM case) but are linearly chirped. The effects of dispersion were included in a simple fashion in a 1996 study [133].

A semi-analytic theory of actively mode-locked fiber lasers that takes into account both the dispersive and nonlinear effects has been developed recently [134]. It assumes that an autosoliton is formed inside the fiber-laser cavity such that

$$A(z,t) = a[\mathrm{sech}(t/\tau)]^{1+iq} \exp(i\phi), \qquad (5.3.12)$$

where the four pulse parameters, a, τ, q, and ϕ, represent the amplitude, width, chirp, and phase associated with the pulse. They vary from one round trip to the next as the laser pulse builds up, but eventually acquire constant values after a steady state has been reached. The moment method is used to find ordinary differential equations governing dynamic evolution of these parameters. These equations reveal how the pulse energy E_p, width τ, and chirp q change from one round trip to the next inside a Yb-doped fiber laser as the FM mode-locked pulse builds up and reaches a steady state.

The semi-analytic approach has been applied to both the AM and FM mode locking in the case of Yb-doped fiber lasers. In both cases, pulse evolves over a large number of round trips. In fact, thousands of round trips are necessary in the case of FM mode locking before pulse parameters converge to their steady-state values [135]. In general, the approach to steady state takes longer when the cavity dispersion is normal. The interesting question is how the pulse width τ and chirp parameter q depend on the average cavity dispersion β_2 and the nonlinear parameter γ. Figure 5.19 shows this dependence in the case of AM mode locking using $g_c = 0.55\ \mathrm{m}^{-1}$, $\alpha = 0.17\ \mathrm{m}^{-1}$, $T_2 = 47$ fs, and a 2-m-long cavity. As seen there, the steady-state values of these parameters also depend on whether the dispersion is normal or anomalous. In particular, the pulse width τ is larger in the case of normal dispersion. Note that τ increases with $|\beta_2|$ in the case of normal dispersion, but it decreases when the net GVD is anomalous. Pulse parameters also depend on the numerical value of γ, indicating that the SPM effects within the doped fiber play an important role. The main conclusion is that the Ginzburg–Landau equation governs the pulse dynamics even in actively mode-locked fiber lasers and that such lasers operate in the autosoliton regime discussed in Section 4.5.

5.3.4 Harmonic Mode Locking

The most common technique for active mode locking of fiber lasers makes use of an amplitude or phase modulator. Both acousto-optic and electro-optic modulators have been used for this purpose. However, most bulk modulators are not suitable for fiber lasers because of their size. They also introduce large coupling losses when light is coupled into and out of the modulator. An exception occurs in the case of LiNbO$_3$ modulators, which are relatively compact and can be integrated within the fiber cavity with reasonable coupling losses. They can also be modulated at speeds as high as 40 GHz [136]. For these reasons, LiNbO$_3$ modulators are commonly used for mode-locking fiber lasers.

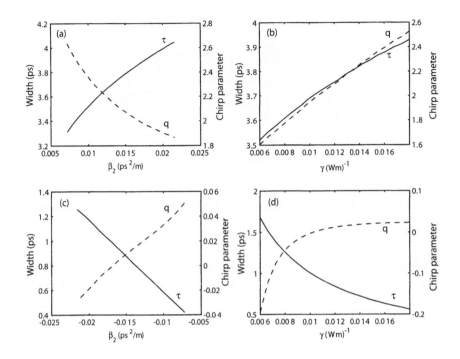

Figure 5.19: Pulse width τ and chirp parameter q as a function of β_2 (left column, $\gamma = 12\ \text{W}^{-1}/\text{km}$) and γ (right column, $\beta_2 = 0.015\ \text{ps}^2/\text{km}$). Cavity dispersion is normal for the top row and anomalous for the bottom row.

Active mode locking of EDFLs initially produced pulses of widths >10 ps. In a 1989 experiment, 4-ps pulses were generated using a ring cavity that included 2 km of standard fiber with large anomalous GVD [137]. Generation of short pulses was attributed to the soliton effects. In fact, the peak power of pulses was in good agreement with the expected peak power for the fundamental soliton from Eq. (5.3.9). The solitonlike nature of emitted pulses was also confirmed by the inferred "sech" shape and by the measured time-bandwidth product of 0.35. The pulse width was reduced to below 2 ps in an FM mode-locking experiment that used a Fabry–Perot cavity [138]. The fiber was only 10 m long, resulting in a longitudinal-mode spacing of about 10 MHz. This laser was referred to as the *fiber-soliton laser* since the "sech" pulses were nearly chirp free, with a time-bandwidth product of only 0.3. The laser wavelength could be tuned over the range of 1.52 to 1.58 μm, indicating that such lasers can serve as a source of tunable picosecond pulses in the 1.55-μm wavelength region of interest for optical communication systems. In this experiment, the LiNbO$_3$ modulator was operated at 420 MHz. This kind of mode locking, where modulation frequency is an integer multiple of the mode spacing, is called *harmonic mode locking* [28].

The performance of harmonically mode-locked EDFLs has continued to improve [139]–[157]. As early as 1990, the pulse-repetition rate was extended to 30 GHz by using a high-speed LiNbO$_3$ modulator [139]. A ring cavity of 30-m length was used

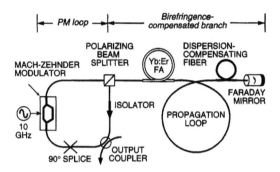

Figure 5.20: A harmonically mode-locked fiber laser employing the sigma configuration. The 10-m-long fiber amplifier (FA) is connected to a 90-m propagation loop in the linear section. (From Ref. [146]; ©1996 OSA.)

with an intracavity isolator for unidirectional operation. In a 1992 experiment, a fiber laser provided transform-limited 3.5- to 10-ps pulses with a time-bandwidth product of 0.32 at repetition rates of up to 20 GHz [140]. The laser was used in a system experiment to demonstrate that such pulses can be used for soliton communication systems at bit rates of up to 8 Gb/s.

A common problem with harmonically mode-locked fiber lasers is that they are unable to produce stable, equal-amplitude pulse trains over extended periods unless a stabilization technique is used. A phase-locking technique in which the optical phase is locked to the electrical drive of the modulator has been used with success [141]. In another approach, the use of a high-finesse Fabry–Perot etalon with a free spectral range equal to the repetition rate has resulted in a stable harmonically mode-locked EDFL suitable for soliton communication systems [142]. Polarization-maintaining fibers have also been used to make actively mode-locked EDFLs. In 1993, such a laser produced 6-ps pulses at repetition rates of up to 40 GHz and at wavelengths tunable over a wide range from 40 to 50 nm [143].

In a different approach, a cavity design known as the *sigma configuration* is used for making environmentally stable fiber lasers [146]. Figure 5.20 shows the σ-shaped cavity schematically consisting of two parts. A loop made of polarization-maintaining fiber contains a LiNbO$_3$ amplitude modulator and an output coupler. The loop is coupled to a linear section through a polarizing beam splitter. This section is made of traditional fibers and does not preserve the polarization state. However, it terminates with a Faraday rotator placed in front of a mirror. Such a Faraday mirror produces orthogonally polarized light on reflection. As a result, all birefringence effects are totally compensated during each round trip in the linear section.

A dispersion-compensating fiber can be used in the linear branch for reducing the average GVD. Such a dispersion-management technique has many advantages discussed later in Chapter 7. It is important to note that the sigma cavity is functionally equivalent to a ring cavity because of the Faraday mirror. In 1996 experiment, such a laser produced 1.3-ps pulses at the 10-GHz repetition rate through soliton-shaping effects while maintaining a negligible pulse-dropout rate and low noise [146]. The pulse shape was close to Gaussian in the center but fitted the "sech" shape better in the wings. This is a well-known feature of dispersion-managed solitons (see Chapter 8). The pulse width also decreased at higher power levels because of the increased nonlinear phase shift produced by SPM. This feature is in agreement with Eq. (5.3.9).

Active mode locking requires the modulation frequency of the $LiNbO_3$ modulator to remain matched to the longitudinal-mode spacing Δv (or a multiple of it) quite precisely. This is difficult to realize in practice because of fluctuations in Δv induced by environmental changes. The matching problem can be solved automatically by using the technique of regenerative mode locking [144]. In this technique, the electrical signal for the modulator at the correct modulation frequency is generated from the laser output using a clock-extraction circuit, a phase controller, and a microwave amplifier. Even though the laser is not initially mode locked, its power spectrum contains the beat signal at frequencies corresponding to multiples of the longitudinal-mode spacing. This signal can be used to produce pulse trains at high repetition rates through harmonic mode locking. As early as 1995, 1.8-ps pulses were produced at the 20-GHz repetition rate using regenerative mode locking of a ring cavity made by using polarization-maintaining fiber components [145]. The output pulses could be compressed to below 0.2 ps in a fiber amplifier made with dispersion-decreasing fiber. The wavelength of the regeneratively mode-locked laser was tunable over a considerable range within the gain spectrum of erbium ions. Moreover, the mode-locked pulse train exhibited low timing jitter (about 120 fs) and small energy fluctuations (about 0.2%) at a repetition rate of 10 GHz [149]. In a 1999 experiment, the technique of regenerative mode locking produced a 40-GHz pulse train tunable over 1530 to 1560 nm while maintaining pulse widths close to 1 ps [150].

A harmonically mode-locked fiber laser can also be stabilized with an electronic feedback loop that is used to adjust the cavity length. Such a scheme has been used for a sigma-configuration laser whose cavity included a piezoelectric transducer for fine adjustment of the cavity length [152]. The cavity also included an optical filter (bandwidth 16 nm). A careful analysis of this sigma laser showed that it has three distinct regions of operation. Figure 5.21 shows how temporal and spectral widths of the mode-locked pulses change with increasing intracavity power. At low power levels, the nonlinear effects (SPM) in silica fibers are negligible, and the laser produces Gaussian-shaped pulses of width close to 5 ps. As the intracavity power increases, the soliton effects become important, and the pulses become narrower, more intense, and attain a certain fixed energy level (as required for autosolitons). If the average power is not large enough to sustain such pulses in all time slots (because of a high repetition rate enforced by the modulator), pulse dropouts occur in a random fashion. Finally, when the intracavity power exceeds a certain value (about 5 mW), the laser emits a train of short optical pulses (width 1–3 ps) with a negligible dropout rate, low noise, and low timing jitter. This behavior is in agreement with numerical simulations based on the Maxwell–Bloch equations [153]. The theory predicts a fourth regime in which more than one pulse may occupy the same time slot at high power levels when GVD is uniform inside the cavity.

In general, the use of dispersion management improves the laser performance considerably. It helps to reduce the timing jitter in the position of mode-locked pulses

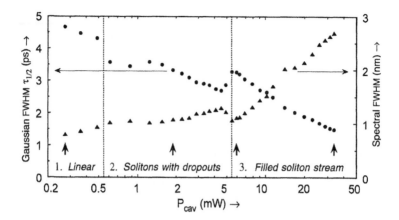

Figure 5.21: Temporal (circles) and spectral (triangles) widths for a sigma laser as a function of intracavity optical power P_{cav}. (From Ref. [152]; ©2000 OSA.)

within the pulse train. Fiber lasers employing dispersion management and polarization-maintaining fibers can be designed to emit 10-GHz pulse trains simultaneously at two different wavelengths [151]. At a single wavelength, the repetition rate of the mode-locked pulse train can be made as high as 40 GHz. The repetition rate of 64 GHz was realized in a FM mode-locked fiber laser in which the optical phase was modulated at 16 GHz, and a Fabry–Perot filter with a 64-GHz free spectral range was used to initiate harmonic mode locking [148]. By 2003, the use of a Fabry–Perot filter inside an EDFL cavity provided mode-locked pulse trains at 24 wavelengths simultaneously [154]. Such sources are useful for WDM applications.

Highly nonlinear fibers were used to make EDFLs soon after they became available. Figure 5.22 shows the design of an EDFL whose cavity contains a 10-m-long polarization-maintaining photonic crystal fiber (PCF), in addition to a 20-m-long piece of erbium-doped active fiber [155]. This passive fiber not only acted as a nonlinear medium but also provided enough anomalous GVD for managing cavity dispersion. This laser also employed the technique of regenerative mode locking. More precisely, the electric signal for driving the Mach–Zehnder modulator was obtained by extracting the 10 GHz clock from the current generated by detecting 30% of the laser output power. The laser produced 1-ps-wide pulses tunable over 1535 to 1560 nm at a repetition rate of 10 GHz. It could also be mode-locked at 40 GHz with a pulse width close to 1.3 ps. Pulses were intense enough that their spectrum could be shifted toward longer wavelengths by as much as 90 nm through the Raman-induced frequency shift. In a 2006 experiment, four-wave mixing inside a 1-km-long highly nonlinear fiber was used to stabilize an EDFL designed to mode lock simultaneously at four equally spaced wavelengths [157].

Although Yb-doped fiber lasers are often mode-locked passively, harmonic mode locking has also been employed for them. In a 2004 experiment, such a laser was FM mode locked at its 281st harmonic, resulting in a repetition rate >10 GHz [158]. Figure 5.23 shows schematically the laser cavity containing a bulk FM modulator. A grating

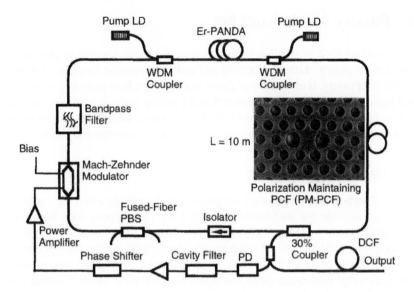

Figure 5.22: Experimental setup of a regeneratively mode-locked EDFL. The inset shows a cross section of the polarization-maintaining photonic crystal fiber. (From Ref. [155]; ©2004 IEEE.)

pair was used to make the average cavity dispersion slightly anomalous. Since Yb-doped fibers exhibit large normal dispersion near 1050 nm, a grating pair or a suitable photonic crystal fiber is needed for such lasers. The laser produced linearly polarized, 2-ps-wide, mode-locked pulses whose center wavelength was tunable over a 58-nm window centered near 1050 nm. The measured temporal and spectral characteristics of the laser indicated that mode-locked pulses were nearly chirp free with a "sech" shape, and they were in good agreement with the predictions of the Ginzburg–Landau equation [159]. These results indicate that such fiber lasers operate in the autosoliton regime discussed in Section 4.5.

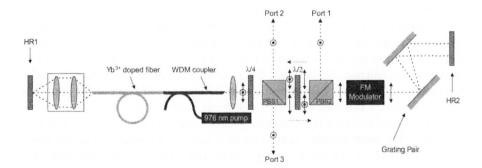

Figure 5.23: The cavity employed for a Yb-doped fiber laser mode locked at its 281st harmonic using a bulk FM modulator. (From Ref. [158]; ©2004 OSA.)

5.4 Passive Mode Locking

Passive mode locking is an all-optical nonlinear technique capable of producing ultrashort optical pulses, without requiring any active component (such as a modulator) inside the laser cavity. It makes use of a nonlinear device whose response to an entering optical pulse is intensity dependent such that the exiting pulse is narrower than the input pulse. Several implementations of this basic idea have been used to make passively mode-locked fiber lasers. This section discusses mostly experimental results.

5.4.1 Saturable Absorbers

Saturable absorbers have been used for passive mode locking since the early 1970s. In fact, their use was the sole method available for this purpose until the advent of additive-pulse mode locking. The basic mechanism behind mode locking is easily understood by considering a fast saturable absorber whose absorption can change on a timescale of the pulse width. When an optical pulse propagates through such an absorber, its wings experience more loss than the central part, which is intense enough to saturate the absorber. The net result is that the pulse is shortened during its passage through the absorber. Pulse shortening provides a mechanism through which a laser can minimize cavity losses by generating intense pulses if the CW radiation is unable to saturate the absorber.

To quantify the extent of pulse shortening in a saturable absorber, we should replace g_0 by $-\alpha_0$ in Eq. (4.1.7), where α_0 is the small-signal absorption coefficient. The resulting equation can be integrated analytically to obtain

$$\ln(P_{out}/P_{in}) + (P_{out} - P_{in})/P_{sa} + \alpha_0 l_a = 0, \qquad (5.4.1)$$

where P_{in} and P_{out} are the input and output powers, P_{sa} is the saturation power, and l_a is the length of the saturable absorber. For a fast-responding saturable absorber, Eq. (5.4.1) applies along the entire pulse and can be used to obtain the output shape $P_{out}(t)$ for a given input shape $P_{in}(t)$. The output pulse is always slightly narrower than the input pulse because of higher absorption in the low-intensity wings.

The pulse-formation process is quite complex in passively mode-locked lasers [28]. Fluctuations induced by spontaneous emission are enhanced by the saturable absorber during multiple round trips inside the laser cavity until an intense pulse capable of saturating the absorber is formed. The pulse continues to shorten until it becomes so short that its spectral width is comparable to the gain bandwidth. The reduced gain in spectral wings then provides the broadening mechanism that stabilizes the pulse width to a specific value. In the case of fiber lasers, GVD and SPM also play an important role in evolution of mode-locked pulses and should be included.

It is not easy to find a fast saturable absorber responding at time scales of 1 ps or shorter. The most suitable material for fiber lasers is a semiconductor absorbing medium [160]–[172]. Its use is more practical with a Fabry–Perot cavity since the absorber can be attached to one of the cavity mirrors. The saturable absorber can be made using either a single or a large stack (>100 layers) of quantum-well layers. In the latter case, it forms a periodic structure called the *superlattice*. Each period of

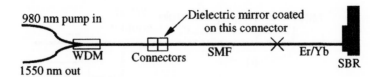

Figure 5.24: A fiber laser that was mode locked using a saturable Bragg reflector. (From Ref. [167]; ©1998 OSA.)

the superlattice consists of alternating absorbing and transparent layers. In the case of EDFLs, all layers are made using the InGaAsP material but the layer composition is altered appropriately. In some cases, the mirror attached to the saturable absorber is also made using a periodic arrangement of quarter-wavelength-thick layers that form a grating and reflect light through Bragg diffraction. Such a device is referred to as a *saturable Bragg reflector* to emphasize the use of a Bragg grating.

A superlattice was first used in 1991 for passive mode locking of a ring-cavity EDFL [160]. It produced mode-locked pulses of 1.2-ps duration with a "sech" shape, as expected from theory. In a 1993 experiment [161], the superlattice saturable absorber consisted of 82 periods, and each period used a 7.8-nm-thick absorbing InGaAs layer and a 6.5-nm-thick transparent InP layer. The absorber was integrated with a Bragg reflector (made of alternating InGaAsP and InP layers) that acted as one of the mirrors of the Fabry–Perot laser cavity. Mode-locked pulses obtained from this laser were relatively broad (≈ 22 ps) for a 6.2-m cavity, but their width could be reduced to 7.6 ps by shortening the doped fiber to 2 m. With further refinements, mode-locked pulses as short as 0.84 ps with pulse energies of 0.85 nJ were obtained at a repetition rate of 22 MHz [162]. The same technique was used for mode locking a Nd-doped fiber laser, and 4-ps pulses were obtained using a heavily doped fiber of 6-cm length [163].

A superlattice saturable absorber integrated with a Bragg reflector requires the growth of hundreds of thin layers using molecular-beam epitaxy. In a different approach, a single 2-μm-thick epitaxial layer of InGaAsP, grown on an InP substrate, acted as a saturable absorber [164]. It was directly mounted on a mirror serving as the output coupler. The 1.2-m-long erbium-doped fiber was the polarization-preserving type. The mode-locked laser produced 320-fs pulses with 40 pJ energy. The laser was self-starting and its output was linearly polarized along a principal axis of the fiber. By codoping the gain-producing fiber with ytterbium, such a laser can be pumped with diode-pumped Nd:YAG or Nd:YLF lasers or directly with a semiconductor laser.

A semiconductor laser amplifier can also be used as a saturable absorber when it is biased below its threshold. Its use allows the construction of a self-starting, passively mode-locked EDFL that can be switched between mode-locked and CW states by simply changing the amplifier bias current. In a 1993 experiment, such a laser produced mode-locked pulses of 1.25-ps width at a repetition rate in the range from 10 to 50 MHz in a ring-cavity configuration [165].

Fiber lasers that are mode locked using a saturable Bragg reflector inside a short Fabry–Perot cavity have quite interesting properties. Figure 5.24 shows the cavity design schematically. A short piece (length 15 cm) of doped fiber is butt coupled to the

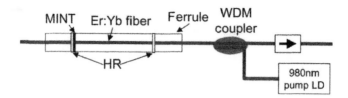

Figure 5.25: A fiber laser that was mode locked at a repetition rate of 5 GHz owing to its 2-cm-long cavity; MINT and HR stand for a mode locker incorporating nanotubes and high-reflectivity mirrors, respectively. (From Ref. [169]; ©2005 IEEE.)

saturable Bragg reflector. Its other end is sliced to a section of standard telecommunication fiber (length about 30 cm) and terminated with a connector on which a high-reflectivity dielectric mirror has been coated. A piece of dispersion-compensating fiber can also be included for dispersion management.

In a series of experiments, total cavity length was changed from 0.5 to 2.5 m, and the average GVD was varied over wide range (normal to anomalous) by using the dispersion-compensating fiber [166]. A mode-locked pulse train could be formed even in the case of normal GVD, but the pulse width was close to 16 ps at a repetition rate of about 40 MHz. This is expected from the results of Section 4.6. Much shorter pulses were observed when the average GVD was anomalous. Pulse widths below 0.5 ps formed over a wide range of average GVD ($\beta_2 = -2$ to -14 ps^2/km) although they were not transform limited. This is expected because of the chirp associated with the autosolitons (see Section 4.5). For short laser cavities (under 50 cm), harmonic mode locking was found to occur. A 45-cm-long laser produced transform-limited, 300-fs pulses at a repetition rate of 2.6 GHz through harmonic passive mode locking [167]. The laser was able to self-organize into a steady state such that 11 pulses with nearly uniform spacing were present simultaneously inside the cavity. Cross-correlation measurements indicated that spacing between pulses was uniform to within 4% of the expected value.

By 2004, a new type of saturable absorber based on carbon nanotubes was employed for mode locking fiber lasers [168]. A carbon nanotube consists of a hexagonal network of carbon atoms in the shape of a cylinder whose diameter is ∼5 nm but length can exceed 10 μm [173]. The ends of the cylinder are capped with a fullerene molecule, forming a closed structure for the carbon nanotube. A thin film of such a polymer can act as a saturable absorber depending on structural details. In a 2005 experiment [169], such a device was used to mode lock an EDFA with the cavity design shown in Figure 5.25. It consists of a 2-cm-long active fiber sandwiched between two high-reflectivity (99.87%) mirrors. One of the mirrors has a thin layer (<1 μm thick) of the nanotube polymer deposited on it. Because of such a short cavity, this laser emitted pulses of widths <1 ps at a repetition rate of 5 GHz.

Mode-locked fiber lasers have continued to evolved in recent years. In a 2006 experiment [170], quantum dots were used for making a superlattice saturable absorber, which was used to mode lock a Yb-doped fiber laser. In another 2006 experiment, a microstructured fiber whose core was doped with Nd to provide gain was

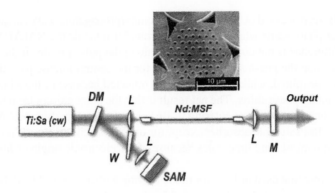

Figure 5.26: A Nd-doped fiber laser that was mode locked with a saturable absorber mirror (SAM). The microstructure of the doped fiber is also shown. DM, L, M, MSF, and W stand for dichroic mirror, lens, mirror, microstructured fiber, and wedge, respectively. (From Ref. [171]; ©2006 IEEE.)

employed together with a saturable-absorber mirror [171]. Figure 5.26 shows the laser cavity schematically. The 56-cm-long Nd-doped microstructured fiber provides not only gain, when pumped suitably, but also anomalous dispersion ($\beta_2 = -20$ ps^2/km) and a relatively large value of the nonlinear parameter ($\gamma = 90$ W^{-1}/km). As a result, this laser was able to generate nearly transform-limited pulses of widths as short as 180 fs in the form of autosolitons.

A similar scheme was adopted for a Yb-doped fiber laser, whose cavity contained a 27-cm-long Yb-doped microstructured fiber [172]. Because of the anomalous dispersion of this fiber, the net dispersion over one round trip was about -0.017 ps^2 at wavelengths near 1035 nm. The laser emitted 335-fs-wide mode-locked pulses at the fundamental repetition rate of 117.5 MHz. The pulses were in the form of autosolitons with a relatively low chirp. The measured 4-nm spectral bandwidth of pulses corresponded to a time-bandwidth product of 0.37 for the emitted pulses.

5.4.2 Nonlinear Fiber-Loop Mirrors

An undesirable aspect of semiconductor-based saturable absorbers is that fiber lasers using them lose their all-fiber nature. A solution is provided by the nonlinear fiber-loop mirrors (Sagnac interferometers) whose power-dependent transmission can shorten an optical pulse just as saturable absorbers do (see Section 3.2). Fiber lasers making use of a Sagnac loop for passive mode locking are referred to as *figure-8 lasers* because of the appearance of their cavity (see Figure 5.4). The physical mechanism responsible for mode locking is known as the *interferometric* or *additive-pulse mode locking*.

The operation of a figure-8 laser can be understood as follows. The central 3-dB coupler in Figure 5.4 splits the entering radiation into two equal counterpropagating parts. The doped fiber providing amplification is placed close to the central coupler such that one wave is amplified at the entrance to the loop while the other experiences amplification just before exiting the loop, resulting in a nonlinear amplifying-loop mirror (NALM). As discussed in Section 3.2, the counterpropagating waves acquire different nonlinear phase shifts while completing a round trip inside the NALM. Moreover, the phase difference is not constant but varies along the pulse profile. If the NALM is adjusted such that the phase shift is close to π for the central intense part, this part of the pulse is transmitted, while pulse wings get reflected because of their lower power levels and smaller phase shifts. The net result is that the pulse exiting from the NALM is narrower compared with that entering it. Because of this property, a NALM behaves similarly to a fast saturable absorber except for one major difference—it is capable of responding at femtosecond timescales because of the electronic origin of fiber nonlinearity.

NALMs were first used in 1991 for mode locking a fiber laser [174]. Pulses shorter than 0.4 ps were generated in the form of fundamental solitons even in early experiments in which the fiber laser was pumped using a Ti:sapphire laser [175]. In a later experiment, 290-fs pulses were produced from an EDFL pumped by 1.48-μm InGaAsP semiconductor lasers [176]. The threshold for mode-locked operation was only 50 mW. Once mode locking initiated, pump power could be decreased to as low as 10 mW.

It is generally difficult to produce pulses shorter than 100 fs from figure-8 lasers. However, mode-locked pulses as short as 30 fs were obtained by amplifying the laser output, and then compressing the amplified pulse in a dispersion-shifted fiber [177]. Pulse shortening inside a fiber amplifier occurs because of adiabatic amplification of fundamental solitons (see Section 4.3). Since the amplified pulse is chirped, it can be further compressed by using a fiber with the appropriate dispersion. By 1993, pulse widths close to 100 fs were generated directly from a figure-8 laser by employing a polarization-sensitive isolator and a short piece of normal-GVD fiber for chirp compensation [178].

Passively mode-locked fiber lasers suffer from a major drawback that has limited their usefulness. It was observed in several experiments that the repetition rate of mode-locked pulses was essentially uncontrollable and could vary over a wide range. Typically, several pulses circulate simultaneously inside the laser cavity, and the spacing among them is not necessarily uniform. In contrast with the case of active mode locking, nothing in the cavity determines the relative location of pulses. As a result, the position of each pulse is determined by various other effects such as fiber birefringence and soliton interactions.

When a single pulse circulates inside the laser cavity, the repetition rate is equal to the round-trip frequency Δv. However, if pulses inside a fiber laser propagate as fundamental solitons, their energy is fixed such that

$$E_s = P_0 \int_{-\infty}^{\infty} |A(z,t)|^2 \, dt = 2P_0 T_s. \tag{5.4.2}$$

As both the soliton width T_s and the peak power P_0 are limited by the laser design, the pulse energy E_s is fixed or quantized [179]. On the other hand, the average intracavity power P_{av} is determined by the pumping level and gain saturation. If $P_{av} > (\Delta v)E_s$, multiple pulses, each of quantized energy E_s, must coexist inside the laser cavity. If these pulses are uniformly spaced, the fiber laser would behave similarly to a harmonically mode-locked laser. However, multiple pulses need not necessarily be uniformly

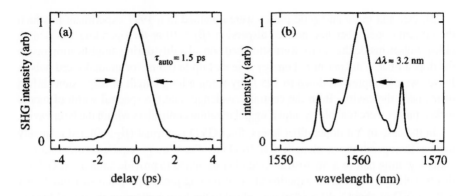

Figure 5.27: (a) Autocorrelation trace and (b) spectrum of mode-locked pulses obtained from a figure-8 laser employing a 94:6 central coupler. (From Ref. [184]; ©1994 OSA.)

spaced. Most fiber lasers emit pulse trains at the fundamental repetition rate Δv such that spacing among pulses in each period is virtually random. Under some operating conditions, the laser emits a train of bunched pulses such that each bunch contains 10 or more closely spaced solitons, each with the quantized energy E_s. The number of pulses within each fundamental period depends on the pumping level, among other things.

The key to stabilizing a figure-8 laser consists of implementing a scheme that can adjust the repetition rate f_r in such a way that $f_r E_s$ nearly equals the average circulating power inside the laser cavity. In a 1992 scheme, a subring was added to the left ring in Figure 5.5 containing the isolator [180]. The subring was only 1.6-m long, while the total loop length was 60.8 m. In this arrangement, the repetition rate of the subring cavity is 38 times that of the main laser cavity. Pulses circulating inside the subring provide a seed and lead to the formation of 38 uniformly spaced pulses in the main laser cavity under mode-locked operation. Such a laser emitted 315-fs pulses at a fixed repetition rate of 125 MHz. A similar control of the repetition rate can be realized by placing a mirror close to one port of the output coupler [181]. In this case, the optical feedback from the external mirror provides the seed and fixes the relative location of pulses in a periodic manner. The mirror distance controls the repetition rate of such fiber lasers.

Figure-8 lasers have continued to improve [182]–[189]. Transform-limited pulses of 1.35-ps duration, at wavelengths continuously tunable over a 20-nm range, were generated in 1993 by using an intracavity Fabry–Perot filter for spectral stabilization and a feedback loop for temporal stabilization [183]. In another experiment, the central coupler (see Figure 5.4) was unbalanced such that 94% of the intracavity power was propagating in the direction in which laser emission occurred [184]. Such a laser had lower cavity losses and was found to be more easily mode locked than figure-8 lasers with balanced (50:50) central couplers. The laser was able to generate pulses shorter than 1 ps. Figure 5.27 shows the autocorrelation trace of 970-fs pulses together with the corresponding spectrum. The origin of spectral sidebands seen in this figure is discussed later.

Shorter and more energetic pulses were obtained in a 1997 experiment in which the erbium-doped fiber had normal dispersion ($\beta_2 > 0$) at the operating wavelength [186]. Pulses inside the cavity were stretched considerably during amplification inside the doped fiber. This permitted energy levels of up to 0.5 nJ for mode-locked pulses. Pulses were compressed down to 125 fs by using a long length of dispersion-shifted fiber inside the cavity. Both the central wavelength and the spectral width of mode-locked pulses were tunable by adjusting polarization controllers within the laser cavity.

In the case of Yd-doped fiber lasers, fiber GVD is normal ($\beta_2 > 0$) and relatively large at the operating wavelength near 1060 nm. Such lasers require incorporation of either a grating pair or a microstructured fiber designed to provide anomalous GVD at this wavelength. In a 2003 experiment, a 20-m-long piece of a microstructured fiber with a GVD of about $-40 \text{ ps}^2/\text{km}$ was added to the loop containing the doped fiber to make the average GVD anomalous [188]. With a suitable design, such a laser emitted a mode-locked train of 850-fs-wide pulses at a wavelength of 1065 nm. Figure-8 fiber lasers can operate at a variety of wavelengths if Raman amplification is used (in place of a doped fiber) to provide the optical gain within the laser cavity. In a 2005 experiment [189], such lasers could be operated at wavelengths of 1330, 1410, and 1570 nm by pumping them with suitable pump lasers and emitted mode-locked pulses of widths <1 ps.

5.4.3   Nonlinear Polarization Rotation

Fiber lasers can also be mode locked by using intensity-dependent changes in the state of polarization (occurring because of SPM and XPM) when the orthogonally polarized components of a single pulse propagate inside an optical fiber. The physical mechanism behind mode locking makes use of the nonlinear birefringence and is similar to that of a Kerr shutter [55]. From a conceptual point of view, the mode-locking mechanism is identical to that used for figure-8 lasers (additive-pulse mode locking) except that the orthogonally polarized components of the same pulse are used in place of two counterpropagating waves. From a practical standpoint, passive mode locking can be accomplished by using a cavity with a single fiber ring.

The mode-locking process can be understood using the ring cavity shown in Figure 5.28. A polarizing isolator placed between two polarization controllers acts as the mode-locking element. It plays the double role of an isolator and a polarizer such that light leaving the isolator is linearly polarized. Consider a linearly polarized pulse just after the isolator. The polarization controller placed after the isolator changes the polarization state to elliptical. The polarization state evolves nonlinearly during propagation of the pulse because of SPM- and XPM-induced phase shifts imposed on the orthogonally polarized components. The state of polarization is nonuniform across the pulse because of the intensity dependence of the nonlinear phase shift. The second polarization controller (one before the isolator) is adjusted such that it forces the polarization to be linear in the central part of the pulse. The polarizing isolator lets the central intense part of the pulse pass but blocks (absorbs) the low-intensity pulse wings. The net result is that the pulse is slightly shortened after one round trip inside the ring cavity, an effect identical to that produced by a fast saturable absorber.

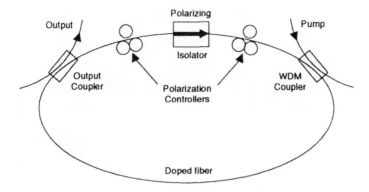

Figure 5.28: Design of a fiber laser passively mode locked via nonlinear polarization-rotation.

The technique of nonlinear polarization rotation was first used in 1992 for passive mode locking of fiber lasers and has resulted in considerable improvement of such lasers [190]–[193]. By the end of 1992, stable, self-starting pulse trains of 452-fs-wide pulses were generated at a 42-MHz repetition rate with this technique [192]. Further improvements occurred when it was realized that the presence of anomalous GVD within the laser cavity is not necessarily beneficial, since it limits both the width and the energy of mode-locked pulses. In a 1993 experiment, 76-fs pulses—with 90-pJ energy and 1 kW of peak power—were generated by using a ring cavity in which the average GVD was normal [193].

Considerable research has been done to understand and to improve fiber lasers making use of nonlinear polarization rotation for passive mode locking [194]–[205]. Pulses with <50 fs widths and high energies (up to 1 nJ) were obtained from an Nd-doped fiber laser in a Fabry–Perot configuration in which a moving mirror was used to start mode locking [194]. In the case of EDFLs, high-energy (>0.5 nJ), ultrashort (<100 fs) pulses at a repetition rate of 48 MHz were obtained in a ring-cavity configuration in which the net dispersion was positive [195]. The ring cavity of this fiber laser consisted of a piece of erbium-doped fiber (of length about 1 m) with normal GVD ($\beta_2 \approx 5$ ps^2/km) and several types of optical fibers (total length 2–6 m) with anomalous GVD in the wavelength region near 1.56 μm. The average dispersion could be changed from anomalous to normal by adjusting the cavity length. Such cavities are called *dispersion managed*, as their net dispersion can be tailored to a desired value. The laser is referred to as a *stretched-pulse fiber laser*, because pulses circulating inside the cavity stretch considerably in the section with normal GVD. It was found that high-energy pulses could be generated only when the average or net dispersion in the cavity was normal. The emitted pulses were relatively broad (>1 ps) but could be compressed down to below 100 fs by using an appropriate length of fiber because of their highly chirped nature. The location of the output coupler plays an important role in such lasers, because pulse width varies by a large factor along the cavity length. Mode-locked pulses as short as 63 fs were generated in 1995 with proper optimization [197].

For practical applications, environmental stability is often an important issue. The

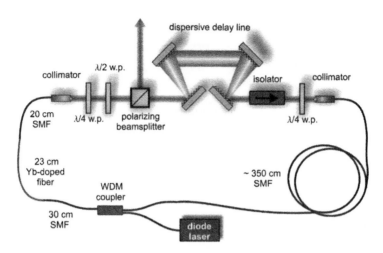

Figure 5.29: A Yb-doped fiber laser that was mode locked passively to generate 36-fs pulses near 1030 nm. The grating pair provides enough anomalous dispersion to offset the normal GVD of various fibers. (From Ref. [199]; ©2003 OSA.)

main source of environmental instability is a relatively long length of the fiber inside the laser cavity required to produce a large enough nonlinear phase shift. Temperature and stress variations can lead to birefringence fluctuations that affect the mode-locking process. The problem can be solved to a large extent by reducing the fiber length to under 10 m and using a fiber with high built-in birefringence (polarization-maintaining fiber) so that linear birefringence is not affected by environmental changes. In a 1994 scheme, a Fabry–Perot cavity in which one of the mirrors acts as a Faraday rotator was used to realize environmentally stable operation [196]. The Faraday mirror rotated the polarization such that the reflected light was orthogonally polarized. As a result, the phase shift induced by linear birefringence was exactly canceled after one round trip, while the nonlinear phase shift remained unaffected. The Faraday mirror also eliminated the walk-off effects induced by the group-velocity mismatch in high-birefringence fibers. Such a laser was capable of producing 360-fs pulses of 60-pJ energy at a stable repetition rate of 27 MHz.

Starting in 1999, considerable attention focused on Nd- and Yb-doped mode-locked fiber lasers making use of a double-clad geometry for efficient pumping [198]–[203]. Such lasers can produce ultrashort pulses with realtively high energies. In one experiment, pulse energy was close to 4 nJ when a Nd-doped fiber laser was mode-locked using the technique of nonlinear polarization rotation, but the pulse width was close to 1 ps [198]. In a 2003 experiment, a Yb-doped fiber laser with the cavity design shown in Figure 5.29 emitted 36-fs mode-locked pulses with 1.5-nJ energy [199]. The grating pair is used to offset the normal GVD of various fibers. The pulse energy for such lasers operating in the autosoliton regime (net dispersion anomalous) can be increased to near 3 nJ but pulses also become wider. As seen in Section 4.5, autosolitons can also form in the normal-dispersion regime. Indeed, 116-fs pulses with 1.7-nJ energy were emitted in a 2004 experiment from such a laser [200]. This laser

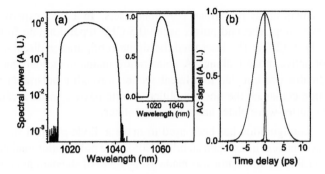

Figure 5.30: (a) Optical spectrum and (b) autocorrelation trace for a Yb-doped fiber laser mode-locked with net normal dispersion to generate parabolic pulses near 1030 nm. The inner narrow trace corresponds to the dechirped pulse. The inset shows the spectrum on a linear scale. (From Ref. [201]; ©2004 APS.)

was passively mode-locked at the fifth harmonic of the fundamental repetition rate of 20.4 MHz.

It turns out that much higher pulse energies can be obtained by operating the laser such that the average dispersion inside the cavity is normal and parabolic-shape pulses are generated in the self-similar regime. Such pulses are chirped linearly, and their width typically exceeds 1 ps, but they can be compressed down to 100 fs externally using a suitable piece of fiber. Figure 5.30 shows the measured spectrum and shape of such parabolic pulses for a Yb-doped mode-locked fiber laser, operated such that the accumulated dispersion over its cavity length was 0.004 ps^2 [201]. The laser emitted 4.2-ps-wide chirped pulses that could be compressed down to 250 fs. Pulse energies in this experiment were close to 2 nJ but could be increased to above 10 nJ with a suitable design. In a 2005 experiment, mode-locked pulses with 14-nJ energy were produced in the self-similar regime [202]. Their width was close to 5 ps but it could be reduced to near 100 fs by dechirping them. The dynamic aspects of passively mode-locked fiber lasers are discussed in Section 5.5.

5.4.4 Hybrid Mode Locking

Hybrid mode locking combines more than one mode-locking technique within the same laser cavity to improve the laser performance. The most obvious combination incorporates an amplitude or phase modulator inside a passively mode-locked fiber laser. The modulator provides periodic timing slots to produce a regular pulse train, while a passive mode-locking technique shortens the pulse compared to that expected from active mode locking alone. An added benefit is that the modulator can be operated at a frequency that is a high multiple of the round-trip frequency, resulting in a well-defined repetition rate that can exceed 10 GHz or more while the mode spacing remains close to 10 MHz.

As early as 1991, the active and passive mode-locking techniques were combined by using a phase modulator [206]. Since then, this combination has led to considerable improvement in the performance of fiber lasers. In a 1994 experiment, it was used to generate subpicosecond pulses at the 0.5-GHz repetition rate from a single-polarization EDFL [207]. The laser used the sigma configuration discussed earlier in the context of active mode locking. A polarization-maintaining loop containing the LiNbO$_3$ modulator was coupled to a linear section through a polarizing beam splitter. This section contained the fiber amplifier and the passive mode-locking element composed of quarter-wave plates and a Faraday rotator.

The sigma configuration has been used to make a diode-pumped stretched-pulse EDFL with excellent environmental stability [208]. The polarization-maintaining loop containing the LiNbO$_3$ modulator was made using 7.5 m of standard fiber with a GVD of -20 ps^2/km. The linear section contained 1 m of erbium-doped fiber with normal GVD ($\beta_2 \approx 100$ ps^2/km), together with a quarter-wave plate, a half-wave plate, and two Faraday rotators (whose presence makes the sigma-laser cavity equivalent to a ring cavity). The net dispersion in the cavity was normal with a value of about 0.02 ps^2. The doped fiber was pumped using 980-nm diode lasers. The laser produced mode-locked pulses with 1.2-nJ energy (average power 20 mW) at a pump power of 200 mW. The pulse width from the laser was about 1.5 ps but could be compressed down to below 100 fs using a dispersive delay line (see Chapter 6).

It is also possible to combine two passive mode-locking techniques within the same fiber laser. In one approach, a superlattice saturable absorber is added to a laser that is passively mode locked via nonlinear polarization rotation. This combination was used in 1996 for a cladding-pumped fiber laser [209]. The laser produced 200-fs pulses with pulse energies of up to 100 pJ at a wavelength near 1560 nm. Pulse energies of up to 1 nJ were obtained by increasing the GVD inside the fiber cavity while maintaining a pulse width close to 3 ps. The intracavity saturable absorber is helpful for initiating mode locking (self-starting capability), whereas the steady-state pulse shape is governed by the nonlinear polarization evolution. The laser is also environmentally stable because of the use of a compensation scheme for linear polarization drifts.

In another implementation of hybrid mode locking, an Nd-doped fiber laser was tuned continuously over a 75-nm bandwidth [210]. Such a wide tuning range (more than twice the FWHM of the gain spectrum) was realized by optimizing the reflection characteristics of the semiconductor saturable absorber. The duration of mode-locked pulses was 0.3 to 0.4 ps over the entire tuning range. A chirped fiber grating was used in 1995 for dispersion compensation inside the cavity of an Nd-doped fiber laser [211]. Such a laser was self-starting and could be passively mode locked by using just the saturable absorber as the mode-locking element. No intracavity polarization controllers were required for its optimization. The laser generated mode-locked pulses of 6-ps duration with output energies as high as 1.25 nJ.

In an interesting application of hybrid mode locking, an Nd-doped fiber laser was mode locked at two wavelengths simultaneously [212]. A prism pair, used for dispersion compensation, also separated the paths taken by the two intracavity beams. The 1.06-μm beam was mode-locked using a saturable absorber, whereas the technique of nonlinear polarization rotation was used for mode locking the 1.1-μm beam. Such a device operates as if the two lasers share the same gain medium, and cross-gain saturation plays an important role in its operation. Indeed, it was necessary to mismatch the

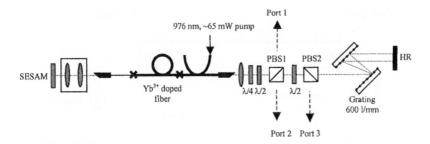

Figure 5.31: A Yb-doped fiber laser mode locked using nonlinear polarization rotation in combination with saturable absorption. The grating pair provides enough anomalous dispersion to offset the normal GVD of fibers. (From Ref. [213]; ©2002 OSA.)

cavity lengths slightly to introduce an offset of at least 0.5 kHz between the repetition rates of the two mode-locked pulse trains. The difference between the repetition rates was tunable from 0.5 kHz to >1 MHz.

In a 2002 experiment, hybrid mode locking was used to realize a low-noise Yb-doped fiber laser [213]. As seen in Figure 5.31, this laser used nonlinear polarization rotation for mode locking in combination with a semiconductor saturable absorber that was attached to a mirror of the Fabry–Perot cavity employed. The grating pair compensates for the normal dispersion of the fiber. Such a laser was self-starting and generated 110-fs pulses of high quality with 60-pJ energy. It also exhibited relatively low intensity noise (<0.05% noise level). In an interesting variation, self-starting mode locking was achieved by combining nonlinear polarization rotation with the frequency-shifted feedback [214]. The laser generated chirped pulses that could be compressed down to 68-fs duration.

5.4.5   Other Mode-Locking Techniques

Several other nonlinear techniques have been suggested for passively mode locking fiber lasers. In one scheme, the nonlinear phenomenon of cross-phase modulation is employed to produce an all-fiber laser [215]–[219]. Figure 5.32 shows the design of such a laser schematically. A relatively long length (several kilometers) of fiber is inserted into the ring cavity through two WDM couplers. Pump pulses from an external laser propagate into this fiber and modulate the phase of laser light through XPM. If the repetition rate of pump pulses is an integer multiple of the mode spacing, XPM forces the fiber laser to produce mode-locked pulses. Pulses shorter than 10 ps have been obtained by this technique at repetition rates up to 40 GHz. Such a laser has been used to transfer an arbitrary bit pattern from the pump-pulse wavelength to the laser wavelength [216], resulting in wavelength conversion. This technique has been used to make an optically programmable mode-locked laser such that emitted pulses represent the result of logic operations between elements of the driving pulse train [217].

In a a variation of this basic idea, synchronous pumping was used for mode locking fiber lasers. Implementation of this technique is extremely simple for lasers pumped with semiconductor lasers because one can simply modulate the pump-laser current

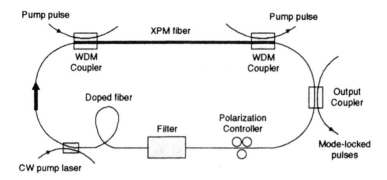

Figure 5.32: Experimental setup for observation of XPM-induced mode locking in fiber lasers. (From Ref. [215]; ©1992 IEE.)

at the appropriate frequency. Because of a relatively long fluorescence time of the dopants, it is not possible to modulate the gain at frequencies in excess of 1 MHz or so. However, pump pulses can modulate the laser field through XPM. Even though the XPM-induced phase shift is expected to be relatively small, it is large enough to initiate mode locking. In a 1992 experiment in which an EDFL was pumped at 980 nm, mode-locked pulses were relatively broad with widths >100 ps [220] because of a group-velocity mismatch. They can be shortened only by decreasing this mismatch. Indeed, pulses shorter than 50 ps were generated when a 1.48-μm pump laser was used [221]. Another way to increase the XPM-induced phase shift is to increase the peak power of pump pulses. Mode-locked pulses as short as 2 ps were generated in an EDFL pumped with a Nd:YAG laser producing 100-ps pulses at the 100-MHz repetition rate [222]. Soliton shaping plays an important role in these experiments since phase-modulated CW laser radiation is converted into nearly chirp-free soliton pulses through the combined action of GVD and SPM.

Several other variations have been used. In one scheme, a semiconductor optical amplifier is used as the mode-locking element [223]. In essence, the long piece of silica fiber in Figure 5.32 is replaced with a pigtailed amplifier. When pump pulses and laser light propagate inside the amplifier, the nonlinear phenomenon of cross-gain saturation modulates both the amplitude and the phase of laser light. GVD and SPM occurring inside the fiber cavity convert the modulated signal into a train of mode-locked soliton pulses. Pulses shorter than 10 ps at repetition rates of up to 20 GHz have been generated by this technique [224].

In another scheme, an acousto-optic modulator and an optical filter are placed inside the laser cavity [225]–[228]. The role of modulator is to shift the laser frequency by a small amount ($\sim$100 MHz). Such a frequency-shifted feedback in combination with the fiber nonlinearity leads to formation of picosecond optical pulses inside the laser cavity. Pulses shorter than 10 ps were generated when a narrowband optical filter was inserted inside the laser cavity [226]. The theory of such lasers is similar to that used for soliton communication systems making use of sliding-frequency guiding filters. In both cases, the soliton maintains itself by changing its frequency adiabatically so that its spectrum remains close to the gain peak. The CW light, in contrast, moves away from the gain peak after a few round trips because of the frequency shift and thus experiences higher losses than the soliton. As a result, the fiber laser emits mode-locked soliton pulses. Such a laser is classified as passively mode locked [228], since nothing modulates the amplitude or phase of laser light at the round-trip frequency or its multiple.

A dual-core fiber, whose one core is doped with erbium ions but the other is left undoped, provides not only gain but also the saturable absorption necessary for mode locking [229]. The operation of such a laser makes use of optical switching in nonlinear directional couplers (see Section 2.3). At low powers, a part of the mode energy is transferred to the undoped core and constitutes a loss mechanism for the laser cavity. At high powers, such an energy transfer ceases to occur, and most of the energy remains confined to the doped core. As a result, a dual-core fiber acts as a fast saturable absorber and shortens an optical pulse propagating through it. Mode locking can also be achieved by using a dual-core fiber as a fiber-loop mirror or simply by placing it inside the Fabry–Perot cavity of a fiber laser [230]. The underlying mode-locking mechanism in all cases is the nonlinear mode coupling that can also be implemented using an array of fibers or planar waveguides [231].

Fiber gratings have turned out to be quite useful for mode-locked lasers. In one scheme, fiber gratings are used to make a coupled-cavity fiber laser that can be mode locked through additive-pulse mode locking. In a simple implementation of this idea, three Bragg gratings were used to form two coupled Fabry–Perot cavities [232]. In one cavity, the fiber was doped with erbium and pumped at 980 nm while the other cavity had the standard undoped fiber. Both cavities had the same nominal length with total length ranging from 1 to 6 m. The laser produced relatively wide mode-locked pulses (width >50 ps) without requiring stabilization of individual cavity lengths. The latter feature is somewhat surprising since additive-pulse mode locking in coupled-cavity lasers normally requires precise matching of the cavity lengths. It can be understood by noting that the effective penetration distance in a fiber grating before light is reflected depends on the wavelength of light. As a result, the laser can adjust its wavelength to match the cavity lengths automatically. The self-matching capability of coupled-cavity fiber lasers can be extended by using chirped gratings [233]. Such a laser produced 5.5-ps mode-locked pulses, which could be compressed down to below 1 ps because of their chirped nature.

The colliding-pulse mode-locking technique can also be adopted for fiber lasers. In a 2001 experiment, it was used to mode lock an EDFL at the 20th harmonic of the fundamental repetition rate, producing 7-ps pulses at a repetition rate of 200 MHz [234]. Since then, this technique has been used to generate 380-fs pulses of high quality at a 605-MHz repetition rate from a Yb-doped fiber laser [235]. A 500-nm-thick InGaAs layer was used as a saturable absorber. The basic idea consists of colliding counter-propagating pulses inside a thin saturable absorber. The thickness of this absorber sets the smallest pulse width that can be generated inside the laser.

The fundamental repetition rate of passively mode-locked fiber lasers is typically limited to <100 MHz because it is difficult to reduce the cavity length much below 1 m. Even harmonic mode locking of such lasers rarely exceeds a repetition rate of 1 GHz. The repetition rate can be increased to 10 GHz and beyond by incorporating an

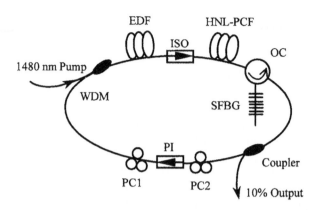

Figure 5.33: Schematic of a fiber laser mode locked passively at very high repetition rates using a sampled fiber Bragg grating (SFBG): EDF, erbium-doped fiber; HNL-PCF, highly nonlinear photonic crystal fiber; ISO, isolator; OC, optical circulator; PC, polarization controller, PI, polarizing isolator. (From Ref. [239]; ©2005 OSA.)

intracavity modulator. In a 2004 experiment [158], a Yb-doped fiber laser was mode locked at 281th harmonic of the fundamental repetition rate using an FM modulator (see Figure 5.23), resulting in mode locking at rates close to 10 GHz. A repetition rate of 40 GHz has been realized in some fiber lasers.

An interesting question is whether high repetition rates can be realized using a passive technique. Several such techniques, developed in recent years for this purpose, make use of chirped or sampled fiber gratings [236]–[240]. The grating can be either placed outside the laser to multiply the repetition rate, or a part of the laser cavity itself. When a chirped grating is employed outside the laser, the repetition rate of the pulse train is enhanced through a time-domain analog of the Talbot effect [236]. A sampled fiber Bragg grating was used in 2000 to multiply of the repetition rate of 1-GHz pulse train by a a factor of 4 [237]. By 2003, the use of superimposed fiber gratings allowed considerable flexibility in tailoring the the repetition rate as well as the shape and duration of individual pulses [238]. In such a device, two or more gratings of different periods are written over the same piece of fiber. Each grating is either uniform or is chirped suitably.

In a 2005 experiment [239], a sampled fiber grating was placed inside the laser cavity, as shown schematically in Figure 5.33, to obtain passive mode locking at repetition rates >100 GHz. The mode-locking mechanism makes use of the so-called dissipative four-wave mixing (FWM) process [241]. More precisely, the sampled fiber grating acts as an optical filter with periodic transmission peaks separated by Δv (100 GHz or so depending on the sample period employed), thereby creating spectral bands separated by this frequency interval. FWM inside the highly nonlinear fiber creates additional spectral bands and provides phase locking among them. The output of such a laser is a pulse train at a repetition rate of Δv or its multiple. Such a laser can easily operate at repetition rates >100 GHz.

The laser cavity in Figure 5.33 contains twos fibers, the one doped with erbium for providing the gain and the other for initiating the FWM process. If the Raman gain

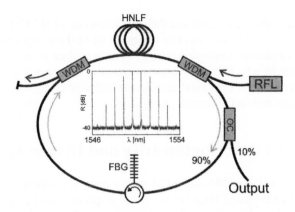

Figure 5.34: A fiber Raman laser mode locked passively at 100 GHz using a 1-km-long highly nonlinear fiber (HNLF) and a sampled fiber Bragg grating (FBG). OC, RFL, and WDM denote the optical circulator, Raman fiber laser used for pumping, and directional couplers. The inset shows the reflectivity spectrum of the grating. (From Ref. [242]; ©2006 OSA.)

within the highly nonlinear fiber is employed by pumping it suitably, one has the simplest design for a passively mode-locked fiber laser capable of generating short optical pulses at ultrahigh repetition rates. This approach was adopted in a 2006 experiment for creating pulses at the 100-GHz repetition rate [242]. As shown schematically in Figure 5.34, the laser cavity contains only a 1-km-long highly nonlinear fiber and a fiber grating. The fiber has its zero-dispersion wavelength at 1555 nm and its nonlinear parameter γ has a value of $14 \ W^{-1}$/km. It is pumped near 1450 nm to provide the maximum Raman gain at wavelengths close to 1550 nm.

The reflectivity spectrum of the sampled fiber grating has multiple peaks separated by 100 GHz around the central wavelength of 1550 nm. This grating acts as a passive mode-locking element by filtering the laser spectrum such that the two or three central peaks suffer the least loss. These filtered frequencies create additional bands, all separated by 100 GHz, through FWM. Phases of multiple bands are locked because of the phase-sensitive nature of this process. Such a laser emits a stable pulse train at the 100-GHz repetition rate, once the pump power is increased above the laser threshold of 150 mW. The autocorrelation trace indicated that 600-fs pulses are separated by 10 ps [242]. Note that the laser has a fundamental repetition rate of only 100 kHz because of the 1-km-long fiber, and it is operating at a harmonic whose number is close to 1 million. This laser shows clearly how the nonlinear effects such as SRS and FWM occurring in optical fibers can be exploited to advantage.

FWM is also the nonlinear process behind the self-induced modulation-instability fiber lasers, in which a mode-locked pulse train is generated passively without any other active or passive mode-locking mechanism [243]–[246]. In a 1997 experiment, a pulse train at a repetition rate of 115 GHz (pulse width about 2 ps) was generated from an EDFL through this process [244]. The laser cavity contained a Fabry–Perot filter whose free spectral range was cosen to coincide with the gain peak of modulation instability [55]. The sigma configuration was employed in a 2001 experiment [245].

Again, pulses were about 2 ps wide and their repetition rate of 107 GHz was set by the Fabry–Perot filter placed inside the laser cavity. In all of these experiments, the laser operated in the anomalous-dispersion regime. A train of dark solitons was created in a later experiment in which an EDFL was operated in the normal-dispersion region [246]. The experiment and numerical simulations indicated that dissipative FWM also played a significant role in the formation of pulses.

5.5 Role of Fiber Nonlinearity and Dispersion

Nonlinear effects such as SPM and XPM play a dominant role in the operation of most passively mode-locked fiber lasers. Fiber dispersion also plays an important role, especially when the soliton effects are relevant. This section is devoted to a theoretical discussion of such effects. Both the numerical and analytical methods were used during the 1990s to understand and to quantify the role of dispersive and nonlinear effects occurring inside both the active and passive fibers [247]–[265].

5.5.1 Saturable-Absorber Mode Locking

The theory of passive mode locking is based on the same Ginzburg–Landau equation used earlier for active mode locking. The main difference is in the functional form of the cavity-loss parameter α appearing in Eq. (5.3.3), which should include the intensity dependence of losses produced by the saturable absorber [130]. More specifically, α is given by

$$\alpha = \alpha_c + \alpha_0(1 + |A|^2/P_{sa})^{-1} \approx \alpha_c + \alpha_0 - \alpha_0|A|^2/P_{sa}, \qquad (5.5.1)$$

where P_{sa} is the saturation power of the absorber, assumed to be much larger than the peak power levels associated with optical pulses circulating inside the laser cavity. This assumption is made only to simplify the following analysis.

If we substitute Eq. (5.5.1) in Eq. (5.3.3), we find that the presence of saturated absorption modifies the parameter α_2 such that $\alpha_2 \approx -\alpha_0/P_{sa}$ if two-photon absorption is negligible. The new value of α_2 is negative. This is understandable from a physical viewpoint since the intensity dependence of a saturable absorber is just the opposite that of a two-photon absorber (absorption decreases with increasing intensity). In the following discussion we use Eq. (5.3.3) with negative values of α_2.

A change in the sign of α_2 does not affect the form of the solution given in Eq. (4.5.4). We can thus conclude that a passively mode-locked fiber laser emits pulses in the form of a chirped soliton whose amplitude is given by [131]

$$u(\xi, \tau) = N_s[\text{sech}(p\tau)]^{1+iq} \exp(iK_s\xi). \qquad (5.5.2)$$

The three parameters N_s, p, and q are determined in terms of the laser parameters as indicated in Eqs. (4.5.5)–(4.5.8). They are in turn related to the soliton width T_s, peak power P_s, and the frequency chirp $\delta\omega$ as (see Section 4.4)

$$T_s = T_2/p, \quad P_s = |\beta_2|N_s^2/(\gamma T_2^2), \quad \delta\omega = q\tanh(p\tau)/T_s. \qquad (5.5.3)$$

Using p from Eq. (4.5.6), the pulse width can be written in terms of the laser parameters (assuming anomalous GVD) as

$$T_s = \left\{ \frac{|\beta_2|[2q+d(q^2-1)]}{g_c - \alpha_c - \alpha_0} \right\}^{1/2}, \tag{5.5.4}$$

where $d = g_c/(\beta_2 \Omega_g^2)$ is related to the gain bandwidth. It is evident that GVD and SPM play a major role in establishing the width of the mode-locked pulse train.

This simple theory needs modification for modeling fiber lasers that are mode locked using semiconductor saturable absorbers. The reason is that a semiconductor does not response instantaneously. In fact, the response time of a quantum well is typically longer than the width of mode-locked pulses. The carrier dynamics can be included by replacing Eq. (5.5.1) with $\alpha = \alpha_c + \alpha_s$. The absorption coefficient α_s of the saturable absorber satisfies the following rate equation:

$$\frac{\partial \alpha_s}{\partial t} = \frac{\alpha_0 - \alpha_s}{\tau_s} - \frac{|A|^2}{E_{sa}} \alpha_s, \tag{5.5.5}$$

where τ_s is the recovery time and $E_{sa} \equiv \tau_s P_{sa}$ is the saturation energy of the absorber. For a fast-responding absorber, α_s is given by the steady-state solution of this equation, and the chirped-soliton solution of Eq. (5.5.2) is recovered.

Equation (5.5.5) can be solved approximately in the opposite limit of a relatively slow absorber and leads to the following expression for α:

$$\alpha = \alpha_c + \alpha_0 \exp\left[-\frac{1}{E_{sa}} \int_0^t |A(z,t)|^2 \, dt \right]. \tag{5.5.6}$$

The use of this equation in Eq. (5.4.1) leads to a modified Ginzburg–Landau equation that can be solved analytically in several important cases [263]. Actual quantum-well absorbers are found to have both fast and slow recovery mechanisms. A realistic model for such saturable absorbers has been developed [262]. The resulting Ginzburg–Landau equation is solved numerically; its predictions agree well with the experimental data.

5.5.2 Additive-Pulse Mode Locking

The Ginzburg–Landau equation can be extended for fiber lasers making use of additive-pulse mode locking [256]. The pulse-shortening effect of the mode-locking element (nonlinear fiber-loop mirror or nonlinear polarization rotation) is included through amplitude and phase changes induced on the pulse circulating inside the cavity.

Consider first the figure-8 laser in which a Sagnac loop imposes amplitude and phase changes on each pulse as it circulates inside it. Transmittivity of a Sagnac loop in which an amplifier is located at the entrance of the loop is given in Section 3.2 for CW beams. For an optical pulse with amplitude $A(t)$, it can be written as

$$T_S(t) = 1 - 2\rho(1-\rho)\{1+\cos[(1-\rho-G\rho)\gamma|A(t)|^2 L_s]\}, \tag{5.5.7}$$

where ρ is the bar-port transmission of the coupler and L_s is the loop length. For a 3-dB coupler, $\rho = 0.5$ and Eq. (5.5.7) reduces to

$$T_S(t) = \sin^2[(G-1)\gamma|A(t)|^2 L_s/4]. \qquad (5.5.8)$$

If loop length L_s is chosen such that $(G-1)\gamma P_0 L_s = 2\pi$, where P_0 is the peak power, the central part of a pulse is transmitted without loss, but the pulse wings experience loss. This intensity-dependent loss is referred to as *self-amplitude modulation* and is similar to that induced by a fast saturable absorber.

We can distribute the intensity-dependent loss introduced by the Sagnac loop over the cavity length and include its effects through the parameter α_2 in the Ginzburg–Landau equation. The effect of loop-induced nonlinear phase shift can also be included by modifying the parameter γ. The steady-state solution of the Ginzburg–Landau equation remains in the form of the chirped soliton of Eq. (5.5.2), but its width and peak power are affected by the loop parameters. This analytic solution can be used to study the effect of fiber dispersion and nonlinearity on the performance of figure-8 fiber lasers [26]. A similar technique can be used for fiber lasers that are mode locked via nonlinear polarization rotation [256].

Modeling of realistic mode-locked fiber lasers requires consideration of several other factors. For example, spontaneous emission seeds the growth of mode-locked pulses and should be included. Another effect that becomes important for ultrashort pulses is the self-frequency shift of solitons resulting from intrapulse Raman scattering. It is common to solve the Ginzburg–Landau equation numerically (with the split-step Fourier method [55], for example) because such an approach automatically includes the effects of SPM, XPM, GVD, and intrapulse Raman scattering [251]–[254]. This equation reduces to a generalized NLS equation in the parts of the laser cavity where the fiber is undoped. In the case of figure-8 lasers, the evolution of counterpropagating pulses should be considered separately inside the Sagnac loop, and the two optical fields should be combined coherently at the central coupler to determine the transmitted field.

For fiber lasers designed to make use of nonlinear polarization rotation, one must consider the evolution of orthogonally polarized components of the optical pulse by solving a set of two coupled Ginzburg–Landau equations, generalized to include the XPM effects. The effects of spontaneous emission can be included approximately by starting numerical simulations with a broadband noise pulse acting as a seed [254]. The noise pulse is propagated around the laser cavity repeatedly until a steady state is reached. Gain saturation is included by considering average power circulating inside the laser cavity. Such numerical simulations are capable of predicting most features observed experimentally.

5.5.3 Spectral Sidebands and Pulse Width

The Ginzburg–Landau equation provides only an approximate description of passively mode-locked fiber lasers. Real lasers show features not explained by this model. For example, pulse spectra of most fiber lasers exhibit sidebands, similar to those seen in Figure 5.27. In fact, several pairs of such sidebands appear under some operating

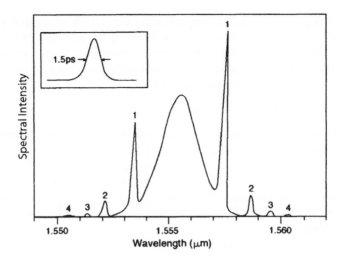

Figure 5.35: Output spectrum of an EDFL mode locked via nonlinear polarization rotation. Inset shows the autocorrelation trace of mode-locked pulses. (From Ref. [191]; ©1992 OSA.)

conditions. Figure 5.35 shows an example of such a pulse spectrum obtained from a fiber laser that was passively mode locked via nonlinear polarization rotation [191]. The ring cavity of this laser incorporated 122 m of standard fiber (undoped) with a total loop length of 148 m.

The origin of spectral sidebands seen in the output of fiber lasers is well understood [266]–[269]. The chirped soliton, found by solving the Ginzburg–Landau equation, represents the average situation since this equation ignores discrete nature of perturbations experienced by such solitons during each round trip. In reality, a part of the soliton energy leaves the cavity at the output coupler and constitutes a loss to the soliton circulating inside the cavity. The energy builds up to its original value as the pulse is amplified during each round trip. The net result is that the soliton energy and the peak power vary periodically, with a period equal to the cavity length. This amounts to creating a nonlinear-index grating that can affect soliton properties through Bragg diffraction, among other things. The situation is similar to that occurring in optical communication systems when pulses are amplified periodically to overcome fiber losses (see Section 4.3.2). In both cases, solitons adjust to perturbations by shading a part of their energy in the form of dispersive waves, also known as the *continuum radiation*.

Normally, dispersive waves produced by perturbations form a low-level, broadband background that accompanies the soliton. However, in the case of periodic perturbations, dispersive waves of certain frequencies can be resonantly enhanced, resulting in the spectral sidebands seen in Figure 5.35. The frequency and the amplitude of sidebands can be calculated using the perturbation theory of solitons [268]. The frequency can also be calculated by using a phase-matching condition, if spectral sidebands are interpreted to result from a four-wave mixing process that is phase matched by the index grating created by periodic perturbations.

A simple physical approach to understanding the growth of spectral sidebands makes use of a constructive interference condition. If the dispersive wave at a frequency $\omega_0 + \delta\omega$, where ω_0 is the soliton carrier frequency, is to grow on successive round trips, the phase difference between the soliton and that dispersive wave must be a multiple of 2π during a single round trip; that is,

$$|\beta(\omega_0) + K_s - \beta(\omega_0 + \delta\omega) - \beta_1\delta\omega|L = 2\pi m, \tag{5.5.9}$$

where m is an integer, $\beta(\omega_0 + \delta\omega)$ is the propagation constant of the dispersive wave, and K_s is the soliton wave number appearing in Eq. (4.5.4).

In general, one must use Eq. (4.5.7) to determine K_s. However, if the soliton is nearly unchirped, $K_s = (2L_D)^{-1}$, where $L_D = T_s^2/|\beta_2|$ is the dispersion length for a soliton of width T_s. By expanding $\beta(\omega_0 + \delta\omega)$ in a Taylor series and retaining terms up to quadratic in $\delta\omega$, Eq. (5.5.9) leads to the following expression for the sideband frequencies [266]:

$$\delta\omega T_s = \pm(8mz_0/L - 1)^{1/2}, \tag{5.5.10}$$

where $z_0 = (\pi/2)L_D$ is the soliton period. The predictions of Eq. (5.5.10) agree quite well with the position of sidebands seen in Figure 5.35 when mode-locked pulses are nearly transform limited. In the case of chirped solitons, the use of Eq. (4.5.5)–(4.5.7) with $s = -1$ ($\beta_2 < 0$) and $p = T_2/T_s$ leads to the result

$$\delta\omega T_s = \pm(8mz_0/L - 1 + q^2 - 2qd)^{1/2}, \tag{5.5.11}$$

where the chirp parameter q is obtained from Eq. (4.5.8). The effect of third-order dispersion on the location of spectral sidebands can also be taken into account [269] by including the cubic term in the Taylor-series expansion of $\beta(\omega_0 + \delta\omega)$ in Eq. (5.5.9).

Periodic perturbations occurring in a fiber laser also limit the duration of mode-locked pulses. This limit is similar to that restricting amplifier spacing in soliton communication systems and has the same origin [25]. If the solitons were to recover from periodic perturbations, they should be perturbed as little as possible during each round trip. In particular, the phase shift K_sL acquired by the soliton over the cavity length L must be a small fraction of 2π. By using $K_s = (2L_D)^{-1}$ with $L_D = T_s^2/|\bar{\beta}_2|$, the soliton width T_s is limited by

$$T_s \gg (|\bar{\beta}_2|L/4\pi)^{1/2}. \tag{5.5.12}$$

Here $\bar{\beta}_2$ represents the average value of GVD inside the dispersion-managed cavity. If we use $L = 20$ m and $\bar{\beta}_2 = -4$ ps^2/km as typical values for a figure-8 laser, $T_s \gg 80$ fs. Indeed, it is difficult to generate pulses much shorter than 100 fs from mode-locked fiber lasers. Equation (5.5.12) also shows that shorter mode-locked pulses can be generated by reducing the cavity length and average GVD inside the laser cavity. If we use $L = 2$ m and $\beta_2 = -1$ ps^2/km as optimized values for a mode-locked fiber laser, the condition (5.5.12) becomes $T_s \gg 25$ fs. Such lasers can generate pulses shorter than 100 fs. This is indeed what has been observed experimentally [193]. Measurements of pulse widths over a wide range of residual dispersion show that the pulse width indeed scales as $(|\bar{\beta}_2|L)^{1/2}$ [269].

The average GVD inside the cavity does not have to be anomalous for mode-locked fiber lasers. The general solution given in Eq. (5.5.2) exists for both normal and anomalous GVD. However, Eqs. (4.5.5)–(4.5.8) show that the chirp is relatively large in the

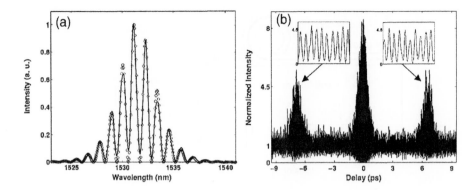

Figure 5.36: (a) Measured spectrum (circles) with a theoretical fit and (b) interferometric auto-correlation of the laser output in the form of soliton pairs. (From Ref. [271]; ©2002 OSA.)

case of normal GVD. Indeed, mode-locked pulses emitted from dispersion-managed fiber lasers with normal average GVD are heavily chirped. For this reason, they can be compressed considerably outside the laser cavity. In a 1994 experiment, the shortest pulse width (about 76 fs) was obtained from a mode-locked EDFL when the output pulse was compressed by using a piece of fiber with appropriate dispersion [195].

5.5.4 Phase Locking and Soliton Collisions

As mentioned earlier, multiple pulses can circulate simultaneously inside the cavity of a passively mode-locked fiber laser, depending on the pump power and other parameters. Under some operating conditions, the laser may even emit two or three pulses bunched together. It is well known from soliton theory that closely spaced solitons interact with each other through XPM such that they attract or repel each other depending on their relative phase difference [55]. Bound states of such solitons can also form under certain conditions. As mode-locked pulses emitted by fiber lasers are in the form of autosolitons, one would expect them to interact through the same mechanism and may even form a bound state through some kind of phase locking. It turns out that this is indeed the case.

A bound state of solitons was observed in a 2002 experiment at the output of a passively mode-locked fiber laser [270]. Since then, several studies have focused on such bound states of two or three solitons that can form in both the anomalous and normal-dispersion regions [271]–[277]. Theoretically, stable soliton pairs were discovered in 1998 by solving the Ginzburg–Landau equation [278]. The relative phase difference of the two pulses was found to be locked at $\pi/2$. This same value was observed in a 2002 experiment in which a fiber-ring laser was passively mode-locked through nonlinear polarization rotation such that the average cavity dispersion was anomalous [271]. Figure 5.36 shows the optical spectrum and autocorrelation of the laser output. The optical spectrum exhibited a multipeak structure that could be fitted well by assuming that the laser output was in the form of two 610-fs soliton pulses, separated by 6.8 ps, and differing in phase by $\pi/2$. The same width was deduced from

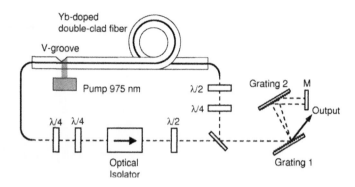

Figure 5.37: A fiber laser mode locked passively to produce a soliton triplet in the normal-dispersion regime. (From Ref. [274]; ©2004 IEEE.)

the autocorrelation trace. The spacing between two pulses of a pair depends on the pump power and increases with it. In this experiment, phase locking of more than 2 pulses was observed when pump power exceeded 80 mW.

It was found in 2003 that soliton pairs and triplets can also form when the average value of cavity dispersion lies in the normal region [272]. The main difference was that a pump power close to 250 mW was needed. Also, each pulse was heavily chirped and was much broader compared with the anomalous case. Indeed, a three-peak autocorrelation trace (which is indicative of two separate pulses) was formed only when they were dechirped down to a width of around 850 fs. This behavior is expected from the autosoliton theory of Section 4.5, based on the Ginzburg–Landau equation. Clear evidence of the bound state of three solitons in the normal-dispersion was seen in a 2004 study using a Yb-doped fiber laser [274]. The spacing among pulses was not necessarily the same under different operating conditions. Figure 5.37 shows the cavity design of this laser. The Yb-doped fiber is of double-clad type, and passive mode locking is realized through nonlinear polarization rotation. The grating pair compensates the normal dispersion of all fibers only partially, leaving a net dispersion of 0.047 ps^2 at the operating wavelength of 1050 nm. The five-peak autocorrelation trace measured at a pump power of 1.85 W and displayed in Figure 5.38 corresponds to a bound state of three 4.3-ps pulses separated by about 20 ps. Each pulse is chirped so much that it could be compressed down to 100 fs with a suitable compression scheme (see Chapter 6).

As discussed earlier, parabolic-shape pulses can also form in the case of normal dispersion inside mode-locked fiber lasers. A bound state of two and three parabolic pulses was observed in a 2006 experiment with a Yb-doped fiber laser [276]. The laser design was similar to that shown in Figure 5.37 but the wave plates were adjusted to operate the laser in the parabolic-pulse regime. The two positively chirped parabolic pulses that formed the bound state were 5.4-ps wide (each with 1.7-nJ energy) and were separated by about 15 ps. They could be compressed down to 100 fs by dechirping them. At a different setting of polarization controllers and a higher pumping power, the same laser generated a bound state of three parabolic pulses.

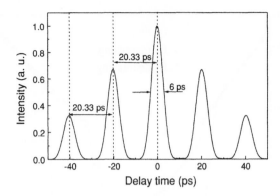

Figure 5.38: Measured autocorrelation trace of a uniformly spaced triplet containing three pulses separated by 20.33 ps. (From Ref. [274]; ©2004 IEEE.)

5.5.5 Polarization Effects

So far in this chapter, we have not addressed the issue of the state of polarization of light emitted from mode-locked fiber lasers. As discussed in Chapter 6 of Ref. [55], optical fibers do not preserve polarization unless they are specifically designed to do so. As a result, the state of polarization of output light may not be constant in time. It may change from pulse to pulse or even over the duration of a single pulse. The situation is quite interesting for short-cavity fiber lasers for which the cavity length is a small fraction of the beat length. In general, polarization evolution is important in all mode-locked fiber lasers and should be included for a proper understanding of such lasers [279]–[286].

The polarization effects were investigated thoroughly in a 1997 experiment in which a fiber laser of cavity length ~1 m or less was mode locked passively using a saturable Bragg reflector [279]. The cavity design was similar to that shown in Figure 5.24, except for the addition of a polarization controller that was made by wrapping standard single-mode fiber on two 5.5-cm-diameter paddles. It allowed continuous adjustment of the linear birefringence within the cavity by changing the azimuthal angles θ_1 and θ_2 of the paddles. A linear polarizer was placed at the output of the laser to analyze the polarization state. It converted polarization changes into periodic amplitude changes and introduced AM sidebands in the optical spectrum around each longitudinal mode. The presence of these sidebands is a sign that the state of polarization is not constant from pulse to pulse. Moreover, their frequency spacing Δ provides a quantitative measure of the temporal period over which polarization evolves. For this reason, this frequency is called the *polarization-evolution frequency*.

It was discovered experimentally that the AM sidebands disappear ($\Delta = 0$) for a certain combination of the angles θ_1 and θ_2. Figure 5.39 shows variations in Δ with θ_1 at a fixed value of θ_2. The polarization-evolution frequency decreases as θ_1 approaches $\pi/2$ and drops to zero in the vicinity of this value. The range of angles over which $\Delta = 0$ depends on the direction from which θ_1 approaches $\pi/2$, indicating that this phenomenon exhibits hysteresis. When $\Delta = 0$, the polarization of mode-locked pulses

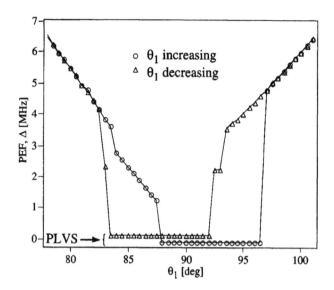

Figure 5.39: Measured variation of polarization-evolution frequency (PEF) Δ with θ_1. Polarization locking (PLVS) occurs when Δ equals 0. (From Ref. [282]; ©2000 OSA.)

is locked in such a way that all pulses have the same state of polarization in spite of the presence of linear birefringence within the laser cavity [280]. Such pulses are referred to as the *polarization-locked vector soliton*.

Properties of the such vector solitons was investigated in a 2000 experiment [282]. It turns out that the polarization state of the soliton can be linear or elliptical. In the case of elliptical polarization, the two linearly polarized components have different amplitudes and phases. The relative phase difference is fixed at $\pm\pi/2$ in all cases but the amplitude difference depends on the linear birefringence within the cavity. In the case of a linearly polarized vector soliton, the total energy of the soliton is carried by the component polarized along the slow axis. The existence of such solitons is related to the polarization instability of light polarized along the fast axis [55]. A theoretical model based on a set of two coupled Ginzburg–Landau equations is able explain most of the experimental data [283].

Vector solitons can also form inside a fiber laser when a birefringent Sagnac loop is used for for mode locking [281]. In the case of a fiber laser mode locked via nonlinear polarization rotation, birefringence effects can lead to pulse-to-pulse periodic variations in both the peak power and the state of polarization [284]. An amplitude- and polarization-locked pulse train is generated only when the axis of polarization of the polarizing isolator is aligned with the slow axis of the fiber.

A ring-cavity design similar to that shown in Figure 5.28 was used in a 2002 experiment for a Yb-doped fiber laser [285]. More specifically, the polarizer placed between two half-wave plates. Depending on the orientation of these phase plates, the fiber laser could be operated in the continuous, Q-switched, or mode-locked regime. An unstable self-pulsing regime was also observed for some orientations. A set of two coupled Ginzburg–Landau equations was used to model the experimental behavior. Numerical solutions of these equations show that the state of polarization (SOP) of the optical field is elliptical in general. Although the SOP evolves over the length of the cavity, it does not change from one round trip to next at a fixed point within the cavity after a steady state is reached [286]. The elliptical SOP of soliton and its continuous evolution occur even in passive fibers whenever a pulse propagates in the form of a vector soliton [287].

Problems

5.1 Why does the gain in fiber lasers vary along the fiber length? Derive the threshold condition by including such axial variations and considering a round trip inside the laser cavity.

5.2 Use the threshold condition [Eq. (5.1.3)] to derive an expression for the pump power required to reach threshold in fiber lasers.

5.3 Why is the gain in a laser clamped at its threshold value? Use this feature to derive expressions for the output power and the slope efficiency of fiber lasers.

5.4 How would you design the Fabry–Perot cavity of a fiber laser without using actual mirrors? Show two such designs and explain their operation.

5.5 Explain how stimulated Brillouin and Raman scattering process affect the output powers of Yb-doped fiber lasers. What steps can be taken in practice to minimize their impact?

5.6 Use the rate equations appropriate for a Q-switched fiber laser and derive an expression for the pulse width. You may consult Refs. [28]–[30] or any other book on lasers.

5.7 Derive an expression for the output intensity by considering N longitudinal modes of the cavity such that the phase difference between two neighboring modes is constant. Estimate the pulse width when 10,000 modes in a fiber laser are mode locked in this way. Assume a ring cavity of 5-m perimeter.

5.8 Explain how XPM can be used to induce mode locking in fiber lasers. Use diagrams as necessary.

5.9 The absorption of a fast saturable absorber saturates with power P as $\alpha = \alpha_0(1 + P/P_{sa})^{-1}$, where P_{sa} is the saturation power. Estimate the extent of pulse shortening occurring when a 1-ps hyperbolic-secant pulse of peak power $P_0 = 100P_{sa}$ passes through the saturable absorber. Assume that only 0.1% of low-power light is transmitted.

5.10 Explain the mode-locking process in a figure-8 fiber laser. What limits the pulse width in such lasers?

5.11 How can nonlinear birefringence be used to advantage for passive mode locking of fiber lasers? Draw the laser cavity schematically and explain the purpose of each component.

5.12 What is the origin of sidebands often seen in the spectrum of pulses emitted from passively mode-locked fiber lasers? Derive an expression for their frequencies.

References

[1] E. Snitzer, *Phys. Rev. Lett.* **7**, 444 (1961).

[2] J. Stone and C. A. Burrus, *Appl. Phys. Lett.* **23**, 388 (1973); *Appl. Opt.* **13**, 1256 (1974).

[3] M. I. Dzhibladze, Z. G. Esiashvili, E. S. Teplitskii, S. K. Isaev, and V. R. Sagaradze, *Sov. J. Quantum Electron.* **13**, 245 (1983).

[4] R. J. Mears, L. Reekie, S. B. Poole, and D. N. Payne, *Electron. Lett.* **21**, 738 (1985).

[5] L. Reekie, R. J. Mears, S. B. Poole, and D. N. Payne, *J. Lightwave Technol.* **4**, 956 (1986).

[6] I. M. Jauncey, L. Reekie, R. J. Mears, D. N. Payne, C. J. Rowe, D. C. J. Reid, I. Bennion, and C. Edge, *Electron. Lett.* **22**, 987 (1986).

[7] I. P. Alcock, A. I. Ferguson, D. C. Hanna, and A. C. Tropper, *Opt. Lett.* **11**, 709 (1986); *Electron. Lett.* **22**, 268 (1986).

[8] M. Shimizu, H. Suda, and M. Horiguchi, *Electron. Lett.* **23**, 768 (1987).

[9] L. Reekie, I. M. Jauncey, S. B. Poole, and D. N. Payne, *Electron. Lett.* **23**, 884 (1987); *Electron. Lett.* **23**, 1078 (1987).

[10] I. M. Jauncey, L. Reekie, J. E. Townsend, D. N. Payne, and C. J. Rowe, *Electron. Lett.* **24**, 24 (1988).

[11] K. Liu, M. Digonnet, K. Fesler, B. Y. Kim, and H. J. Shaw, *Electron. Lett.* **24**, 838 (1988).

[12] D. C. Hanna, R. M. Percival, I. M. Perry, R. G. Smart, and A. C. Trooper, *Electron. Lett.* **24**, 1068 (1988).

[13] G. T. Maker and A. I. Ferguson, *Electron. Lett.* **24**, 1160 (1988).

[14] I. N. Duling, L. Goldberg, and J. F. Weller, *Electron. Lett.* **24**, 1333 (1988).

[15] M. C. Brierley, P. W. France, and C. A. Miller, *Electron. Lett.* **24**, 539 (1988).

[16] M. C. Farries, P. R. Morkel, and J. E. Townsend, *Electron. Lett.* **24**, 709 (1988).

[17] D. C. Hanna, R. M. Percival, I. R. Perry, R. G. Smart, P. J. Suni, J. E. Townsend, and A. C. Trooper, *Electron. Lett.* **24**, 1111 (1988).

[18] R. Wyatt, B. J. Ainslie, and S. P. Craig, *Electron. Lett.* **24**, 1362 (1988).

[19] M. S. O'Sullivan, J. Chrostowski, E. Desurvire, and J. R. Simpson, *Opt. Lett.* **14**, 438 (1989).

[20] W. J. Barnes, S. B. Poole, J. E. Townsend, L. Reekie, D. J. Taylor, and D. N. Payne, *J. Lightwave Technol.* **7**, 1461 (1989).

[21] M. C. Farries, P. R. Morkel, R. I. Laming, T. A. Birks, D. N. Payne, and E. J. Tarbox, *J. Lightwave Technol.* **7**, 1473 (1989).

[22] Y. Kimura, K. Susuki, and M. Nakazawa, *Opt. Lett.* **14**, 999 (1989).

[23] R. Wyatt, *Electron. Lett.* **26**, 1498 (1989).

[24] M. J. F. Digonnet, Ed., *Rare-Earth Doped Fiber Lasers and Amplifiers* (Marcel Dekker, New York, 2001).

[25] G. P. Agrawal, *Fiber-Optic Communication Systems*, 3rd ed. (Wiley, Hoboken, NJ, 2002).

[26] I. N. Duling III, Ed., *Compact Sources of Ultrashort Pulses* (Cambridge University Press, New York, 2006).

[27] A. J. Brown, J. Nilsson, D. J. Harter, and A. Tünnermann, Eds., *Fiber Lasers III: Technology, Systems, and Applications*, Vol. 6102 (SPIE Press, Bellingham, WA, 2006).

[28] A. E. Siegman, *Lasers* (University Science Books, Mill Valley, CA, 1986).

[29] W. T. Silfvast, *Laser Fundamentals*, 2nd ed. (Cambridge University Press, New York, 2004).

[30] O. Svelto, *Principles of Lasers*, 4th ed. (Plenum, New York, 1998).

[31] D. C. Hanna, R. M. Percival, I. R. Perry, R. G. Smart, J. E. Townsend, and A. C. Tropper, *Opt. Commun.* **78**, 187 (1990).

[32] J. Y. Allain, M. Monerie, and H. Poignant, *Electron. Lett.* **26**, 166 (1990); *Electron. Lett.* **26**, 261(1990); *Electron. Lett.* **27**, 189 (1991).

[33] S. Sanders, R. G. Waarts, D. G. Mehuys, and D. F. Welch, *Appl. Phys. Lett.* **67**, 1815 (1995).

[34] R. Scheps, *Prog. Quantum Electron.* **20**, 271 (1996).

[35] M. Zeller, H. G. Limberger, and T. Lasser, *IEEE Photon. Technol. Lett.* **15**, 194 (2003).

[36] R. P. Kashyap, I. R. Armitage, R. Wyatt, S. T. Davey, and D. L. Williams, *Electron. Lett.* **26**, 730 (1990).

[37] G. A. Ball, W. W. Morey, and W. H. Glenn, *IEEE Photon. Technol. Lett.* **3**, 613 (1991).

[38] D. B. Mortimore, *J. Lightwave Technol.* **6**, 1217 (1988).

[39] P. Barnsley, P. Urquhart, C. A. Miller, and M. C. Brierley, *J. Opt. Soc. Am. A* **5**, 1339 (1988).

[40] P. R. Morkel, in *Rare Earth Doped Fiber Lasers and Amplifiers*, M. J. F. Digonnet, Ed. (Marcel Dekker, New York, 1993), Chap. 6.

[41] Y. Chaoyu, P. Jiangde, and Z. Bingkun, *Electron. Lett.* **25**, 101 (1989).

[42] F. Hakimi, H. Po, T. Tumminelli, B. C. McCollum, L. Zenteno, N. M. Cho, and E. Snitzer, *Opt. Lett.* **14**, 1060 (1989).

[43] H. Po, J. D. Cao, B. M. Laliberte, R. A. Minns, R. F. Robinson, B. H. Rockney, R. R. Tricca, and Y. H. Zhang, *Electron. Lett.* **29**, 1500 (1993).

[44] L. Zenteno, *J. Lightwave Technol.* **11**, 1435 (1993).

[45] H. Zellmer, U. Willamowski, A. Tünnermann, H. Welling, S. Unger, V. Reichel, H.-R. Muller, T. Kirchhof, and P. Albers, *Opt. Lett.* **20**, 578 (1995).

[46] H. M. Pask, R. J. Carman, D. C. Hanna, A. C. Tropper, C. J. Mackechnie, P. R. Barber, and J. M. Dawes, *IEEE J. Sel. Topics Quantum Electron.* **1**, 2 (1995).

[47] V. Dominic, S. MacCormack, R. Waarts, S. Sanders, S. Bicknese, R. Dohle, E. Wolak, P. S. Yeh and E. Zucker, *Electron. Lett.* **35**, 1158 (1999).

[48] J. M. Sousa, J. Nilsson, C. C. Renaud, J. A. Alvarez-Chavez, A. B. Grudinin, and J. D. Minelly, *IEEE Photon. Technol. Lett.* **11**, 39 (1999).

[49] J. Limpert, A. Liem, H. Zellmer, and A. Tünnermann, *Electron. Lett.* **39**, 645 (2003).

[50] J. Nilsson, W. A. Clarkson, R. Selvas, J. K. Sahu, P. W. Turner, S. U. Alam, and A. B. Grudinin, *Opt. Fiber Technol.* **10**, 5 (2004).

[51] A. S. Kurkov and E. M. Dianov, *Quantum Electron.* **34**, 8811 (2004).

[52] C. H. Liu, B. Ehlers, F. Doerfel, S. Heinemann, A. Carter, K. Tankala, J. Farroni, and A. Galvanauskas, *Electron. Lett.* **23**, 1471 (2004).

[53] Y. Jeong, J. K. Sahu, D. N. Payne, and J. Nilsson, *Opt. Express* **12**, 6088 (2004).

[54] J. Limpert, O. Schmidt, J. Rothhardt, F. Röser, T. Schreiber, A. Tünnermann, S. Ermeneux, P. Yvernault, and F. Salin, *Opt. Express* **14**, 2715 (2006).

[55] G. P. Agrawal, *Nonlinear Fiber Optics*, 4th ed. (Academic Press, San Diego, CA, 2007).

[56] J. P. Koplow, D. A. V. Kliner, and L. Goldberg, *Opt. Lett.* **25**, 442 (2000).

[57] P. Wang, L. J. Cooper, W. A. Clarkson, J. Nilsson, R. B. Williams, J. Sahu, and A. K. Vogel, *Electron. Lett.* **40**, 1325 (2004).

[58] P. Wang, L. J. Cooper, J. K. Sahu, and W. A. Clarkson, *Opt. Lett.* **31**, 2265 (2006).

[59] Z. Jiang and J. R. Marciante, *J. Opt. Soc. Am. B* **23**, 2051 (2006).

[60] G. C. Valley, *IEEE J. Quantum Electron.* **22**, 704 (1986).

[61] W. L. Barnes, P. R. Morkel, L. Reekie, and D. N. Payne, *Opt. Lett.* **14**, 1002 (1989).

[62] Y. Kimura, K. Susuki, and M. Nakazawa, *Opt. Lett.* **14**, 999 (1989).

[63] K. Iwatsuki, H. Okumura, and M. Saruwatari, *Electron. Lett.* **26**, 2033 (1990).

[64] J. L. Zyskind, J. W. Sulhoff, J. W. Stone, D. J. DiGiovanni, L. W. Stulz, H. M. Presby, A. Piccirilli, and P. E. Pramayon, *Electron. Lett.* **27**, 2148 (1991).

[65] D. A. Smith, M. W. Maeda, J. J. Johnson, J. S. Patel, M. A. Saifi, and A. V. Lehman, *Opt. Lett.* **16**, 387 (1991).

[66] P. D. Humphrey and J. E. Bowers, *IEEE Photon. Technol. Lett.* **5**, 32 (1993).

[67] C. V. Poulsen and M. Sejka, *IEEE Photon. Technol. Lett.* **5**, 646 (1993).

[68] Y. T. Chieng and R. A. Minasian, *IEEE Photon. Technol. Lett.* **6**, 153 (1994).

[69] R. Kashyap, *Fiber Bragg Gratings* (Academic Press, San Diego, CA, 1999), Chap. 8.

[70] G. A. Ball and W. H. Glenn, *J. Lightwave Technol.* **10**, 1338 (1982).

[71] S. V. Chernikov, R. Kashyap, P. F. McKee, and J. R. Taylor, *Electron. Lett.* **29**, 1089 (1993).

[72] S. V. Chernikov, J. R. Taylor, and R. Kashyap, *Opt. Lett.* **18**, 2024 (1993); *Electron. Lett.* **29**, 1788 (1993).

[73] J. T. Kringlebton, P. R. Morkel, L. Reekie, J. L. Archambault, and D. N. Payne, *IEEE Photon. Technol. Lett.* **5**, 1162 (1993).

[74] G. A. Ball, C. E. Holton, G. Hull-Allen, and W. W. Morey, *IEEE Photon. Technol. Lett.* **6**, 192 (1994).

[75] J. L. Archambault and S. G. Grubb, *J. Lightwave Technol.* **15**, 1378 (1997).

[76] K. Hsu, W. H. Loh, L. Dong, and C. M. Miller, *J. Lightwave Technol.* **15**, 1438 (1997).

[77] W. H. Loh, B. N. Samson, L. Dong, G. J. Cowle, and K. Hsu, *J. Lightwave Technol.* **16**, 114 (1998).

[78] G. P. Agrawal and N. K. Dutta, *Semiconductor Lasers*, 2nd ed. (Van Nostrand Reinhold, New York, 1993).

[79] W. H. Loh and R. I. Lamming, *Electron. Lett.* **31**, 1440 (1995).

[80] M. Sejka, P. Varming, J. Hübner, and M. Kristensen, *Electron. Lett.* **31**, 1445 (1995).

[81] J. Hübner, P. Varming, and M. Kristensen, *Electron. Lett.* **33**, 139 (1997).

[82] M. Ibsen, S.-U. Alam, M. N. Zervas, A. B. Grudinin, and D. N. Payne, *IEEE Photon. Technol. Lett.* **11**, 1114 (1999).

[83] M. Ibsen, S. Y. Set, C. S. Goh, and K. Kikuchi, *IEEE Photon. Technol. Lett.* **14**, 21 (2002).

[84] L. B. Fu, R. Selvas, M. Ibsen, J. K. Sahu, J. N. Jang, S.-U. Alam, J. Nilsson, D. J. Richardson, D. N. Payne, C. Codemard, S. Goncharov, I. Zalevsky, and A. B. Grudinin, *IEEE Photon. Technol. Lett.* **15**, 655 (2003).

[85] P. Horak, N. Y. Voo, M. Ibsen, and W. H. Loh, *IEEE Photon. Technol. Lett.* **18**, 998 (2006).

[86] Y. Dai, X. Chen, J. Sun, Y. Yao, and S. Xie, *IEEE Photon. Technol. Lett.* **18**, 1964 (2006).

[87] J. Chow, G. Town, B. Eggleton, M. Isben, K. Sugden, and I. Bennion, *IEEE Photon. Technol. Lett.* **8**, 60 (1996).

[88] N. Park and P. F. Wysocki, *IEEE Photon. Technol. Lett.* **8**, 1459 (1996).

[89] S. Yamashita, K. Hsu, and W. H. Loh, *IEEE J. Sel. Topics Quantum Electron.* **3**, 1058 (1997).

[90] N. J. C. Libatique and R. K. Jain, *IEEE Photon. Technol. Lett.* **11**, 1584 (1999).

[91] G. Brochu, S. LaRochelle, and R. Slavík, *J. Lightwave Technol.* **23**, 44 (2005).

[92] J. J. Pan and Y. Shi, *IEEE Photon. Technol. Lett.* **11**, 36 (1999).

[93] K. H. Ylä-Jarkko A. B. Grudinin, *IEEE Photon. Technol. Lett.* **15**, 191 (2003).

[94] A. Liem, J. Limpert, H. Zellmer, and A. Tnnermann, *Opt. Lett.* **28**, 1537 (2003).

[95] C. Alegria, Y. Jeong, C. Codemard, J. K. Sahu, J. A. Alvarez-Chavez, L. Fu, M. Ibsen, and J. Nilsson, *IEEE Photon. Technol. Lett.* **16**, 1825 (2004).

[96] Y. Jeong, J. Nilsson, D. B. S. Soh, C. A. Codemard, P. Dupriez, C. Farrell, J. K. Sahu, J. Kim, S. Yoo, D. J. Richardson, and D. N. Payne, in *Proc. Optical Fiber Commun.*, Paper OThW7 (Optical Society of America, Washington, DC, 2006).

[97] G. H. M. van Tartwijk and G. P. Agrawal, *Prog. Quant. Electron.* **22**, 43 (1998).

[98] F. Sanchez, P. LeBoudec, P. L. François, and G. M. Stefan, *Phys. Rev. A* **48**, 2220 (1993).

[99] E. Lacot, F. Stoeckel, and M. Chenevier, *Phys. Rev. A* **49**, 3997 (1994).

[100] M. Dinand and C. Schütte, *J. Lightwave Technol.* **13**, 14 (1995).

[101] F. Sanchez, M. LeFlohic G. M. Stefan, P. LeBoudec, and P. L. François, *IEEE J. Quantum Electron.* **31**, 481 (1995).

[102] R. Rangel-Rojo and M. Mohebi, *Opt. Commun.* **137**, 98 (1997).

[103] S. Bielawski, D. Derozier, and P. Glorieux, *Phys. Rev. A* **46**, 2811 (1992).

[104] M. Haeltermann, S. Trillo, and S. Wabnitz, *Phys. Rev. A* **47**, 2344 (1993).

[105] F. Sanchez and G. Stefan, *Phys. Rev. E* **53**, 2110 (1996).

[106] Q. L. Williams, J. Garcia-Ojalvo, and R. Roy, *Phys. Rev. A* **55**, 2376 (1997).

[107] L. G. Luo, T. J. Lee, and P. L. Chu, *J. Opt. Soc. Am. B* **15**, 972 (1998).

[108] H. D. I. Abarbanel, M. B. Kennel, M. Buhl, and C. T. Lewis, *Phys. Rev. A* **60**, 2360 (1999).

[109] G. D. van Wiggeren and R. Roy, *Science* **279**, 2524 (1998); *Phys. Rev. Lett.* **81**, 3547 (1998).

[110] L. G. Luo, P. L. Chu, and H. F. Liu, *IEEE Photon. Technol. Lett.* **12**, 269 (2000).

[111] A. Uchida, F. Rogister, J. García-Ojalvo, and R. Roy, in *Progrss in Optics*, Vol. 48, E. Wolf, Ed. (Elevier, New York, 2005), Chap. 5.

[112] R. Paschotta, R. Haring, E. Gini. H. Melchior, U. Keller, H. L. Offerhaus, and D. J. Richardson, *Opt. Lett.* **24**, 388 (1999).

[113] C. C. Renaud, R. J. Selvas-Aguilar, J. Nilsson, P. W. Turner, and A. B. Grudinin, *IEEE Photon. Technol. Lett.* **8**, 976 (1999).

[114] J. A. Alvarez-Chavez, H. L. Offerhaus, J. Nilsson, P. W. Turner, W. A. Clarkson, and D. J. Richardson, *Opt. Lett.* **25**, 37 (2000).

[115] C. C. Renaud, H. L. Offerhaus, J. A. Alvarez-Chavez, J. Nilsson, W. A. Clarkson, P. W. Turner, D. J. Richardson, and A. B. Grudinin, *IEEE J. Quantum Electron.* **37**, 199 (2001).

[116] M. Laroche, A. M. Chardon, J. Nilsson, D. P. Shepherd, and W. A. Clarkson, *Opt. Lett.* **27**, 1980 (2002).

[117] Y. Wang and C. Xu, *Opt. Lett.* **29**, 1060 (2004).

[118] A. Piper, A. Malinowski, K. Furusawa, and D. J. Richardson, *Electron. Lett.* **40**, 928 (2004).

[119] O. Schmidt, J. Rothhardt, F. Röser, S. Linke, T. Schreiber, K. Rademaker, J. Limpert, S. Ermeneux, P. Yvernault, F. Salin, and A. Tünnermann, *Opt. Lett.* **32**, 1551 (2000).

[120] S. V. Chernikov, Y. Zhu, J. R. Taylor, and V. P. Gapontsev, *Opt. Lett.* **22**, 298 (1997).

[121] Z. J. Chen, A. B. Grudinin, J. Porta, and J. D. Minelly, *Opt. Lett.* **23**, 454 (1998).

[122] A. Hideur, T. Chartier, C. Özkul, and F. Sanchez, *Opt. Commun.* **186**, 311 (2000).

[123] Y. X. Fan, F. Y. Lu, S. L. Hu, K. C. Lu, H. J. Wang, X. Y. Dong, J. L. He, and H. T. Wang, *Opt. Lett.* **29**, 724 (2004).

[124] A. A. Fotiadi, P. Mégret, and M. Blondel, *Opt. Lett.* **29**, 1078 (2004).

[125] Y. Zhao and S. D. Jackson, *Opt. Lett.* **31**, 751 (2006); *Opt. Lett.* **31**, 2739 (2006).

[126] M. Y. Cheng, Y. C. Chang, A. Galvanauskas, P. Mamidipudi, R. Changkakoti, and P. Gatchell, *Opt. Lett.* **30**, 358 (2005).

[127] L. Farrow, D. A. V. Kliner, P. E. Schrader, A. A. Hoops, S. W. Moore, G. R. Hadley, and R. L. Schmitt, *Proc. SPIE* **6102**, 61020L (2006).

[128] D. Brooks and F. Di Teodoro, *Appl. Phys. Lett.* **89**, 111119 (2006).

[129] D. J. Kuizenga and A. E. Siegman, *IEEE J. Quantum Electron.* **6**, 694 (1970).

[130] H. A. Haus, *J. Appl. Phys.* **46**, 3049 (1975); *IEEE J. Quantum Electron.* **11**, 323 (1975).

[131] H. A. Haus, in *Compact Sources of Ultrashort Pulses*, I. N. Duling III, Ed. (Cambridge University Press, New York, 2006).

[132] F. X. Kärtner, D. Kopf, and U. Keller, *J. Opt. Soc. Am. B* **12**, 486 (1995).

[133] K. Tamura and M. Nakazawa, *Opt. Lett.* **21**, 1930 (1996).

[134] N. G. Usechak and G. P. Agrawal, *Opt. Express* **13**, 2075 (2005).

[135] N. G. Usechak and G. P. Agrawal, *J. Opt. Soc. Am. B* **22**, 2570 (2005).

[136] K. Noguchi, O. Mitomi, and H. Miyazawa, *J. Lightwave Technol.* **16**, 615 (1998).

[137] J. D. Kafka, T. Baer, and D. W. Hall, *Opt. Lett.* **14**, 1269 (1989).

[138] K. Smith, J. R. Armitage, R. Wyatt, N. J. Doran, and S. M. J. Kelly, *Electron. Lett.* **26**, 1149 (1990).

[139] A. Takada and H. Miyazawa, *Electron. Lett.* **26**, 216 (1990).

[140] H. Takara, S. Kawanishi, M. Saruwatari, and K. Noguchi, *Electron. Lett.* **28**, 2095 (1992).

[141] X. Shan, D. Cleland, and A. Elis, *Electron. Lett.* **28**, 182 (1992).

[142] G. T. Harvey and L. F. Mollenauer, *Opt. Lett.* **18**, 107 (1993).

[143] T. Pfeiffer and G. Veith, *Electron. Lett.* **29**, 1849 (1993).

[144] M. Nakazawa, E. Yoshida, and Y. Kimura, *Electron. Lett.* **30**, 1603 (1994).

[145] E. Yoshida, Y. Kimura, and M. Nakazawa, *Electron. Lett.* **31**, 377 (1995).

[146] T. F. Carruthers and I. N. Duling III, *Opt. Lett.* **21**, 1927 (1996).

[147] R. Kiyan, O. Deparis, O. Pottiez, P. Mégret, and M. Blondel, *Opt. Lett.* **24**, 1029 (1999).

[148] K. S. Abedin, N. Onodera, and M. Hyodo, *Opt. Lett.* **24**, 1564 (1999).

[149] E. Yoshida and M. Nakazawa, *IEEE Photon. Technol. Lett.* **11**, 548 (1999).

[150] E. Yoshida, N. Shimizu, and M. Nakazawa, *IEEE Photon. Technol. Lett.* **11**, 1587 (1999).

[151] B. Bakshi and P. A. Andrekson, *IEEE Photon. Technol. Lett.* **11**, 1387 (1999).

[152] T. F. Carruthers, I. N. Duling III, M. Horowitz, and C. R. Menyuk, *Opt. Lett.* **25**, 153 (2000).

[153] M. Horowitz, C. R. Menyuk, T. F. Carruthers, and I. N. Duling III, *IEEE Photon. Technol. Lett.* **12**, 266 (2000); *J. Lightwave Technol.* **18**, 1565 (2000).

[154] J.-N. Maran, S. LaRochelle, and P. Besnard, *Opt. Lett.* **28**, 2082 (2003).

[155] K. S. Abedin and F. Kubota, *IEEE J. Sel. Topics Quantum Electron.* **10**, 1203 (2004).

[156] J. Vasseur, M. Hanna, and J. M. Dudley, *Opt. Commun.* **256**, 394 (2005).

[157] M. Tang, X. L. Tian, P. Shum, S. N. Fu, and H. Dong, *Opt. Express* **14**, 1726 (2006).

[158] N. G. Usechak, J. D. Zuegel, and G. P. Agrawal, *Opt. Lett.* **29**, 1360 (2004).

[159] N. G. Usechak, G. P. Agrawal, and J. D. Zuegel, *IEEE J. Quantum Electron.* **41**, 753 (2005).

[160] M. Zirngibl, L. W. Stulz, J. Stone, J. Hugi, D. DiGiovanni, and P. B. Hansen, *Electron. Lett.* **27**, 1734 (1991).

[161] W. H. Loh, D. Atkinson, P. R. Morkel, M. Hopkinson, A. Rivers, A. J. Seeds, and D. N. Payne, *IEEE Photon. Technol. Lett.* **5**, 35 (1993).

[162] W. H. Loh, D. Atkinson, P. R. Morkel, M. Hopkinson, A. Rivers, A. J. Seeds, and D. N. Payne, *Appl. Phys. Lett.* **63**, 4 (1993).

[163] W. H. Loh, D. Atkinson, P. R. Morkel, R. Grey, A. J. Seeds, and D. N. Payne, *Electron. Lett.* **29**, 808 (1993).

[164] E. A. De Souza, M. N. Islam, C. E. Soccolich, W. Pleibel, R. H. Stolen, J. R. Simpson, and D. J. DiGiovanni, *Electron. Lett.* **29**, 447 (1993).

[165] D. Abraham, R. Nagar, V. Mikhelashvili, and G. Eisenstein, *Appl. Phys. Lett.* **63**, 2857 (1993).

[166] B. C. Collings, K. Bergman, S. T. Cundiff, S. Tsuda, J. N. Kutz, J. E. Cunningham, W. Y. Jan, M. Koch, and W. H. Knox, *IEEE J. Sel. Topics Quantum Electron.* **3**, 1065 (1997).

[167] B. C. Collings, K. Bergman, and W. H. Knox, *Opt. Lett.* **23**, 123, (1998).

[168] S. Y. Set, H. Yaguchi, Y. Tanaka, and M. Jablonski, *IEEE J. Sel. Topics Quantum Electron.* **10**, 137 (2004).

[169] S. Yamashita, Y. Inoue, K. Hsu, T. Kotake, H. Yaguchi, D. Tanaka, M. Jablonski, and S. Y. Set, *IEEE Photon. Technol. Lett.* **17**, 750 (2005).

[170] R. Herda, O. G. Okhotnikov, E. U. Rafailov, W. Sibbett, P. Crittenden, and A, Starodumov, *IEEE Photon. Technol. Lett.* **18**, 157 (2006).

[171] M. Moenster, P. Glas, R. Iliew, R. Wedell, and G. Steinmeyer, *IEEE Photon. Technol. Lett.* **18**, 2502 (2006).

[172] A. Isomäki and O. G. Okhotnikov, *Opt. Express* **14**, 9238 (2006).

[173] M. J. O'Connell, Ed., *Carbon Nanotubes: Properties and Applications* (CRC Press, Boca Raton, FL, 2006).

[174] D. J. Richardson, R. I. Laming, D. N. Payne, V. J. Matsas, and M. W. Phillips, *Electron. Lett.* **27**, 542 (1991); *Electron. Lett.* **27**, 1451 (1991).

[175] I. N. Duling III, *Electron. Lett.* **27**, 544 (1991); *Opt. Lett.* **16**, 539 (1991).

[176] M. Nakazawa, E. Yoshida, and Y. Kimura, *Appl. Phys. Lett.* **59**, 2073 (1991).

[177] D. J. Richardson, A. B. Grudinin, and D. N. Payne, *Electron. Lett.* **28**, 778 (1992).

[178] M. Nakazawa, E. Yoshida, and Y. Kimura, *Electron. Lett.* **29**, 63 (1993).

[179] A. B. Grudinin, D. J. Richardson, and D. N. Payne, *Electron. Lett.* **28**, 67 (1992).

[180] E. Yoshida, Y. Kimura, and M. Nakazawa, *Appl. Phys. Lett.* **60**, 932 (1992).

[181] M. L. Dennis and I. N. Duling III, *Electron. Lett.* **28**, 1894 (1992); *Appl. Phys. Lett.* **62**, 2911 (1993).

[182] S. Wu, J. Strait, R. L. Fork, and T. F. Morse, *Opt. Lett.* **18**, 1444 (1993).

[183] D. U. Noske and J. R. Taylor, *Electron. Lett.* **29**, 2200 (1993).

[184] A. J. Stentz and R. W. Boyd, *Electron. Lett.* **30**, 1302 (1994).

[185] W. Y. Oh, B. Y. Kim, and H. W. Lee, *IEEE J. Quantum Electron.* **32**, 333 (1996).

[186] T. O. Tsun, M. K. Islam, and P. L. Chu, *Opt. Commun.* **141**, 65 (1997).

[187] N. H. Seong and D. Y. Kim, *IEEE Photon. Technol. Lett.* **14**, 459 (2002); *Opt. Lett.* **27**, 1521 (2002).

[188] A. V. Avdokhin, S. V. Popov and J. R. Taylor, *Opt. Express* **11**, 265 (2003).

[189] D. A. Chestnut and J. R. Taylor, *Opt. Lett.* **30**, 2982 (2005).

[190] V. J. Matsas, T. P. Newson, D. J. Richardson, and D. N. Payne, *Electron. Lett.* **28**, 1391 (1992).

[191] D. U. Noske, N. Pandit, and J. R. Taylor, *Opt. Lett.* **17**, 1515 (1992).

[192] K. Tamura, H. A. Haus, and E. P. Ippen, *Electron. Lett.* **28**, 2226 (1992).

[193] K. Tamura, E. P. Ippen, H. A. Haus, and L. E. Nelson, *Opt. Lett.* **18**, 1080 (1993).

[194] M. H. Ober, M. Hofer, and M. E. Fermann, *Opt. Lett.* **18**, 367 (1993).

[195] K. Tamura, L. E. Nelson, H. A. Haus, and E. P. Ippen, *Appl. Phys. Lett.* **64**, 149 (1994).

[196] M. E. Fermann, L.-M. Yang, M. L. Stock, and M. J. Andrejco, *Opt. Lett.* **19**, 43 (1994).

[197] K. Tamura, E. P. Ippen, and H. A. Haus, *Appl. Phys. Lett.* **67**, 158 (1995).

[198] R. Hofer, M. Hofer, and G. A. Reider, *Opt. Commun.* **169**, 135 (1999).

[199] F. Ö. Ilday, J. Buckley, L. Kuznetsova, and F. W. Wise, *Opt. Express* **11**, 3550 (2003).

[200] B. Ortaç, A. Hideur, M. Brunel, *Opt. Lett.* **29**, 1995 (2004).

[201] F. Ö Ilday, J. R. Buckley, W. G. Clark, and F. W. Wise, *Phys. Rev. Lett.* **92**, 213902 (2004).

[202] J. R. Buckley, F. W. Wise, F. Ö. Ilday, and T. Sosnowski, *Opt. Lett.* **30**, 1888 (2005).

[203] J. R. Buckley, S. W. Clark, and F. W. Wise, *Opt. Lett.* **31**, 1340 (2006).

[204] C. K. Nielsen and S. R. Keiding, *Opt. Lett.* **32**, 1474 (2007).

[205] J. Chen, J. W. Sickler, E. P. Ippen, and F. X. Kärtner, *Opt. Lett.* **32**, 1566 (2007).

[206] R. P. Davey, R. P. E. Fleming, K. Smith, R. Kashyap, and J. R. Armitage, *Electron. Lett.* **27**, 2087(1991).

[207] T. F. Carruthes, I. N. Duling III, and M. L. Dennis, *Electron. Lett.* **30**, 1051 (1994).

[208] D. J. Jones, L. E. Nelson, H. A. Haus, and E. P. Ippen, *IEEE J. Sel. Topics Quantum Electron.* **3**, 1076 (1997).

[209] M. E. Fermann, D. Harter, J. D. Minelly, and G. G. Vienne, *Opt. Lett.* **21**, 967 (1996).

[210] M. H. Ober, M. Hofer, G. A. Reider, G. D. Sucha, M. E. Fermann, D. Harter, C. A. C. Mendonca, and T. H. Chiu, *Opt. Lett.* **20**, 2305 (1995).

[211] M. Hofer, M. H. Ober, R. Hofer, M. E. Fermann. G. D. Sucha, D. Harter, K. Sugden, I. Bennion, C. A. C. Mendonca, and T. H. Chiu, *Opt. Lett.* **20**, 1701 (1995).

[212] M. H. Ober, G. D. Sucha, and M. E. Fermann, *Opt. Lett.* **20**, 195 (1995).

[213] L. Lefort, J. H. V. Price, and D. J. Richardson, G. J. Spühler, R. Paschotta, U. Keller, A. R. Fry, and J. Weston, *Opt. Lett.* **27**, 291 (2002).

[214] L. Lefort, A. Albert, V. Couderc, and A. Barthelemy, *IEEE Photon. Technol. Lett.* **14**, 1674 (2002).

[215] E. J. Greer and K. Smith, *Electron. Lett.* **28**, 1741 (1992).

[216] J. K. Lucek and K. Smith, *Opt. Lett.* **18**, 12,226 (1993).

[217] M. Obro, K. Lucek, K. Smith, and K. J. Blow, *IEEE Photon. Technol. Lett.* **6**, 799 (1994).

[218] D. S. Peter, G. Onishchukov, W. Hodel, and H. P. Weber, *Electron. Lett.* **30**, 1595 (1994).

[219] T. Aakjer and J. H. Povlsen, *Opt. Commun.* **112**, 315 (1994).

[220] E. M. Dianov, T. R. Martirosian, O. G. Okhotnikov, V. M. Paramonov, and A. M. Prokhorov, *Sov. Lightwave Commun.* **2**, 275 (1992).

[221] O. G. Okhotnikov, F. M. Araujo, and J. R. Salcedo, *IEEE Photon. Technol. Lett.* **6**, 933 (1994).

[222] D. U. Noske, A. Boskovic, M. J. Guy, and J. R. Taylor, *Electron. Lett.* **29**, 1863 (1993).

[223] C. R. Ó. Cochláin, R. J. Mears, and G. Sherlock, *IEEE Photon. Technol. Lett.* **5**, 25 (1993).

[224] D. M. Patrick, *Electron. Lett.* **30**, 43 (1994).

[225] H. Sabert and E. Brinkmeyer, *Electron. Lett.* **29**, 2124 (1993); *J. Lightwave Technol.* **12**, 1360 (1994).

[226] M. Romagnoli, S. Wabnitz, P. Franco, M. Midrio, F. Fontana, and G. E. Town, *J. Opt. Soc. Am. B* **12**, 72 (1995).

[227] D. O. Culverhouse, D. J. Richardson, T. A. Birks, and P. St. J. Russell, *Opt. Lett.* **20**, 2381 (1995).

[228] J. Porta, A. B. Grudinin, Z. J. Chen, J. D. Minelly, and N. J. Traynor, *Opt. Lett.* **23**, 615 (1998).

[229] H. G. Winful and D. T. Walton, *Opt. Lett.* **17**, 1688 (1992).

[230] R.-J. Essiambre and R. Vallé, *Opt. Commun.* **105**, 142 (1994).

[231] J. Proctor and J. N. Kutz, *Opt. Express* **13**, 8933 (2005).

[232] P. K. Cheo, L. Wang, and M. Ding, *IEEE Photon. Technol. Lett.* **8**, 66 (1996).

[233] D. W. Huang, G. C. Lin, and C. C. Yang, *IEEE J. Quantum Electron.* **35**, 138 (1999).

[234] O. G. Okhotnikov and M. Guina, *Appl. Phys. B* **72**, 381 (2001).

[235] Y. Deng, M. W. Koch, F. Lu, G. W. Wicks, W. H. Knox, *Opt. Express* **12**, 3872 (2004).

[236] J. Azaña and M. A. Muriel, *Opt. Lett.* **24**, 1672 (1999); *IEEE J. Sel. Topics Quantum Electron.* **7**, 728 (2001).

[237] P. Petropoulos, M. Ibsen, M. N. Zervas, and D. J. Richardson, *Opt. Lett.* **25**, 521 (2000).

[238] J. Azaña, R. Slavík, P. Kockaert, L. R. Chen, and S. LaRochelle, *J. Lightwave Technol.* **21**, 1490 (2003).

[239] S. Zhang, F. Lu, X. Dong, P. Shum, X. Yang, X. Zhou, Y. Gong, and C. Lu, *Opt. Lett.* **30**, 2852 (2005).

[240] J. Magné, J. Bolger, M. Rochette, S. LaRochelle, L. R. Chen, B. J. Eggleton, and J. Azaña, *J. Lightwave Technol.* **24**, 2091 (2006).

[241] M. Quiroga-Teixeiro, C. B. Clausen, M. P. Sørensen, P. L. Christiansen, and P. A. Andrekson, *J. Opt. Soc. Am. B* **15**, 1315 (1998).

[242] J. Schröder, S. Coen, F. Vanholsbeeck, and T. Sylvestre, *Opt. Lett.* **31**, 3489 (2006).

[243] P. Franco, F. Fontana, I. Cristiani, M. Midrio, and M. Romagnoli, *Opt. Lett.* **20**, 2009 (1995).

[244] E. Yoshida and M. Nakazawa, *Opt. Lett.* **22**, 1409 (1997).

[245] P. Honzatko, P. Peterka, and J. Kanka, *Opt. Lett.* **26**, 810 (2001).

[246] T. Sylvestre, S. Coen, P. Emplit, and M. Haelterman, *Opt. Lett.* **27**, 482 (2002).

[247] A. G. Bulushev, E. M. Dianov, and O. G. Okhotnikov, *Opt. Lett.* **15**, 968 (1990); *Opt. Lett.* **16**, 88 (1991).

[248] H. A. Haus, J. G. Fujimoto, and E. P. Ippen, *J. Opt. Soc. Am. B* **8**, 2068 (1991).

[249] P. A. Bélanger, *J. Opt. Soc. Am. B* **8**, 2077 (1991).

[250] S. M. J. Kelly, K. Smith, K. J. Blow, and N. J. Doran, *Opt. Lett.* **16**, 1337 (1991).

[251] M. P. Soerensen, K. A. Shore, T. Geisler, P. L. Christiansen, J. Mork, and J. Mark, *Opt. Commun.* **90**, 65 (1992).

[252] T. Geisler, K. A. Shore, M. P. Soerensen, P. L. Christiansen, J. Mork, and J. Mark, *J. Opt. Soc. Am. B* **10**, 1166 (1993).

[253] V. Tzelepis, S. Markatos, S. Kalpogiannis, T. Sphicopoulos, and C. Caroubalos, *J. Lightwave Technol.* **11**, 1729 (1993).

[254] V. Tzelepis, T. Sphicopoulos, and C. Caroubalos, *IEEE Photon. Technol. Lett.* **6**, 47 (1994).

[255] I. N. Duling III, C. J. Chen, P. K. A. Wai, and C. R. Menyuk, *IEEE J. Quantum Electron.* **30**, 194 (1994).

[256] H. A. Haus, E. P. Ippen, and K. Tamura, *IEEE J. Quantum Electron.* **30**, 200 (1994).

[257] C. J. Chen, P. K. A. Wai, and C. R. Menyuk, *Opt. Lett.* **19**, 198 (1994); *Opt. Lett.* **20**, 350 (1995).

[258] G. Sucha, S. R. Bolton, S. Weiss, and D. S. Chemla, *Opt. Lett.* **20**, 1794 (1995).

[259] M. Margalit and M. Orenstein, *Opt. Commun.* **124**, 475 (1996).

[260] S. Namiki and H. A. Haus, *IEEE J. Quantum Electron.* **33**, 649 (1997).

[261] G. H. M. van Tartwijk and G. P. Agrawal, *J. Opt. Soc. Am. B* **14**, 2618 (1997).

[262] J. N. Kutz, B. C. Collings, K. Bergman, S. Tsuda, S. T. Cundiff, W. H. Knox, P. Holmes, and M. Weinstein, *J. Opt. Soc. Am. B* **14**, 2681 (1997).

[263] N. N. Akhmediev, A. Ankiewicz, M. J. Lederer, and B. Luther-Davies, *Opt. Lett.* **23**, 280 (1988).

[264] G. H. M. van Tartwijk and G. P. Agrawal, *IEEE J. Quantum Electron.* **34**, 1854 (1998).

[265] D. Arbel and M. Orenstein, *IEEE J. Quantum Electron.* **35**, 977 (1999).

[266] S. M. J. Kelly, *Electron. Lett.* **28**, 806 (1992).

[267] N. Pandit, D. U. Noske, S. M. J. Kelly, and J. R. Taylor, *Electron. Lett.* **28**, 455 (1992).

[268] J. P. Gordon, *J. Opt. Soc. Am. B* **9**, 91 (1992).

[269] M. L. Dennis and I. N. Duling III, *IEEE J. Quantum Electron.* **30**, 1469 (1994).

[270] D. Y. Tang, B. Zhao, D. Y. Shen, C. Lu, W. S. Man, and H. Y. Tam, *Phys. Rev. A* **66**, 033806 (2002).

[271] P. Grelu, F. Belhache, F. Gutty, and J. M. Soto-Crespo, *Opt. Lett.* **27**, 966 (2002); *J. Opt. Soc. Am. B* **20**, 863 (2003).

[272] P. Grelu, J. Béal, and J. M. Soto-Crespo, *Opt. Express* **11**, 2238 (2003).

[273] M. Olivier, V. Roy, M. Pich, and F. Babin, *Opt. Lett.* **29**, 1461 (2004).

[274] B. Ortaç, A. Hideur, T. Chartier, M. Brunel, P. Grelu, H. Leblond, and F. Sanchez, *IEEE Photon. Technol. Lett.* **16**, 1274 (2004).

[275] V. Roy, M. Olivier, and M. Piché, *Opt. Express* **13**, 2716 (2005).

[276] B. Ortaç, A. Hideur, M. Brunel, C. Chédot, J. Limpert, A. Tünnermann, and F. Ö. Ilday, *Opt. Express* **14**, 6075 (2006).

[277] G. Martel, C. Chédot, V. Réglier, A. Hideur, B. Ortaç, and P. Grelu, *Opt. Lett.* **32**, 343 (2007).

[278] N. N. Akhmediev, A. Ankiewicz and J. M. Soto-Crespo, *J. Opt. Soc. Am. B* **15**, 515 (1998). *Opt. Lett.* **23**, 280 (1998).

[279] S. T. Cundiff, B. C. Collings, and W. H. Knox, *Opt. Express* **1**, 12 (1997).

[280] S. T. Cundiff, B. C. Collings, N. N. Akhmediev, J. M. Soto-Crespo, K. Bergman, and W. H. Knox, *Phys. Rev. Lett.* **82**, 3988 (1999).

[281] J. W. Haus, G. Shaulov, E. A. Kuzin, and J. Sanchez-Mondragon, *Opt. Lett.* **24**, 376 (1999).

[282] B. C. Collings, S. T. Cundiff, N. N. Akhmediev, J. M. Soto-Crespo, K. Bergman, and W. H. Knox, *J. Opt. Soc. Am. B* **17**, 354 (2000).

[283] J. M. Soto-Crespo, N. N. Akhmediev, B. C. Collings, S. T. Cundiff, K. Bergman, and W. H. Knox, *J. Opt. Soc. Am. B* **17**, 366 (2000).

[284] A. D. Kim, J. N. Kutz, and D. J. Muraki, *IEEE J. Quantum Electron.* **36**, 465 (2000).

[285] H. Leblond, M. Salhi, A. Hideur, T. Chartier, M. Brunel, and F. Sanchez, *Phys. Rev. A* **65**, 063811 (2002).

[286] J. Wu, D. Y. Tang, L. M. Zhao, and C. C. Chan, *Phys. Rev. E* **74**, 046605 (2006).

[287] F. Lu, Q. Lin, W. H. Knox, and G. P. Agrawal, *Phys. Rev. Lett.* **93**, 183901 (2004).

Chapter 6

Pulse Compression

An important application of nonlinear fiber optics consists of compressing optical pulses. Pulses shorter than 5 fs have been produced by using the nonlinear and dispersive effects occurring simultaneously inside silica fibers. This chapter is devoted to the study of pulse-compression techniques, the theory behind them, and the experimental issues related to them. Section 6.1 presents the basic idea and introduces the two kinds of compressors commonly used for pulse compression. The grating-fiber compressors discussed in Section 6.2 use a fiber with normal group-velocity dispersion (GVD) followed by a grating pair. The soliton-effect compressors described in Section 6.3 make use of higher-order solitons forming when self-phase modulation (SPM) and anomalous GVD occur simultaneously. The use of fiber gratings for pulse compression is discussed in Section 6.4, whereas Section 6.5 focuses on the technique of chirped-pulse amplification. Section 6.6 is devoted to dispersion-managed fiber compressors. Section 6.7 describes several other nonlinear techniques that use optical fibers for pulse compression including cross-phase modulation and four-wave mixing.

6.1 Physical Mechanism

The basic idea behind optical pulse compression is borrowed from *chirp radar*, where chirped pulses at microwave frequencies are compressed by passing them through a dispersive delay line [1]. The physical mechanism can be understood from Section 3.2 of Ref. [2], where propagation of chirped optical pulses in a linear dispersive medium is discussed. Such a medium imposes a dispersion-induced chirp on the pulse during its propagation. If the initial chirp is in the opposite direction of that imposed by GVD, the two tend to cancel each other, resulting in an output pulse that is narrower than the input pulse.

To see how such cancelation can produce shorter pulses, consider the propagation of a chirped Gaussian pulse inside an optical fiber. In the absence of the nonlinear effects, we should solve the linear equation

$$i\frac{\partial U}{\partial z} = \frac{\beta_2}{2}\frac{\partial^2 U}{\partial T^2},\tag{6.1.1}$$

with the input field

$$U(0,T) = \exp\left[-\frac{(1+iC)}{2}\frac{T^2}{T_0^2}\right], \tag{6.1.2}$$

where C is the chirp parameter and T_0 is the pulse width. Solving Eq. (6.1.1) with the Fourier transform method, we obtain

$$U(z,T) = \frac{1}{2\pi}\int_{-\infty}^{\infty}\tilde{U}(0,\omega)\exp\left(\frac{i}{2}\beta_2\omega^2 z - i\omega T\right)d\omega, \tag{6.1.3}$$

where $\tilde{U}(0,\omega)$ is the Fourier transform of the input field. Using $U(0,T)$ from Eq. (6.1.2), we obtain

$$U(z,T) = [1-i\xi(1+iC)]^{-1/2}\exp\left\{-\frac{(1+iC)T^2}{2T_0^2[1-i\xi(1+iC)]}\right\}, \tag{6.1.4}$$

where the propagation distance $\xi = z/L_D$ is normalized to the dispersion length $L_D = T_0^2/|\beta_2|$.

The preceding equation can be written in the form of Eq. (6.1.2) with the new pulse parameters T_1 and C_1 that change with ξ as

$$T_1 = T_0[(1+sC\xi)^2 + \xi^2]^{1/2}, \qquad C_1(z) = C + s(1+C^2)\xi, \tag{6.1.5}$$

where $s = \text{sgn}(\beta_2) = \pm 1$, depending on the nature of GVD. Note that even an initially unchirped pulse $(C = 0)$ becomes chirped inside the fiber. What is more important from the standpoint of pulse compression is that a fiber of suitable length can unchirp an initially chirped pulse if β_2 and C have opposite signs. Equation (6.1.5) shows that the pulse width is reduced initially under such conditions. It can be used to find the compression factor F_c in the form

$$F_c(\xi) = T_0/T_1 = [(1+sC\xi)^2 + \xi^2]^{-1/2}. \tag{6.1.6}$$

A pulse is compressed $(F_c > 1)$ only if $sC < 0$; that is, $\beta_2 C < 0$. This condition just states that chirp cancelation occurs only if the initial chirp and GVD-induced chirp are of opposite kinds. Positively chirped pulses $(C > 0)$ require anomalous GVD for compression (and vice versa).

Equation (6.1.6) also shows that the shortest pulse is obtained at a specific distance given by $\xi = |C|/(1+C^2)$. The maximum compression factor at that distance is also fixed by the input chirp and is $F_c = (1+C^2)$. This limit is easily understood by noting that spectrum of a chirped input pulse is broader by a factor of $1+C^2$ compared with that of an unchirped pulse. In the time domain, the compression process can be visualized as follows. Different frequency components of the pulse travel at different speeds in the presence of GVD. If the leading edge of the pulse is delayed by just the right amount to arrive nearly with the trailing edge, the output pulse is compressed. Positively chirped pulses (frequency increasing toward the trailing side) require anomalous or negative GVD in order to slow down the red-shifted leading edge. By contrast, negatively chirped pulses require normal or positive GVD to slow down the blue-shifted leading edge.

Early pulse-compression studies made use of both normal and anomalous GVD, depending on the technique through which frequency chirp was initially imposed on the pulse [3]–[7]. In the case of negatively chirped pulses, pulses were transmitted through liquids or gases such that they experienced normal GVD [4]. In the case of positively chirped pulses, a grating pair was found to be most suitable for providing anomalous GVD [6]. In these early experiments, pulse compression did not make use of any nonlinear optical effects. Although the use of the nonlinear process of SPM for pulse compression was suggested [8] as early as 1969, the experimental work on SPM-based pulse compression took off only during the 1980s, when the use of single-mode silica fibers as a nonlinear medium became widespread [9]–[24]. It led in 1987 to the creation of optical pulses as short as 6 fs in the 620-nm wavelength region [14]. By 1988, compression factors as large as 5000 had been attained. Such advances were possible only after the evolution of optical pulses in silica fibers was properly understood.

Pulse compressors based on nonlinear fiber optics can be classified into two broad categories: *grating-fiber* and *soliton-effect* compressors. In a grating-fiber compressor, the input pulse is propagated in the normal-dispersion regime of a nonlinear fiber; it is then compressed externally using a grating pair. The role of fiber is to impose a nearly linear, positive chirp on the pulse through the combination of SPM and GVD [2]. The grating pair provides the anomalous GVD required for compression of positively chirped pulses [6].

A soliton-effect compressor, in contrast, consists of only a piece of fiber whose length is suitably chosen. The input pulse propagates in the anomalous-GVD regime of the fiber and is compressed through an interplay between SPM and GVD. Compression occurs because of an initial pulse-narrowing phase through which all higher-order solitons go before the input shape is restored after one soliton period [2]. The compression factor depends on the peak power of the pulse, which determines the soliton order. The two types of compressors are complementary and generally operate in different regions of the optical spectrum. Grating-fiber compressors are useful for compressing pulses in the visible and near-infrared regions while soliton-effect compressors work typically in the range from 1.3 to 1.6 μm. The wavelength region near 1.3 μm offers special opportunities, since both kinds of compressors can be combined to yield large compression factors by using dispersion-shifted fibers.

6.2 Grating-Fiber Compressors

In the visible and near-infrared regions ($\lambda < 1.3$ μm), a grating-fiber compressor is commonly used for pulse compression [10]–[24]. Figure 6.1 shows such a compressor schematically in the double-pass configuration [15]. The input pulse is coupled into a single-mode fiber, where it broadens spectrally and develops a positive chirp across its entire width. The output pulse is then sent through a grating pair, where it experiences anomalous GVD and gets compressed. The optical beam is sent back through the grating pair to reconvert it to the original cross section. The mirror M_1 is slightly tilted to separate the outgoing beam from the incoming one. The mirror M_2 deflects the compressed pulse out of the compressor without introducing any additional losses.

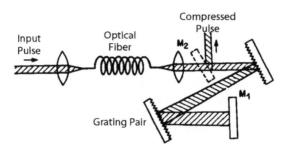

Figure 6.1: A grating-fiber compressor in the double-pass configuration. Mirror M_2 (shown dashed) is located above the plane of the figure. Mirror M_1 is slightly tilted to separate the outgoing beam from the incoming beam.

6.2.1   Grating Pair

A pair of parallel gratings acts as a dispersive delay line. Optical pulses propagating through such a grating pair behave as if they were transmitted through an optical fiber with anomalous GVD [6]. In this subsection, we focus on the theory behind a grating pair [25]–[28].

   When an optical pulse is incident at one grating of a pair of parallel gratings, different frequency components associated with the pulse are diffracted at slightly different angles. As a result, they experience different time delays during their passage through the grating pair. It turns out that the blue-shifted components arrive earlier than the red-shifted components. In a positively chirped pulse, blue-shifted components occur near the trailing edge of the pulse whereas the leading edge consists of red-shifted components. Thus, the trailing edge catches up with the leading edge during passage of the pulse through the grating pair, and the pulse is compressed.

   Mathematically, the phase shift acquired by a specific spectral component of the pulse at the frequency ω passing through the grating pair is given by $\phi_g(\omega) = \omega l_p(\omega)/c$, where the optical path length $l_p(\omega)$ is obtained from Figure 6.2 using simple geometrical arguments and is given by [27]

$$l_p(\omega) = l_1 + l_2 = d_0 \sec\theta_r [1 + \cos(\theta_r - \theta_i)], \qquad (6.2.1)$$

where d_0 is the grating separation (see Figure 6.2). The diffraction theory of gratings shows that, when light is incident at an angle θ_i, the diffraction angle θ_r is given by Eq. (1.1.1). We use this relation and assume first-order diffraction ($m = 1$). The frequency dependence of θ_r is responsible for the dispersion induced by a grating pair.

   If the spectral width of the optical pulse is a small fraction of its center frequency ω_0, it is useful to expand $\phi_g(\omega)$ in a Taylor series around ω_0 as

$$\phi_g(\omega) = \phi_0 + \phi_1(\omega - \omega_0) + \tfrac{1}{2}\phi_2(\omega - \omega_0)^2 + \tfrac{1}{6}\phi_3(\omega - \omega_0)^3 + \cdots, \qquad (6.2.2)$$

where ϕ_0 is a constant and ϕ_1 is related to the transit time through the grating pair. The parameters ϕ_2 and ϕ_3 take into account the dispersive effects associated with the grating pair and can be obtained by expanding $l_p(\omega)$ in a Taylor series and using Eq.

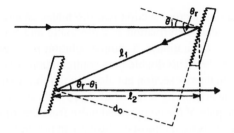

Figure 6.2: Geometry of a grating pair used as a dispersive delay line.

(1.1.1) for θ_r. The result is given by

$$\phi_2 = \frac{-8\pi^2 c d_g}{\omega_0^3 \Lambda^2 \cos^2 \theta_{r0}}, \qquad \phi_3 = \frac{24\pi^2 c d_g (1 + \sin \theta_i \sin \theta_{r0})}{\omega_0^4 \Lambda^2 \cos^4 \theta_{r0}}, \qquad (6.2.3)$$

where θ_{r0} is the diffraction angle for ω_0 and d_g is the center-to-center spacing between the gratings $(d_g = d_0 \sec \theta_{r0})$.

In most cases of practical interest, the spectral width of the pulse satisfies the condition $\Delta\omega \ll \omega_0$, and the cubic and higher-order terms in the expansion (6.2.2) can be neglected. If we ignore the unimportant constant and linear terms, the frequency-dependent part of the phase shift is governed by ϕ_2. Since ϕ_2 is negative from Eq. (6.2.3), a grating pair introduces anomalous GVD. This can be seen more clearly by considering the optical field at the output end of the grating pair. If $U_{\text{in}}(T)$ is the input field, the output field is given by

$$U_{\text{out}}(T) = \frac{1}{2\pi} \int_{-\infty}^{\infty} \tilde{U}_{\text{in}}(\omega - \omega_0) \exp\left[\frac{i}{2}\phi_2(\omega - \omega_0)^2 - i\omega T\right] d\omega, \qquad (6.2.4)$$

where $\tilde{U}_{\text{in}}$ is the Fourier transform of U_{in}. Comparing Eq. (6.2.4) with Eq. (6.1.3), the effective GVD parameter for a grating pair is given by $\beta_2^{\text{eff}} = \phi_2/d_g$. It is also possible to introduce an effective dispersion length as $L_D^{\text{eff}} = T_0^2/|\beta_2^{\text{eff}}|$, where T_0 is the input pulse width. An order of magnitude estimate of β_2^{eff} is obtained from Eq. (6.2.3). In the visible region ($\lambda_0 \approx 0.5\ \mu$m), $\beta_2^{\text{eff}} \sim 1000$ ps^2/km if we use $\Lambda \sim 1\ \mu$m. This corresponds to $L_D^{\text{eff}} \sim 1$ m for $T_0 = 1$ ps.

The grating separation required for pulse compression depends on the amount of positive chirp; typically d_g is a fraction of L_D^{eff} for femtosecond pulses ($d_g \sim 10$ cm). However, it becomes impractically large (>10 m) for $T_0 > 10$ ps. It is possible to increase β_2^{eff} by letting the pulse be incident at the grazing angle so that θ_{r0} in Eq. (6.2.3) approaches $\pi/2$. However, as is evident from this equation, such a scheme increases the contribution of third-order dispersion (TOD), and the cubic term in Eq. (6.2.2) must be included. Inclusion of the TOD term becomes necessary for ultrashort pulses ($T_0 \sim 10$ fs) whose bandwidth $\Delta\omega$ is comparable to ω_0. The effects of TOD can be minimized by using special gratings that are engraved directly on a prism [28].

A drawback of the grating pair is that spectral components of a pulse are dispersed not only temporally but also spatially. As a result, the optical beam diverges between the two gratings, acquiring a cross section that resembles an elongated ellipse rather than a circle. Such a beam deformation is undesirable and becomes intolerable in the case of large grating separations. A simple solution is to reflect the beam back through the grating pair [29]. This double-pass configuration not only recollimates the beam back into its original cross section but also doubles the amount of GVD, thereby reducing the grating separation by a factor of 2 [15]. A slight tilt of the reflecting mirror can separate the path of the compressed pulse from that of the input pulse. The double-pass configuration is used almost exclusively in practice.

Another disadvantage of the grating pair is related to the diffraction losses associated with it. Typically, 60 to 80% of the pulse energy remains in the pulse during first-order diffraction at a grating. This results in an energy loss of about a factor of 2 during a single pass through the grating pair or a factor of 4 in the double-pass configuration. Two alternative schemes can produce anomalous GVD with significantly smaller losses. One scheme makes use of a Gires–Tournois interferometer for the purpose of pulse compression [3]. Such an interferometer can reflect almost all of the pulse energy while imposing a dispersive phase shift of the quadratic form on various spectral components. In another scheme, a pair of two prisms provides anomalous GVD through refraction [30]. However, the required prism spacing is typically quite large (>10 m) because of the relatively small dispersion of fused quartz.

The prism spacing can be reduced by using materials such as dense flint glass and TeO_2 crystals [31]. In the case of TeO_2 crystal prisms, the spacing becomes comparable to that of a grating pair. In a 1988 experiment, 800-fs pulses were compressed to 120 fs by using a prism-pair spacing of 25 cm [32]. The energy loss of a prism pair can be reduced to 2% or less. A phase grating induced in a crystal by a chirped ultrasonic wave provides an alternative to the grating pair [33]. The use of fiber gratings is also quite attractive for this purpose [34]. As discussed in Chapter 1, a fiber grating can provide anomalous GVD even when it is fabricated within the core of a normal-GVD fiber.

6.2.2   Optimum Compressor Design

Several important questions need to be answered for optimum performance of grating-fiber compressors. The most important among them are (i) What is the optimum fiber length for given values of input pulse parameters? (ii) How far apart should the two gratings be to obtain high-quality pulses with maximum compression? To answer these questions, we should consider how an input pulse of a certain width and peak power evolves inside optical fibers in the presence of both SPM and GVD [35]–[45].

It is useful to employ a normalized form of the nonlinear Schrödinger (NLS) equation for this purpose. In the case of positive GVD ($\beta_2 > 0$), the NLS equation can be written as [2]

$$i\frac{\partial U}{\partial \xi} - \frac{1}{2}\frac{\partial^2 U}{\partial \tau^2} + N^2 \exp(-\alpha L_D \xi)|U|^2 U = 0, \tag{6.2.5}$$

where $\tau = T/T_0$, $\xi = z/L_D$, α accounts for fiber losses, and N^2 is defined as

$$N^2 = \frac{L_D}{L_{NL}} = \frac{\gamma P_0 T_0^2}{|\beta_2|}. \tag{6.2.6}$$

In these equations, $U = Ae^{-\alpha z}/P_0^{1/2}$ is the normalized amplitude, P_0 is the peak power of input pulses of width T_0, and γ is the nonlinear parameter. The nonlinear length $L_{NL} \equiv (\gamma P_0)^{-1}$ is defined in the usual manner. The soliton period, $z_0 = (\pi/2)\xi$, can also be used in place of ξ. It serves as a useful length scale even in the normal-dispersion regime with the interpretation that the pulse width nearly doubles at $z = z_0$ in the absence of SPM [37].

The performance of a grating-fiber compressor can be simulated by solving Eq. (6.2.5) numerically to obtain $U(z, \tau)$ at the fiber output then using it as the input field in Eq. (6.2.4) to find the compressed pulse shape. The parameter ϕ_2 in Eq. (6.2.3) can be adjusted to optimize the compressor performance. The optimum compressor is one for which the grating separation corresponds to an optimum value of ϕ_2 such that the peak power of the compressed pulse is largest. This is precisely how a grating-fiber compressor is optimized in practice. In the following discussion, fiber losses are neglected since the fiber lengths used in practice are relatively short ($\alpha L \ll 1$).

Consider first the case of pure SPM by neglecting GVD. In the absence of GVD, the shape of an input pulse remains unchanged, but its spectrum broadens with propagation [2]. More important, however, from the standpoint of pulse compression, is the SPM-induced frequency chirp. For a Gaussian pulse, the chirp is linear only over the central part of the pulse. When such a pulse is passed through a grating pair, only the central part is compressed. Since a significant amount of pulse energy remains in the wings, the compressed pulse is not of high quality.

It turns out that GVD of fibers helps improve the pulse quality considerably [35]. Normal GVD broadens the pulse and reshapes it to become nearly rectangular. At the same time, the pulse develops a nearly linear chirp across its entire width. As a result of this linear chirp, the grating pair can compress most of the pulse energy into a narrow pulse. Figure 6.3 shows the pulse shape at the fiber output, the frequency chirp across the pulse, and the compressed pulse for $N = 5$ and $z/z_0 = 0.5$ [37]. For comparison, the upper row shows the corresponding plots in the absence of GVD for a fiber length chosen such that the pulse is compressed by about the same factor in both cases ($N^2 z/z_0 = 4.5$). Even though neither N^2 nor z_0 is finite in the limit $\beta_2 = 0$, their ratio remains finite and can be used to compare the two cases. A comparison of the two rows in Figure 6.3 reveals the beneficial effect of GVD on the pulse quality when $\beta_2 > 0$ for the fiber used for chirping the pulse. However, this benefit is realized only at the expense of reduced compression at a given value of the input peak power [37].

To quantify the performance of grating-fiber compressors, it is useful to introduce two parameters:

$$F_c = T_{FWHM}/T_{comp}, \qquad Q_c = |U_{out}(0)|^2/F_c, \qquad (6.2.7)$$

where T_{comp} is the full width at half maximum (FWHM) of the compressed pulse. Clearly, F_c is the compression factor. The parameter Q_c is a measure of the quality of the compressed pulse. Its value at the fiber input is 1, and $Q_c \approx 1$ is desirable for the compressed pulse if nearly all of the pulse energy has to reappear in it.

Numerical simulations based on the NLS equation show that an optimum value of the fiber length exists for which both F_c and Q_c are maximum [37]. Figure 6.4 shows variations in F_c and Q_c with z/z_0 for values of N in the range from 1 to 20,

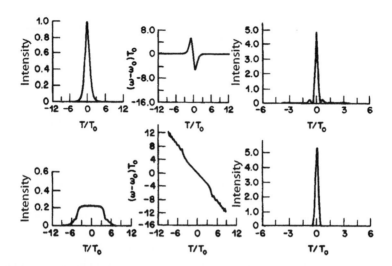

Figure 6.3: Pulse shape and chirp profile at fiber output and compressed pulse after the grating pair. The effects of GVD are ignored in the upper row and $N^2 z/z_0 = 4.5$. In the lower row, $N = 5$, $z/z_0 = 0.5$, and GVD is included. (From Ref. [37]; ©1984 OSA.)

assuming a "sech" shape for input pulses. For values of $N > 5$, the maxima of F_c and Q_c are evident, indicating the need to optimize the fiber length. The existence of an optimum fiber length z_{opt} can be understood qualitatively as follows. For $z < z_{opt}$, the SPM-induced chirp has not yet been linearized, whereas for $z > z_{opt}$ the GVD effects broaden the pulse so much that SPM loses its effectiveness. Indeed, z_{opt} is well approximated by $(6L_D L_{NL})^{1/2}$, showing the relative importance of both the GVD and SPM effects for pulse compression.

From the standpoint of compressor design, it is useful to provide simple design rules that govern the optimum fiber length and the optimum grating separation for realizing maximum compression for given values of the pulse and fiber parameters. The numerical results of Figure 6.4 can be used to obtain the following relations valid for $N \gg 1$:

$$z_{opt}/z_0 \approx 1.6/N, \qquad (6.2.8)$$

$$\tfrac{1}{2}|\phi_2|/T_{FWHM}^2 \approx 1.6/N, \qquad (6.2.9)$$

$$1/F_c \approx 1.6/N, \qquad (6.2.10)$$

where the grating parameter ϕ_2 is related to the optimum grating separation through Eq. (6.2.3). The numerical factor depends on the input pulse shape and would be slightly different than 1.6 for shapes other than a hyperbolic secant. Equations (6.2.8)–(6.2.10) are fairly accurate for all pulse shapes as long as $N > 10$. Similar relations have been obtained using the inverse scattering method and making certain approximations about the pulse shape and the chirp [36]. In applying Eqs. (6.2.8)–(6.2.10) in practice, the parameter N is first estimated from Eq. (6.2.6) for given values of the peak power P_0 and the width T_0 associated with a pulse. Then, the fiber length z_{opt} is obtained from

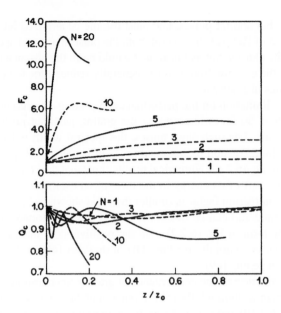

Figure 6.4: Compression factor F_c and quality factor Q_c as a function of fiber length for values of N in the range from 1 to 20. Grating separation is optimized in each case to maximize the peak power of the compressed pulse. (From Ref. [37]; ©1984 OSA.)

Eq. (6.2.8) while Eqs. (6.2.3) and (6.2.9) provide the grating separation. Finally, the compression factor is estimated from Eq. (6.2.10).

6.2.3 Practical Limitations

Although the preceding theory is applicable in most practical situations, it is important to keep in mind its limitations [46]–[53]. First, input pulses were assumed to be unchirped. It is easy to include the effect of a linear initial chirp by solving Eq. (6.2.5) numerically [46]. For down-chirped pulses ($C < 0$), the optimum fiber length increases, since the positive chirp provided by the fiber has to compensate for the initial negative chirp. At the same time, the compression factor is slightly reduced because such a compensation is not perfect all along the pulse width. The opposite occurs in the case of up-chirped pulses ($C > 0$). However, for large values of N ($N > 10$), the changes in z_{opt} and F_c are relatively small ($<10\%$) for pulses whose spectral width is up to twice of that expected in the absence of chirp. A related issue is the effect of a random chirp on pulse compression because of phase fluctuations associated with some input pulses. Numerical simulations show that the average compression factor is reduced by an amount that depends on the noise variance but the optimum value of the fiber length remains relatively unaffected [47].

Another limitation of the results shown in Figure 6.4 is that they are based on Eq. (6.2.5), which neglects the higher-order nonlinear and dispersive effects. This is justified as long as the spectral width $\Delta\omega \ll \omega_0$, and the results are fairly accurate for pulse widths $T_0 > 1$ ps. For shorter pulses, one must use the generalized NLS equation derived in Section 2.3 of Ref. [2]. In general, both the pulse shape and spectrum become asymmetric. As the chirp is not as linear as it would be in the absence of higher-order nonlinear effects, the compression factor is generally reduced for femtosecond pulses from the predictions of Figure 6.4.

A more severe limitation on the performance of grating-fiber compressors for ultrashort pulses ($T_0 < 50$ fs) is imposed by the grating pair that no longer acts as a quadratic compressor. For such short pulses, the spectral width is large enough that the cubic term in the expansion (6.2.2) becomes comparable to the quadratic term and must be included in Eq. (6.2.4). Numerical results show that the compressed pulse then carries a significant part of its energy in the form of an oscillatory trailing edge [48]. As a result, the compression factor is smaller than that shown in Figure 6.4. This limitation is fundamental and can be overcome only if a way is found to counteract the effect of the cubic term in Eq. (6.2.2). On the positive side, the cubic term can be exploited to compensate partially for the TOD of the fiber [42] or the nonlinear chirp induced by self-steepening [49].

An ultimate limitation on the performance of grating-fiber compressors is imposed by stimulated Raman scattering [50]–[53]. Even though the compression factor $F_c \propto N$, according to Eq. (6.2.10) and can in theory be increased by increasing the peak power of the incident pulse, it is limited in practice, since the peak power must be kept below the Raman threshold to avoid the transfer of pulse energy to the Raman pulse. Furthermore, even if some energy loss is acceptable, the Raman pulse can interact with the pump pulse through cross-phase modulation and deform the linear nature of the frequency chirp. It is possible to achieve large compression factors even in the Raman regime with an optimization of the design parameters [53]. Numerical simulations show that a significant part of the pulse energy remains uncompressed because of mutual interaction between the pump and Raman pulses. For highly energetic pulses, parametric processes such as four-wave mixing can suppress the Raman process to some extent, but they eventually limit the extent of pulse compression [54].

The performance of a grating-fiber compressor can be improved with the spectral-window method [18], in which a suitable aperture is placed near the mirror M_1 in Figure 6.1 to filter the pulse spectrum selectively. The technique of spectral filtering is a powerful technique that can be used not only to improve the performance of a grating-fiber compressor but also to control the pulse shape through spectral modifications inside the compressor [55]–[57]. This is possible because a grating pair separates the spectral components spatially and allows one to modify them (both in amplitude and phase) by using masks placed near the mirror M_1 in Figure 6.1.

6.2.4 Experimental Results

In the 1981 experiment in which an optical fiber was first used for pulse compression, 5.5-ps input pulses at 587 nm, with peak powers of 10 W, were propagated through a 70-m-long fiber [9]. The 20-ps output pulses were nearly rectangular in shape and had an SPM-broadened spectrum with a nearly linear chirp across the entire pulse. This experiment used sodium vapor, instead of a grating pair, as a dispersive-delay line. The compressed pulse was 1.5-ps wide. Compression factor of 3.7 is in agreement

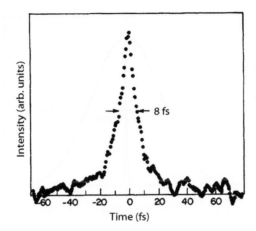

Figure 6.5: Autocorrelation trace of a 40-fs input pulse compressed by a grating-fiber compressor. (From Ref. [13]; ©1985 AIP.)

with the results of Figure 6.4, if we note that the experimental values of the parameters correspond to $N \approx 7$ and $z/z_0 \approx 0.25$. Even the pulse shape at the fiber output was in close agreement with the numerical simulations based on the NLS equation.

The compression technique was extended in 1982 to the femtosecond domain by using a grating pair as a dispersive delay line [10]. In this experiment, 90-fs pulses at 619 nm were passed through a 15-cm-long fiber and were compressed to about 30 fs after passing through the grating pair. The fiber and pulse parameters were such that $N \approx 3$ and $z/z_0 \approx 1.5$. The compression factor of about 3 is expected from Figure 6.4. This experiment led to a series of experiments [12]–[14] in which the pulse width was reduced in succession to about 6 fs. In one experiment, 40-fs pulses at 620 nm, with a peak intensity of about 1 TW/cm^2, were passed through a 7-mm-long fiber then compressed to 8 fs by using a grating pair [13]. Figure 6.5 shows the autocorrelation trace of the compressed pulse. The corresponding spectrum had a spectral width of about 70 nm, indicating that a pulse width of 6 fs would be possible if the compressed pulse were transform limited. The most important factor that limited compression was the TOD of the grating pair resulting from the ϕ_3 term in Eq. (6.2.2). In a later experiment, the effects of TOD were compensated for by using a combination of gratings and prisms, and the pulse indeed compressed to 6 fs [14]. Such a pulse consists of only three optical cycles at a wavelength of 620 nm.

In a different set of experiments, the objective was to maximize the compression factor. Compression by a factor of 12 was achieved in a 1983 experiment in which 5.4-ps input pulses from a dye laser were compressed to 0.45 ps using a 30-m-long fiber [11]. A higher compression factor of 65 was obtained using a two-stage compression scheme in which the pulse was passed through two grating pairs in succession. In a 1984 experiment, single-stage compression by a factor of 80 was realized using 33-ps pulses at 532 nm from a frequency-doubled Nd:YAG laser [15]. Passage of these pulses through a 105-m-long fiber, followed by a grating pair (separation $d_g = 7.24$ m), resulted in compressed pulses of 0.41-ps duration. The experiment utilized a double-

256   Chapter 6. Pulse Compression

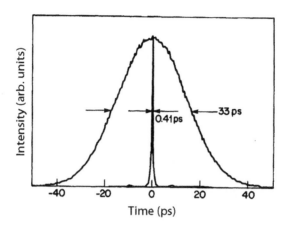

Figure 6.6: Measured autocorrelation traces of the input and compressed pulses showing a single-stage compression by a factor of 80. (From Ref. [15]; ©1984 AIP.)

pass configuration. Figure 6.6 shows the compressed pulse and compares it to the input pulse. The input peak power of 240 W corresponds to $N \approx 145$. Equation (6.2.10) predicts a compression factor of about 90 for this value of N, in reasonable agreement with the experimental value of 80. Although larger values of the compression factor are possible in principle, peak powers cannot be increased much more in practice because of the onset of stimulated Raman scattering.

The experiments just described were performed in the visible region of the optical spectrum. The grating-fiber compression technique has been extended to the near-infrared region to obtain ultrashort pulses at 1.06 and 1.32 μm. Input pulses at these wavelengths are generally obtained from mode-locked Nd:YAG lasers and are typically 100-ps wide. As a result, the dispersion length and the parameter z_0 are relatively large ($\sim$100 km). Equation (6.2.8) indicates that the optimum fiber length exceeds 1 km even for values of $N \sim 100$. The optimum value of the grating separation is also relatively large ($d_g > 1$ m), as seen from Eqs. (6.2.3) and (6.2.9).

In a 1984 experiment, 60-ps pulses at 1.06 μm were compressed by a factor of 15 after passing through a 10-m-long fiber and a grating pair with spacing of about 2.5 m [16]. In a different experiment, a compression factor of 45 was achieved by using a 300-m-long fiber and a compact grating pair [17]. The compressed pulses at 1.06 μm generally carry a significant amount of energy in the uncompressed wings because a smaller fiber length than that dictated by Eq. (6.2.8) is often used in practice to reduce optical losses. In the absence of fully developed GVD effects only the central part of the pulse is linearly chirped, and the energy in the wings remains uncompressed. The energy in uncompressed pulse wings can also be reduced by using the nonlinear birefringence in such a way that the fiber acts as an intensity discriminator [58].

The technique of spectral windowing has been used to remove energy from pulse wings [18]. It makes use of the observation that the wings contain spectral components at the extreme end of the pulse spectrum, which can be filtered by placing an aperture (or window) near the mirror M_1 in Figure 6.1. Figure 6.7 compares the autocorrelation

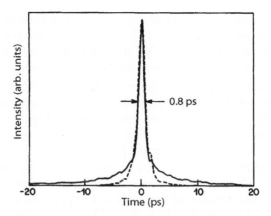

Figure 6.7: Autocorrelation traces of compressed pulses with (dashed curve) and without (solid line) spectral windowing. (From Ref. [56]; ©1986 OSA.)

traces of compressed pulses obtained with and without spectral windowing [56]. The 75-ps input pulses were compressed to about 0.8 ps in a conventional grating-fiber compressor, resulting in a compression factor of more than 90. The use of spectral windowing nearly eliminated the pulse wings while the pulse width increased slightly to 0.9 ps. This technique can also be used to modify the pulse shape by using suitable masks in place of a simple aperture [55]–[57]. Temporal modulation of the chirped pulses at the fiber output (before entering the grating pair) has also been used for this purpose [59]. Such techniques convert the grating-fiber compressor into a versatile tool that can be used for pulse synthesis.

It is generally difficult to achieve compression factors larger than 100 for 1.06-μm pulses because of the onset of stimulated Raman scattering. Compression by a factor of 110 was realized in an experiment in which 60-ps pulses were propagated through a 880-m-long fiber [24]. Even higher compression factors can be obtained by using two grating-fiber compressors in series [19]. In one experiment, 90-ps pulses were compressed to 0.2 ps, resulting in an overall compression factor of 450 [22]. At the same time, the peak power increased from 480 W to 8 kW. Two compressors provided the same compression factor of about 21. It was noticed that, even though pulses after the first compressor had significant amounts of energy in their wings, pulses emerging from the second compressor were of high quality. The reason is related to the different input pulse widths. The 4.2-ps pulses fed to the second compressor were short enough that GVD was able to linearize the chirp across the whole pulse. The experimental results were in close agreement with theory.

The grating-fiber compression technique has been extended to 1.32 μm, a wavelength at which mode-locked Nd:YAG lasers can provide powerful pulses of about 100-ps duration [20]. However, since standard fibers provide normal dispersion only for $\lambda < 1.3$ μm, it is necessary to use a dispersion-shifted fiber with its zero-dispersion wavelength around 1.55 μm. The optimum fiber length typically exceeds 2 km. This is not, however, a limiting factor because of smaller fiber losses ($\approx$0.4 dB/km) near 1.32 μm. Compression by a factor of 50 was realized when 100-ps pulses were chirped using a 2-km-long dispersion-shifted fiber [21] (zero dispersion at 1.59 μm). The fiber length was less than optimum ($z_{opt} \approx 3.3$ km) to reduce the grating separation to manageable dimensions. Equation (6.2.10) predicts a compression factor of 80 if the optimum fiber length with optimum grating separation were employed ($N \approx 130$).

An advantage of operating at the 1.32-μm wavelength is that the grating pair can be replaced by a piece of fiber, making it possible to realize a compact all-fiber compressor. Two fibers with positive and negative values of the GVD parameter β_2 are fused together to make the compressor. The fiber with positive β_2 produces linear chirp across the pulse while the fiber with negative β_2 compresses it. The lengths of the two fibers need to be optimized using Eqs. (6.2.8) and (6.2.9). The grating parameter ϕ_2 is replaced by $\beta_2 L_2$, where L_2 is the optimum length of the second fiber with negative β_2. In a feasibility demonstration of this concept, 130-ps pulses were compressed to about 50 ps using a 2-km-long fiber ($\beta_2 \approx 18.4$ ps^2/km) followed by a 8-km-long fiber with $\beta_2 \approx -4.6$ ps^2/km [20]. In later experiments, a two-stage compression technique in which a grating-fiber compressor was followed by an anomalous-GVD fiber was used to obtain compression factors of up to 5000. The second-stage compression in these experiments results from the effects of higher-order solitons, a topic covered in the next section.

With the advent of mode-locked Ti:sapphire lasers, considerable attention focused during the 1990s on reducing the pulse width to below 5 fs. By 1996, optical pulses shorter than 8 fs were generated directly from a Ti:sapphire laser [60]. If a 10-fs pulse could be compressed even by a factor of 3 using a grating-fiber compressor, one would be able to attain pulses shorter than 4 fs in the 800-nm wavelength region. Such a pulse would contain less than two optical cycles! It is not easy to realize such short pulses in practice because of several higher-order nonlinear effects that limit the extent of pulse compression. Nonetheless, pulse widths in the range of 3 to 5 fs have been obtained in several experiments [61]–[69]. The spectral width of a 5-fs pulse exceeds 100 THz. It is hard to find a grating or prism pair whose GVD is constant over such a large spectral range. As a result, the TOD limits the compressor performance considerably.

A number of techniques are employed to circumvent the limitations imposed by the higher-order dispersive and nonlinear effects. In a 1997 experiment, a polarization-maintaining fiber with a 2.75-μm-diameter core was used to chirp 13-fs input pulses [61]. Although the optimum fiber length was estimated to be only 1 mm [see Eq. (6.2.8)], practical considerations forced the use of 3-mm-long fiber. The spectrum of the chirped pulse was more than 250 nm wide. A grating pair, followed by a four-prism combination, was used for pulse compression. The width of the compressed pulse was 4.9 fs and was found to be limited by the TOD effects. The use of a Gires–Tournois interferometer formed using a *chirped mirror* reduced the pulse width to 4.6 fs [63]. A chirped mirror is made by depositing multiple layers of two different dielectrics on a substrate, similar to the saturable Bragg mirror used for mode locking of fiber lasers (see Section 5.4). Just as a chirped fiber grating provides large GVD, a chirped mirror can introduce large GVD on reflection. Such mirrors can be designed, by varying layer thicknesses, in such a way that their GVD is uniform over a large bandwidth (>150 THz) with little residual TOD [70].

A shortcoming of using optical fibers for chirping the input pulse is that peak powers of compressed pulses are limited by the damage threshold of silica. This problem can be solved by using a hollow fiber with a relatively large diameter that is filled with a noble gas. In a 1997 set of experiments, a 60-cm-long hollow fiber (diameter 80 μm) was filled with argon or krypton [62]. The nonlinear and dispersive effects of the gas were used to chirp 20-fs pulses with 40-μJ energy. Pulses as short as 4.5 fs were produced by compressing the chirped pulses using a chirped mirror in combination with two pairs of fused silica prisms. By 2003, pulse width could be reduced to 3.4 fs (about one and a half optical cycle) by first chirping the pulse through SPM inside an argon-filled hollow fiber, then compensating it over a wide bandwidth extending from 495 to 1090 nm [69].

Highly nonlinear fibers, such as photonic crystal and other microstructured fibers, have been used in recent years for pulse compression because of their unusual dispersive and nonlinear characteristics [71]–[73]. The nonlinear parameter γ is relatively large ($\gamma > 20$ W^{-1}/km) in such fibers owing to their much smaller effective mode area compared with traditional fibers. As a result, pulses can be chirped through SPM by a large amount without requiring long fiber lengths. In a 2004 experiment, 250-fs pulses were compressed down to below 25 fs using a short section ($\sim$1 cm) of a photonic crystal fiber with a 2.6-μm-diameter core, followed with a grating pair [72]. In another experiment, a 20-cm-long fiber with a 5-μm-diameter core was used in combination with a grating pair to compress 100-fs pulses down to 20 fs [73]. The zero-dispersion wavelength of this fiber was matched to the laser wavelength to ensure that dispersive effects played a minor role during SPM-induced spectral broadening of pulses. A large-mode-area fiber must be employed when pulses requiring compensation have high average and peak power levels. In a 2003 experiment, the spectrum of 810-fs pulses was broadened inside a 17-cm-long microstructured fiber with an effective mode area of 200 μm^2 [71]. The pulses were unchirped using a prism pair to obtain 33-fs pulses with a peak power of 12 MW.

6.3 Soliton-Effect Compressors

Optical pulses at wavelengths exceeding 1.3 μm generally experience both SPM and anomalous GVD during their propagation in silica fibers. Thus, a single piece of fiber can act as a compressor by itself, without requiring an external grating pair, and such an approach has been used since 1983 for this purpose [74]–[91]. The compression mechanism is related to a fundamental property of higher-order solitons. As discussed in Section 5.2 of Ref. [2], these solitons follow a periodic evolution pattern such that they undergo an initial narrowing phase at the beginning of each period. Because of this property, with an appropriate choice of the fiber length, input pulses can be compressed by a factor that depends on the soliton order N. Such a compressor is referred to as the *soliton-effect compressor* to emphasize the role of solitons.

6.3.1 Compressor Optimization

The evolution of a soliton of order N inside optical fibers is governed by the NLS equation. One can neglect fiber losses ($\alpha = 0$), since fiber lengths employed in practice

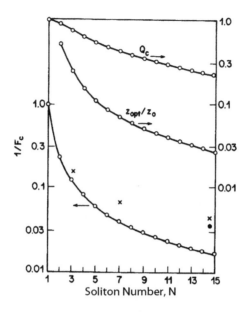

Figure 6.8: Variation of compression factor F_c, optimum fiber length z_{opt}, and quality factor Q_c with the parameter N. Data points correspond to experiments performed with 320-m (crosses) and 100-m (solid dots) fibers. (From Ref. [74]; ©1983 OSA.)

are relatively small ($\alpha L \ll 1$). In the case of anomalous GVD ($\beta_2 < 0$), Eq. (6.2.5) becomes

$$i\frac{\partial U}{\partial \xi} + \frac{1}{2}\frac{\partial^2 U}{\partial \tau^2} + N^2|U|^2U = 0, \tag{6.3.1}$$

where the parameter N is given by Eq. (6.2.6). Even though higher-order solitons follow an exact periodic pattern only for integer values of N, Eq. (6.3.1) can be used to describe pulse evolution for arbitrary values of N. In general, the input pulse goes through an initial narrowing phase for all values of $N > 1$. The optimum fiber length z_{opt} corresponds to the location at which the width of the central spike is minimum. The compression factor is the ratio of the FWHM of the compressed pulse to that of the input pulse.

Numerical techniques have been used to obtain the compression factor F_c and the optimum fiber length z_{opt} as a function of N [74]. The inverse scattering method can also be used to obtain these quantities for integer values of N. Figure 6.8 shows the variation of F_c^{-1} and z_{opt}/z_0 with N for values of N from 1 to 15. Also shown is the quality factor Q_c, defined as the fraction of input pulse energy appearing in the compressed pulse. In contrast to the case of a grating-fiber compressor, Q_c is significantly smaller than its ideal value of unity and decreases monotonically as N increases. This drawback is inherent in all soliton-effect compressors. The remaining pulse energy appears in the form of a broad pedestal around the compressed pulse. The physical origin of the pedestal can be understood as follows. During the initial narrowing stage, the evolution of higher-order solitons is dominated by SPM. Since the SPM-induced chirp is linear only over the central part of the pulse, only the central part is compressed by anomalous GVD. Energy in the pulse wings remains uncompressed and appears as a broad pedestal.

Numerical simulations performed for values of N up to 50 show that the compression factor F_c and the optimum fiber length of a soliton-effect compressor are well approximated by the empirical relations [38]

$$F_c \approx 4.1N, \tag{6.3.2}$$

$$\frac{z_{opt}}{z_0} \approx \frac{0.32}{N} + \frac{1.1}{N^2}. \tag{6.3.3}$$

These relations are accurate to within a few percent for $N > 10$ and can serve as simple design rules, similar to those given by Eqs. (6.2.8)–(6.2.10) for grating-fiber compressors. A direct comparison shows that, for the same values of N and z_0, a soliton-effect compressor provides pulse compression that is larger by a factor of 6.5 with a fiber that is shorter by a factor of 5. However, the pulse quality is poorer since the compressed pulse carries only a fraction of the input energy, with the remaining energy appearing in the form of a broad pedestal. The results of Figure 6.8 assume an unchirped input pulse with "sech" shape. Much higher compression factors are possible for chirped input pulses having specific pulse shapes [84].

6.3.2 Experimental Results

In a 1983 experiment [74], 7-ps pulses from a color-center laser operating near 1.5 μm were propagated through a 320-m-long fiber ($z/z_0 \approx 0.25$). As the input peak power was increased beyond 1.2 W (the power level corresponding to a fundamental soliton), the output pulse became narrower than the input pulse by a factor that increased with increasing N. The observed values of the compression factor are shown in Figure 6.8 (crosses) for three values of N. The compression factor was close to the theoretical value of 8 for $N = 3$ but became significantly smaller for larger values of N. This can be understood by noting that the fiber length of 320 m was close to optimum for $N = 3$ but became far too large for $N > 3$. Indeed, a reduction in the fiber length to 100 m ($z/z_0 \approx 0.077$) increased the compression factor to 27 for $N = 13$. The autocorrelation trace of the 0.26-ps compressed pulse is shown in Figure 6.9.

It was observed experimentally that the broad pedestal associated with the compressed pulse could be partially suppressed under certain experimental conditions. As discussed in Chapter 6.2 of Ref. [2], the origin of pedestal suppression is related to the nonlinear birefringence of optical fibers that can make the fiber act as an intensity discriminator [58]. This mechanism can, in principle, eliminate the pedestal almost completely. Another possibility for removing the pedestal is to filter out the low-frequency components of the compressed pulse that are associated with the pedestal. The numerical results show that the bandwidth $\Delta\nu_f$ of such a filter is related to the parameter N and the input pulse width T_{FWHM} by the relation [76]

$$\Delta\nu_f \approx 0.2(N/T_{FWHM}), \tag{6.3.4}$$

where the numerical factor depends slightly on the input pulse shape.

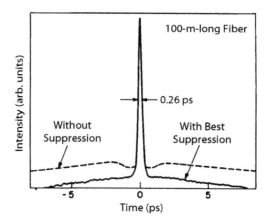

Figure 6.9: Autocorrelation trace of a 7-ps input pulse compressed to 0.26 ps by using a soliton-effect compressor. Dashed and solid curves compare the pedestal with and without the nonlinear birefringence effect. (From Ref. [74]; ©1983 OSA.)

Soliton-effect compressors can provide quite high compression factors. A compression factor of 110 was realized when 30-ps pulses were compressed to 275 fs by passing them through a 250-m-long fiber [75]. The fiber length is nearly optimum in this experiment if we take $N \approx 28$ (corresponding to a peak power of 0.6 kW) and note that $z_0 \approx 20$ km for 30-ps input pulses. The observed compression is also in agreement with Eq. (6.3.2). Compression factors >1000 were realized using a two-stage scheme in which a grating-fiber compressor was followed with a soliton-effect compressor [78]–[80]. These experiments used 100-ps input pulses, emitted by a mode-locked Nd:YAG laser operating at 1.32 μm. In the first stage, a grating-fiber compressor was used to obtain compressed pulses of widths in the range of 1 to 2 ps. These pulses were fed into a soliton-effect compressor whose fiber length was carefully optimized to achieve compression factors of about 50. In a 1988 experiment [80], the initial 90-ps pulse was compressed to only 18 fs (consisting of only four optical cycles) by such a two-stage scheme, resulting in a net compression factor of 5000. Figure 6.10 shows the autocorrelation trace and the spectrum of the 18-fs pulse. The narrow central feature in the spectrum corresponds to the pedestal seen in the autocorrelation trace that carries 69% of the total energy.

Several experiments have employed multiple cascaded soliton-effect compressors to realize shortest pulses. In a 1999 experiment [89], 7.5-ps optical pulses were compressed down to 20 fs using four stages, resulting in a total compression factor of more than 3700. Each stage employed a different fiber with optimized dispersion characteristics. In particular, the third stage used a 4-m-long fiber whose dispersion was decreasing along its length (a dispersion-decreasing fiber), followed with a fourth stage whose 60-cm-long fiber had the same value of dispersion over a wide wavelength range (a dispersion-flattened fiber). This experiment indicated that a higher-order nonlinear effect that shifts the pulse spectrum toward long wavelengths (the Raman-induced frequency shift discussed in the following section) played an important role during the

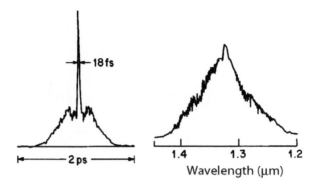

Figure 6.10: Autocorrelation trace and spectrum of an 18-fs pulse obtained by compressing 90-ps input pulses using a two-stage compression scheme. (From Ref. [80]; ©1988 Taylor & Francis.)

last two stages. The same effect also helped in minimizing the undesirable pedestal component seen in Figure 6.10. In a 2006 experiment [91], 5.4-ps parabolic pulses, generated through Raman amplification in the normal-dispersion regime (see Section 4.6.2), were compressed down to 20 fs using two stages of soliton-effect compression. The first stage employed a 3.5-m-long photonic bandgap fiber (with air core) that was followed with a 50-cm-long highly nonlinear fiber. The use of highly nonlinear fibers has become quite common in recent years because of the relatively high values of the nonlinear parameter ($\gamma > 50$ W^{-1}/km) offered by them.

6.3.3 Higher-Order Nonlinear Effects

In pulse-compression experiments involving femtosecond pulses, the optimum fiber length was found to be larger [79] by more than a factor of 2.5 than that predicted by Eq. (6.3.3). This is not unexpected, since Eq. (6.3.3) is based on the numerical solution of Eq. (6.3.1), which neglects the higher-order dispersive and nonlinear effects that become increasingly more important as pulses become shorter than 100 fs. For an accurate prediction of the optimum fiber length, one must include the effects of TOD, self-steepening, and intrapulse Raman scattering by solving the following generalized NLS equation:

$$i\frac{\partial U}{\partial \xi} + \frac{1}{2}\frac{\partial^2 U}{\partial \tau^2} - i\delta_3 \frac{\partial^3 U}{\partial \tau^3} + N^2 \left(|U|^2 U + is_0 \frac{\partial |U|^2 U}{\partial \tau} - \tau_R U \frac{\partial |U|^2}{\partial \tau} \right) = 0, \quad (6.3.5)$$

where the three parameters δ_3, s_0, and τ_R govern, respectively, the effects of TOD, self-steepening, and intrapulse Raman scattering (see Section 5.5 of Ref. [2]).

For not-too-short pulses (width >50 fs) propagating not too close to the zero-dispersion wavelength of the fiber, the dominant contribution comes from intrapulse Raman scattering. It manifests as a shift of the pulse spectrum toward the red side, a phenomenon referred to as the *soliton self-frequency shift* or the *Raman-induced frequency shift*. Associated with this shift is a delay of the optical pulse because of a

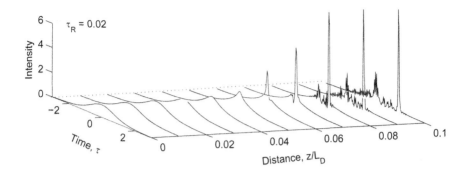

Figure 6.11: Evolution of a 10th-order soliton in the presence of intrapulse Raman scattering. The compressed pulse separates from the pedestal because of Raman-induced frequency shift. (From Ref. [82]; ©1990 OSA.)

change in the group velocity of the pulse. Such a delay affects substantially the interplay between GVD and SPM that is responsible for pulse compression. Numerical simulations indeed show that the optimum fiber length is longer than that predicted by Eq. (6.3.1) when the higher-order nonlinear effects are included in the analysis.

Interestingly, intrapulse Raman scattering improves the quality of the compressed pulse by producing pedestal-free pulses [82]. Figure 6.11 shows the evolution of the $N = 10$ soliton using $\tau_R = 0.02$, a value appropriate for 0.2-ps input pulses. The pulse begins to compress, and a narrow spike riding on a broad pedestal is formed near $\xi = 0.06$. This behavior is similar to that occurring when $\tau_R = 0$. However, the evolution becomes qualitatively different for $\xi > 0.06$ because of intrapulse Raman scattering. More specifically, the narrow spike travels slower than the pedestal and separates from it because of the change in its group velocity induced by the Raman-induced frequency shift. Moreover, the pedestal can be removed by spectral filtering. The net result is to produce a red-shifted, pedestal-free, compressed pulse. At the same time, the optimum fiber length is longer and the compression factor is larger compared to the values obtained from Figure 6.8.

Intrapulse Raman scattering, in combination with the induced modulation instability, can be used to obtain a train of pedestal-free ultrashort optical pulses at high repetition rates [83]. The basic idea consists of injecting a sinusoidally modulated CW beam into an optical fiber. Weak AM sidebands are amplified through the gain provided by modulation instability if the modulation frequency falls within the bandwidth of the instability gain. At the same time, the pulse compresses through the soliton-effect compression if the peak power is large enough to excite a higher-order soliton. In the absence of intrapulse Raman scattering, compressed pulses ride on a broad pedestal forming from the CW background. However, because of the Raman-induced frequency shift, the spectrum of the pulse train separates from the pedestal spectrum. A bandpass filter can be used to remove the pedestal and obtain a train of ultrashort optical pulses at a repetition rate determined by the initial modulation frequency. Numerical simulations reveal that a pulse train of width $\sim$100 fs at repetition rates $\sim$100 GHz can be generated by this technique [86].

The TOD, in general, degrades the quality of compressed pulses when femtosecond pulses propagate close to the zero-dispersion wavelength of the fiber used for soliton-effect compression [87]. However, if the TOD parameter δ_3 is negative, the combination of TOD and intrapulse Raman scattering can improve the performance of a soliton-effect compressor. Numerical simulations based on Eq. (6.3.5) show that negative values of β_3 result in larger compression factors and higher peak powers [88]. Although β_3 is positive for most fibers, dispersion-compensating fibers with negative values of β_3 were developed during the late 1990s. The use of dispersion management, a technique in which two or more fibers with different dispersion characteristics are spliced together, can provide soliton-effect compressors for which the average GVD is small but anomalous while the average value of β_3 is negative. Such compressors should prove useful for compressing femtosecond pulses.

In recent years, highly nonlinear fibers have been used to generate a supercontinuum by broadening the pulse spectrum though the fission of higher-order solitons (see Chapter 12 of Ref. [2]). Higher-order nonlinear and dispersive effects play a dominant role in supercontinuum generation. In fact, the input pulse breaks up into a number of subpulses though soliton fission, and the spectrum of each subpulse shifts rapidly toward the longer-wavelength side through intrapulse Raman scattering. It was realized soon after the discovery of supercontinuum that this process can be used to create ultrashort compressed pulses [92]–[98]. In a 2002 experiment, 2.4-ps pulses from a mode-locked, Yb-doped fiber laser could be compressed down to 110 fs by launching them into a Yb-doped microstructured fiber, pumped suitably to act as an amplifier [92]. The pulses were amplified and compressed simultaneously, as they formed a Raman soliton whose spectrum shifted continuously toward the red side.

The important question is how short pulses can be produced with this approach. In practice, one deals with a pulse train that exhibits small pulse-to-pulse variations. The supercontinuum generation process is extremely sensitive to the pulse peak power, as it is driven by the fiber nonlinearity. Even a difference of 0.2% in the peak power can change significantly the phases associated with different spectral components of the pulse [93]. Thus, a single compressor can not provide optimum compression for all pulses of a pulse train. A stochastic approach was used in a 2004 study to reveal the fundamental limitations imposed by pulse-to-pulse fluctuations [94]. It indicated that the average duration of compressed pulses can still be close to 5 fs under optimized conditions. Indeed, 5.5-fs compressed pulses were produced in a 2005 experiment in which a supercontinuum was generated by launching 15-fs pulses into a 5-mm-long microstructured fiber and dechirping them with a dispersive delay line [96].

A dispersive delay line can be avoided if compression is realized through the soliton effects alone. Numerical simulations show that the use of fibers with core diameters of <1 μm (called photonic nanowires) can compress pulses down to 2 fs with suitable launch conditions [97]. Experimentally, 70-fs pulses could be compressed down to 6.8 fs inside a microstructured fiber with the 980-nm-diameter core. Only a 2-mm-long piece of such a fiber was needed to compress the pulse by a factor of 10.3. Figure 6.12 shows the intensity and spectral profiles of the compressed pulse, together with the corresponding phase profiles, all deduced from the cross-correlation FROG (frequency-resolved optical gating) traces. Even though the pulse rides on a pedestal,

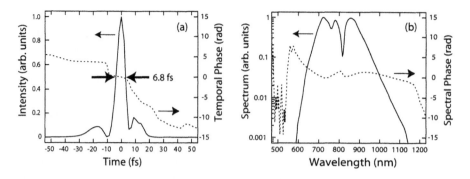

Figure 6.12: (a) Shape and (b) spectrum of the compressed pulse together with the corresponding phase profiles (dotted curves), as retrieved from the measured FROG traces (From Ref. [97]; ©2005 OSA.)

the quality factor F_c is 0.73 in this case.

Pedestal-free compressed pulses were obtained in a 2006 experiment in which several fibers were combined to compensate for the higher-order dispersive effects [98]. More specifically, 260-fs pulses from a mode-locked fiber laser operating at 1.55 μm were compressed down to 22 fs using two stages of pulse compression. During the first stage, an erbium-doped fiber amplifier (EDFA) was used in combination with standard fibers to generate a 110-fs pulse with a relatively smooth, 40-nm-wide, single-peak spectrum. This pulse was then launched into a normally dispersive, highly nonlinear fiber, whose optimum length was found to be 40 cm. Although the pulse spectrum shifted toward the red side (through intrapulse Raman scattering) and broadened considerably in this fiber (with a FWHM of 250 nm), it maintained its single-peak nature because of the presence of normal dispersion that also increased the pulse width to 820 fs. This stretched pulse was dechirped using two pieces of standard-dispersion and reverse-dispersion fibers, resulting in the final pedestal-free 22-fs pulse. Such a performance was possible because the experiment employed the nonlinear effects within optical fibers but avoided the fission of higher-order solitons leading to supercontinuum generation.

6.4   Fiber Bragg Gratings

As discussed in Chapter 1, fiber Bragg gratings exhibit large GVD in the vicinity of the stop-band edges. The grating-induced GVD can be varied from normal to anomalous, and its magnitude can be tailored over a wide range by chirping the grating period. Since a fiber grating acts as a dispersive delay line, it can be used for compressing chirped pulses in place of the bulk-grating pair [99]. Moreover, the onset of various nonlinear effects within the fiber grating points to the possibility of realizing pulse compression using a compact, all-fiber device [100]. For these reasons, the use of fiber gratings for pulse compression attracted attention soon after such gratings became

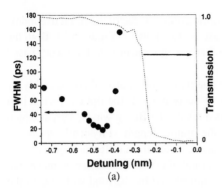 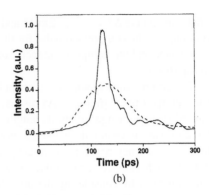

(a) (b)

Figure 6.13: (a) Changes in width (FWHM) of pulses compressed by a fiber grating as a function of wavelength detuning $\Delta\lambda$. Dotted curve shows changes in transmittivity. (b) Shape of compressed pulse for $\Delta\lambda = -0.45$ nm. Dashed curve shows the input pulse. (From Ref. [115]; ©1998 OSA.)

available [101]–[118]. In this section we discuss the important role played by fiber gratings in pulse compressors.

6.4.1 Gratings as a Compact Dispersive Element

A uniform grating reflects light whose wavelength falls within the stop band centered at the Bragg wavelength λ_B. Outside but close to the stop-band edges, the grating provides large dispersion. The effective values of the grating-induced GVD and TOD depend on the detuning δ and are given by [see Eq. (1.3.26)]

$$\beta_2^g = -\frac{\text{sgn}(\delta)\kappa^2/v_g^2}{(\delta^2 - \kappa^2)^{3/2}}, \qquad \beta_3^g = \frac{3|\delta|\kappa^2/v_g^3}{(\delta^2 - \kappa^2)^{5/2}}. \qquad (6.4.1)$$

where κ is the coupling coefficient as defined in Section 1.3. The GVD parameter β_2^g depends on the sign of the detuning δ. The GVD is anomalous on the high-frequency side of the stop band where δ is positive. In contrast, GVD becomes normal ($\beta_2^g > 0$) on the low-frequency side of the stop band ($\delta < 0$). The TOD remains positive in all cases. Both β_2^g and β_3^g become infinitely large if the optical frequency falls close to the edges of the stop band such that $\delta = \kappa$. Figure 1.7 showed how GVD varies in the vicinity of the stop-band edges of a grating. Typical values of $|\beta_2^g|$ can easily exceed 10^7 ps^2/km. As a result, a 1-cm-long fiber grating may provide as much dispersion as 10-km of silica fiber or a bulk-grating pair with more than one meter spacing.

A simple application thus consists of replacing the bulk-grating pair in a grating-fiber compressor with a fiber grating. The resulting all-fiber device can be quite compact. Unfortunately, the TOD affects the quality of compressed pulses significantly since β_3^g increases rapidly as δ approaches the stop-band edges located at $\delta = \pm\kappa$. Figure 6.13 shows how the pulse width changes with wavelength detuning when 80-ps (FWHM) pulses—obtained from a Q-switched, mode-locked Nd:YLF laser and chirped through SPM within the laser—are transmitted through a 6.5-cm-long apodized grating. The shortest pulse width of about 15 ps is obtained for $\Delta\lambda =$

−0.45 nm, where $\Delta\lambda = -(\lambda_B^2/2\pi\bar{n})\delta$ and $\bar{n}$ is the average value of the refractive index. The shape of the compressed pulse for this case is also shown in Figure 6.13. Both the pulse shape and the compression factor of 5.3 are in agreement with the theory based on Eq. (6.4.1).

The compression factor as well as pulse quality can be improved considerably by using chirped fiber gratings. As discussed in Section 1.7.2, the optical period in a chirped grating changes along its length. As a result, the Bragg wavelength at which the stop band is centered also shifts along the grating length. Physically speaking, different frequency components of the pulse are reflected from different regions of the grating. Such a device can introduce a large amount of GVD in the reflected pulse. We can estimate it by considering the time delay introduced by the total shift $\Delta\lambda_t$ in the Bragg wavelength. With $T_r = D_g L_g \Delta\lambda_t = 2\bar{n}L_g/c$, where T_r is the round-trip time for a grating of length L_g, the dispersion parameter is given by

$$D_g = -(2\pi c/\lambda^2)\beta_2^g = 2\bar{n}/(c\Delta\lambda_t). \tag{6.4.2}$$

Values of $|\beta_2^g|$ can exceed 5×10^7 ps^2/km for $\Delta\lambda_t = 0.2$ nm. Such shifts in the Bragg wavelength can be realized by chirping the grating period linearly such that it changes by only 0.1% at the two ends of the grating.

Several experiments in 1994 used chirped gratings for pulse compression [102]–[104]. A major motivation was to compensate for dispersion-induced broadening of pulses in fiber-optic communication systems [119]. In a 1995 experiment, dispersion compensation at 10 Gb/s over 270 km of standard fiber ($\beta_2 \approx -20$ ps^2/km) was realized using a 12-cm-long chirped grating [105]. Compression factors in excess of 100 can be realized by this technique. The only disadvantage of a chirped fiber grating from a practical standpoint is that the compressed pulse is reflected rather than transmitted. An optical circulator is commonly used to separate the reflected pulse from the incident pulse because of its relatively low insertion losses.

In most grating-fiber compressors, a long piece of normal-GVD fiber is used to chirp the input pulse before it can be compressed with a grating. The required fiber length can be shortened considerably by employing a narrow-core microstructured silica fiber. A fiber made of a different glass can also help if its nonlinear parameter n_2 is relatively large. In a 2006 experiment [118], a 4.1-m-long chalcogenide fiber was used to broaden the spectrum of a 6-ps pulse with a carrier wavelength of 1550 nm. The fiber exhibited normal dispersion near this wavelength (a measured β_2 of 710 ps^2/km) and its n_2 value is more than 300 times than that of silica. Because of such large values of β_2 and n_2, the 6-ps pulse with only 35-W peak power could be compressed down to 420 fs when a chirped fiber grating was used to dechirp the pulse. These results show that chalcogenide fibers can be useful for pulse compression, provided peaks powers are kept low enough that two-photon absorption (nearly negligible in silica fibers) does not limit their performance.

6.4.2 Grating-Induced Nonlinear Chirp

A fiber grating can also be used to chirp an input pulse and broaden its spectrum if it exhibits normal GVD and pulse energy is high enough that the SPM effects are

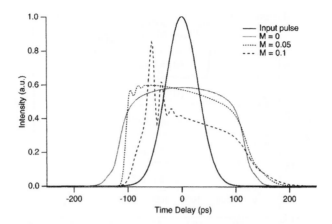

Figure 6.14: Effect of TOD on a 70-ps Gaussian pulse chirped using a fiber grating. Parameter $M = 3.6\delta_3$ is a measure of the relative strength of TOD. (From Ref. [113]; ©1998 OSA.)

relatively large. A dispersive delay with anomalous GVD can then be used to dechirp the pulse. Such a device results in a compact compressor, but it requires high input pulse energies. Moreover, the fiber grating should be designed carefully to minimize the effects of TOD and to avoid the onset of stimulated Raman scattering [113].

The effect of TOD on the chirping process within the grating can be studied numerically by solving the nonlinear coupled-mode equations of Section 1.3. However, as pointed out there, these equations reduce to a modified NLS equation of the form

$$i\frac{\partial U}{\partial \xi} - \frac{1}{2}\frac{\partial^2 U}{\partial \tau^2} - \delta_3 \frac{\partial^3 U}{\partial \tau^3} + N^2|U|^2 U = 0, \qquad (6.4.3)$$

where $\delta_3 = \beta_3^g/(6\beta_2^g T_0)$ is the effective TOD parameter for a grating. Fiber losses within the grating can be neglected because of its short length. Figure 6.14 shows how the shape of a 70-ps Gaussian pulse is affected by TOD when it is chirped using a 10-cm-long fiber grating with normal GVD ($\beta_2 = 50$ ps^2/cm). The peak intensity of the pulse is taken to be 170 GW/cm^2. The effects of TOD are included by changing δ_3. In the absence of TOD ($\delta_3 = 0$), the chirped pulse is nearly rectangular, as found in Section 6.2. However, the pulse becomes asymmetric and develops considerable internal structure as δ_3 increases. This structure affects the quality of the compressed pulse and should be minimized when designing the fiber grating [113].

The TOD parameter δ_3 depends on the detuning parameter δ and can be reduced by moving the stop-band edge of the grating away from the optical wavelength so that the pulse is not too close to the edge. However, the GVD parameter β_2^g becomes smaller as $|\delta|$ increases, resulting in a longer dispersion length. Since the optimum length of the grating is about $z_{opt} = (6L_D L_{NL})^{1/2}$, a longer grating is needed. The compression factor is limited by the onset of stimulated Raman scattering, since the parameter N cannot be made very large by increasing input peak powers. A careful consideration of various dispersive and nonlinear effects shows that the maximum compression factor is limited to a value of about 6 [113]. It should also be stressed that pulses chirped

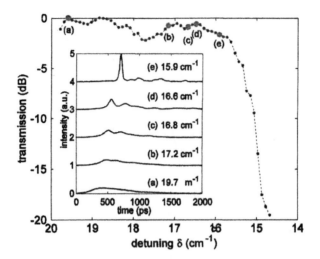

Figure 6.15: Measured transmission and output pulse shapes for several values of detuning from the Bragg wavelength in the case of 1.4-kW input pulses. (From Ref. [117]; ©2005 OSA.)

nonlinearly by a grating cannot be compressed by another fiber grating acting as a dispersive delay line, because of their high peak power levels. The nonlinear effects in the second grating can be avoided by reducing the pulse energy. Alternatively, a grating or prism pair can be used for pulse compression.

6.4.3 Bragg-Soliton Compression

As discussed in Section 1.6, fiber gratings support Bragg solitons. These solitons can be used for pulse compression in the same way that higher-order solitons produce soliton-effect compression in fibers without a grating. The advantage of a fiber grating is that the compressor length can be reduced from hundreds of meters to a few centimeters.

Since the nonlinear coupled-mode equations describing pulse propagation in fiber gratings reduce to an effective NLS equation under appropriate conditions (see Section 1.5.2), the analysis of Section 6.3 applies to fiber gratings as well, as long as the GVD parameter β_2 and the nonlinear parameter γ are replaced by their equivalent values given in Eq. (1.5.10). An estimate of the compression factor for values of N in the range from 2 to 15 is provided by $F_c \approx 4.6(N-1)$. The pulse shapes shown in Figure 1.17 were obtained at the output end of a 7.5-cm-long apodized fiber grating for 80-ps input pulses having a peak intensity of 11 GW/cm^2. The soliton order N is different for different curves, seen in Figure 1.17, since the GVD parameter β_2^g changes with detuning δ. Choosing the values of β_2^g and the pulse peak power P_0 such that the input pulse corresponds to a $N = 2$ soliton while the grating length $L = z_0/2$, the compression factor is expected to be 4.6. Indeed, pulse compression by a factor of 4 was observed experimentally under such conditions [115].

In a 2005 experiment, 580-ps pulses obtained from a microchip Q-switched laser were compressed down to 45 ps by forming solitons through modulation instability inside a 10-cm-long fiber grating. The peak power of pulses exceeded 1 kW and was large enough to create multiple subpulses within the Q-switched pulse envelope. Figure 6.15 shows the measured transmission and the output pulse shapes in the case of 1.4-kW input pulses for several values of detuning from the Bragg wavelength near the short-wavelength edge of the stop band. As the pulse wavelength is tuned closer to the stop-band edge by stretching the grating and shifting its Bragg wavelength, the output pulses become narrower. For a detuning of $\delta = 15.9$ cm^{-1}, most of the output energy is contained in a single narrow pulse that is shorter by a factor of 12 compared with the input pulse. This narrowing is similar to soliton-effect compression discussed earlier. The formation of a single dominant pulse depends strongly on the peak power of input pulses. For example, 2.5-kW input pulses formed a train of 5 narrow pulses of different amplitudes at the grating output, as expected from the onset of modulation instability at high power levels. Numerical simulations based on the nonlinear coupled-mode equations as well as the effective NLS equation (see Section 1.6) agree well with the experimental data.

Another grating-based nonlinear scheme for compressing optical pulses makes use of the push-broom effect in which a weak broad pulse is swept by a strong pump pulse such that most of the energy of the broad pulse piles up at the front end of the pump pulse [106]. The physical mechanism behind optical push broom was discussed in Section 1.6.4. The nonlinear chirp in this case is generated not by SPM but by cross-phase modulation (XPM). Pulse compression induced by the push-broom effect was seen in a 1997 experiment [112]. XPM-induced pulse compression is discussed in more detail later in this chapter.

Although soliton-based compression has been observed in fiber gratings, its use is likely to be limited in practice. The reason is related to the combination of a relatively low value of n_2 in silica glasses and relatively short lengths of fiber gratings. A nonlinear phase shift ($\phi_{NL} = \gamma P_0 L$) of π requires values of P_0 in excess of 1 kW even for a relatively long grating ($L = 50$ cm). The power levels can be reduced by more than a factor of 100 if chalcogenide glasses are used for making the fiber grating.

6.5 Chirped-Pulse Amplification

Many applications require optical pulses with high energies (1 mJ or more) that are not readily available from the laser producing the pulse train. It is then necessary to amplify the pulse externally by using one or more amplifiers. However, as the peak power of a pulse increases, the nonlinear effects within the amplifier begin to distort the pulse. The technique of chirped-pulse amplification (CPA) has been found extremely useful to solve this problem. This technique was used as early as 1974 [120] but drew widespread attention only after 1985 when it was used to obtain ultrashort pulses with terawatt peak powers [121]–[125].

The basic idea behind CPA consists of chirping and stretching the pulse before it is amplified. The nonlinear and dispersive effects of optical fibers are often used for this purpose. During its passage through the fiber, the pulse is not only chirped but it also broadens, as discussed in Section 6.2. Pulse stretching (by a factor of 100 or more) reduces the peak power substantially, making it possible to amplify the pulse by a large amount before gain saturation limits the energy-extraction efficiency. The amplified pulse is then compressed back by using a grating pair, fiber grating, or simply a piece of fiber acting as a dispersive delay line. In essence, the scheme is similar to that shown in Figure 6.1 except that an amplifier is inserted between the fiber and the grating pair.

The use of CPA resulted by 1990 in the advent of tabletop terawatt laser systems in which mode-locked pulses emitted from solid-state lasers are amplified to obtain picosecond, or even femtosecond, pulses with terawatt peak powers. In a 1991 experiment, this technique was used to produce 1052-nm pulses of 3.5-ps duration, 28-J energy, and 8-TW peak power [125]. The quality of the compressed pulse was improved by selecting only a central portion of the pulse where the frequency chirp was linear; a saturable absorber was used for this purpose.

Although the use of optical fibers is not essential for CPA, their dispersive and nonlinear properties provide a simple way to stretch the pulse while chirping it. Pulse energy can be increased by using any fiber amplifier. If fibers can be used for compressing the amplified pulse as well, one can develop an all-fiber source of high-energy ultrashort pulses. However, the amplified pulses are often so intense that the nonlinear effects within the compressor fiber are difficult to avoid in practice. Two possible solutions to this problem are based on the use of chirped fiber gratings and hollow-core photonic bandgap fibers. The former provides high dispersion over such a short length that the nonlinear effects can remain negligible. The latter usees a core filled with air, a relatively weak nonlinear medium.

6.5.1 Chirped Fiber Gratings

Starting in 1993, the technique of CPA was applied to realize all-fiber sources of high-energy optical pulses [126]–[135]. The initial experiments employed EDFAs operating in the 1.55-μm spectral region for pulse amplification, in combination with chirped fiber gratings for stretching and compressing optical pulses. As discussed in Chapter 5, mode-locked fiber lasers can produce pulses shorter than 1 ps but their energy is typically below 0.1 nJ. The energy level can be increased considerably by amplifying such pulses in an erbium-doped fiber amplifier, but the nonlinear effects occurring during amplification limit the pulse quality. Therefore, it is useful to stretch the pulse by a factor of 100 or so (while chirping it) before amplification. The amplified pulse is then compressed back using a suitable compressor. By 1994, this technique produced 800-fs pulses with pulse energies of 100 nJ at a repetition rate of 200 kHz by using bulk gratings [128]. Bulk gratings were later replaced by chirped fiber gratings to realize a compact, all-fiber device. Figure 6.16 shows the experimental setup schematically for this technique. In one experiment, a chirped fiber grating stretched 330-fs pulses to 30 ps, while the second grating recompressed the amplified pulse back to 408 fs [130]. In another experiment, the pulse energy was boosted to 20 nJ by using a cladding-pumped fiber amplifier [131]. In both experiments, the quality of compressed amplified pulses deteriorated at energy levels beyond 5 nJ because of the onset of nonlinear effects inside the compressor grating.

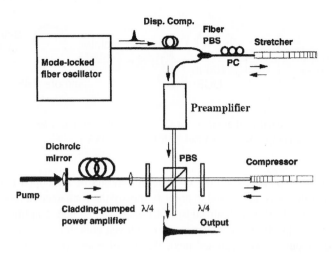

Figure 6.16: A chirped-pulse amplification technique making use of two chirped fiber gratings for stretching and compressing the pulses. (PBS: polarization beam splitter; PC: polarization controller) (From Ref. [131]; ©1995 OSA.)

Higher pulse energies can be obtained if the pulse width is in the picosecond regime. In a 1996 experiment, 1.9-ps pulses with 300-nJ energy levels were obtained using the technique of CPA [133]. Linearly chirped pulses were obtained using a semiconductor laser and were relatively wide (>1 ns) with low energy levels (<10 pJ). They were amplified to energy levels as high as 3 μJ using two fiber amplifiers. A 12-cm-long chirped fiber grating compressed the pulse to under 2 ps. Nonlinear effects inside the grating were relatively small, since the pulse was compressed fully near the output end of the grating. In fact, the onset of nonlinear effects in the 5-cm-long pigtail (attached to the grating) was believed to limit pulse energies in this experiment.

In a 1999 experiment, the pulse energy was increased to beyond 1 μJ (peak powers up to 500 kW) by forming a chirped grating inside the core of a large-mode-area fiber [136]. Figure 6.16 shows the experimental setup schematically. Optical pulses (width 1.5 ps) from a 1.53-μm fiber laser were first stretched to 600 ps using a chirped fiber grating, before being amplified to an energy of 15 μJ using three fiber amplifiers. They were then compressed back to below 4 ps with a second chirped grating made using a fiber with a large mode area of 450 μm^2. As the nonlinear parameter γ scales inversely with the effective mode area, nonlinear effects are reduced considerably in the second grating at the same pulse-energy level. Indeed, the nonlinear effects were found to be negligible at energies of up to 0.9 μJ in this experiment.

Much higher pulse energies can be realized if a cladding-pumped fiber amplifier with a relatively wide core is used to overcome the peak-power limitations of single-mode fibers [137]–[139]. The only problem is the multimode nature of such double-clad fibers. In principle, if only the fundamental mode is excited by the input beam, only that mode would propagate and be amplified. In practice, it is difficult to launch light only in the fundamental mode of a multimode fiber and to prevent excitation of higher-order modes through mode coupling. However, a number of techniques,

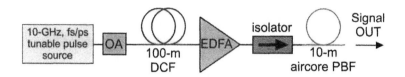

Figure 6.17: Experimental configuration for CPA in which a hollow-core photonic bandgap fiber (PBF) acts as a compressor. (From Ref. [140]; ©2003 OSA.)

such as selective doping of the fiber core, tight bending of the doped fiber, and an optimized fiber design, can be used to amplify mostly the fundamental mode. Indeed, pulse energies of >1 mJ were realized by 2001 using such a fiber amplifier with a 50-μm-diameter core [138]. These experiments employed bulk gratings for compressing the high-energy pulse. As discussed next, the use of microstructured fibers in later experiments permitted an all-fiber configuration [139].

6.5.2 Photonic Crystal Fibers

The CPA technique has benefitted considerably with the advent of photonic crystal and other microstructured fibers. The dispersive effects in such fibers depend on the cladding design and can be controlled by modifying the size and spacing of air holes. The nonlinear effects in such fibers can be enhanced or reduced by changing the size of the core. In fact, the core may even consist of air in the so-called *photonic bandgap fibers* that confine light though a periodic array of air holes inside the silica cladding and do not depend on the phenomenon of total internal reflection. Such a fiber is well suited for compressing high-energy pulses simply because the onset of the nonlinear effects in air occurs at much higher power levels compared with silica.

Starting in 2003, hollow-core, photonic bandgap fibers were used for CPA in several experiments involving Er-doped or Yb-doped fiber amplifiers [140]–[145]. Figure 6.17 shows a typical experimental configuration with an EDFA providing the amplification. In this experiment [140], a 100-m-long dispersion-compensating fiber (DCF) providing normal dispersion ($\beta_2 \approx 165$ ps^2/km near 1550 nm) was used for chirping and stretching input pulses of about 0.5-ps width. After amplification, the pulses were compressed back to 1 ps or so using a 10-m-long photonic bandgap fiber. This fiber provided large anomalous dispersion ($\beta_2 \approx -1460$ ps^2/km near 1560 nm) and acted as a nearly linear dispersive delay line.

A similar approach was used to amplify 250-fs pules at 1040 nm with 1-nJ energy using an Yb-doped fiber amplifier and to produce 100-fs pulses with 82-nJ energy and a peak power close to 0.82 MW [141]. In this case, a 1.9-m-long piece of standard single-mode fiber was used to stretch the pulse to 1.9 ps and chirp it at the same time. The amplifier employed a 2.1-m-long large-mode-area photonic crystal fiber especially designed with an air cladding; it increased the average power by a factor of about 200. The width of the pulse increased to 4 ps at the amplifier output and its spectrum was 28.5-nm wide. The 2-m long photonic bandgap fiber provided sufficient anomalous dispersion to compress the pulse down to 100 fs.

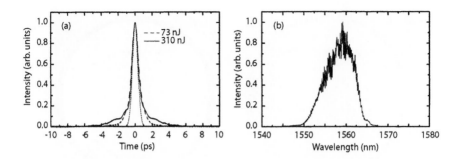

Figure 6.18: (a) Autocorrelation traces of output pulses with 310-nJ energy (solid line) and 73-nJ energy (dashed line). Dotted line shows the transform-limited pulse. (b) Measured spectrum of the 310-nJ pulse. (From Ref. [143]; ©2004 OSA.)

Even higher pulse energies have been realized using CPA with a photonic bandgap fiber acting as a compressor. In a 2004 experiment, 570-fs pulse at 1560 nm with an energy of 310-nJ were obtained, starting from 400-fs pulses of 60-pJ energy that were stretched using 60-m of standard fiber and amplified using three EDFAs [143]. A 9.3-m-long photonic bandgap fiber was used in a double-pass configuration in combination with a Faraday rotator to compress amplified pulses. Figure 6.18 shows the 310-nJ output pulse (solid trace) and its spectrum. The pedestal seen there is due to the SPM effects in the last amplifier that produce a chirp that is not linear in the pulse wings. Indeed, the pedestal is reduced considerably for 73-nJ pulses (dashed trace).

In an interesting approach, short 1-ps pulses with 20-kW peak powers were generated starting from the continuous-wave (CW) light emitted by a semiconductor laser operating at 1566 nm [144]. A $LiNbO_3$ modulator, driven by 35-ps electrical pulses, was used to create 40-ps optical pulses (at a 50-MHz repetition rate) that were chirped inside a 3.5-km-long dispersion-shifted fiber. The chirped pulses were amplified then compressed in 110-m-long of photonic bandgap fiber exhibiting a nearly constant anomalous dispersion of -115 ps^2/km over the entire pulse bandwidth. The measured autocorrelation trace indicated that 80% of pulse energy was contained within a compressed 1-ps pulse, with a peak power of about 20 kW.

Although most experiments have employed a doped-fiber amplifier, any kind of optical amplifier can be used for CPA. A fiber-based Raman amplifier was used in a 2005 experiment in which Raman gain was obtained by pumping a 9-km-long dispersion-shifted fiber [146]. Even more interesting is to employ a fiber-optic parametric amplifier. Solid-state parametric amplifiers have been used in an extension of the basic CPA technique, referred to as *optical parametric CPA*, and have produced pulses as short as 3.9 fs and peak powers as high as 350 TW [147]. The use of a fiber-optic parametric amplifier for this purpose is also attractive [148]. Such amplifiers can provide wide-band amplification in a single- or dual-pump configuration, and they offer considerable flexibility in the choice of signal and pump wavelengths [149].

The CPA technique is actively being pursued for developing high-power Yb-doped fiber sources capable of emitting high-energy femtosecond pulses [139]. Two approaches are being used for such lasers. In one, the core size of the fiber amplifier

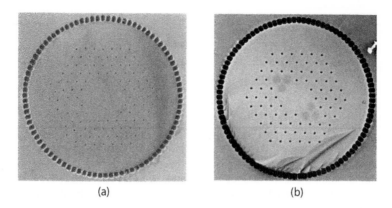

(a)   (b)

Figure 6.19: Cross section of two photonic crystal fibers with a Yb-doped core developed for high-power fiber lasers. (From Ref. [139]; ©2006 IEEE.)

is increased to minimize the nonlinear effects within the double-clad fiber. Figure 6.19 shows two photonic crystal fibers used for this purpose in recent experiments with core diameters as large as 60 μm. Both of them have a central core, an inner cladding with air holes, and a circular outer cladding in the form of a ring containing mostly air. The fiber on right also contains six stress-applying elements to induce birefringence within the fiber. Pulse energies of up to 100 μJ were realized by 2006 with such fiber amplifiers within a CPA configuration. The main limitation comes from the undesirable SPM-induced chirp within the amplifier.

In the second approach, the nonlinear effects occurring within the fiber amplifier are used to advantage by operating it in the normal-dispersion regime. As discussed in Section 4.6.2, pulses of any shape evolve toward similaritons whose shape is parabolic and whose width and peak power increase exponentially along the fiber length. This approach does not require stretching of an optical pulse before it enters the amplifier, because it is stretched automatically as it is amplified. The most useful feature of this approach is that the resulting high-energy pulses are chirped linearly across their entire temporal profile. As a result they can be compressed with a grating pair, or a suitable piece of hollow-core fiber, without exhibiting a residual pedestal. In one such experiment, high-quality 80-fs pulses with 140-nJ energy were obtained by using a 9-m-long Yb-doped fiber amplifier with only a 30-μm-diameter core [139].

6.6   Dispersion-Managed Fibers

In several pulse-compression schemes, multiple fibers with different dispersion characteristics are employed in a judicial manner. Dispersion management is an established technique in optical communication systems and its use for pulse compression turns out to be quite beneficial. This section focuses on several types of dispersion-managed fiber compressors.

6.6.1   Dispersion-Decreasing Fibers

An interesting pulse-compression technique makes use of a single piece of optical fiber in which the magnitude of the GVD parameter $|\beta_2|$ decreases along the fiber length [150]–[154]. Such fibers are referred to as *dispersion-decreasing fibers* (DDFs) and can be made by tapering the core diameter of the fiber at the preform-drawing stage. As the magnitude of $|\beta_2|$ depends on the core size, its value decreases along the fiber length. The basic idea was proposed in 1988, but it was not until 1992 that compression factors as large as 16 were realized experimentally [152].

The physical mechanism behind DDF-induced compression can be understood from Eq. (6.2.6), which shows how the soliton order N depends on the GVD parameter β_2. Consider a fundamental soliton launched at the fiber input so that $N = 1$ initially. If the numerical value of $|\beta_2|$ decreases along the fiber length, N can be preserved at its input value $N = 1$ only if the pulse width decreases as $|\beta_2|^{1/2}$. In essence, the situation is similar to that occurring in fiber amplifiers, where an increase in the peak power of a pulse results in its compression. The analogy between a fiber amplifier and a DDF can be established mathematically by using the NLS equation (6.3.1). In the case of a DDF, this equation can be written as

$$i\frac{\partial U}{\partial \xi} + d(\xi)\frac{1}{2}\frac{\partial^2 U}{\partial \tau^2} + N^2|U|^2 U = 0, \tag{6.6.1}$$

where the parameter $d(\xi) = |\beta_2(\xi)/\beta_2(0)|$ governs dispersion variations along the fiber length. If we make the transformation [155]

$$\eta = \int_0^\xi d(y)dy, \qquad u = NU/\sqrt{d}, \tag{6.6.2}$$

Eq. (6.6.1) is reduced to

$$i\frac{\partial u}{\partial \eta} + \frac{1}{2}\frac{\partial^2 u}{\partial \tau^2} + |u|^2 u = i\Gamma u. \tag{6.6.3}$$

The parameter Γ depends on the nature of dispersion variations along the fiber length as $\Gamma(\eta) = -d_\eta/(2d)$, where d_η stands for the derivative of d with respect to η.

Equation (6.6.3) shows that the effect of decreasing dispersion along the fiber length is mathematically equivalent to adding a gain term to the NLS equation. The effective gain coefficient Γ becomes constant when GVD decreases exponentially along the fiber length. Equation (6.6.3) can be used to study pulse compression in DDFs [86]. It is important to include the contribution of Raman-induced frequency shift in the femtosecond regime, because changes in GVD affect the frequency shift considerably [152].

Starting in 1991, DDFs were used for pulse compression in several experiments. In one experiment, 130-fs pulses were compressed down to 50 fs by using a 10-m section of a DDF [151]. Much larger compression factors were realized in 1992 by using 3.5-ps pulses from a mode-locked fiber laser operating near 1.55 μm [152]. Such pulses were transmitted through a fiber whose dispersion decreased from 10 to 0.5 ps/(km-nm) over a length of 1.6 km. The input pulse was compressed by a factor of 16 (to

230 fs), while its spectrum shifted by 10 nm due to the frequency shift induced by intrapulse Raman scattering. In another experiment, 630-fs pulses were compressed down to 115 fs by using a fiber whose dispersion decreased from 10 to 1.45 ps/(km-nm) over a length of 100 m [153].

Pulse compression has also been realized in a fiber whose dispersion was constant over the fiber length but decreased with wavelength in the 1.55- to 1.65-μm wavelength region [154]. A 95-fs input pulse at 1.57 μm compressed to 55 fs over 65 m of such fiber. The Raman-induced frequency shift plays a crucial role in this experiment. In effect, the GVD decreased along the fiber length in the reference frame of the pulse because of the frequency shift induced by intrapulse Raman scattering. The pulse spectrum was indeed observed to shift from 1.57 to 1.62 μm as the pulse compressed.

The DDF pulse-compression mechanism has been used to generate a train of ultrashort pulses [155]–[159]. The basic idea consists of injecting a CW beam, with weak sinusoidal modulation imposed on it, into an optical fiber exhibiting gain [155]. Since decreasing dispersion is equivalent to an effective gain, such fibers can be used in place of a fiber amplifier [156]. As the sinusoidal signal propagates, individual pulses within each modulation cycle are compressed. The combined effect of GVD, SPM, and decreasing GVD is to convert a nearly CW signal into a high-quality train of ultrashort solitons [86]. The repetition rate of pulses is governed by the frequency of initial sinusoidal modulation.

Several experiments have used DDFs to generate ultrashort pulses at high repetition rates. Sinusoidal modulation in these experiments is imposed by beating two optical signals. In a 1992 experiment [157], the outputs of two distributed-feedback (DFB) semiconductor lasers, operating continuously at slightly different wavelengths near 1.55 μm, were combined in a fiber coupler to produce a sinusoidally modulated signal at a beat frequency that could be varied in the 70 to 90 GHz range by controlling the laser temperature. The beat signal was amplified to power levels of about 0.3 W by using a fiber amplifier. It was then propagated through a 1-km dispersion-shifted fiber, followed by a DDF whose dispersion decreased from 10 to 0.5 ps/(km-nm) over a length of 1.6 km. The output consisted of a high-quality pulse train at 70 GHz with individual pulses of 1.3-ps width. By 1993, this technique had led to the generation of a 250-fs soliton train at 80- to 120-GHz repetition rates when the peak power of the beat signal was enhanced to about 0.8 W by synchronous modulation of the laser current [158]. Figure 6.20 shows the spectrum and the autocorrelation trace of such a soliton train at the 114-GHz repetition rate. Even though the laser wavelengths are near 1.5 μm, the spectrum of the soliton train is centered at 1.565 μm because of the Raman-induced frequency shift.

During the 1990s, pulse compression is DDFs remained an active topic of interest [160]–[169]. In one set of experiments, the objective was to produce a train of ultrashort pulses at high repetition rates. A 40-GHz train of 3.4-ps pulses was generated by compressing the sinusoidal signal (obtained beating two DFB lasers) using a combination of standard, dispersion-shifted, and dispersion-decreasing fibers [160]. In another experiment, 0.8-ps pulses at a repetition rate of 160 GHz were produced starting with a 10-GHz train of relatively wide (>10 ps) pulses obtained from a gain-switched semiconductor laser [161]. Chirped pulses were first compressed down to

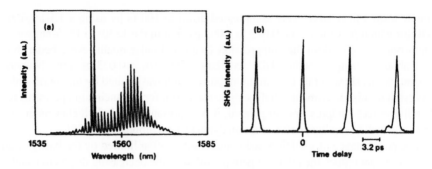

Figure 6.20: (a) Spectrum and (b) autocorrelation trace of a 114-GHz pulse train generated by using a dispersion-decreasing fiber. (From Ref. [158]; ©1993 AIP.)

6.3 ps in a normal-GVD fiber. They were then amplified and compressed to 0.8 ps using two sections (2 and 5 km long) of DDFs. Finally, time-division multiplexing was used to increase the repetition rate to 160 GHz. Pulses as short as 170 fs were produced by combining decreasing GVD with amplification [162]. Such sources of ultrashort optical pulses are useful for fiber-optic communication systems [119].

In another set of experiments, the DDF was used for transmission of ultrashort pulses over relatively long lengths. The objective in this case was to maintain the width of a fundamental soliton in spite of fiber losses. In a constant-dispersion fiber, solitons broaden as they loose energy because of weakening of the nonlinear effects. However, soliton width can be maintained if GVD decreases exponentially. This behavior was indeed observed in a 40-km DDF [163]. The Raman-induced spectral shift can disturb the balance between the GVD and SPM because GVD changes with frequency. Soliton width can still be maintained if the dispersion profile is designed such that it remains exponential in the reference frame of the spectrally shifting soliton.

The optimum GVD profile is not exponential in the presence of the Raman-induced spectral shift. Numerical simulations indicate that linear and Gaussian dispersion profiles result in better-quality compressed pulses and larger compression factors for sub-picosecond pulses [165]. An analytic approach can also be used to find the optimum GVD profile [164]. It shows that, in a long DDF, the GVD should be normal far from the input end to counteract the effect of the Raman-induced spectral shift. A variational approach has also been developed to determine the optimum GVD profile and its dependence on the width and peak power of input pulses [168]. Compression factors larger than 50 are possible by launching input pulses (with peak powers such that the soliton order $N \approx 1$) into a DDF whose length is about one soliton period [166]. This technique takes advantage of soliton-effect compression but requires lower peak powers and produces compressed pulses of better quality.

The TOD becomes the ultimate limiting factor for compression of ultrashort pulses in DDFs. Dispersion-flattened fibers, in which the TOD parameter β_3 is reduced considerably from its standard value (about 0.08 ps^3/km), became available in the late 1990s. The use of such a DDF has resulted in improved compression characteristics. In one set of experiments, 3-ps pulses at the 10-GHz repetition rate, obtained from a mode-locked fiber laser, were compressed down to 100 fs by using a 1.1-km-long DDF for which β_3 varied from 0.023 to 0.003 ps^3/km in the 1530- to 1565-nm wavelength range [169]. Both the compression factor and pulse quality were better than those obtained using a standard DDF with larger TOD ($\beta_3 = 0.073$ ps^3/km). Numerical simulations showed that the minimum pulse width was limited by the fourth-order dispersion in this experiment. DDFs can be used even for enhancing the performance of soliton-effect compressors. In general, it is important to include higher-order dispersive as well as nonlinear effects for analyzing such compressors [170].

The performance of a DDF-based compressor is often limited by the background fiber losses that reduce the effective gain provided by the decreasing dispersion, and it can be improved if the DDF is pumped suitably to provide gain. As discussed in Section 4.6 and later in Section 6.7.3, fiber amplifiers can also compress optical pulses. Indeed, distributed Raman amplification inside constant-dispersion fibers was used in several recent experiments to compress pulses propagating inside them [171]–[173]. It turns out that distributed Raman amplification inside a DDF enhances the compression factor as well as the quality of compressed pulses. In a 2004 experiment, 13-ps pulses, generated at a repetition rate of 10 GHz at wavelengths near 1550 nm, were compressed down to 1.3 ps using a 20-km-long DDF whose effective gain resulting from dispersion changes was just large enough to cancel the fiber loss [174]. Three semiconductor lasers, operating at wavelengths of 1430, 1460, and 1490 nm, launched 500 mW of total pump power and amplified pulse energies by a a factor of 6.3 through distributed Raman amplification. No compression was observed when pump lasers were turned off. The same setup was used to generate 2.2-ps pulses at a 40-GHz repetition rate by compressing the output of a DFB laser that was first modulated sinusoidally using a LiNbO$_3$ modulator [175]. Figure 6.21 shows the numerically simulated temporal and spectral profiles at the input and output ends of the fiber compressor. The experimental results agree well with the theory.

Dispersion of a microstructured fiber can also be varied along its length by tapering its diameter. Such DDFs can be used for pulse compression. However, there is a practical limitation because the diameter of air holes within the cladding also decreases with tapering, resulting in high losses because of the weaker mode confinement. Nevertheless, a compression factor of 10 should be possible for 3-ps pulses propagating inside a 28-m-long tapered fiber [176]. Such large compression factors have not yet been observed. In a recent experiment, compression of 130 fs pulses down to 60 fs was observed inside a holey fiber whose dispersion near the 1.06-μm wavelength changed from 5.4 to 2.3 ps/(km-nm) over its 8.1-m length [177]. The theoretically predicted compression factor is also close to 2 for this fiber after the wavelength dependence of the GVD and fiber losses are taken into account.

6.6.2 Comblike Dispersion Profiles

Several experiments have employed a comblike dispersion profile for producing ultrashort pulse trains [178]–[183]. In most cases, a comblike dispersion profile is produced by alternating pieces of low- and high-dispersion fibers in series. A dual-frequency fiber laser was used in a 1993 experiment to generate the high-power beat signal that was launched into such a sequence of fibers [178]. The output consisted of a 2.2-ps

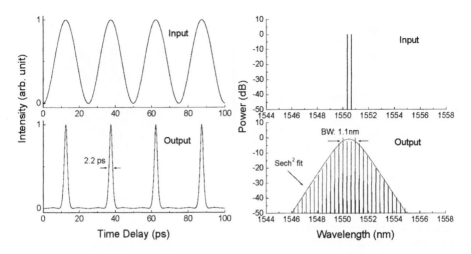

Figure 6.21: (a) Temporal and (b) spectral profiles at the input (top row) and output (bottom row) ends of a 20-km-long DDF compressor pumped to provide 8-dB Raman amplification. (From Ref. [175]; ©2004 OSA.)

pulse train at the 59-GHz repetition rate. Such a device can be used as an all-fiber source of pulse trains suitable for high-speed communication systems. It operates by separating the effects of SPM and GVD in low- and high-dispersion fibers. Specifically, SPM dominates in low-dispersion sections and induces a frequency chirp on the beat signal. By contrast, GVD dominates in high-dispersion sections and it compresses the chirped pulse.

Highly nonlinear fibers have been used recently in combination with standard fibers to produce fiber compressors in which both the dispersion and nonlinear parameters (β_2 and γ) vary in a comblike fashion [180]–[183]. In a 2005 experiment, such a fiber chain was used to create a train of 730-fs soliton pulses at a repetition rate of 160 GHz by launching an optical beat signal with an average power of about 200 mW [180]. The fiber chain was 520 m long and contained six stages of alternating standard and highly nonlinear fibers of different lengths. In another experiment, a 1.4-km-long chain consisting of 12 stages of such fiber pairs was used to compress a 40-GHz optical beat signal to form a train of 500-fs compressed pulses [181]. In each case, pulses become chirped in fiber sections with high γ and low β_2 and are compressed in fiber sections with low γ and high β_2. Thus, each successive fiber pair increases the total compression factor.

Section lengths for the comblike dispersion profile used in these experiments were not uniform and were optimized to yield the best performance. It is intuitively clear that a uniform section length is not the optimum choice because pulses entering each section of the fiber chain have different widths, chirps, and peak powers. Since the dispersion length changes from section to section, fiber length L_f of each section should be adjusted to keep the normalized length L_f/L_D the same along the whole chain, as seen from Eq. (6.6.1) written in the soliton units.

One can combine the idea of DDF with the comblike dispersion profile by de-

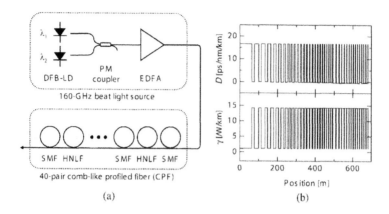

Figure 6.22: (a) Experimental setup used for compressing a 160-GHz beat signal using a fiber chain with a comblike dispersion profile. (b) Changes in the dispersion and nonlinear parameters along the fiber chain. (From Ref. [182]; ©2005 IEEE.)

manding that the average dispersion of sections decrease along the fiber chain in a certain fashion. The decrease may be linear, quadratic, or exponential, and the optimum choice is not obvious. An iterative procedure was used in 2005 for this purpose [182]. The basic idea is to approximate the dispersion profile $d(\xi)$ in Eq. (6.6.1) with a polynomial as $d(\xi) = \sum_{m=1}^{M} d_m \xi^m$, where M is an integer, and find the optimum value of the expansion coefficients d_m that yield the target pulse at the output end. Figure 6.22(a) shows the experimental setup and the dispersion profile, optimized using $M = 5$ for converting a 160-GHz optical beat signal to a soliton pulse train. The 684-m-long chain consisted of 40 alternating sections of a standard fiber ($\gamma < 2$ W^{-1}/km; $\beta_2 = -21$ ps^2/km) and a highly nonlinear fiber ($\gamma = 14.2$ W^{-1}/km; $\beta_2 = 0.32$ ps^2/km). Such a fiber chain compressed the 160-GHz sinusoidal signal into a pulse train consisting of 324-fs solitons. Both the shape and the spectrum corresponded to a nearly perfect train of sech-shape pulses.

The number of required fiber sections can be reduced considerably using the concept of self-similarity, called *rescaled pulse propagation* in a 2006 study [183]; such a pulse is also referred to as a *similariton*. The basic idea consists of designing the comblike dispersion profile of the fiber chain such that each stage consisting of standard and highly nonlinear fibers rescales the pulse by the same factor while maintaining its shape. Mathematically, the solution of Eq. (6.6.1) after the nth stage at a distance ξ_n has the form

$$U(\xi_n, \tau) = R^{-n/2} U(0, \tau/R^n), \qquad (6.6.4)$$

where the rescaling parameter R represents the factor by which the pulse compresses during each stage. Figure 6.23(a) shows the simulated and measured autocorrelation traces after each stage for a four-stage compressor designed with $R = 2.1$. The dispersion and nonlinear parameters for each fiber section and section lengths are shown in part (b). Such a 1.1-km-long eight-fiber chain compresses 7.2-ps input pulses down to 370-fs in a self-similar fashion. The same concept of self-similarity can also be applied to DDFs [184]. It shows that a linearly chirped, self-similar pulse can form in a DDF

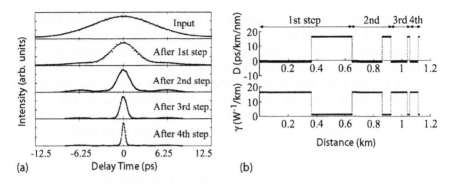

Figure 6.23: (a) Simulated and measured (dots) autocorrelation traces at the input and after each stage for a four-stage compressor. (b) Dispersion and nonlinear parameters versus section lengths for the comblike fiber chain. (From Ref. [183]; ©2006 IEEE.)

whose dispersion decreases along its length. This solution can be used in practice to obtain pedestal-free compressed pulses if the input pulse is first chirped suitably.

Recently a comblike dispersion profile, with decreasing values of the average anomalous dispersion (increasing $|\beta_2|$) along the fiber chain, was used as a distributed Raman amplifier [185]. Input pulses were first chirped so that they propagated as similaritons inside the amplifier whose widths decreased as they were amplified. When the resulting 17.8-ps pulses were amplified in the 16-section 640-m-long fiber chain, they were compressed down to 1.47 ps. A similar second stage reduced the pulse width close to 0.5 ps. This behavior is similar to the parabolic-shape pulses found in Section 4.6.2, whose width increased in a self-similar fashion as they experienced normal dispersion during amplification. It turns out that sech-shape, chirped, similaritons can exist even when the gain, dispersion, and nonlinear parameters change with distance, provided they satisfy a compatibility condition [186]. For certain specific dispersion and gain profiles, similaritons compress as they are amplified.

6.7 Other Compression Techniques

The pulse compression techniques described so far show how the interplay between SPM and GVD can be used to compress optical pulses by using fibers. Several other techniques based on this basic idea have been developed. This section discusses how these methods use nonlinear effects in optical fibers to produce ultrashort optical pulses.

6.7.1 Cross-Phase Modulation

The use of SPM-induced chirp for pulse compression requires input pulses to be intense enough that their spectrum broadens considerably during propagation inside an optical fiber. Clearly, such a technique cannot be used for compressing low-energy pulses. Since cross-phase modulation (XPM) also imposes a frequency chirp on optical pulses, it can be used for compressing weak optical pulses [187]–[200]. Of course, the use of XPM requires an intense pump pulse that must be copropagated with the weak input pulse (referred to as the *probe pulse* in pump-probe experiments). However, the pump pulse is allowed to have a different wavelength. As discussed in Section 7.5 of Ref. [2], the XPM-induced chirp is affected by pulse walk-off and depends critically on the initial pump-probe delay. As a result, the practical use of XPM-induced pulse compression requires a careful control of the pump-pulse parameters such as its width, peak power, wavelength, and synchronization with the probe pulse.

Two cases must be distinguished depending on the relative magnitudes of the walk-off length L_W and the dispersion length L_D. If $L_D \gg L_W$ throughout the fiber, the GVD effects are negligible. In that case, the fiber imposes the chirp through XPM, and a dispersive delay line (a grating pair or another type of fiber) is needed to compress the chirped pulse. A nearly linear chirp can be imposed across the entire probe pulse when the pump pulse is much wider compared with it [189]. The compression factor depends on the energy of pump pulses and can easily exceed 10.

Another pulse-compression mechanism can be used when L_D and L_W are comparable in magnitude. In this case, the same piece of fiber that is used to impose the XPM-induced chirp also compresses the pulse through the GVD. In some sense, this scheme is analogous to the soliton-effect compressor for the XPM case. However, in contrast with the SPM case where compression can occur only in the anomalous-GVD regime, the XPM offers the possibility of pulse compression even in the visible region (normal GVD) without the need of a grating pair. The performance of such a compressor can be studied by solving numerically the following set of two coupled NLS equations (see Section 7.4 of Ref. [2]):

$$\frac{\partial U_1}{\partial \xi} + \text{sgn}(\beta_{21})\frac{i}{2}\frac{\partial^2 U_1}{\partial \tau^2} = iN^2(|U_1|^2 + 2|U_2|^2)U_1, \tag{6.7.1}$$

$$\frac{\partial U_2}{\partial \xi} \pm \frac{L_D}{L_W}\frac{\partial U_2}{\partial \tau} + \frac{i}{2}\frac{\beta_{22}}{\beta_{21}}\frac{\partial^2 U_2}{\partial \tau^2} = iN^2\frac{\omega_2}{\omega_1}(|U_2|^2 + 2|U_1|^2)U_2, \tag{6.7.2}$$

where $U_j = A_j/\sqrt{P_j}$ is the normalized amplitude, ω_j is the carrier frequency, P_j is the peak power, and β_{2j} is the GVD parameter for the pump ($j = 1$) and probe ($j = 2$) pulses. The soliton order N and the walk-off length L_W are introduced using

$$N^2 = \frac{\gamma_1 P_1 T_0^2}{|\beta_{21}|}, \qquad L_W = \frac{v_{g1}v_{g2}T_0}{|v_{g1} - v_{g2}|}, \tag{6.7.3}$$

where v_{g1} and v_{g2} are the group velocities of the pump and probe pulses. These equations govern the evolution of pump and probe pulses in the presence of SPM, XPM, and group-velocity mismatch [189]. One can introduce a relative time delay T_d between the input pump and probe pulses such that the faster-moving pulse overtakes the slower pulse and passes through it. In general, a trade-off exists between the magnitude and the quality of compression.

XPM-induced pulse compression in the normal-GVD region of a fiber can also occur when the XPM coupling is due to interaction between the orthogonally polarized components of a single beam [190]. Indeed, an experiment in 1990 demonstrated

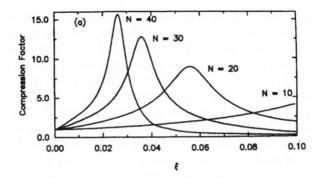

Figure 6.24: Compression factor as a function of fiber length for a Raman amplifier pumped by a pulse propagating as an Nth-order soliton. (From Ref. [193]; ©1993 OSA.)

pulse compression by using just such a technique [191]. A polarizing Michelson interferometer was used to launch 2-ps pulses in a 1.4-m birefringent fiber (with a 2.1-mm beat length) such that the peak power and the relative delay of the two polarization components were adjustable. For a relative delay of 1.2 ps, the weak component was compressed by a factor of about 6.7 when the peak power of the other polarization component was 1.5 kW.

When both the pump and signal pulses propagate in the normal-GVD region of the fiber, the compressed pulse is necessarily asymmetric because of the group-velocity mismatch and the associated walk-off effects. The group velocities can be made nearly equal when wavelengths of the two pulses lie on opposite sides of the zero-dispersion wavelength (about 1.3 μm in standard silica fibers). One possibility consists of compressing 1.55-μm pulses by using 1.06-μm pump pulses. The probe pulse by itself is too weak to form an optical soliton. However, the XPM-induced chirp imposed on it by a copropagating pump pulse may be strong enough that the probe pulse goes through an initial compression phase associated with the higher-order solitons. As an example, weak picosecond pulses can be compressed by a factor of 10 using pump pulses with $N = 30$ by optimizing the fiber length. This method of pulse compression is similar to that provided by higher-order solitons even though the compressed pulse never forms a soliton. Using dispersion-shifted fibers, the technique can be used even when both pump and probe wavelengths are in the 1.55-μm region as long as the zero-dispersion wavelength of the fiber lies in the middle. In a 1993 experiment [192], 10.6-ps signal pulses were compressed to 4.6 ps by using 12-ps pump pulses. Pump and signal pulses were obtained from mode-locked semiconductor lasers operating at 1.56 and 1.54 μm, respectively, with a 5-GHz repetition rate. Pump pulses were amplified to an average power of 17 mW by using a fiber amplifier. This experiment demonstrates that XPM-induced pulse compression can occur at power levels available with semiconductor lasers.

An extension of this idea makes use of Raman amplification for simultaneous amplification and compression of picosecond optical pulses [193]. The probe pulse extracts energy from the pump pulse through stimulated Raman scattering and is amplified. At the same time, it interacts with the pump pulse through XPM, which imposes a nearly linear frequency chirp on it and compresses it in the presence of anomalous GVD. Equations (6.7.1) and (6.7.2) can be used to study this case, provided a Raman term is added to these equations (see Section 8.3 of Ref. [2]). Numerical simulations show that compression factors as large as 15 can be realized while the pulse is amplified a millionfold [193]. The quality of the compressed pulses is also quite good, with no pedestal and little ringing. Figure 6.24 shows the compression factor as the probe pulse is amplified by a pump pulse intense enough to form a soliton of order N. Pulse compression is maximum for an optimum fiber length, a feature similar to that of soliton-effect compressors. This behavior is easily understood by noting that GVD reduces the XPM-induced chirp to nearly zero at the point of maximum compression. Simultaneous Raman-induced amplification and XPM-induced compression of picosecond optical pulses was observed in a 1996 experiment [194].

Several other XPM-based techniques can be used for pulse compression. In one scheme, a probe pulse is compressed as it travels with a pump pulse that is launched as a higher-order soliton and is thus compressed through soliton-effect compression [196]. Compression factors as large as 25 are predicted by the coupled NLS equations. In this technique, a pump pulse transfers its compression to the copropagating probe pulse through XPM-induced coupling between the two pulses. Both pulses must propagate in the anomalous-GVD regime of the optical fiber and their wavelengths should not be too far apart. This is often the situation for WDM lightwave systems with channel spacing of 1 nm or so.

Another technique is based on propagation of two ultrashort pulses (widths below 100 fs) inside an optical fiber [197]. The second pulse is generated through second-harmonic generation. As a result, carrier frequencies of the two pulses are far apart but their relative phase is locked. Both pulses are launched into an optical fiber where they interact through XPM. The pulse spectra are relatively broad to begin with for ultrashort pulses. They broaden further through SPM and XPM to the extent that the two spectra merge at the fiber output, producing an extremely broad continuum. A grating pair, in combination with a spatial phase modulator, can then be used to compensate the chirp and realize a single supershort pulse whose spectrum corresponds to the merge spectra of the two input pulses. Pulses shorter than 3 fs are predicted theoretically.

In a 1999 experiment, a three-wave mixing technique was used to generate 8-fs pulses in the ultraviolet region near 270 nm [198]. Ultrashort 35-fs pulses, from a Ti:sapphire laser operating near 800 nm, and second-harmonic pulses were coupled together into a hollow silica fiber containing argon gas. The sum-frequency pulse generated was chirped because of XPM and could be compressed to 8 fs using a grating pair.

6.7.2 Gain Switching in Semiconductor Lasers

The technique of gain switching can produce pulses of about 20-ps duration from semiconductor lasers directly. As a rule, pulses emitted from gain-switched semiconductor lasers are chirped because of refractive-index variations occurring during pulse generation. In contrast with the positive chirp induced through SPM in optical fibers, the frequency chirp of gain-switched pulses is negative; that is, the frequency decreases toward the trailing edge. Such pulses with negative values of the chirp parameter C can be compressed if an optical fiber having positive GVD is used and its length is suitably optimized. This technique has been used in many experiments to generate trains of ultrashort optical pulses suitable for optical communications [201]–[211]. Nonlinear properties of optical fibers are not used in this technique, since laser pulses entering the fiber are already chirped. The role of the fiber is to provide positive dispersion. The output pulse is not only compressed but also becomes nearly transform limited. The use of dispersion-shifted fibers is necessary in the 1.5-μm wavelength region.

As early as 1986, the gain-switching technique was used to obtain 6-ps optical pulses at a repetition rate of 12 GHz [201]. The repetition rate was increased to 100 GHz by 1988 through time-division multiplexing of compressed pulses. Since the frequency chirp imposed by the gain-switched laser is not perfectly linear, the compressed pulse is not generally transform limited. The quality of the compressed pulse can be significantly improved by using a bandpass filter that passes the central part of the pulse spectrum where the chirp is nearly linear [202]. By 1990, transform-limited optical pulses of 3-ps duration were generated in this way [203].

Even shorter pulses can be obtained if the compressed pulse is further compressed in a fiber with anomalous GVD by using the soliton-effect compression technique of Section 6.4. In one experiment [204], 1.26-ps optical pulses were generated by compressing a 17.5-ps gain-switched pulse through such a two-stage compression scheme. It was necessary to boost the pulse energy by using semiconductor laser amplifiers before launching the pulse into the second fiber so that the pulse energy would be large enough to excite a higher-order soliton. By 1993, the pulse width was reduced to below 1 ps [205]. Pulses as short as 230 fs were obtained by using a tunable distributed Bragg reflector (DBR) laser [206]. The ultimate performance of the gain-switching technique is limited by SPM occurring in the normal-GVD fiber used for pulse compression, since the resulting positive chirp tends to cancel the negative chirp of gain-switched pulses. Numerical simulations are often used to optimize the performance [207].

Much shorter pulses were obtained in 1995 by using fiber birefringence during pulse compression in such a way that pulses were reshaped as they were being compressed [208]. A train of 185-fs high-quality pedestal-free pulses were successfully generated by this technique. In a later experiment, 16-ps pulses from a gain-switched laser were compressed down to 110 fs [209]. Linear compression of chirped pulses in a normal-GVD fiber resulted in transform-limited 4.2-ps pulses. Such pulses were then amplified and compressed to below 0.2 ps by launching them into a dispersion-shifted fiber as a higher-order soliton ($N = 10$ to 12) to use soliton-effect compression. The effects of TOD as well as intrapulse Raman scattering became important for such short pulses. Experimental results were in agreement with the predictions of Eq. (6.3.5). The shortest optical pulses (about 21 fs) were obtained by pumping a surface-emitting laser synchronously in the external-cavity configuration [85]. Pulses emitted from such a laser were heavily chirped and could be compressed down to 21 fs by using a grating pair followed with a soliton-effect compressor.

A chirped fiber grating can also be used to compress gain-switched pulses [211]. The role of the grating is to provide dispersion that induces a chirp of the opposite nature on the chirped input pulse. If the two chirps cancel perfectly, a transform-limited narrower pulse is reflected back from the grating. High-quality 3.5-ps pulses have been produced with this approach by launching 10.5-ps gain-switched pulses and using a nonlinearly chirped fiber grating.

6.7.3 Optical Amplifiers

Under certain conditions, amplification of optical pulses can chirp the pulse such that it can be compressed if it is subsequently propagated through an optical fiber having appropriate GVD. An example is provided by semiconductor optical amplifiers. When picosecond optical pulses are amplified in such amplifiers, gain saturation leads to non-linear changes in the refractive index of the semiconductor gain medium [212]–[215]. In essence, the amplifier imposes a frequency chirp on the amplified pulse through the process of SPM [213]. However, in contrast with the gain-switched semiconductor lasers, the chirp imposed on the pulse is such that the frequency increases with time over a large portion of the pulse (similar to the SPM-induced chirp in optical fibers). As a result, the amplified pulse can be compressed if it is passed through a fiber having anomalous GVD ($\beta_2 < 0$). The compression mechanism is similar to the soliton-effect compression scheme of Section 6.4 with the difference that the SPM-induced chirp is imposed by the amplifier instead of the fiber. The main advantage of this technique stems from the fact that low-energy pulses, which cannot be compressed in fibers directly because their peak power is below the $N = 1$ level, can be amplified and compressed by using an amplifier followed by a piece of optical fiber.

Amplifier-induced pulse compression was observed in a 1989 experiment in which 40-ps pulses, emitted from a 1.52-μm mode-locked semiconductor laser, were first amplified in a semiconductor laser amplifier then propagated through 18 km of optical fiber having $\beta_2 = -18$ ps^2/km [213]. The compression factor was about 2 because of relatively low pulse energies (~ 0.1 pJ). Pulse-shape measurements through a streak camera were in good agreement with the theoretical prediction [214]. The technique can be used for simultaneous compensation of fiber loss and dispersion in fiber-optic communication systems. In a demonstration of the basic concept [215], a 16-Gb/s signal could be transmitted over 70 km of conventional fiber having large dispersion when a semiconductor laser amplifier was used as an in-line amplifier. In the absence of amplifier-induced chirp, the signal could not be transmitted over more than 15 km since optical pulses experienced excessive broadening.

As discussed in Section 4.6, fiber amplifiers compress optical pulses propagating in the anomalous-GVD regime as they are amplified [216]–[225]. The compression mechanism is similar to that associated with higher-order solitons. Specifically, the amplifying pulse forms a fundamental soliton ($N = 1$) when its peak power becomes large enough. With further increase in the peak power, N begins to exceed 1. As seen from Eq. (6.2.6), the soliton order can be maintained ($N = 1$) if pulse width decreases with amplification. Thus, the amplified pulse keeps on compressing as long as the amplification process remains adiabatic. This compression mechanism is evident in Figure 4.12, which was obtained by solving the Ginzburg–Landau equation that governs the amplification process in fiber amplifiers. The compression stops eventually because of a finite gain bandwidth associated with fiber amplifiers. Compression of femtosecond pulses was achieved in erbium-doped fiber amplifiers soon after they became available [219]. In a 1990 experiment, 240-fs input pulses were compressed by about a factor of 4 [220]. In a later experiment, 124-fs pulses emitted from a mode-locked fiber laser were shortened to 50 fs by using a 6-m-long fiber amplifier [223].

Amplification can also improve the performance of standard grating-fiber compressors [225]. It turns out that, if the normal-GVD fiber—used for imposing SPM-induced chirp—is doped and pumped to provide gain, the pulse acquires a nearly parabolic shape. As seen in Figure 6.3, the pulse shape is close to being rectangular in the absence of gain. Sharp edges of such pulses leads to optical wave breaking, a phenomenon that does not occur for parabolic-shaped pulses. The SPM-induced chirp remains linear even when pulses are amplified in a normal-GVD fiber amplifier. As a result, the amplified pulse can be compressed with a dispersive delay line without forming a pedestal. In the 1996 experiment, 350-fs pulses were compressed down to 77 fs using a 4-m-long fiber amplifier with 18-dB gain, followed by a prism pair. An added benefit of this technique is that it can be used for relatively weak input pulses as long as the amplified pulse becomes intense enough to undergo SPM-induced phase shift.

The creation of parabolic-shape pulses in the normal-GVD regime of fiber amplifiers attracted considerable attention after 2000 because of the self-similar nature of such pulses. As discussed in Section 4.4.2, both the pulse width and peak power keep increasing in the self-similar regime, while a linear chirp is maintained across the entire pulse. As a result, such amplifiers produce high-energy pulses that can be compressed using any dispersive element (grating pairs, chirped fiber gratings, or photonic crystal fibers). This approach has already been discussed in Section 6.5 in the context of CPA. The formation and compression of such pulses was analyzed in a 2006 study in the presence of stimulated Raman scattering [226].

A Raman fiber amplifier can also be used for compressing optical pulses. If amplification occurs in the anomalous-GVD regime, pulses are compressed through the soliton-effect compression [171]–[173]. In contrast, if amplification occurs in the anomalous-GVD regime, pulses become broad, acquire a parabolic shape, and are chirped linearly [227]. Such pulses can thus be compressed by passing them using a dispersive delay line.

Optical pulses can also be compressed through four-wave mixing (FWM) inside a fiber-optic parametric amplifier [228]–[230]. In a 1997 experiment, 93-ps signal pulses were compressed down to 20 ps when 50-ps pump pulses (wavelength separation 4.9 nm) were copropagated with them inside a 5-km-long fiber [228] and were amplified by 29 dB. An idler pulse was also generated during parametric amplification. Its width and peak powers were close to those of the signal pulse. It was found later that a multiple FWM process can be used to generate a train of short pulses at ultrahigh repetition rates. A 160-GHz train of 1.3-ps pulses was generated in a 2002 experiment when CW light from two lasers, separated in frequency by 160 GHz, was launched inside in the naomalous-GVD regime of a 1-km-long fiber. optical beat signal at the 160-GHz frequency [229]. Figure 6.25 shows the measured FROG traces, autocorrelation, and spectrum that were used to characterize the pulse train.

The underlying physical mechanism that initiates the formation of the pulse train through multiple FWM is known as the *induced modulation instability* (see Section 5.1

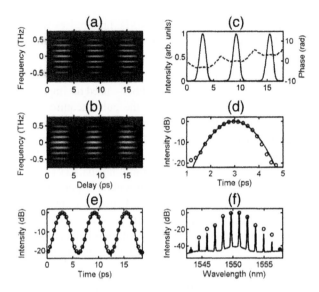

Figure 6.25: (a) Measured and (b) retrieved FROG traces of the 160-GHz pulse; (c) intensity and phase profiles deduced from FROG traces and (d) a Gaussian fit to the pulse shape; (e) autocorrelation and (f) spectrum of the pulse train at fiber output together with the values (circles) deduced from the FROG data. (From Ref. [229]; ©2002 OSA.)

of Ref. [2]). This instability can be interpreted as a FWM process and is responsible for providing the gain to the 160-GHz AM sidebands of the input optical beat signal. As these sidebands grow in amplitude, they can act interact with the input pumps through another FWM process and create additional sidebands in the optical spectrum of Figure 6.25(f). The buildup of these sidebands is accompanied with the compression of pulses and results in the 1.3-ps pulse train appearing in Figure 6.25(c). Notice that this entire process is closely related to the formation of pulse trains inside a DDF (see Section 6.6.1). By 2006, the use of multiple FWM processes produced 170-fs pulses at a repetition rate of 1 THz [230].

6.7.4 Fiber-Loop Mirrors and Other Devices

The use of nonlinear fiber-loop mirrors and interferometers for optical switching has been discussed in Chapter 3. The intensity-dependent transmission of these devices can be used for pulse shaping and compression [231]–[238]. For example, compressed pulses are often accompanied by a broad pedestal (see Section 6.3). If such a low-quality pulse is passed through a nonlinear fiber-loop mirror, the central, intense part of the pulse can be separated from the low-power pedestal [231]. In general, any nonsoliton pulse is compressed by a Sagnac interferometer designed to transmit the central part while blocking the low-intensity pulse wings. A fiber amplifier is sometimes incorporated within the fiber loop if the input pulse energy is below the switching threshold. Compression factors are relatively small (two or less) since the technique works by clipping pulse wings [232].

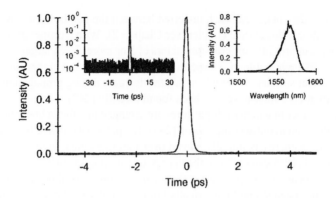

Figure 6.26: Autocorrelation trace of 170-fs pulses compressed using a fiber-loop mirror. Long-range autocorrelation on a log scale (left inset) and pulse spectrum (right inset) are also shown. (From Ref. [236]; ©2001 IEEE.)

The use of multiple sections with different dispersion characteristics inside the Sagnac loop (dispersion management) can improve the performance of such pulse compressors [233]. In a 1998 experiment, the use of a dispersion-imbalanced fiber-loop mirror provided a net compression factor of 80 and resulted in the formation of nearly unchirped 270-fs pulses with a negligible pedestal [234]. The loop employed two 15-m-long pieces of standard and fiber and dispersion-shifted fibers with a net normal dispersion of about 0.84 ps^2/km. It compressed 22-ps pulses that were first chirped through SPM to 2.5 ps. As these pulses were still linearly chirped, they could be compressed further down to 270 fs using a fiber with large normal dispersion.

Several variations of this approach are possible. In a 1999 experiment, a dispersion-flattened fiber was employed inside a dispersion-imbalanced fiber-loop mirror to realize pulse compression over a wide wavelength range extending from 1530 to 1565 nm [236]. In a later experiment, the performance was improved further by using several different types of polarization-maintaining fibers. Such a fiber-loop mirror compressed 3-ps pulses down to 140 fs, a value that was limited by the higher-order dispersive effects within the loop. These pulse were unchirped and amplified using an optical filer with an EDFA, resulting in 170-fs transform-limited pulses. Figure 6.26 shows the autocorrelation and the optical spectrum at the amplifier output. The peak-to-pedestal ratio of more than 35 dB indicates the high quality of compressed pulses. The use of a DDF can inside a fiber-loop mirror has also been proposed for the soliton-effect compression to obtain high-quality compressed pulses [237].

A dual-core fiber also exhibits nonlinear switching between its two cores (see Section 2.3), which can be exploited for pulse shaping and compression [239]–[242]. The use of such a fiber in a Sagnac-loop configuration offers several advantages since it combines two nonlinear mechanisms in a single device. Numerical simulations show that it can be used for pulse shaping, pedestal suppression, and pulse compression [239]. An interesting feature of such a device is that one can introduce additional coupling between the two cores by twisting the dual-core fiber within the loop. The operation of the device is quite complex and requires the solution of coupled NLS equations, since the energy can be transferred between the two cores in a way similar to that of a nonlinear directional coupler (see Chapter 2). With a proper design, a dual-core fiber loop can provide compression factors of 5 or more. Even larger compression factors (up to 20) are possible in symmetric dual-core fibers using higher-order solitons [241]. Self-similar solitary waves whose widths decrease during propagation inside an asymmetric twin-core fiber have also been found recently [242].

A multicore fiber in which multiple cores are arranged in either a linear or a circular fashion can also lead to pulse compression [243]. If a pulse is launched into such a fiber array such that its energy is initially spread over multiple cores, the nonlinear effects such as SPM and XPM can collapse the energy distribution in such a way that almost all the energy appears in a single core in the form of a compressed pulse. Compression in fiber arrays has been studied numerically but not yet observed experimentally.

Problems

6.1 Explain the operation of a grating-fiber compressor. Use diagrams as necessary.

6.2 Derive an expression for the effective GVD coefficient of a grating pair.

6.3 Develop a computer simulation program capable of modeling the performance of a grating-fiber compressor. Use the split-step Fourier method for solving Eq. (6.2.5) and implement the action of a grating pair through Eq. (6.2.4). Comment on the validity of Eqs. (6.2.8)–(6.2.10) for $N = 10$

6.4 A 100-ps pulse with 1-μJ energy (emitted from a 1.06-μm Nd:YAG laser) is compressed by using a grating-fiber compressor. Estimate the maximum compression factor, optimum fiber length, and optimum grating separation for typical parameter values.

6.5 Solve numerically the NLS equation that models the performance of a soliton-effect compressor. Find the maximum compression factor and the optimum fiber length when a 10th-order soliton is launched and the soliton period is 10 km.

6.6 What is the origin of pedestal formation in soliton-effect compressors? How can the pedestal be removed from the compressed pulse?

6.7 Solve numerically the generalized NLS equation and reproduce the results shown in Figure 6.11 for the 10th-order soliton.

6.8 How can a chirped fiber grating be used for pulse compression? Estimate the dispersion provided by such a grating if its period changes by 0.1% over a 10-cm length. Assume $\lambda = 1.55 \ \mu$m and $\bar{n} = 1.45$.

6.9 What is meant by chirped-pulse amplification? How can this technique be used to produce high-energy ultrashort pulses?

6.10 Explain how a hollow-core microstructured silica fiber can confine light to its core containing air. Why is such a fiber useful for chirped-pulse amplification?

6.11 Solve Eq. (6.6.1) in the case of a fundamental soliton ($N = 1$) propagating inside a DDF with $d(\xi) = \exp(-\xi)$ over a distance up to $\xi = 3$. Explain why the pulse compresses inside such a fiber.

6.12 Dispersion of a fiber decreases exponentially from 20 to 1 ps/(km-nm) over a length of 1 km. Estimate the compression factor for a 1-ps pulse launched as a fundamental soliton at the high-GVD end.

6.13 Solve Eqs. (6.7.1) and (6.7.2) numerically when the pump pulse, launched with $N = 30$ and experiencing normal dispersion, is used to compress a weak probe pulse experiencing anomalous dispersion. Assume that the two pulses have the same width initially and propagate at the same speed. Assume $\omega_2 = 0.8\omega_1$ and $\beta_{22} = -\beta_{21}$. Estimate the compression factor at $\xi = 0.05$.

6.14 Explain how a pulse can be compressed using the technique of Raman amplification. Reproduce the results shown in Figure 6.24 by solving the appropriate equations from Chapter 8 of Ref. [2].

References

[1] J. R. Klauder, A. C. Price, S. Darlington, and W. J. Albersheim, *Bell Syst. Tech. J.* **39**, 745 (1960).

[2] G. P. Agrawal, *Nonlinear Fiber Optics*, 4th ed. (Academic Press, San Diego, CA, 2007).

[3] F. Gires and P. Tournois, *Compt. Rend. Acad. Sci.* **258**, 6112 (1964).

[4] J. A. Giordmaine, M. A. Duguay, and J. W. Hansen, *IEEE J. Quantum Electron.* **4**, 252 (1968).

[5] M. A. Duguay and J. W. Hansen, *Appl. Phys. Lett.* **14**, 14 (1969).

[6] E. B. Treacy, *IEEE J. Quantum Electron.* **5**, 454 (1969).

[7] J. K. Wigmore and D. R. Grischkowsky, *IEEE J. Quantum Electron.* **14**, 310 (1978).

[8] R. A. Fisher, P. L. Kelley, and T. K. Gustafson, *Appl. Phys. Lett.* **14**, 140 (1969).

[9] H. Nakatsuka, D. Grischkowsky, and A. C. Balant, *Phys. Rev. Lett.* **47**, 910 (1981).

[10] C. V. Shank, R. L. Fork, R. Yen, R. H. Stolen, and W. J. Tomlinson, *Appl. Phys. Lett.* **40**, 761 (1982).

[11] B. Nikolaus and D. Grischkowsky, *Appl. Phys. Lett.* **42**, 1 (1983).

[12] J. G. Fujimoto, A. M. Weiner, and E. P. Ippen, *Appl. Phys. Lett.* **44**, 832 (1984).

[13] W. H. Knox, R. L. Fork, M. C. Downer, R. H. Stolen, C. V. Shank, and J. A. Valdmanis, *Appl. Phys. Lett.* **46**, 1120 (1985).

[14] R. L. Fork, C. H. Brito Cruz, P. C. Becker, and C. V. Shank, *Opt. Lett.* **12**, 483 (1987).

[15] A. M. Johnson, R. H. Stolen, and W. M. Simpson, *Appl. Phys. Lett.* **44**, 729 (1984).

[16] E. M. Dianov, A. Y. Karasik, P. V. Mamyshev, G. I. Onischukov, A. M. Prokhorov, M. F. Stel'makh, and A. A. Fomichev, *Sov. J. Quantum Electron.* **14**, 726 (1984).

[17] J. D. Kafka, B. H. Kolner, T. Baer, and D. M. Bloom, *Opt. Lett.* **9**, 505 (1984).

[18] J. P. Heritage, R. N. Thurston, W. J. Tomlinson, A. M. Weiner, and R. H. Stolen, *Appl. Phys. Lett.* **47**, 87 (1985).

[19] A. S. L. Gomes, W. Sibbett, and J. R. Taylor, *Opt. Lett.* **10**, 338 (1985).

[20] K. J. Blow, N. J. Doran, and B. P. Nelson, *Opt. Lett.* **10**, 393 (1985).

[21] K. Tai and A. Tomita, *Appl. Phys. Lett.* **48**, 309 (1986).

[22] B. Zysset, W. Hodel, P. Beaud, and H. P. Weber, *Opt. Lett.* **11** 156 (1986).

[23] B. Valk, K. Vilhelmsson, and M. M. Salour, *Appl. Phys. Lett.* **50**, 656 (1987).

[24] E. M. Dianov, A. Y. Karasik, P. V. Mamyshev, A. M. Prokhorov, and D. G. Fursa, *Sov. J. Quantum Electron.* **17**, 415 (1987).

[25] J. M. McMullen, *Appl. Opt.* **18**, 737 (1979).

[26] I. P. Cristov and I. V. Tomov, *Opt. Commun.* **58**, 338 (1986).

[27] S. D. Brorson and H. A. Haus, *Appl. Opt.* **27**, 23 (1988); *J. Opt. Soc. Am. B* **5**, 247 (1988).

[28] P. Tournois, *Electron. Lett.* **29**, 1414 (1993); *Opt. Commun.* **106**, 253 (1994).

[29] J. Debois, F. Gires, and P. Tournois, *IEEE J. Quantum Electron.* **9**, 213 (1973).

[30] O. E. Martinez, J. P. Gordon, and R. L. Fork, *J. Opt. Soc. Am. A* **1**, 1003 (1984).

[31] J. D. Kafka and T. Baer, *Opt. Lett.* **12**, 401 (1987).

[32] M. Nakazawa, T. Nakashima, H. Kubota, and S. Seikai, *J. Opt. Soc. Am. B* **5**, 215 (1988).

[33] V. E. Pozhar and V. I. Pustovoit, *Sov. J. Quantum Electron.* **17**, 509 (1987).

[34] R. Kashyap, *Fiber Bragg Gratings* (Academic Press, San Diego, CA, 1999).

[35] D. Grischkowsky and A. C. Balant, *Appl. Phys. Lett.* **41**, 1 (1982).

[36] R. Meinel, *Opt. Commun.* **47**, 343 (1983).

[37] W. J. Tomlinson, R. H. Stolen, and C. V. Shank, *J. Opt. Soc. Am. B* **1**, 139 (1984).

[38] E. M. Dianov, Z. S. Nikonova, A. M. Prokhorov, and V. N. Serkin, *Sov. Tech. Phys. Lett.* **12**, 311 (1986).

[39] E. A. Golovchenko, E. M. Dianov, P. V. Mamyshev, and A. M. Prokhorov, *Opt. Quantum Electron.* **20**, 343 (1988).

[40] E. M. Dianov, L. M. Ivanov, P. V. Mamyshev, and A. M. Prokhorov, *Sov. J. Quantum Electron.* **19**, 197 (1989).

[41] D. G. Fursa, P. V. Mamyshev, and A. M. Prokhorov, *Sov. Lightwave Commun.* **2**, 59 (1992).

[42] M. Stern, J. P. Heritage, and E. W. Chase, *IEEE J. Quantum Electron.* **28**, 2742 (1992).

[43] R. F. Mols and G. J. Ernst, *Opt. Commun.* **94**, 509 (1992).

[44] E. M. Dianov, A. B. Grudinin, A. M. Prokhorov, and V. N. Serkin, in *Optical Solitons—Theory and Experiment*, J. R. Taylor, Ed. (Cambridge University Press, Cambridge, UK, 1992), Chap. 7.

[45] M. Karlsson, *Opt. Commun.* **112**, 48 (1994).

[46] D. Mestdagh, *Appl. Opt.* **26**, 5234 (1987).

[47] V. A. Vysloukh and L. K. Muradyan, *Sov. J. Quantum Electron.* **17**, 915 (1987).

[48] W. J. Tomlinson and W. H. Knox, *J. Opt. Soc. Am. B* **4**, 1404 (1987).

[49] H. Kubota and M. Nakazawa, *Opt. Commun.* **66**, 79 (1988).

[50] T. Nakashima, M. Nakazawa, K. Nishi, and H. Kubota, *Opt. Lett.* **12**, 404 (1987).

[51] A. M. Weiner, J. P. Heritage, and R. H. Stolen, *J. Opt. Soc. Am. B* **5**, 364 (1988).

[52] A. S. L. Gomes, A. S. Gouveia-Neto, and J. R. Taylor, *Opt. Quantum Electron.* **20**, 95 (1988).

[53] M. Kuckartz, R. Schulz, and H. Harde, *J. Opt. Soc. Am. B* **5**, 1353 (1988).

[54] A. P. Vertikov, P. V. Mamyshev, and A. M. Prokhorov, *Sov. Lightwave Commun.* **1**, 363 (1991).

[55] J. P. Heritage, A. M. Weiner, and R. H. Thurston, *Opt. Lett.* **10**, 609 (1985).

[56] A. M. Weiner, J. P. Heritage, and R. N. Thurston, *Opt. Lett.* **11**, 153 (1986).

[57] R. N. Thurston, J. P. Heritage, A. M. Weiner, and W. J. Tomlinson, *IEEE J. Quantum Electron.* **22**, 682 (1986).

[58] R. H. Stolen, J. Botineau, and A. Ashkin, *Opt. Lett.* **7**, 512 (1982).

[59] M. Haner and W. S. Warren, *Appl. Phys. Lett.* **52**, 1458 (1988).

[60] L. Xu, C. Spielmann, F. Krausz, and R. Szipöcs, *Opt. Lett.* **21**, 1259 (1996).

[61] A. Baltuska, Z. Wei, M. S. Pshenichnikov, and D. A. Wiersma, *Opt. Lett.* **22**, 102 (1997).

[62] M. Nisoli, S. De Silvestri, O. Svelto, R. Szipcs, K. Ferencz, Ch. Spielmann, S. Sartania, and F. Krausz, *Opt. Lett.* **22**, 1562 (1997).

[63] A. Baltuska, Z. Wei, M. S. Pshenichnikov, D. A. Wiersma, and R. Szipöcs, *Appl. Phys. B* **65**, 175 (1997).

[64] L. Xu, N. Karasawa, N. Nakagawa, R. Morita, H. Shigekawa, and M. Yamashita, *Opt. Commun.* **162**, 256 (1999).

[65] G. Steinmeyer, D. H. Sutter, L. Gallmann, N. Matuschek, and U. Keller, *Science* **286**, 1507 (1999).

[66] T. Brabec and F. Krausz, *Rev. Mod. Phys.* **72**, 545 (2000).

[67] G. Cerullo, S. De Silvestri, M. Nisoli, S. Sartania, S. Stagira, and O. Svelto, *IEEE J. Sel. Topics Quantum Electron.* **6**, 948 (2000).

[68] B. Schenkel, J. Biegert, U. Keller, C. Vozzi, M. Nisoli, G. Sansone, S. Stagira, S. De Silvestri, and O. Svelto, *Opt. Lett.* **28**, 1987 (2003).

[69] K. Yamane, Z. Zhang, K. Oka, R. Morita, M. Yamashita, and A. Suguro, *Opt. Lett.* **28**, 2258 (2003).

[70] G. Tempea, F. Krausz, C. Spielmann, and T. Barbec, *IEEE J. Sel. Topics Quantum Electron.* **4**, 193 (1998).

[71] T. Sudmeyer, F. Brunner, E. Innerhofer, R. Paschotta, K. Furusawa, J. C. Baggett, T. M. Monro, D. J. Richardson, and U. Keller, *Opt. Lett.* **28**, 1951 (2003).

[72] G. McConnell and E. Riis, *Appl. Phys. B* **78**, 557 (2004).

[73] F. Druon and P. Georges, *Opt. Express* **13**, 3383 (2005).

[74] L. F. Mollenauer, R. H. Stolen, J. P. Gordon, and W. J. Tomlinson, *Opt. Lett.* **8**, 289 (1983).

[75] E. M. Dianov, A. Y. Karasik, P. V. Mamyshev, G. I. Onischukov, A. M. Prokhorov, M. F. Stel'makh, and A. A. Fomichev, *JETP Lett.* **40**, 903 (1984).

[76] E. M. Dianov, Z. S. Nikonova, A. M. Prokhorov, and V. N. Serkin, *Sov. Tech. Phys. Lett.* **12**, 310 (1986).

[77] P. Beaud, W. Hodel, B. Zysset, and H. P. Weber, *IEEE J. Quantum Electron.* **23**, 1938 (1987).

[78] K. Tai and A. Tomita, *Appl. Phys. Lett.* **48**, 1033 (1986).

[79] A. S. Gouveia-Neto, A. S. L. Gomes, and J. R. Taylor, *Opt. Lett.* **12**, 395 (1987).

[80] A. S. Gouveia-Neto, A. S. L. Gomes, and J. R. Taylor, *J. Mod. Opt.* **35**, 7 (1988).

[81] E. M. Dianov, Z. S. Nikinova, and V. N. Serkin, *Sov. J. Quantum Electron.* **19**, 937 (1989).

[82] G. P. Agrawal, *Opt. Lett.* **15**, 224 (1990).

[83] P. V. Mamyshev, S. V. Chernikov, E. M. Dianov, and A. M. Prokhorov, *Opt. Lett.* **15**, 1365 (1990).

[84] N. N. Akhmediev and N. V. Mitzkevich, *IEEE J. Quantum Electron.* **27**, 849 (1991).

[85] W. H. Xiang, S. R. Friberg, K. Watanabe, S. Machida, Y. Sakai, H. Iwamura, and Y. Yamamoto, *Appl. Phys. Lett.* **59**, 2076 (1991).

[86] P. V. Mamyshev, in *Optical Solitons—Theory and Experiment*, J. R. Taylor, Ed. (Cambridge University Press, Cambridge, UK, 1992), Chap. 8.

[87] K. C. Chan and H. F. Liu, *IEEE J. Quantum Electron.* **31**, 2226 (1995).

[88] K. C. Chan and W. H. Cao, *J. Opt. Soc. Am. B* **15**, 2371 (1998).

[89] Y. Matsui, M. D. Pelusi, and A. Suzuki, *IEEE Photon. Technol. Lett.* **11**, 1217 (1999).

[90] D. Huhse, O. Reimann, E. H. Bottcher, and D. Bimberg, *Appl. Phys. Lett.* **75**, 2530 (1999).

[91] B. Kibler, C. Billet, P.-A. Lacourt, R. Ferriere, and J. M. Dudley, *IEEE Photon. Technol. Lett.* **18**, 1831 (2006).

[92] J. H. V. Price, K. Furasawa, T. M. Monro, L. Lefort, and D. J. Richardson, *J. Opt. Soc. Am. B* **19**,1286 (2002).

[93] G. Chang, T. B. Norris, and H. G. Winful, *Opt. Lett.* **28**, 546 (2003).

[94] J. Dudley and S. Coen, *Opt. Express* **12**, 2423 (2004).

[95] J. W. Nicholson, A. D. Yablon, P. S. Westbrook, K. S. Feder, and M. F. Yan, *Opt. Express* **12**, 3025 (2004).

[96] B. Schenkel, R. Paschotta, and U. Keller, *J. Opt. Soc. Am. B* **22**, 687 (2005).

[97] M. A. Foster, A. L. Gaeta, Q. Cao, and R. Trebino, *Opt. Express* **13**, 6848 (2005).

[98] J. Takayanagi, N. Nishizawa, T. Sugiura, M. Yoshida, and T. Goto, *IEEE J. Quantum Electron.* **42**, 287 (2006).

[99] F. Ouellette, *Opt. Lett.* **12**, 847 (1987); *Appl. Opt.* **29**, 4826 (1990); *Opt. Lett.* **16**, 303 (1991).

[100] H. G. Winful, *Appl. Phys. Lett.* **46**, 527 (1985).

[101] D. S. Peter, W. Hodel, and H. P. Weber, *Opt. Commun.* **112**, 59 (1994).

[102] B. J. Eggleton, K. A. Ahmed, L. Poladian, K. A. Ahmed, and H. F. Liu, *Opt. Lett.* **19**, 877 (1994).

[103] J. A. R. William, I. Bennion, K. Sugden, and N. Doran, *Electron. Lett.* **30**, 985 (1994).

[104] R. Kashyap, S. V. Chernikov, P. F. McKee, and J. R. Taylor, *Electron. Lett.* **30**, 1078 (1994).

[105] P. A. Krug, T. Stefans, G. Yoffe, F. Ouellette, P. Hill, and G. Dhosi, *Electron. Lett.* **31**, 1091 (1995).

[106] C. M. de Sterke, *Opt. Lett.* **17**, 914 (1992); M. J. Steel and C. M. de Sterke, *Phys. Rev. A* **49**, 5048 (1994).

[107] S. V. Chernikov, J. R. Taylor, and R. Kashyap, *Opt. Lett.* **20**, 1586 (1995).

[108] D. Taverner, D. J. Richardson, M. N. Zervas, L. Reekie, L. Dong, and J. L. Cruz, *IEEE Photon. Technol. Lett.* **7**, 1436 (1995).

[109] K. Tamura, T. Komukai, T. Yamamoto, T. Imai, E. Yoshida, amd M. Nakazawa, *Electron. Lett.* **31**, 2194 (1995).

[110] C. M. de Sterke, N. G. R. Broderick, B. J. Eggleton, and M. J. Steel, *Opt. Fiber Technol.* **2**, 253 (1996).

[111] C. M. de Sterke, B. J. Eggleton, and P. A. Krug, *J. Lightwave Technol.* **15**, 1494 (1997).

[112] N. G. R. Broderick, D. Taverner, D. J. Richardson, M. Isben, and R. I. Laming, *Phys. Rev. Lett.* **79**, 4566 (1997); *Opt. Lett.* **22**, 1837 (1997).

[113] G. Lenz, B. J. Eggleton, and N. M. Litchinitser, *J. Opt. Soc. Am. B* **15**, 715 (1998).

[114] G. Lenz and B. J. Eggleton, *J. Opt. Soc. Am. B* **15**, 2979 (1998).

[115] B. J. Eggleton, G. Lenz, R. E. Slusher, and N. M. Litchinitser, *Appl. Opt.* **37**, 7055 (1998).

[116] B. J. Eggleton, G. Lenz, and N. M. Litchinitser, *Fiber Integ. Opt.* **19**, 383 (2000).

[117] J. T. Mok, I. C. M. Littler, E. Tsoy, and B. J. Eggleton, *Opt. Lett.* **30**, 2457 (2005).

[118] L. Fu, A. Fuerbach, I. C. M. Littler, and B. J. Eggleton, *Appl. Phys. Lett.* **88**, 081116 (2006).

[119] G. P. Agrawal, *Fiber-Optic Communication Systems*, 3rd ed. (Wiley, Hoboken, NJ, 2002).

[120] R. A. Fisher and W. K. Bischel, *Appl. Phys. Lett.* **24**, 468 (1974); *IEEE J. Quantum Electron.* **11**, 46 (1975).

[121] B. Strickland and G. Mourou, *Opt. Commun.* **55**, 447 (1985).

[122] P. Maine, D. Strickland, P. Bado, M. Pessot, and G. Mourou, *IEEE J. Quantum Electron.* **24**, 398 (1988).

[123] M. Pessot, J. Squier, G. Mourou, and D. J. Harter, *Opt. Lett.* **14**, 797 (1989).

[124] M. Ferray, L. A. Lompre, O. Gobert, A. L. Huillier, G. Mainfray, C. Manus, A. Sanchez, and A. S. Gomes, *Opt. Commun.* **75**, 278 (1990).

[125] K. I. Yamakawa, C. P. J. Barty, H. Shiraga, and Y. Kato, *IEEE J. Quantum Electron.* **27**, 288 (1991).

[126] W. Hodel, D. S. Peter, H. P. Weber, *Opt. Commun.* **97**, 233 (1993).

[127] M. L. Stock and G. Mourou, *Opt. Commun.* **106**, 249 (1994).

[128] M. E. Fermann, A. Galvanauskas, and D. Harter, *Appl. Phys. Lett.* **64**, 1315 (1994).

[129] L. M. Yang, T. Sosnowski, M. L. Stock, T. B. Norris, J. Squier, G. Mourou, M. L. Dennis, and I. N. Duling III, *Opt. Lett.* **20**, 1044 (1995).

[130] A. Galvanauskas, M. E. Fermann, D. Harter, K. Sugden, and I. Bennion, *Appl. Phys. Lett.* **66**, 1053 (1995).

[131] J. D. Minelly, A. Galvanauskas, M. E. Fermann, D. Harter, J. E. Caplan, Z. J. Chen, and D. N. Payne, *Opt. Lett.* **20**, 1797 (1995).

[132] S. Kane and J. Squier, *IEEE J. Quantum Electron.* **31**, 2052 (1995).

[133] A. Galvanauskas, P. A. Krug, and D. Harter, *Opt. Lett.* **21**, 1049 (1996).

[134] D. S. Peter, W. Hodel, and H. P. Weber, *Opt. Commun.* **130**, 75 (1996).

[135] A. Galvanauskas, D. Harter, M. A. Arbore, M. H. Chou, and M. M. Fejer, *Opt. Lett.* **23**, 1695 (1998).

[136] N. G. R. Broderick, D. J. Richardson, D. Taverner, J. E. Caplen, L. Dong, and M. Ibsen, *Opt. Lett.* **24**, 566 (1999).

[137] M. Hofer, M. E. Fermann, A. Galvanauskas, D. Harter, and R. S. Windeler, *IEEE Photon. Technol. Lett.* **11**, 650 (1999).

[138] A. Galvanauskas, *IEEE J. Sel. Topics Quantum Electron.* **7**, 504 (2001).

[139] J. Limpert, F. Roser, T. Schreiber, and A. Tünnermann, *IEEE J. Sel. Topics Quantum Electron.* **12**, 233 (2006).

[140] C. J. S. de Matos, J. R. Taylor, T. P. Hansen, K. P. Hansen, and J. Broeng, *Opt. Express* **11**, 2832 (2003).

[141] J. Limpert, T. Schreiber, S. Nolte, H. Zellmer, and A. Tünnermann, *Opt. Express* **11**, 3332 (2003).

[142] C. J. S. de Matos and J. R. Taylor, *Opt. Express* **12**, 405 (2004).

[143] G. Imeshev, I. Hartl, and M. E. Fermann, *Opt. Express* **12**, 6508 (2004).

[144] C. J. S. de Matos, R. E. Kennedy, S. V. Popov, and J. R. Taylor, *Opt. Lett.* **30**, 436 (2005).

[145] A. Shirakawa, M. Tanisho, and K. Ueda, *Opt. Express* **14**, 12039 (2006).

[146] C. J. S. de Matos and J. R. Taylor, *Opt. Express* **13**, 2828 (2005).

[147] A. Dubietis, R. Butkus, and A. P. Piskarskas, *IEEE J. Sel. Topics Quantum Electron.* **12**, 163 (2006).

[148] M. Hanna, F. Druon, P. Georges, *Opt. Express* **14**, 2783 (2006).

[149] F. Yaman, Q. Lin, and G. P. Agrawal, in *Guided Wave Optical Components and Devices*, B. P. Pal, Ed. (Academic Press, Boston, 2005), Chap. 7.

[150] H. H. Kuehl, *J. Opt. Soc. Am. B* **5**, 709 (1988).

[151] S. V. Chernikov and P. V. Mamyshev, *J. Opt. Soc. Am. B* **8**, 1633 (1991).

[152] S. V. Chernikov, D. J. Richardson, E. M. Dianov, and D. N. Payne, *Electron. Lett.* **28**, 1842 (1992).

[153] S. V. Chernikov, E. M. Dianov, D. J. Richardson, and D. N. Payne, *Opt. Lett.* **18**, 476 (1993).

[154] P. V. Mamyshev, P. G. J. Wigley, J. Wilson, G. I. Stegeman, V. A. Smenov, E. M. Dianov, and S. I. Miroshnichenko, *Phys. Rev. Lett.* **71**, 73 (1993).

[155] E. M. Dianov, P. V. Mamyshev, A. M. Prokhorov, and S. V. Chernikov, *Opt. Lett.* **14**, 1008 (1989).

[156] P. V. Mamyshev, S. V. Chernikov, and E. M. Dianov, *IEEE J. Quantum Electron.* **27**, 2347 (1991).

[157] S. V. Chernikov, J. R. Taylor, P. V. Mamyshev, and E. M. Dianov, *Electron. Lett.* **28**, 931 (1992).

[158] S. V. Chernikov, D. J. Richardson, and R. I. Laming, *Appl. Phys. Lett.* **63**, 293 (1993).

[159] P. Schell, D. Bimberg, V. A. Bogatyrjov, E. M. Dianov, A. S. Kurkov, V. A. Semenov, and A. A. Sysoliatin, *IEEE Photon. Technol. Lett.* **6**, 1191 (1994).

[160] A. V. Shipulin, E. M. Dianov, R. J. Richardson, and D. N. Payne, *IEEE Photon. Technol. Lett.* **6**, 1380 (1994).

[161] K. Suzuki, K. Iwatsuki, S. Nishi, M. Samwatari, and T. Kitoh, *IEEE Photon. Technol. Lett.* **6**, 352 (1994).

[162] M. Nakazawa, E. Yoshida, K. Kubota, and Y. Kimura, *Electron. Lett.* **30**, 2038 (1994).

[163] A. J. Stentz, R. W. Boyd, and A. F. Evans, *Opt. Lett.* **20**, 1770 (1995).

[164] R. J. Essiambre and G. P. Agrawal, *Opt. Lett.* **21**, 116 (1996).

[165] A. Mostofi, H. Hatami-Hanza, and P. L. Chu, *IEEE J. Quantum Electron.* **33**, 620 (1997).

[166] M. D. Pelusi and H. F. Liu, *IEEE J. Quantum Electron.* **33**, 1430 (1997).

[167] J. Wu, C. Lou, Y. Gao, W. Xu, Q. Guo, C. Liao, and S. Liu, *High Technol. Lett.* **8**, 8 (1998).

[168] K. I. M. McKinnon, N. F. Smyth, and A. L. Worthy, *J. Opt. Soc. Am. B* **16**, 441 (1999).

[169] K. R. Tamura and M. Nakazawa, *IEEE Photon. Technol. Lett.* **11**, 319 (1999).

[170] Z. Shumin, L. Fuyuna, X. Wenchengc, Y. Shipingb, W. Jiana, and D. Xiaoyi, *Opt. Commun.* **237**, 1 (2004).

[171] P. C. Reeves-Hall, S. A. E. Lewis, S. V. Chernikov, and J. R. Taylor, *Electron. Lett.* **36**, 622 (2000).

[172] P. C. Reeves-Hall and J. R. Taylor, *Electron. Lett.* **37**, 417 (2001).

[173] T. E. Murphy, *IEEE Photon. Technol. Lett.* **14**, 1424 (2002).

[174] T. Kogure, J. H. Lee, and D. J. Richardson, *IEEE Photon. Technol. Lett.* **16**, 1167 (2004).

[175] J. H. Lee, Y. G. Han, S. B. Lee, T. Kogure, and D. Richardson, *Opt. Express* **12**, 2187 (2004).

[176] J. Hu, B. S. Marks, C. R. Menyuk, J. Kim, T. F. Carruthers, B. M. Wright, T. F. Taunay, and E. J. Friebele, *Opt. Express* **14**, 4026 (2006).

[177] M. L. V. Tse, P. Horak, J. H. V. Price, F. Poletti, F. He, and D. J. Richardson, *Opt. Lett.* **31**, 3504 (2006).

[178] S. V. Chernikov, J. R. Taylor, and R. Kashyap, *Electron. Lett.* **29**, 1788 (1993); *Electron. Lett.* **30**, 433 (1994); *Opt. Lett.* **19**, 539 (1994).

[179] J. P. Wang, Y. Wu, C. Y. Lou, and Y. Z. Gao, *Opt. Eng.* **42**, 2380 (2003).

[180] K. Igarashi, J. Hiroishi, T. Yagi and S. Namiki, *Electron. Lett.* **41**, 688 (2005).

[181] K. Igarashi, H. Tobioka, H. Takahashi, T. Yagi, and S. Namiki, *Electron. Lett.* **41**, 797 (2005).

[182] Y. Ozeki, S. Takasaka, T. Inoue, K. Igarashi, J. Hiroishi, R. Sugizaki, M. Sakano, and S. Namiki, *IEEE Photon. Technol. Lett.* **17**, 1698 (2005).

[183] T. Inoue, H. Tobioka, K. Igarashi, and S. Namiki, *J. Lightwave Technol.* **24**, 2510 (2006).

[184] Y. Ozeki and T. Inoue, *Opt. Lett.* **31**, 1606 (2006).

[185] D. Méchin, S. H. Im, V. I. Kruglov, and J. D. Harvey, *Opt. Lett.* **31**, 2106 (2006).

[186] V. I. Kruglov, A. C. Peacock, and J. D. Harvey, *Phys. Rev. E* **71**, 056619 (2005).

[187] E. M. Dianov, P. V. Mamyshev, A. M. Prokhorov, and S. V. Chernikov, *Sov. J. Quantum Electron.* **18**, 1211 (1988).

[188] J. T. Manassah, *Opt. Lett.* **13**, 755 (1988).

[189] G. P. Agrawal, P. L. Baldeck, and R. R. Alfano, *Opt. Lett.* **14**, 137 (1989); *Phys. Rev. A* **40**, 5063 (1989).

[190] Q. Z. Wang, P. P. Ho, and R. R. Alfano, *Opt. Lett.* **15**, 1023 (1990); *Opt. Lett.* **16**, 496 (1991).

[191] J. E. Rothenberg, *Opt. Lett.* **15**, 495 (1990).

[192] A. D. Ellis and D. M. Patrick, *Electron. Lett.* **29**, 149 (1993).

[193] C. Headley III and G. P. Agrawal, *J. Opt. Soc. Am. B* **10**, 2383 (1993); *IEEE J. Quantum Electron.* **31**, 2058 (1995).

[194] R. F. de Souza, E. J. S. Fonseca, M. J. Hickmann, and A. S. Gouveia-Neto, *Opt. Commun.* **124**, 79 (1996).

[195] W. H. Cao and Y. W. Zhang, *Opt. Commun.* **128**, 23 (1996).

[196] C. Yeh and L. Bergman, *Phys. Rev. E* **57**, 2398 (1998).

[197] M. Yamashita, H. Sone, R. Morita, and H. Shigekawa, *IEEE J. Quantum Electron.* **34**, 2145 (1998).

[198] C. G. Durfee III, S. Backus, H. C. Kapteyn, and M. M. Murnane, *Opt. Lett.* **24**, 697 (1999).

[199] K. C. Chan and W. H. Cao, *Opt. Commun.* **178**, 79 (2000).

[200] L. Guo and C. Zhou, *Opt. Commun.* **257**, 180 (2006).

[201] A. Takada, T. Suigi, and M. Saruwatari, *Electron. Lett.* **22**, 1347 (1986).

[202] M. Nakazawa, K. Suzuki, and Y. Kimura, *Opt. Lett.* **15**, 588 (1990); *IEEE Photon. Technol. Lett.* **2**, 216 (1990).

[203] R. T. Hawkins, *Electron. Lett.* **26**, 292 (1990).

[204] H. F. Liu, Y. Ogawa, S. Oshiba, and T. Tonaka, *IEEE J. Quantum Electron.* **27**, 1655 (1991).

[205] J. T. Ong, R. Takahashi, M. Tsuchiya, S. H. Wong, R. T. Sahara, Y. Ogawa, and T. Kamiya, *IEEE J. Quantum Electron.* **29**, 1701 (1993).

[206] A. Galvanauskas, P. Blixt, and J. A. Tellefsen, Jr., *Appl. Phys. Lett.* **63**, 1742 (1993).

[207] L. Chusseau, *IEEE J. Quantum Electron.* **30**, 2711 (1994).

[208] K. A. Ahmed, K. C. Chan, and H. F. Liu, *IEEE J. Sel. Topics Quantum Electron.* **1**, 592 (1995).

[209] L. Chusseau and E. Delevaque, *IEEE J. Sel. Topics Quantum Electron.* **2**, 500 (1996).

[210] R. Yatsu, K. Taira, and M. Tsuchiya, *Opt. Lett.* **24**, 1172 (1999).

[211] P. M. Anandarajah, C. Guignard, A. Clarke, D. Reid, M. Rensing, L. P. Barry, G. Edvell, and J. D. Harvey, *IEEE J. Sel. Topics Quantum Electron.* **12**, 255 (2006).

[212] G. P. Agrawal and N. A. Olsson, *Opt. Lett.* **14**, 500 (1989).

[213] N. A. Olsson and G. P. Agrawal, *Appl. Phys. Lett.* **55**, 13 (1989).

[214] G. P. Agrawal and N. A. Olsson, *IEEE J. Quantum Electron.* **25**, 2297 (1989).

[215] N. A. Olsson, G. P. Agrawal, and K. W. Wecht, *Electron. Lett.* **25**, 603 (1989).

[216] K. J. Blow, N. J. Doran, and D. Wood, *J. Opt. Soc. Am. B* **5**, 381 (1988); *J. Opt. Soc. Am. B* **5**, 1301 (1988).

[217] P. A. Bélanger, L. Gagnon, and C. Paré, *Opt. Lett.* **14** 943 (1989); L. Gagnon and P. A. Bélanger, *Phys. Rev. A* **43**, 6187 (1991).

[218] G. P. Agrawal, *IEEE Photon. Technol. Lett.* **2**, 216 (1990); *Opt. Lett.* **16**, 226 (1991); *Phys. Rev. A* **44**, 7493 (1991).

[219] I. Y. Khrushchev, A. B. Grudinin, E. M. Dianov, D. V. Korobkin, V. A. Semenov, and A. M. Prokhorov, *Electron. Lett.* **26**, 456 (1990).

[220] M. Nakazawa, K. Kurokawa, H. Kobota, K. Suzuki, and Y. Kimura, *Appl. Phys. Lett.* **57**, 653 (1990).

[221] V. S. Grigoryan and T. S. Muradyan, *J. Opt. Soc. Am. B* **8**, 1757 (1991).

[222] V. V. Afanasjev, V. N. Serkin, and V. A. Vysloukh, *Sov. Lightwave Commun.* **2**, 35 (1992).

[223] E. Yoshida, Y. Kimura, and M. Nakazawa, *Jpn. J. Appl. Phys.* **32**, 3461 (1993).

[224] M. L. Quiroga-Teixeiro, D. Anderson, P. A. Andrekson, A. Berntson, and M. Lisak, *J. Opt. Soc. Am. B* **13**, 687 (1996).

[225] K. Tamura and M. Nakazawa, *Opt. Lett.* **21**, 68 (1996).

[226] D. B. S. Soh, J. Nilsson, and A. B. Grudinin, *J. Opt. Soc. Am. B* **23**, 1 (2006).

[227] C. Finot, G. Millot, S. Pitois, C. Billet, and J. M. Dudley, *IEEE J. Sel. Topics Quantum Electron.* **10**, 1211 (2004).

[228] Y. Yamamoto and M. Nakazawa, *IEEE Photon. Technol. Lett.* **9**, 1595 (1997).

[229] S. Pitois, J. Fatome, and G. Millot, *Opt. Lett.* **27**, 1729 (2002).

[230] J. Fatome, S. Pitois, and G. Millot, *IEEE J. Quantum Electron.* **42**, 1038 (2006).

[231] K. Smith, N. J. Doran, and P. G. J. Wigley, *Opt. Lett.* **15**, 1294 (1990).

[232] R. A. Betts, S. J. Frisken, C. A. Telford, and P. S. Atherton, *Electron. Lett.* **27**, 858 (1991).

[233] A. L. Steele, *Electron. Lett.* **29**, 1971 (1993).

[234] I. Y. Khrushchev, I. H. White, and R. V. Penty, *Electron. Lett.* **34**, 1009 (1998).

[235] M. D. Pelusi, Y. Matsui, and A. Suzuki, *IEEE J. Quantum Electron.* **35**, 867 (1999).

[236] K. R. Tamura and M. Nakazawa, *IEEE Photon. Technol. Lett.* **11**, 230 (1999); *IEEE Photon. Technol. Lett.* **13**, 526 (2001).

[237] P. K. A. Wai and W. Cao, *J. Opt. Soc. Am. B* **20**, 1346 (2003).

[238] J. H. Lee, T. Kogure, and D. J. Richardson, *IEEE J. Sel. Topics Quantum Electron.* **10**, 181 (2004).

[239] R. J. Essiambre and A. Vallé, *Can. J. Phys.* **71** , 11 (1993).

[240] J. H. Xy, Z. J. Fang, W. Z. Zhang, and F. X. Gan, *Fiber Integ. Opt.* **13**, 365 (1994).

[241] H. Hatami-Hanza, P. L. Chu, M. A. Malomed, and G. D. Peng, *Opt. Commun.* **134**, 59 (1997).

[242] T. S. Raju, P. K. Panigrahi, and K. Porsezian, *Phys. Rev. E* **71**, 026608 (2005); *Phys. Rev. E* **72**, 046612 (2005).

[243] A. B. Aceves, G. G. Luther, C. De Angelis, A. M. Rubenchik, and S. K. Turitsyn, *Opt. Fiber Technol.* **1**, 244 (1995); *Phys. Rev. Lett.* **75**, 73 (1995).

Chapter 7

Fiber-Optic Communications

Soon after the nonlinear effects in optical fibers were observed experimentally, it was realized that they would limit the performance of fiber-optic communication systems [1]. However, nonlinear effects were found to be mostly irrelevant for system design in the 1980s, because both the bit rate and link lengths were limited by fiber losses and group-velocity dispersion (GVD). The situation changed dramatically during the 1990s with the advent of optical amplification, dispersion management, and wavelength-division multiplexing (WDM). These advances increased link lengths to beyond 1000 km and single-channel bit rates to beyond 10 Gb/s. As a result, the nonlinear effects in optical fibers became of paramount concern for optimizing a lightwave system [2]–[5]. This chapter focuses on how the nonlinear effects influence the design of WDM lightwave systems. Loss- and dispersion-management techniques are discussed in Section 7.1 as an introduction to system-related issues. Section 7.2 is devoted to the impact of five major nonlinear effects: stimulated Brillouin and Raman scattering (SBS and SRS), self- and cross-phase modulation (SPM and XPM), and four-wave mixing (FWM). Section 7.3 focuses on optical solitons that employ SPM to advantage and can be used for designing WDM systems. Section 7.4 describes intrachannel nonlinear effects that become important for high-speed channels in pseudo-linear lightwave systems.

7.1 System Basics

All digital lightwave systems transmit information as a continuous stream of 1 and 0 bits. The bit rate B determines the duration of each bit, or the bit slot, as $T_B = 1/B$. An optical pulse is present in the time slot allocated to each 1 bit. It occupies a fraction of the bit slot when the return-to-zero (RZ) format is used. In the case of the nonreturn-to-zero (NRZ) format, the optical pulse occupies the entire bit slot. The most important issue for lightwave systems is how an optical bit stream is affected by losses, GVD, and various nonlinear effects as it propagates down the fiber link [6]–[8]. In this section the focus is on the effects of fiber losses and dispersion; the nonlinear effects are considered in later sections.

301

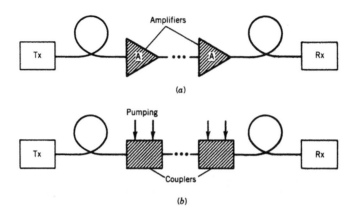

Figure 7.1: Fiber-optic links making use of (a) lumped and (b) distributed amplification schemes for compensating fiber losses.

7.1.1   Loss Management

Silica fibers exhibit minimum losses (about 0.2 dB/km) in the wavelength region near 1.55 μm [6]. For this reason, most modern lightwave systems operate near this wavelength to minimize the impact of fiber losses. Even then, the optical signal is attenuated by a factor of 100 or more over a link length of only 100 km. Since long-haul lightwave systems typically extend over thousands of kilometers, it is evident that fiber losses must be compensated periodically to boost the signal power back to its original value.

A common technique, used exclusively during the 1980s, regenerates the optical signal using a "repeater" in which the bit stream is first converted to the electric domain using an optical receiver and then regenerated with the help of an optical transmitter [6]. This technique becomes quite cumbersome (and expensive) for WDM systems as it requires demultiplexing of all channels at each repeater. As discussed in Chapter 4, fiber amplifiers can amplify multiple WDM channels simultaneously. For this reason, almost all WDM lightwave systems use optical amplifiers for compensating fiber losses. Figure 7.1 shows how amplifiers can be cascaded in a periodic manner to form a chain to extend transmission distances, while retaining the signal in its original optical form. Most systems employ erbium-doped fiber amplifiers (EDFAs) in which losses accumulated over 60–80 km are compensated by using short lengths ($\sim$10 m) of erbium-doped fibers. A distributed amplification scheme is also used for advanced systems. In this case, the Raman gain of the transmission fiber itself is used for compensating losses all along the fiber length. This scheme requires periodic injection of the pump power through fiber couplers as indicated in Figure 7.1(b).

The loss-management technique based on optical amplification degrades the signal-to-noise ratio (SNR) of the optical bit stream, because all amplifiers add noise to the signal through spontaneous emission. Mathematically, this noise can be included by adding a Langevin noise term to the Ginzburg–Landau equation of Section 4.3. If the two-photon absorption term in Eq. (4.3.15) is neglected because of its smallness, the amplification process is governed by

$$\frac{\partial A}{\partial z} + \frac{i}{2}(\beta_2 + ig_0 T_2^2)\frac{\partial^2 A}{\partial T^2} = i\gamma|A|^2 A + \frac{1}{2}(g_0 - \alpha)A + f_n(z, T), \qquad (7.1.1)$$

where g_0 is the gain coefficient and the T_2 term accounts for the decrease in the gain for spectral components of an optical pulse located far from the gain peak. The noise induced by spontaneous emission vanishes on average; that is, $\langle f_n(z, T) \rangle = 0$. If noise is modeled as a Markovian stochastic process with Gaussian statistics, its second moment can be written as

$$\langle f_n(z, T) f_n(z', T') \rangle = n_{sp} h v_0 g_0 \delta(z - z') \delta(T - T'), \qquad (7.1.2)$$

where n_{sp} is the spontaneous-emission factor introduced in Section 4.1 and $h v_0$ is the average photon energy. The two delta functions ensure that all spontaneous-emission events are independent of each other. The noise variance is higher by factor of 2 when both polarization components are considered [9].

In the case of distributed Raman amplification, Eq. (7.1.1) should be solved along the entire fiber link. However, when lumped amplifiers are used periodically, their length is so much shorter than the dispersion and nonlinear length scales that one can set α, β_2, and γ to zero in Eq. (7.1.1). If gain dispersion is also ignored by setting $T_2 = 0$, this equation can be integrated over the amplifier length l_a with the result

$$A_{out}(T) = \sqrt{G}A_{in}(T) + a_n(T), \qquad (7.1.3)$$

where $G = \exp(g_0 l_a)$ is the amplification factor of the lumped amplifier. The amplified spontaneous emission (ASE) noise added by the amplifier is given by

$$a_n(T) = \int_0^{l_a} f_n(z, T) \exp\left[\frac{1}{2}g_0(l_a - z)\right] dz. \qquad (7.1.4)$$

If we use Eq. (7.1.2), the second moment of $a_n(T)$ satisfies

$$\langle a_n(T) a_n(T') \rangle = S_{sp} \delta(T - T'), \qquad (7.1.5)$$

where $S_{sp} = (G - 1)n_{sp}h v_0$ is the spectral density of ASE, introduced earlier in Section 4.1.3. The delta function indicates that the ASE spectral density is frequency independent (white noise). In real amplifiers, the gain spectrum sets the bandwidth over which ASE occurs. An optical filter is often placed just after the amplifier to reduce the noise. If that is the case, noise is added only over the filter bandwidth, and the ASE power becomes

$$P_{sp} = \int_{-\infty}^{\infty} S_{sp} H_f(v - v_0) \, dv \approx S_{sp} \Delta v_f, \qquad (7.1.6)$$

where H_f is the transfer function, v_0 is the center frequency, and Δv_f is the 3-dB bandwidth (FWHM) of the filter.

In a chain of cascaded lumped amplifiers (see Figure 7.1), ASE accumulates from amplifier to amplifier and can build up to high levels [9]. If we assume that all amplifiers are spaced apart by a constant distance L_A and the amplifier gain $G \equiv \exp(\alpha L_A)$

is just large enough to compensate for fiber losses in each fiber section, the total ASE power for a chain of N_A amplifiers is given by

$$P_{\mathrm{sp}} = N_A S_{\mathrm{sp}} \Delta v_f = n_{\mathrm{sp}} h v_0 N_A (G-1) \Delta v_f. \qquad (7.1.7)$$

Clearly, ASE power can become quite large for large values of G and N_A. A side effect of high ASE levels is that at some point ASE begins to saturate amplifiers. Then, signal power begins to decrease while, at the same time, noise power keeps on increasing, resulting in severe degradation of the SNR. The ASE power can be controlled by reducing the amplifier spacing L_A. At first sight, this approach may appear counter-intuitive since it increases N_A. However, noting that $N_A = L/L_A = \alpha L / \ln G$, we find that P_{sp} scales with G as $(G-1)/\ln G$ and can be reduced by lowering the gain of each amplifier. The limit $L_A \to 0$ corresponds to the technique of distributed amplification. In practice, the amplifier spacing L_A cannot be made too small. Typically, L_A is below 50 km for undersea applications but can be increased to 80 km or so for terrestrial systems with link lengths under 3000 km.

7.1.2   Dispersion Management

The performance of any lightwave system is affected by dispersion-induced pulse broadening. The GVD effects can be minimized by operating the system close to the zero-dispersion wavelength λ_{ZD} of the fiber. However, it is not always practical to match the operating wavelength λ with λ_{ZD}. An example is provided by terrestrial operating near 1.55 μm. Such systems often use "standard" telecommunication fibers for which $\lambda_{\mathrm{ZD}} \approx 1.31$ μm. Since the GVD parameter $\beta_2 \approx -20$ ps²/km for such fibers in the 1.55-μm region, dispersion-induced pulse broadening limits the system performance severely. In the case of directly modulated distributed feedback (DFB) lasers, the transmission distance L is limited as $L < (4B|D|\sigma_\lambda)^{-1}$, where σ_λ is the root-mean-square (RMS) spectral width, with a value around 0.15 nm for directly modulated DFB lasers [6]. Using $D = 16$ ps/(km-nm), lightwave systems operating at 2.5 Gb/s are limited to $L \approx 42$ km. Indeed, such systems employ electronic regenerators every 40 km or so and cannot make use of optical amplifiers.

System performance can be improved considerably using external modulators. The transmission distance in this case is limited to $L < (16|\beta_2|B^2)^{-1}$ [8]. Using $\beta_2 = -20$ ps²/km at 1.55 μm, link length is limited to 500 km at 2.5 Gb/s. Although improved considerably compared with the case of directly modulated DFB lasers, this dispersion limit becomes of concern when optical amplifiers are used for loss compensation. Moreover, if the bit rate is increased to 10 Gb/s, the GVD-limited transmission distance drops to 30 km, a value so low that optical amplifiers cannot be used in designing such lightwave systems. Clearly, the relatively large GVD of standard telecommunication fibers severely limits the performance of 1.55-μm systems designed to use the existing telecommunication network at a bit rate of 10 Gb/s or more.

The dispersion-management technique is aimed at solving this practical problem. The basic idea behind dispersion management is quite simple and can be understood using the pulse propagation equation derived in Section 2.3 of Ref. [5]. If nonlinear effects and fiber losses are ignored, this equation can be written as

$$\frac{\partial A}{\partial z} + \frac{i\beta_2}{2}\frac{\partial^2 A}{\partial T^2} - \frac{\beta_3}{6}\frac{\partial^3 A}{\partial T^3} = 0, \tag{7.1.8}$$

where the effect of third-order dispersion (TOD) is included by the β_3 term. In practice, this term can be neglected as long as $|\beta_2|$ is not close to zero. Equation (7.1.8) can be solved with the Fourier-transform method. In the specific case of $\beta_3 = 0$, the solution is given by

$$A(L,T) = \frac{1}{2\pi}\int_{-\infty}^{\infty}\tilde{A}(0,\omega)\exp\left(\frac{i}{2}\beta_2 L\omega^2 - i\omega T\right)d\omega, \tag{7.1.9}$$

where $\tilde{A}(0,\omega)$ is the Fourier transform of $A(0,T)$ and L is the length of a fiber with uniform GVD β_2.

Dispersion-induced degradation of optical signals is caused by different phase shifts ($\phi_s = \beta_2 L\omega^2/2$) acquired by different spectral components as a pulse propagates down the fiber. All dispersion-management schemes attempt to cancel this phase shift so that the input signal can be restored. In the simplest approach, the optical signal is propagated over multiple fiber segments with different dispersion characteristics. The basic idea can be understood by considering just two segments whose GVD parameters are chosen such that

$$\beta_{21}L_1 + \beta_{22}L_2 = 0, \tag{7.1.10}$$

where $L = L_1 + L_2$ and β_{2j} is the dispersion parameter of the fiber segment of length L_j ($j = 1, 2$). If we apply Eq. (7.1.9) to each fiber section and use the condition (7.1.10), it is easy to verify that $A(L,t) = A(0,T)$. As a result, the pulse shape is restored to its input form after traversing the two segments. The second segment is made of dispersion-compensating fiber (DCF) designed to have normal GVD near 1.55 μm ($\beta_{22} > 0$). Its length should be chosen such that $L_2 = -(\beta_{21}/\beta_{22})L_1$. For practical reasons, L_2 should be as small as possible. Commercial DCFs have values of β_{22} in excess of 100 ps^2/km and are designed with a relatively small value of the V parameter.

WDM systems benefit from dispersion management enormously because its use avoids interchannel crosstalk induced by FWM (see Section 7.2.5). However, the GVD cannot be compensated fully for all channels because of the wavelength dependence of β_2. The plot of the accumulated value of dispersion, $\int_0^z D(z)dz$, along the fiber link in Figure 7.2 shows the typical situation for WDM systems for the central and two boundary channels. The GVD effects can be eliminated for the central channel but they remain finite for all other channels. In practice, the accumulated dispersion may exceed 1000 ps/nm for the boundary channels in long-haul WDM systems. Pre- or postcompensation techniques are often used by adding different DCF lengths for different channels at the transmitter or receiver end.

7.2 Impact of Fiber Nonlinearities

As seen in Section 5.2.2, the nonlinear effects occurring inside optical fibers limit the maximum power levels that can be realized with Yb-doped fiber lasers and amplifiers.

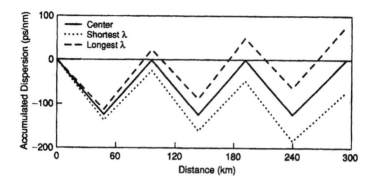

Figure 7.2: Dispersion map for center, shortest-, and longest-wavelength channels of a WDM system. (From Ref. [3]; ©1997 Academic Press)

They also affect the performance of modern long-haul WDM lightwave systems. This section focuses on the five major nonlinear phenomena in this context.

7.2.1 Stimulated Brillouin Scattering

SBS in optical fibers was first observed in 1972 and has been studied extensively since then because of its implications for lightwave systems [10]–[16]. As discussed in Chapter 9 of Ref. [5], SBS generates a Stokes wave propagating in the backward direction. The frequency of the Stokes wave is downshifted by an amount that depends on the wavelength of incident signal. This shift is known as the *Brillouin shift* and is about 11 GHz in the wavelength region near 1.55 μm.

The intensity of the Stokes wave grows exponentially once the input power exceeds a threshold value. For narrowband signals, the threshold power P_{th} can be estimated from the relation [11]

$$g_B P_{th} L_{eff}/A_{eff} \approx 21, \tag{7.2.1}$$

where g_B is the Brillouin gain coefficient, A_{eff} is the effective mode area, and the effective length, $L_{eff} = (1 - e^{-\alpha L})/\alpha$ is smaller than the actual length L because of fiber losses governed by α. For long fibers such that $\alpha L \gg 1$, one can use $L_{eff} \approx 1/\alpha$ (21.7 km for $\alpha = 0.2$ dB/km). Using $g_B \approx 5 \times 10^{-11}$ m/W and $A_{eff} = 50$ μm^2 as typical values, P_{th} can be as low as 1 mW for continuous-wave (CW) signals in the wavelength region near 1.55 μm [12]. Figure 7.3 shows variations in the transmitted reflected power (through SBS) for a 13-km-long dispersion-shifted fiber as the injected CW power is increased from 0.5 to 50 mW. No more than 3 mW could be transmitted through the fiber in this experiment after the onset of SBS.

The SBS threshold increases for CW beams whose spectral width $\Delta \nu_p$ is larger than the Brillouin-gain line width ($\Delta \nu_B \sim 20$ MHz). It also increases when short optical pulses propagate through the fiber because of their relatively wide bandwidths. In lightwave systems, the optical signal is in the form of a time-dependent signal composed of an arbitrary sequence of 1 and 0 bits. One would expect the Brillouin threshold of such a signal to be higher than that of a CW beam. Considerable attention

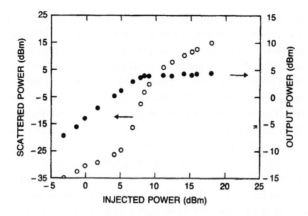

Figure 7.3: Output signal power (solid circles) and reflected SBS power (empty circles) as a function of power injected into a 13-km-long fiber. (From Ref. [14]; ©1992 IEEE.)

has been paid to estimating the Brillouin threshold and quantifying the SBS limitations for practical lightwave systems [17]–[25]. The amount by which the threshold
power increases depends on the modulation format used for data transmission. In the
case of a coherent transmission scheme, the SBS threshold also depends on whether
the amplitude, phase, or frequency of the optical carrier is modulated for information
coding. Most lightwave systems modulate amplitude of the optical carrier and use the
so-called on–off keying scheme.

Calculation of the Brillouin threshold for such systems is quite involved as it requires a time-dependent analysis [17]. Considerable simplification occurs if the bit rate
B is assumed to be much larger than the Brillouin-gain line width Δv_B. Even with this
assumption, the analysis is complicated by the fact that the 1 and 0 bits do not follow
a fixed pattern in realistic communication systems. A simple approach assumes that
the situation is equivalent to that of a CW pump whose spectrum corresponds to that
caused by a random bit pattern. This is justified by noting that the backward nature of
SBS would tend to average out time-dependent fluctuations. An interesting result of
such an approximate analysis is that the Brillouin threshold increases by about a factor
of 2 irrespective of the actual bit rate. As a result, input powers of about 6 mW can be
injected into a fiber without performance degradation resulting from SBS.

In modern WDM systems, fiber losses are compensated periodically using optical
amplifiers. An important question is how amplifiers affect the SBS process. If the
Stokes wave were amplified by amplifiers, it would accumulate over the entire link
and grow enormously. Fortunately, periodically amplified lightwave systems typically
employ an optical isolator within each amplifier that blocks the passage of the Stokes
wave. However, the SBS growth between two amplifiers is still undesirable for two
reasons. First, it removes power from the signal once the signal power exceeds the
threshold level. Second, it induces large fluctuations in the remaining signal, resulting
in degradation of both the SNR and the bit-error rate [15]. For these reasons, single-
channel powers are invariably kept below the SBS threshold and are limited in practice
to below 10 mW.

Some applications require launch powers in excess of 10 mW. An example is provided by the shore-to-island fiber links designed to transmit information over several hundred kilometers, without using in-line amplifiers or repeaters. One way to increase the launched power is to control the SBS by raising its threshold, and several schemes have been proposed for this purpose [26]–[34]. Several among them rely on increasing either the Brillouin-gain bandwidth Δv_B or the effective spectral width of the optical carrier. The former has a value of about 20 MHz for silica fibers, while the latter is typically <10 MHz for DFB lasers. The bandwidth of an optical carrier can be increased, without affecting the system performance, by modulating its phase at a frequency much lower than the bit rate. Typically, the modulation frequency Δv_m is in the range of 200 to 400 MHz. Since the effective Brillouin gain is reduced by a factor of $(1 + \Delta v_m / \Delta v_B)$ in Eq. (7.2.1), the SBS threshold increases by the same factor. The launched power can be increased by more than a factor of 10 by the phase-modulation technique.

If the Brillouin-gain bandwidth Δv_B itself can be increased from its nominal value of 20 MHz to beyond 200 MHz, the SBS threshold can be increased without requiring a phase modulator. One technique uses sinusoidal strain along the fiber length for this purpose. The applied strain changes the Brillouin shift v_B by a few percent in a periodic manner. The resulting Brillouin-gain spectrum is much broader than that occurring for a fixed value of v_B. The strain can be applied during cabling of the fiber. In one fiber cable, Δv_B was found to increase from 50 MHz to 400 MHz [28]. The Brillouin shift v_B can also be changed by making the core radius nonuniform along the fiber length since the longitudinal acoustic frequency depends on the core radius [29]. The same effect can be realized by changing the dopant concentration along the fiber length. This technique increased the SBS threshold of one fiber by 7 dB [30]. A temperature gradient along the fiber length also increases the SBS threshold by shifting v_B in a distributed fashion. A threefold increase in the SBS threshold was observed in a 2001 experiment with a 140°C temperature difference across a 100-m-long fiber [32].

Fiber gratings can also be used to increase the SBS threshold. The Bragg grating is designed such that it is transparent to the forward-propagating pump beam, but the spectrum of the Stokes wave generated through SBS falls entirely within its stop band [33]. A single grating, placed suitably in the middle, may be sufficient for relatively short fibers. Multiple gratings need to be used for long fibers. Another approach tries to minimize the overlap between the optical and acoustic modes through suitable dopants [34].

7.2.2 Stimulated Raman Scattering

As discussed in Chapter 8 of Ref. [5], SRS differs from SBS in several ways. First, it generates a forward-propagating Stokes wave. Second, the Raman shift v_R by which the frequency of the Stokes wave is downshifted is close to 13 THz. Third, the Raman-gain spectrum is extremely broad and extends over a frequency range wider than 20 THz. Fourth, the peak value of the Raman gain g_R is lower by more than a factor of 100 compared with that of the Brillouin gain. SRS was first observed in optical fibers in 1972 [35]. Since then, the impact of SRS on the performance of lightwave systems has been studied extensively [36]–[50].

The Raman threshold, the power level at which the Raman process becomes stimulated and transfers most of the signal power to the Stokes wave, can be written as [11]

$$P_{th} \approx 16A_{eff}/(g_R L_{eff}). \tag{7.2.2}$$

As before, we can replace L_{eff} with $1/\alpha$ for long fiber lengths used in lightwave systems. Using $g_R \approx 1 \times 10^{-13}$ m/W, P_{th} is about 500 mW at wavelengths near 1.55 μm. Since input powers are limited to below 10 mW because of SBS, SRS is not of concern for single-channel lightwave systems.

The situation is quite different for WDM systems, which transmit simultaneously multiple channels spaced 100 GHz or so apart. The fiber link in this case acts as a Raman amplifier such that longer-wavelength channels are amplified by shorter-wavelength channels as long as their wavelength difference is within the Raman-gain bandwidth. The shortest-wavelength channel is depleted most as it can pump all other channels simultaneously. Such an energy transfer among channels can be detrimental for system performance as it depends on the bit pattern, occurring only when 1 bits are present simultaneously in the two channels acting as the pump and signal channels. The signal-dependent amplification leads to power fluctuations, which add to receiver noise and degrade the receiver performance.

The Raman crosstalk can be avoided if channel powers are made so small that Raman amplification is negligible over the fiber length. A simple model considered depletion of the highest-frequency channel in the worst case in which all channels transmit 1 bits simultaneously [36]. A more accurate analysis should consider not only depletion of each channel but also its own amplification by shorter-wavelength channels. If all other nonlinear effects are neglected along with GVD, evolution of the power P_n associated with the nth channel is governed by [8]

$$\frac{dP_n}{dz} + \alpha P_n = C_R P_n \sum_{m=1}^{M} (n-m)P_m, \tag{7.2.3}$$

where $C_R = (dg_R/dv)\Delta v_{ch}/(2A_{eff})$ and α is assumed to be the same for all channels. This set of M coupled nonlinear equations can be solved analytically [42]. The solution shows that channel powers follow an exponential distribution because of Raman-induced coupling among all channels. The depletion factor for the shortest-wavelength channel ($n = 1$) is obtained using $D_R = (P_{10} - P_1)/P_{10}$, where $P_{10} = P_1(0)\exp(-\alpha L)$ is the channel power expected in the absence of SRS. In the case of equal input powers in all channels, D_R is given by

$$D_R = 1 - \exp\left[-\frac{1}{2}M(M-1)C_R P_{ch}L_{eff}\right] \frac{M\sinh(\frac{1}{2}MC_R P_{ch}L_{eff})}{\sinh(\frac{1}{2}M^2 C_R P_{ch}L_{eff})}. \tag{7.2.4}$$

The Raman-induced power penalty is obtained using $\delta_R = -10\log(1 - D_R)$ because the input channel power must be increased by a factor of $(1 - D_R)^{-1}$ to maintain the same system performance. Figure 7.4 shows how the power penalty increases with an increase in the channel power and the number of channels. The channel spacing is

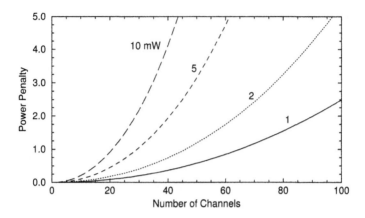

Figure 7.4: Raman-induced power penalty as a function of number of channels for several values of P_{ch}. Channels are 100-GHz apart and launched with equal powers.

assumed to be 100 GHz. As seen from Figure 7.4, the power penalty becomes quite large for WDM systems with a large number of channels. If a value of at most 1 dB is considered acceptable, the limiting channel power is reduced to below 1 mW when the number of WDM channels is larger than 70.

Equation (7.2.4) overestimates the Raman crosstalk because it ignores the fact that signals in each channel consist of a random sequence of 0 and 1 bits. A statistical analysis shows that Raman crosstalk is lower by about a factor of 2 when signal modulation is taken into account [40]. The inclusion of GVD effects that were neglected in this analysis also reduces Raman crosstalk since pulses in different channels travel at different speeds because of the group-velocity mismatch [43]. Figure 7.5 shows the the Raman-induced power penalty for a 100-km-long WDM system with 1-nm channel spacing after including the effects of both GVD and modulation statistics. Each channel operates at 10 Gb/s and is launched with same input peak power of 10 mW. The shortest-wavelength channel is assumed to be located at 1530 nm. The power penalty is smaller by a factor of 2 for standard fibers with $D = 16$ ps/(km-nm) compared with the $D = 0$ case. It exceeds 1 dB when the number of channels becomes more than 25 for $D = 2$ ps/(km-nm).

The effects of Raman crosstalk in a WDM system were quantified in a 1999 experiment by transmitting 32 channels, with 100-GHz spacing, over 100 km [45]. At low input powers ($P_{\text{ch}} = 0.1$ mW), SRS effects were relatively small and channel powers differed by only a few percent after 100 km. However, when the input power for each channel was increased to 3.6 mW, the longest-wavelength channel had 70% more power than the shortest-wavelength channel. Moreover, the channel powers were distributed in an exponential fashion as predicted by Eq. (7.2.3).

In long-haul lightwave systems, the crosstalk is also affected by the use of loss- and dispersion-management schemes. Dispersion management permits high values of GVD locally while reducing it globally. Since the group-velocity mismatch among different channels is quite large in such systems, the Raman crosstalk should be re-

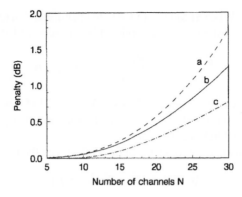

Figure 7.5: Raman-induced power penalty for a 100-km-long WDM system with 1-nm channel spacing. Dispersion parameter is 0, 2, and 16 ps/(km-nm) for curves (a)–(c), respectively. (From Ref. [43]; ©1998 IEEE.)

duced in a dispersion-managed system. In contrast, the use of optical amplifiers for loss management magnifies the impact of SRS-induced degradation. The reason is that in-line amplifiers add broadband noise, which can seed the SRS process. As a result, noise is amplified along the link and results in degradation of the SNR. The SNR can be maintained if the channel power is increased as the number of amplifiers increases. The Raman-limited capacity of long-haul WDM systems depends on a large number of design parameters such as amplifier spacing, optical-filter bandwidth, bit rate, channel spacing, and total transmission distance [39].

Can Raman crosstalk be avoided by a proper system design? Clearly, reducing the channel power is the simplest solution but it may not always be practical. An alternative scheme lets SRS occur over the whole link but cancels the Raman crosstalk by using the technique of spectral inversion [39]. As the name suggests, if the spectrum of the WDM signal were inverted at some appropriate distance, short-wavelength channels would become long-wavelength channels and vice versa. As a result, the direction of Raman-induced power transfer will be reversed such that channel powers become nearly equal at the end of the fiber link. Complete cancelation of Raman crosstalk for a two-channel system requires spectral inversion at mid-span if GVD effects are negligible or compensated [44]. Equation (7.2.3) can be used to show that the spectral-inversion technique should work for an arbitrary number of channels [46]. The location of spectral inversion is not necessarily in the middle of the fiber span but changes depending on gain–loss variations. Spectral inversion can be accomplished through FWM inside a fiber to realize phase conjugation; the same technique is also useful for dispersion compensation [8].

7.2.3 Self-Phase Modulation

As discussed in Chapter 4 of Ref. [5], the intensity dependence of the refractive index leads to an SPM-induced nonlinear phase shift, resulting in chirping and spectral broadening of optical pulses. Clearly, SPM can affect the performance of a lightwave system considerably. When SPM is included with fiber dispersion and losses, the propagation of an optical bit stream through optical fibers is governed by the NLS equation

$$i\frac{\partial A}{\partial z} + \frac{i\alpha}{2}A - \frac{\beta_2}{2}\frac{\partial^2 A}{\partial T^2} + \gamma|A|^2 A = 0, \qquad (7.2.5)$$

where α, β_2, and γ govern the effects of loss, GVD, and SPM, respectively. All three parameters become functions of z when loss- and dispersion-management schemes are employed for long-haul lightwave systems.

It is useful to eliminate the loss term in Eq. (7.2.5) with the transformation

$$A(z,T) = \sqrt{P_0}\, e^{-\alpha z/2} U(z,T), \qquad (7.2.6)$$

where P_0 is the peak power of input pulses. Equation (7.2.5) then takes the form

$$i\frac{\partial U}{\partial z} - \frac{\beta_2(z)}{2}\frac{\partial^2 U}{\partial T^2} + \gamma P_0 p(z)|U|^2 U = 0, \qquad (7.2.7)$$

where power variations along a loss-managed fiber link are included through the periodic function $p(z)$. In the case of lumped amplifiers, $p(z) = e^{-\alpha z}$ between two amplifiers but equals 1 at the location of each amplifier. It is not easy to solve Eq. (7.2.7) analytically except in some simple cases. In the specific case of $p = 1$ and β_2 constant but negative, this equation reduces to the standard NLS equation and has solutions in the form of solitons discussed in Section 7.3.

From a practical standpoint, the effect of SPM is to chirp the pulse and broaden its spectrum. In the absence of dispersion ($\beta_2 = 0$), Eq. (7.2.7) can be solved analytically to study the extent of frequency chirping and spectral broadening. The solution is of the form $U(z,T) = U(0,T)\exp(i\phi_{\mathrm{NL}})$, where the SPM-induced phase shift is given by

$$\phi_{\mathrm{NL}} = \gamma P_0 L_{\mathrm{eff}}|U(0,T)|^2. \qquad (7.2.8)$$

The maximum phase shift, $\phi_{\max} = \gamma P_0 L_{\mathrm{eff}}$, determines the amount of frequency chirp. As a rough design guideline, the SPM effects are negligible when $\phi_{\max} < 1$ or $P_0 < \alpha/\gamma$, where we used $L_{\mathrm{eff}} \approx 1/\alpha$. For typical values of α and γ, SPM becomes important at peak power levels above 25 mW. Since SBS limits power levels to below 10 mW, SPM is of little concern for loss-limited lightwave systems. The situation changes when fiber losses are compensated using optical amplifiers. The SPM effects can then accumulate over the entire link. If N_A amplifiers are used, the maximum phase shift becomes $\phi_{\max} = \gamma P_0 N_A L_{\mathrm{eff}}$. As a result, the peak power is limited to $P_0 < \alpha/\gamma N_A$ or to below 3 mW for links with only 10 amplifiers. Clearly, SPM can be a major limiting factor for long-haul lightwave systems.

The important question is how the SPM-induced chirp affects broadening of optical pulses in the presence of dispersion. The broadening factor can be estimated, without requiring a complete solution of Eq. (7.2.7), using various approximations [51]–[57]. A variational approach was used as early as 1983 [51]. A split-step approach, in which the effects of SPM and GVD are considered separately, also provides a reasonable estimate of pulse broadening [52]. In an extension of this technique, the SPM-induced chirp is treated as an effective chirp parameter at the input end [55]. A perturbation approach, in which the nonlinear term in Eq. (7.2.7) is treated as being relatively small, is also quite useful [57]. We focus on this approach since it can be used for systems with loss and dispersion management.

Following Section 3.3 of Ref. [5], the root-mean-square value of the pulse width can be calculated using $\sigma = [\langle T^2 \rangle - \langle T \rangle^2]^{1/2}$, where

$$\langle T^m \rangle = \frac{\int_{-\infty}^{\infty} T^m |U(z,T)|^2 \, dT}{\int_{-\infty}^{\infty} |U(z,T)|^2 \, dT}. \tag{7.2.9}$$

For a symmetric pulse, $\langle T \rangle = 0$ and σ^2 is approximately given by

$$\sigma^2(z) = \sigma_L^2(z) + \gamma P_0 f_s \int_0^z \beta_2(z_1) \left[\int_0^{z_1} p(z_2) dz_2 \right] dz_1, \tag{7.2.10}$$

where σ_L^2 is the RMS width expected in the linear case ($\gamma = 0$). The shape of the input pulse enters through the parameter f_s, defined as

$$f_s = \frac{\int_{-\infty}^{\infty} |U(0,T)|^4 \, dT}{\int_{-\infty}^{\infty} |U(0,T)|^2 \, dT}. \tag{7.2.11}$$

For a Gaussian pulse with $U(0,T) = \exp[-\frac{1}{2}(T/T_0)^2]$, $f_s = 1/\sqrt{2} \approx 0.7$. For a square pulse, $f_s = 1$.

As an example, consider the case of a uniform-GVD fiber with distributed amplification such that the pulse energy remains nearly constant. Using $p(z) = 1$ with constant β_2 in Eq. (7.2.11), we obtain the simple expression

$$\sigma^2(z) = \sigma_L^2(z) + \frac{1}{2} f_s \gamma P_0 \beta_2 z^2. \tag{7.2.12}$$

This equation shows that the SPM enhances pulse broadening in the normal-GVD regime but leads to pulse compression in the anomalous-GVD regime. This behavior can be understood by noting that the SPM-induced chirp is positive in nature. As a result, the pulse goes through a contraction phase when $\beta_2 < 0$. This is the origin of the existence of solitons in the anomalous-GVD regime. Equation (7.2.10) shows that the soliton effects are beneficial for all pulse shapes and can improve the performance of even NRZ-format systems using nearly square-shaped pulses. This improvement was predicted in the 1980s [58] and has been seen in several experiments.

Consider long-haul lightwave systems in which lumped amplifiers are used periodically for compensation of fiber losses. If dispersion management is also employed, σ_L returns to its input value σ_0 at the end of each fiber section in between two amplifiers. It is evident from Eq. (7.2.10) that such a dispersion-compensation scheme will not work perfectly when SPM effects are significant. Even though the second term in this equation changes sign when β_2 changes sign, power variations in the two sections are different. As a result, the contribution of SPM does not cancel perfectly. Indeed, it has been noticed experimentally that system performance is better when GVD is undercompensated [59]–[61]. Equation (7.2.10) provides a simple explanation of this behavior. By optimizing the amount of GVD in the dispersion-compensating fiber, one can adjust the two terms in this equation and minimize the pulse width.

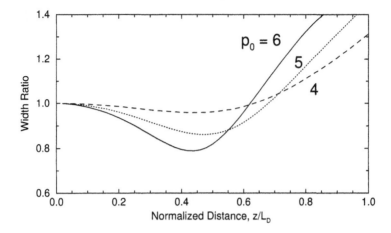

Figure 7.6: Width ratio σ/σ_0 as a function of propagation distance for a super-Gaussian pulse ($m = 2$) at three input peak powers, labeled using $p_0 = \gamma P_0 L_D$.

Equation (7.2.10) remains accurate as long as the second term is small compared with the first since the SPM-induced reduction in pulse width for $\beta_2 < 0$ cannot be expected to be larger than σ_L. As a rough estimate of the validity of this equation, we can use Eq. (7.2.12) to conclude that the simple analysis is valid as long as $z < (L_D L_{NL})^{1/2}$, where $L_D = T_0^2/|\beta_2|$ is the dispersion length and $L_{NL} = (\gamma P_0)^{-1}$ is the nonlinear length [5]. Numerical simulations show that, at a fixed power level, the pulse width reaches a minimum value at some distance and then begins to increase. We show this behavior in Figure 7.6 by solving Eq. (7.2.7) for a super-Gaussian input $U(0, T) = \exp[-\frac{1}{2}(T/T_0)^{2m}]$ with $m = 2$ and $p = 1$. The fiber was assumed to have uniform GVD (no dispersion management). A similar behavior is observed as a function of input peak power when the distance z is kept fixed. As the peak power increases, the pulse initially contracts because of the SPM effects, attains a minimum value at a certain value of the peak power, then begins to increase rapidly. In practical terms, the input power should be optimized properly if we want to take advantage of the solitonlike effects for NRZ systems [58].

Another SPM-induced limitation results from the phenomenon of modulation instability occurring when the signal travels in the anomalous-GVD regime of the transmission fiber. At first sight, it may appear that modulation instability is not likely to occur for a signal in the form of a pulse train. In fact, it affects the performance of periodically amplified lightwave systems considerably. This can be understood by noting that optical pulses in an NRZ-format system occupy the entire time slot and can be several bits long depending on the bit pattern. As a result, the situation is quasi-CW-like. As early as 1990, numerical simulations indicated that system performance of a 6000-km fiber link, operating at bit rates > 1 Gb/s with 100-km amplifier spacing, would be severely affected by modulation instability if the signal propagates in the anomalous-GVD regime and is launched with peak power levels in excess of a few milliwatts [62].

SPM can lead to the degradation of the SNR when optical amplifiers are used for loss compensation [63]–[73]. Such amplifiers add to a signal broadband noise that extends over the entire bandwidth of amplifiers (or optical filters when they are used to reduce noise). Even close to the zero-dispersion wavelength, amplifier noise is enhanced considerably by SPM [67]. In the case of anomalous GVD, spectral components of noise falling within the gain spectrum of modulation instability will be enhanced by this nonlinear process, resulting in further degradation of the SNR [70]. Moreover, periodic power variations occurring in long-haul systems create a nonlinear index grating that can lead to modulation instability even in the normal-GVD regime [66]. Both of these effects have been observed experimentally. In a 10-Gb/s system, considerable degradation in system performance was noticed after a transmission distance of only 455 km [71]. In general, long-haul systems perform better when the average GVD of the fiber link is kept positive ($\bar{\beta}_2 > 0$).

7.2.4 Cross-Phase Modulation

As discussed in Chapter 7 of Ref. [5], when two pulses of different wavelengths propagate simultaneously inside optical fibers, their optical phases are affected not only by SPM but also by XPM. The XPM effects are quite important for WDM lightwave systems, because the phase of each optical channel depends on the bit patterns of all other channels [3]. Fiber dispersion converts phase variations into amplitude fluctuations, affecting the SNR considerably. A proper understanding of the interplay between XPM and GVD is of considerable importance for WDM systems [74]–[96].

Consider an M-channel WDM system with the total optical field

$$A(z,T) = \sum_{m=1}^{M} A_m(z,T) \exp[i(\omega_m - \omega_0)T], \qquad (7.2.13)$$

where ω_m is the carrier frequency of the mth channel and ω_0 is a reference frequency chosen, in practice, to coincide with the frequency of the central channel. Following the method outlined in Section 7.1 of Ref. [5], we obtain a set of M coupled NLS equations:

$$i\frac{\partial A_j}{\partial z} + \frac{i\alpha}{2}A_j + \frac{i}{v_{gj}}\frac{\partial A_j}{\partial T} - \frac{\beta_{2j}}{2}\frac{\partial^2 A_j}{\partial T^2} + \gamma\left(|A_j|^2 + 2\sum_{m\neq j}^{M}|A_m|^2\right)A_j = 0, \quad (7.2.14)$$

where $j = 1$ to M, v_{gj} is the group velocity, and β_{2j} is the GVD parameter. The loss parameter α and the nonlinear parameter γ are assumed to be the same for all channels. The contribution of FWM is neglected in these equations but is discussed later in this section.

In general, the set of M equations should be solved numerically. It can be solved analytically in the CW case with the result $A_j(L) = \sqrt{P_j}\exp(i\phi_j)$, where P_j is the input power and the nonlinear phase shift resulting from a combination of SPM and XPM is given by

$$\phi_j = \gamma L_{\text{eff}}\left(P_j + 2\sum_{m\neq j}^{M}P_m\right). \qquad (7.2.15)$$

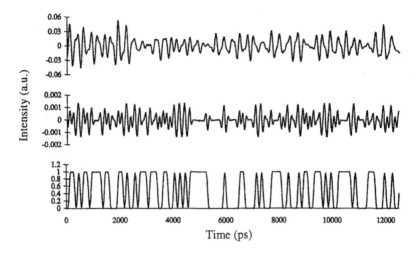

Figure 7.7: XPM-induced power fluctuations on a CW probe for a 130-km link (middle) and a 320-km link (top) with dispersion management. An NRZ bit stream in the pump channel is shown at the bottom. (From Ref. [80]; ©1999 IEEE.)

The CW solution can be applied approximately for NRZ-format systems operating at relatively low bit rates. The phase ϕ_j of a specific channel would vary from bit to bit depending on the bit patterns of neighboring channels. In the worst case, in which all channels have 1 bits in their time slots simultaneously, the XPM-induced phase shift is largest. If the input power is assumed to be the same for each channel, this maximum value is given by

$$\phi_{max} = (\gamma/\alpha)(2M - 1)P_{ch}, \qquad (7.2.16)$$

where L_{eff} was replaced by $1/\alpha$ assuming $\alpha L \gg 1$. The XPM-induced phase shift increases linearly with M and can become quite large. It was measured in 1984 for the two-channel case [74]. Light from two semiconductor lasers operating near 1.3 and 1.5 μm was injected into a 15-km-long fiber. The phase shift at 1.5 μm, induced by the copropagating 1.3-μm wave, was measured using an interferometer. A value of $\phi_{max} = 0.024$ was found for $P_{ch} = 1$ mW. This value is in good agreement with the predicted value of 0.022 from Eq. (7.2.16).

Strictly speaking, the XPM-induced phase shift should not affect system performance if the GVD effects were negligible. However, any dispersion in fiber converts pattern-dependent phase shifts to power fluctuations, reducing the SNR at the receiver. This conversion can be understood by noting that time-dependent phase changes lead to frequency chirping, which affects the dispersion-induced broadening of the signal. Figure 7.7 shows XPM-induced fluctuations for a CW probe launched with a 10-Gb/s pump channel modulated using the NRZ format. The probe power fluctuates by as much as 6% after 320 km of dispersive fiber. The RMS value of fluctuations depends on the channel power and can be reduced by lowering it. As a rough estimate, if we use the condition $\phi_{max} < 1$ in Eq. (7.2.16), the channel power is restricted to

$$P_{ch} < \alpha/[\gamma(2M - 1)]. \qquad (7.2.17)$$

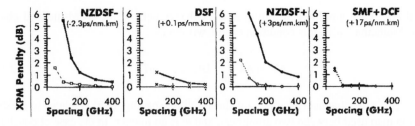

Figure 7.8: XPM-induced power penalty as a function of channel spacing for four fiber links with different dispersion. Thin and thick lines correspond to best and worst cases, respectively. (From Ref. [81]; ©1999 IEEE.)

For typical values of α and γ, P_{ch} should be below 10 mW even for five channels and reduces to below 1 mW for more than 50 channels.

The preceding analysis ignores the effects of group-velocity mismatch. In reality, pulses belonging to different channels travel at different speeds and walk through each other at a rate that depends on their wavelength difference. Since XPM can occur only when pulses overlap in the time domain, its impact is reduced considerably by the walk-off effects. As a faster-moving pulse belonging to one channel collides with and passes through a specific pulse in another channel, the XPM-induced chirp shifts the pulse spectrum first toward red and then toward blue. In a lossless fiber, collisions of two pulses are perfectly symmetric, resulting in no net spectral shift at the end of the collision. In a loss-managed system, with lumped amplifiers placed periodically along the link, power variations make collisions between pulses of different channels asymmetric, resulting in a net frequency shift that depends on the channel spacing. Such frequency shifts lead to timing jitter (the speed of a channel depends on its frequency because of GVD), since their magnitude depends on the bit pattern as well as on channel wavelengths. The combination of amplitude and timing jitter degrades the SNR at the receiver considerably, especially for closely spaced channels, and leads to an XPM-induced power penalty [75].

Figure 7.8 shows how the power penalty depends on channel spacing and fiber dispersion at a bit-error rate of 10^{-10}. These results were obtained by launching two 10-Gb/s channels into a 200-km fiber link with one amplifier located midway [81]. Four types of fibers were used to change the GVD. The pump-channel power was 8 dBm (6.3 mW) while the signal-channel power was kept at 2 dBm. The penalty depends on the relative pump–signal delay and the state of polarizations; thin and thick lines in Figure 7.8 show the best and worst cases. The XPM-induced penalty becomes quite large for large-GVD links and small channel spacing, as expected. It can be reduced to negligible levels for fiber links with small average GVD and relatively large channel spacing (> 50 GHz). A negligible penalty occurs for the dispersion-managed link in which GVD is compensated using a DCF.

In periodically amplified lightwave systems, power variations along the fiber link affect the XPM interaction among channels. If two channels are spaced such that the relative propagation delay ΔT between them over each amplifier span is equal to a multiple of the bit slot T_B, the pulse trains in the two channels will become synchronized after each amplifier, resulting in the enhancement of the XPM-induced phase shift. Mathematically, this condition can be written as

$$\Delta T = \Delta\lambda \int_0^{L_A} D(z)dz = mT_B \qquad (7.2.18)$$

where $\Delta\lambda$ is the channel spacing, L_A is the amplifier spacing, $D(z)$ is related to the dispersion map used between two amplifiers, and m is an integer. In the case of constant-dispersion fibers, this condition becomes $BDL_A\Delta\lambda = m$, where $B = 1/T_B$. System performance is expected to degrade whenever channel spacing $\Delta\lambda$ satisfies Eq. (7.2.18). This was indeed observed in the experiment in which the bit-error rate of a weak probe channel exhibited resonances (it increased significantly) whenever wavelength spacing of the pump channel responsible for XPM satisfied the resonance condition [83].

The XPM effects occurring within a fiber amplifier are normally negligible because of a small length of doped fiber used. The situation changes for the L-band amplifiers, which operate in the 1570- to 1610-nm wavelength region and require fiber lengths in excess of 100 m. The effective mode area of doped fibers used in such amplifiers is relatively small, resulting in larger values of the nonlinear parameter γ and enhanced XPM-induced phase shifts. As a result, the XPM can lead to considerable power fluctuations within an L-band amplifier [77]. A new feature is that such XPM effects are independent of the channel spacing and can occur over the entire bandwidth of the amplifier [84]. The reason for this behavior is that all XPM effects occur before pulses walk off because of group-velocity mismatch.

The XPM effects can be reduced in modern WDM systems through the use of differential phase-shift keying (DPSK) format [94]–[96]. The DPSK is often combined with the RZ format such that a pulse is present in every bit slot and the information is encoded only through phase variations. It is easy to understand qualitatively why XPM-induced penalties are reduced for such a lightwave system. The main reason why XPM leads to amplitude fluctuations and timing jitter when an amplitude-shift keying (ASK) format is used is related to the random power variations that mimic the bit pattern. It is easy to see that the XPM will be totally harmless if channel powers were constant in time, because all XPM-induced phase shifts will be time-independent, producing no frequency and temporal shifts. Although this is not the case for an RZ-DPSK system, the XPM effects are considerably reduced because of a strictly periodic nature of the power variations. Physically speaking, all bits undergo nearly identical collision histories, especially if the average channel power does not vary too much along the link, resulting in negligible XPM-induced power penalties.

7.2.5 Four-Wave Mixing

Four-wave mixing is a major source of nonlinear crosstalk for WDM lightwave systems [97]–[119]. The physical origin of FWM-induced crosstalk, and the resulting system degradation, can be understood by noting that FWM can generate a new wave at the frequency $\omega_F = \omega_i + \omega_j - \omega_k$, whenever three waves of frequencies ω_i, ω_j, and ω_k copropagate inside the fiber. For an M-channel system, i, j, and k vary from 1 to M, resulting in a large combination of new frequencies generated by FWM. In the case of equally spaced channels, most new frequencies coincide with the existing channel frequencies and interfere coherently with the signals in those channels. This interference depends on the bit pattern and leads to considerable fluctuations in the detected signal at the receiver. When channels are not equally spaced, most FWM components fall in between the channels and add to overall noise. In both cases, system performance is affected by the loss in channel powers, but the degradation is much more severe for equally spaced channels because of the coherent nature of crosstalk.

As discussed in Chapter 10 of Ref. [5], the FWM process in optical fibers is governed by a set of four coupled equations whose general solution requires a numerical approach. If we neglect the phase shifts induced by SPM and XPM, assume that the three channels participating in the FWM process remain nearly undepleted, and include fiber losses, the amplitude A_F of the FWM component at the frequency ω_F is governed by

$$\frac{dA_F}{dz} = -\frac{\alpha}{2}A_F + d_F \gamma A_i A_j A_k^* \exp(-i\Delta k z), \qquad (7.2.19)$$

where $A_m(z) = A_m(0)\exp(-\alpha z/2)$ for $m = i,j,k$ and $d_F = 2 - \delta_{ij}$ is the degeneracy factor defined such that its value is 1 when $i = j$ but doubles when $i \neq j$. This equation can be easily integrated to obtain $A_f(z)$. The power transferred to the FWM component in a fiber of length L is given by [97]

$$P_F = |A_F(L)|^2 = \eta_F (d_F \gamma L)^2 P_i P_j P_k e^{-\alpha L}, \qquad (7.2.20)$$

where $P_m = |A_m(0)|^2$ is the launched power in mth channel and η_F is a measure of the FWM efficiency defined as

$$\eta_F = \left| \frac{1 - \exp[-(\alpha + i\Delta k)L]}{(\alpha + i\Delta k)L} \right|^2. \qquad (7.2.21)$$

The FWM efficiency η_F depends on the channel spacing through the phase mismatch governed by

$$\Delta k = \beta_F + \beta_k - \beta_i - \beta_j \approx \beta_2(\omega_i - \omega_k)(\omega_j - \omega_k), \qquad (7.2.22)$$

where the propagation constants were expanded in a Taylor series around $\omega_c = (\omega_i + \omega_j)/2$ and β_2 is the GVD parameter at that frequency. If the GVD of the transmission fiber is relatively large, $(|\beta_2| > 5 \text{ ps}^2/\text{km})$, η_F nearly vanishes for typical channel spacings of 50 GHz or more. In contrast, $\eta_F \approx 1$ close to the zero-dispersion wavelength of the fiber, resulting in considerable power in the FWM component, especially at high channel powers. In the case of equal channel powers, P_F increases as P_{ch}^3. This cubic dependence of the FWM component limits the channel powers to below 1 mW if FWM is nearly phase matched. Since the number of FWM components for an M-channel WDM system increases as $M^2(M-1)/2$, the total power in all FWM components can be quite large. Figure 7.9 shows, as an example, the optical spectrum measured at the output of a 25-km-long dispersion-shifted fiber [$D = -0.2$ ps/(km-nm) for the central channel] when three 3-mW channels are launched into it. The nine FWM components can be seen clearly. None of them coincides with the channel wavelengths because of the unequal channel spacing used in this experiment.

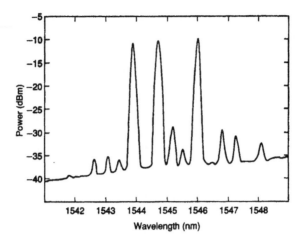

Figure 7.9: Optical spectrum measured at the output of a 25-km-long fiber when three channels, each with 3-mW average power, are launched into it. (From Ref. [3]; ©1997 Academic Press.)

In the case of equal channel spacing, most FWM components fall within the channel spectra and cannot be seen as clearly as in Figure 7.9 in the spectral domain. However, their presence is easily noticed in the time domain as they interfere with the signal coherently. Since the FWM power depends on the bit patterns of three channels, the signal power fluctuates considerably. Figure 7.10 shows the bit patterns observed for the central channel using three fibers with different GVD values. The central channel in this case is located exactly in the middle (see Figure 7.9) such that the channel spacing is constant and equal to 1 nm. The FWM-induced noise is quite large for low GVD values because of the quasi-phase-matched nature of the FWM process.

Modulation instability can enhance the effects of FWM for certain specific values of channel spacing [108]. The reason can be understood by noting that SPM and XPM, ignored in deriving Eq. (7.2.20), can produce phase matching even when $\beta_2 \neq 0$. We can follow Section 10.2 of Ref. [5] to include the SPM and XPM phase shifts. It turns out that Eq. (7.2.20) can still be used but the phase-mismatch factor Δk in Eq. (7.2.22) is replaced with [110]

$$\Delta k \approx \beta_2(\omega_i - \omega_k)(\omega_j - \omega_k) + \gamma(P_i + P_j - P_k)[1 - \exp(-\alpha L_{\text{eff}})]/(\alpha L_{\text{eff}}). \quad (7.2.23)$$

Clearly, Δk may become close to zero for some FWM terms, depending on the channel powers and spacings, when β_2 is in the anomalous-GVD regime of the fiber. The corresponding FWM process will then become phase-matched, resulting in significant power-conversion efficiency. Physically speaking, if the frequency at which the gain peak of modulation instability nearly coincides with the channel spacing in a WDM system, modulation-instability sidebands will overlap with the channel wavelengths. As a result, the FWM process will become enhanced resonantly in spite of the GVD. We can estimate the channel spacing δv_{ch} for which such resonant FWM is expected to occur as

$$\Omega_s = 2\pi \delta v_{\text{ch}} = (2\gamma P_{\text{ch}}/|\beta_2|)^{1/2}. \quad (7.2.24)$$

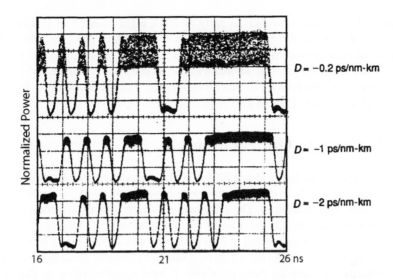

Figure 7.10: Effect of fiber dispersion on the central channel when three 3-mW channels are launched with equal channel spacing among them. (From Ref. [3]; ©1997 Academic Press.)

As a rough estimate, $\delta v_{ch} \approx 50$ GHz when $P_{ch} = 5$ mW, $\beta_2 = -0.2$ ps^2/km, and $\gamma = 2$ W^{-1}/km. Since channel spacing in modern WDM systems is typically 100 GHz or less, resonance enhancement of FWM can easily occur.

A simple scheme for reducing the FWM-induced degradation consists of designing WDM systems with unequal channel spacings [103]. The main impact of FWM in this case is to reduce the channel power. This power depletion results in a power penalty at the receiver, whose magnitude can be controlled by varying the launched power and fiber dispersion. Experimental measurements on a WDM system, in which eight 10-Gb/s channels were transmitted over 137 km of dispersion-shifted fiber, confirm the advantage of unequal channel spacings. In a 1999 experiment, this technique was used to transmit 22 channels, each operating at 10 Gb/s, over 320 km of dispersion-shifted fiber with 80-km amplifier spacing [111]. Channel spacings ranged from 125 to 275 GHz in the 1532- to 1562-nm wavelength region and were determined using a periodic allocation scheme [107]. The zero-dispersion wavelength of the fiber was close to 1548 nm, resulting in near phase matching of many FWM components. Nonetheless, the system performed quite well (because of unequal channel spacings) with less than 1.5-dB power penalty for all channels.

The use of a nonuniform channel spacing is not always practical because many WDM components require equal channel spacings [6]. Also, this scheme is spectrally inefficient since the bandwidth of the resulting WDM signal is considerably larger compared with the case of equally spaced channels [103]. An alternative is offered by the dispersion-management technique discussed earlier. In this case, fibers with normal and anomalous GVD are combined to form a periodic dispersion map such that GVD is locally high all along the fiber even though its average value is quite low. As a result, the FWM efficiency η_F is negligible throughout the fiber, resulting in little FWM-induced crosstalk. As early as 1993, eight channels at 10 Gb/s could be transmitted over 280 km by using dispersion management [100]. By 1996, the use of dispersion management had become quite common for FWM suppression in WDM systems because of its practical simplicity. FWM can also be suppressed by using fibers whose GVD varies along the fiber length [112]. The use of chirped pulses [118] or carrier-phase locking [119] is also helpful for this purpose.

7.3 Solitons in Optical Fibers

We have seen that both the GVD and SPM limit the performance of a lightwave system when acting individually. It turns out that the two can be forced to cooperate such that the pulses form optical solitons. The existence of solitons in optical fibers is the result of a balance between the chirps induced by GVD and SPM [120]–[122]. One can develop an intuitive understanding of how such a balance is possible by following the discussion in Section 6.1. As shown there, a chirped pulse can be compressed inside a fiber whenever β_2 and the chirp parameter C have opposite signs, so that $\beta_2 C$ is negative. As seen in Section 7.2.3, SPM imposes a chirp on the optical pulse such that $C > 0$. If $\beta_2 < 0$, the condition $\beta_2 C < 0$ is readily satisfied. Moreover, as the SPM-induced chirp is power dependent, it is not difficult to imagine that, under certain conditions, SPM and GVD may cooperate in such a way that the SPM-induced chirp is just right to cancel the GVD-induced broadening of the pulse. The optical pulse would then propagate undistorted in the form of a soliton.

7.3.1 Properties of Optical Solitons

To find the conditions under which solitons can form, we set $p(z) = 1$ in Eq. (7.2.7) and assume that pulses propagate in the region of anomalous GVD ($\beta_2 < 0$). Introducing the normalized variables as $\xi = z/L_D$ and $\tau = T/T_0$, Eq. (7.2.7) can be written as

$$i\frac{\partial U}{\partial \xi} + \frac{1}{2}\frac{\partial^2 U}{\partial \tau^2} + N^2 |U|^2 U = 0, \tag{7.3.1}$$

where $L_D = T_0^2/|\beta_2|$, $L_{NL} = (\gamma P_0)^{-1}$, and

$$N^2 = \frac{L_D}{L_{NL}} = \frac{\gamma P_0 T_0^2}{|\beta_2|} \tag{7.3.2}$$

represents a dimensionless combination of the pulse and fiber parameters. Even the parameter N appearing in Eq. (7.3.1) can be removed by introducing $u = NU$ as a renormalized amplitude. With this change, the NLS equation takes on its canonical form

$$i\frac{\partial u}{\partial \xi} + \frac{1}{2}\frac{\partial^2 u}{\partial \tau^2} + |u|^2 u = 0. \tag{7.3.3}$$

The NLS equation (7.3.3) belongs to a special class of nonlinear partial differential equations that can be solved exactly with a mathematical technique known as the *inverse scattering method* [123]–[125]. It was first solved in 1971 by this method [126].

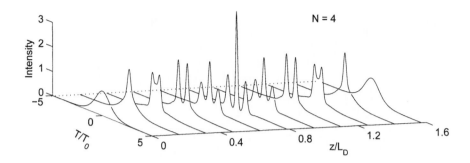

Figure 7.11: Evolution of a fourth-order soliton over one soliton period. The power profile $|u|^2$ is plotted as a function of z/L_D.

The main result can be summarized as follows. When an input pulse having an initial amplitude

$$u(0, \tau) = N \operatorname{sech}(\tau) \tag{7.3.4}$$

is launched into the fiber, its shape remains unchanged during propagation when $N = 1$ but follows a periodic pattern for integer values of $N > 1$ such that the input shape is recovered at $\xi = m\pi/2$, where m is an integer.

An optical pulse whose parameters satisfy the condition $N = 1$ represents a fundamental soliton. Pulses corresponding to other integer values of N are called *higher-order solitons*; that is, N represents the order of the soliton. Noting that $\xi = z/L_D$, the soliton period z_0, defined as the distance over which higher-order solitons recover their original shape, is given by

$$z_0 = \frac{\pi}{2}L_D = \frac{\pi}{2} \frac{T_0^2}{|\beta_2|}. \tag{7.3.5}$$

Figure 7.11 shows the evolution of a fourth-order soliton over one soliton period by solving the NLS equation (7.3.1) numerically with $N = 4$. The pulse shape changes considerably but returns to its original form at $z = z_0$. Only a fundamental soliton maintains its shape during propagation inside optical fibers. In a 1973 study, Eq. (7.3.1) was solved numerically to show that optical solitons can form inside optical fibers [127].

The solution corresponding to the fundamental soliton can be obtained by solving Eq. (7.3.3) directly, without recourse to the inverse scattering method. The approach consists of assuming that a solution of the form

$$u(\xi, \tau) = V(\tau) \exp[i\phi(\xi)] \tag{7.3.6}$$

exists, where V must be independent of ξ for Eq. (7.3.6) to represent a fundamental soliton that maintains its shape during propagation. The phase ϕ can depend on ξ but is assumed to be time independent. When Eq. (7.3.6) is substituted in Eq. (7.3.3) and the real and imaginary parts are separated, we obtain two real equations for V and ϕ. These equations show that ϕ should be of the form $\phi(\xi) = K\xi$, where K is a constant. The function $V(\tau)$ is then found to satisfy the nonlinear differential equation

$$\frac{d^2V}{d\tau^2} = 2V(K - V^2). \tag{7.3.7}$$

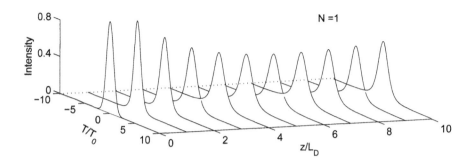

Figure 7.12: Evolution of a Gaussian pulse with $N = 1$ over the range $\xi = 0$ to 10. The pulse evolves toward the fundamental soliton by changing its shape, width, and peak power.

This equation can be solved by multiplying it with $2\,(dV/d\tau)$ and integrating over τ. The result is found to be

$$(dV/d\tau)^2 = 2KV^2 - V^4 + C, \qquad (7.3.8)$$

where C is a constant of integration. Using the boundary condition that both V and $dV/d\tau$ should vanish for any optical pulse at $|\tau| \to \infty$, C can be set to zero.

The constant K in Eq. (7.3.8) is determined using the boundary condition that $V = 1$ and $dV/d\tau = 0$ at the soliton peak, assumed to occur at $\tau = 0$. Its use provides $K = \frac{1}{2}$, resulting in $\phi = \xi/2$. Equation (7.3.8) is easily integrated to obtain $V(\tau) = \mathrm{sech}(\tau)$. We have thus found the well-known "sech" solution [120]–[125]

$$u(\xi, \tau) = \mathrm{sech}(\tau)\exp(i\xi/2) \qquad (7.3.9)$$

for the fundamental soliton by integrating the NLS equation directly. It shows that the input pulse acquires a phase shift $\xi/2$ as it propagates inside the fiber, but its amplitude remains unchanged. It is this property of a fundamental soliton that makes it an ideal candidate for optical communications. In essence, the effects of fiber dispersion are exactly compensated by the fiber nonlinearity when the input pulse has a "sech" shape and its width and peak power are related by Eq. (7.3.2) in such a way that $N = 1$.

An important property of optical solitons is that they are remarkably stable against perturbations. Thus, even though the fundamental soliton requires a specific shape and a certain peak power corresponding to $N = 1$ in Eq. (7.3.2), it can be created even when the pulse shape and the peak power deviate from the ideal conditions. Figure 7.12 shows the numerically simulated evolution of a Gaussian input pulse for which $N = 1$ but $u(0, \tau) = \exp(-\tau^2/2)$. As seen there, the pulse adjusts its shape and width as it propagates down the fiber in an attempt to become a fundamental soliton and attains a "sech" profile for $\xi \gg 1$. Similar behavior is observed when N deviates from 1. It turns out that the Nth-order soliton can form when the input value of N is in the range $N - \frac{1}{2}$ to $N + \frac{1}{2}$ [128]. In particular, the fundamental soliton can be excited for values of N in the range of 0.5 to 1.5.

It may seem mysterious that an optical fiber can force any input pulse to evolve toward a soliton. A simple way to understand this behavior is to think of optical solitons as the temporal modes of a nonlinear waveguide. Higher intensities in the pulse center create a temporal waveguide by increasing the refractive index only in the central part of the pulse. Such a waveguide supports temporal modes just as the core-cladding index difference leads to spatial modes of optical fibers. When the input pulse does not match a temporal mode precisely but is close to it, most of the pulse energy can still be coupled to that temporal mode. The rest of the energy spreads in the form of *dispersive waves*. It will be seen later that such dispersive waves affect system performance and should be minimized by matching the input conditions as close to the ideal requirements as possible. When solitons adapt to perturbations adiabatically, perturbation theory developed specifically for solitons can be used to study how the soliton amplitude, width, frequency, speed, and phase evolve along the fiber.

The NLS equation can be solved with the inverse scattering method even when an optical fiber exhibits normal dispersion [129]. The intensity profile of the resulting solutions exhibits a dip in a uniform background, and it is the dip that remains unchanged during propagation inside an optical fiber [130]. For this reason, such solutions of the NLS equation are called *dark* solitons. Even though dark solitons were observed during the 1980s and their properties have been studied thoroughly [131]–[133], most experiments have employed bright solitons with a "sech" shape. In the following discussion we focus on the fundamental soliton given in Eq. (7.3.9).

7.3.2 Loss-Managed Solitons

As seen in the preceding section, solitons use SPM to maintain their width even in the presence of fiber dispersion. However, this property holds only if soliton energy is maintained inside the fiber. It is not difficult to see that a decrease in pulse energy because of fiber losses would produce pulse broadening simply because a reduced peak power weakens the SPM effect necessary to counteract the GVD. When optical amplifiers are used periodically for compensating fiber losses, soliton energy changes in a periodic fashion. Such energy variations are included in the NLS equation (7.2.7) through the periodic function $p(z)$. In the case of lumped amplifiers, $p(z)$ varies as $e^{-\alpha z}$ between two amplifiers and can change by 20 dB or more over each period. The important question is whether solitons can maintain their shape and width in spite of such large energy fluctuations. It turns out that solitons can remain stable over long distances, provided amplifier spacing L_A is kept much smaller than the dispersion length L_D [134].

In general, changes in soliton energy are accompanied by changes in pulse width. Large rapid variations in $p(z)$ can destroy a soliton if its width changes rapidly through the emission of dispersive waves. The concept of the *path-averaged* soliton [134] makes use of the fact that solitons evolve little over a distance that is short compared with the dispersion length (or soliton period). Thus, when $L_A \ll L_D$, the width of a soliton remains virtually unchanged even if its peak power varies considerably in each section between two amplifiers. In effect, one can replace $p(z)$ by its average value $\bar{p}$ in Eq. (7.2.7) when $L_A \ll L_D$. Noting that $\bar{p}$ is just a constant that modifies γP_0, we recover the standard NLS equation.

From a practical viewpoint, a fundamental soliton can be excited if the input peak power P_s (or energy) of the path-averaged soliton is chosen to be larger by a factor of

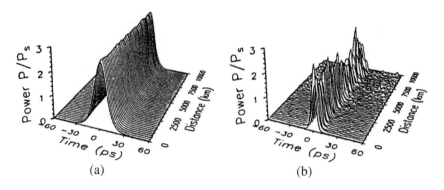

Figure 7.13: Evolution of loss-managed solitons over 10,000 km for (a) $L_D = 200$ km and (b) 25 km with $L_A = 50$ km, $\alpha = 0.22$ dB/km, and $\beta_2 = -0.5$ ps^2/km.

$1/\bar{p}$. If we introduce the amplifier gain as $G = \exp(\alpha L_A)$ and use $\bar{p} = L_A^{-1} \int_0^{L_A} e^{-\alpha z} dz$, the energy enhancement factor for loss-managed solitons is given by

$$f_{\text{LM}} = \frac{P_s}{P_0} = \frac{1}{\bar{p}} = \frac{\alpha L_A}{1 - \exp(-\alpha L_A)} = \frac{G \ln G}{G - 1}. \qquad (7.3.10)$$

Thus, soliton evolution in lossy fibers with periodic lumped amplification is identical to that in lossless fibers, provided amplifiers are spaced such that $L_A \ll L_D$ and the launched peak power is larger by a factor f_{LM}. As an example, $G = 10$ and $f_{\text{LM}} \approx 2.56$ when $L_A = 50$ km and $\alpha = 0.2$ dB/km.

Numerical simulations confirm that the condition $L_A \ll L_D$ must be satisfied for propagating loss-managed solitons over long lengths. Figure 7.13 shows the evolution of a loss-managed soliton over a distance of 10,000 km, assuming that solitons are amplified every 50 km. When the input pulse width corresponds to a dispersion length of 200 km, the soliton is preserved quite well even after 10,000 km because the condition $L_A \ll L_D$ is well satisfied. However, if the dispersion length is reduced to 25 km, the soliton is unable to sustain itself because of the emission of dispersive waves.

How does the limitation on amplifier spacing affect the design of soliton systems? The condition $L_A < L_D$ can be related to the width T_0 through $L_D = T_0^2/|\beta_2|$. The resulting condition is

$$T_0 > \sqrt{|\beta_2|L_A}. \qquad (7.3.11)$$

The pulse width T_0 must be a small fraction of the bit slot $T_b = 1/B$ to ensure that the neighboring solitons are well separated. Mathematically, the soliton solution in Eq. (7.3.9) is valid only when a single pulse propagates by itself. It remains approximately valid for a train of pulses only when individual solitons are well isolated. This requirement can be used to relate the soliton width T_0 to the bit rate B using $T_b = 2q_0 T_0$, where $2q_0$ is a measure of separation between two neighboring pulses in an optical bit stream. Typically, q_0 exceeds 4 to ensure that pulse tails do not overlap significantly. Using $T_0 = (2q_0 B)^{-1}$ in Eq. (7.3.11), we obtain the following design criterion:

$$B^2 L_A < (4q_0^2 |\beta_2|)^{-1}. \qquad (7.3.12)$$

Choosing typical values, $\beta_2 = -2$ ps^2/km, $L_A = 50$ km, and $q_0 = 5$, we obtain $T_0 >$ 10 ps and $B < 10$ GHz. Clearly, the use of path-averaged solitons imposes a severe limitation on both the bit rate and the amplifier spacing for soliton communication systems. To operate even at 10 Gb/s, one must reduce either q_0 or L_A if β_2 is kept fixed. Both of these parameters cannot be reduced much below the values used in obtaining the preceding estimate. A partial solution to this problem was suggested in 2000 when it was proposed that one could prechirp the soliton to relax the condition $L_A \ll L_D$ [135].

The condition $L_A \ll L_D$ can also be relaxed considerably by employing distributed amplification. As discussed in Section 3.2, a distributed-amplification scheme is superior to lumped amplification because its use provides a nearly lossless fiber by compensating losses locally at every point along the fiber link. Historically, distributed Raman amplification of solitons was proposed in 1984 [136] and was used by 1985 in an experiment [137]. A 1988 experiment transmitted solitons over 4000 km using periodic Raman amplification [138]. This experiment used a 42-km recirculating fiber loop whose loss was exactly compensated by injecting pump light from a 1.46-μm color-center laser. Solitons were allowed to circulate many times along the fiber loop and their width was monitored after each round trip. The 55-ps solitons could be circulated along the loop up to 96 times without a significant increase in their pulse width, indicating soliton recovery over 4000 km. The distance could be increased to 6000 km with further optimization. This experiment was the first to demonstrate that solitons could be transmitted over transoceanic distances in principle. The main drawback was that Raman amplification required pump lasers emitting more than 500 mW of power near 1.46 μm. It was not possible to obtain such high powers from semiconductor lasers in 1988, and the color-center lasers used in the experiment were too bulky to be useful for practical lightwave systems.

Starting in 1989, lumped amplifiers were used for loss-managed soliton systems. In a 1991 experiment, 2.5-Gb/s solitons were transmitted over 12,000 km by using a 75-km recirculating loop containing three amplifiers, spaced apart by 25 km [139]. In this experiment, the bit rate–distance product of $BL = 30$ (Tb/s)-km was limited mainly by the timing jitter induced by amplifier noise. During the 1990s, several schemes for reducing the timing jitter were discovered and employed for improving the performance of soliton systems [120]–[122]. Even the technique of distributed Raman amplification was revived around 2000 and is used often for long-haul lightwave systems.

7.3.3 Dispersion-Managed Solitons

The preceding discussion of solitons assumed that the GVD was constant along the fiber link. However, dispersion management is employed commonly for modern WDM lightwave systems as it helps in suppressing FWM among channels. It turns out that solitons can form even when the GVD parameter β_2 varies along the link length but their properties are quite different. In this subsection, we focus on the soliton solutions of Eq. (7.2.7). It includes the effects of loss and dispersion management along the link length through $p(z)$ and $\beta_2(z)$.

Consider first the dispersion-decreasing fibers (DDFs) encountered earlier in Section 6.6.1. It turns out that the restriction $L_A \ll L_D$ imposed normally on loss-managed solitons is relaxed completely in such fibers [140]. Physically, the decreasing GVD counteracts the reduced SPM experienced by solitons weakened from fiber losses. This can be seen from Eq. (7.2.7) by introducing the normalized distance and time variables as

$$\xi = T_0^{-2} \int_0^z |\beta_2(z)| \, dz, \qquad \tau = t/T_0, \qquad (7.3.13)$$

resulting in the following equation:

$$i\frac{\partial U}{\partial \xi} + \frac{1}{2}\frac{\partial^2 U}{\partial \tau^2} + N^2(z)|U|^2 U = 0, \qquad (7.3.14)$$

where $N^2(z) = \gamma P_0 T_0^2 p(z)/|\beta_2(z)|$. If the GVD profile is chosen such that $|\beta_2(z)| = |\beta_2(0)|p(z)$, N becomes a constant and Eq. (7.3.14) reduces to the standard NLS equation obtained earlier. As a result, fiber losses have no effect on a soliton in spite of its reduced energy when DDFs are used.

The main disadvantage of DDFs from a system standpoint is that the average dispersion along the link remains relatively large. Dispersion maps consisting of alternating-GVD fibers are attractive because their use lowers the average dispersion of the entire link, while keeping the GVD of each section large enough that the FWM crosstalk remains negligible in WDM systems.

The use of dispersion management forces each soliton to propagate in the normal-dispersion regime of a fiber during each map period. At first sight, such a scheme should not even work because the normal-GVD fibers do not support solitons and lead to considerable broadening and chirping of the pulse. So, why should solitons survive in a dispersion-managed fiber link? An intense theoretical effort devoted to this issue has led to the discovery of dispersion-managed (DM) solitons [141]–[164]. Physically speaking, if the dispersion length associated with each fiber section used to form the map is a fraction of the nonlinear length, the pulse would evolve in a linear fashion over a single map period. On a longer length scale, solitons can still form if the SPM effects are balanced by the average dispersion. As a result, solitons can survive in an average sense, even though not only the peak power but also the width and shape of such solitons oscillate periodically.

Consider a simple dispersion map consisting of two fibers with opposite GVD characteristics. Soliton evolution is governed by Eq. (7.2.7) in which β_2 is a piece-wise continuous function of z taking values β_{2a} and β_{2n} in the anomalous and normal GVD sections of lengths l_a and l_n, respectively. The map period $L_{map} = l_a + l_n$ can be different from the amplifier spacing L_A. As is evident, the properties of DM solitons will depend on several map parameters even when only two types of fibers are used in each map period. Numerical simulations show that a nearly periodic solution can often be found by adjusting input pulse parameters (width, chirp, and peak power) even though these parameters vary considerably in each map period. The shape of such DM solitons is closer to a Gaussian profile rather than the "sech" shape associated with standard solitons [142]–[144].

Numerical solutions, although essential, do not lead to much physical insight. Several techniques have been used to solve the NLS equation (7.2.7) approximately. One approach makes use of the variational method [145]–[147] or the moment method.

Another expands $B(z,t)$ in terms of a complete set of the Hermite–Gauss functions that are solutions of the linear problem [148]. Still another approach solves an integral equation, derived in the spectral domain using perturbation theory [150]–[152].

Both the variational and moment methods assume that each optical pulse maintains its shape even though its amplitude, width, and chirp may change during propagation. In the case of a chirped Gaussian pulse, we assume that an approximate solution of Eq. (7.2.7) can be written in the form

$$U(z,t) = a \exp[-\tfrac{1}{2}(1+iC)t^2/T^2 + i\phi], \qquad (7.3.15)$$

where a is the amplitude, T is the width, C is the chirp, and ϕ is the phase. All four parameters vary with z. The variational and the moment method then lead to four ordinary differential equations governing the evolution of these four parameters with z. The phase equation can be ignored as it is not coupled to the other three equations. The amplitude equation can be integrated to find that the product a^2T does not vary with z and is related to the input pulse energy E_0 as $E_0 = \sqrt{\pi}P_0 a^2(z)T(z) = \sqrt{\pi}P_0 T(0)$ as $a(0) = 1$. Thus, we need to solve only the following two coupled equations:

$$\frac{dT}{dz} = \frac{\beta_2(z)C}{T}, \qquad (7.3.16)$$

$$\frac{dC}{dz} = (1+C^2)\frac{\beta_2(z)}{T^2} + \frac{\gamma(z)p(z)E_0}{\sqrt{2\pi}\,T}. \qquad (7.3.17)$$

Details of loss and dispersion managements appear in these equations through the z dependence of three parameters β_2, γ, and p.

Equations (7.3.16) and (7.3.17) require values of three pulse parameters at the input end, namely, the width T_0, chirp C_0, and energy E_0, before they can be solved. The pulse energy E_0 is related to the average power launched into the fiber link through the relation $P_{av} = \tfrac{1}{2}BE_0 = (\sqrt{\pi}/2)P_0(T_0/T_b)$, where T_b is the duration of bit slot at the bit rate B. The other two parameters, T_0 and C_0, are found by solving Eqs/ (7.3.16) and (7.3.17) with the periodic boundary conditions to ensure that the DM soliton recovers its initial state after each amplifier. Thus, a new feature of the DM solitons is that the input pulse width and chirp depend on the dispersion map and cannot be chosen arbitrarily.

Figure 7.14 shows how the pulse width T_0 and the chirp C_0 of allowed periodic solutions vary with input pulse energy for a specific dispersion map. The minimum value T_m of the pulse width occurring in the middle of the anomalous-GVD section of the map is also shown. The map is suitable for 40-Gb/s systems and consists of alternating fibers with GVD of -4 and 4 ps^2/km and lengths $l_a \approx l_n = 5$ km such that the average GVD is -0.01 ps^2/km. The solid lines show the case of ideal distributed amplification for which $p(z) = 1$ in Eq. (7.1.10). The lumped-amplification case is shown by the dashed lines in Figure 7.14, assuming 80-km amplifier spacing and 0.25 dB/km losses in each fiber section.

Several conclusions can be drawn from Figure 7.14. First, both T_0 and T_m decrease rapidly as pulse energy is increased. Second, T_0 attains its minimum value at a certain pulse energy E_c while T_m keeps decreasing slowly. Third, T_0 and T_m differ by a large factor for $E_0 \gg E_c$. This behavior indicates that pulse width changes considerably in

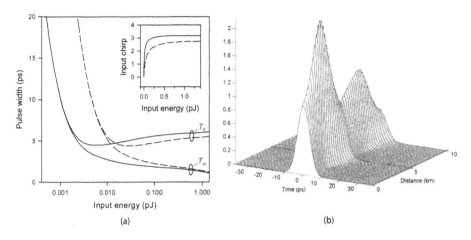

Figure 7.14: (a) Changes in T_0 (upper curve) and T_m (lower curve) with input pulse energy E_0 for $\alpha = 0$ (solid lines) and 0.25 dB/km (dashed lines). The inset shows the input chirp C_0 in the two cases. (b) Evolution of the DM soliton over one map period for $E_0 = 0.1$ pJ and $L_A = 80$ km.

each fiber section when this regime is approached. An example of pulse breathing is shown in Figure 7.14(b) for $E_0 = 0.1$ pJ in the case of lumped amplification. The input chirp C_0 is relatively large ($C_0 \approx 1.8$) in this case. The most important feature of Figure 7.14 is the existence of a minimum value of T_0 for a specific value of the pulse energy. The input chirp $C_0 = 1$ at that point. It is interesting to note that the minimum value of T_0 does not depend much on fiber losses and is about the same for the solid and dashed curves, although the value of E_c is much larger in the lumped amplification case because of fiber losses.

As seen from Figure 7.14, both the pulse width and the peak power of DM solitons vary considerably within each map period. Figure 7.15(a) shows the width and chirp variations over one map period for the DM soliton of Figure 7.14(b). The pulse width varies by more than a factor of 2 and becomes minimum nearly in the middle of each fiber section where frequency chirp vanishes. The shortest pulse occurs in the middle of the anomalous-GVD section in the case of ideal distributed amplification in which fiber losses are compensated fully at every point along the fiber link. For comparison, Figure 7.15(b) shows the width and chirp variations for a DM soliton whose input energy is close to E_c where the input pulse is shortest. Breathing of the pulse is reduced considerably together with the range of chirp variations. In both cases, the DM soliton is quite different from a standard fundamental soliton as it does not maintain its shape, width, or peak power. Nevertheless, its parameters repeat from period to period at any location within the map. For this reason, DM solitons can be used for optical communications in spite of oscillations in the pulse width. Moreover, such solitons perform better from a system standpoint.

Since 1996, a large number of experiments have shown the benefits of DM solitons for lightwave systems [165]–[173]. In one experiment, the use of a periodic dispersion map enabled the transmission of a 20-Gb/s soliton bit stream over 5520 km of a fiber link containing amplifiers at 40-km intervals [165]. In another 20-Gb/s experiment

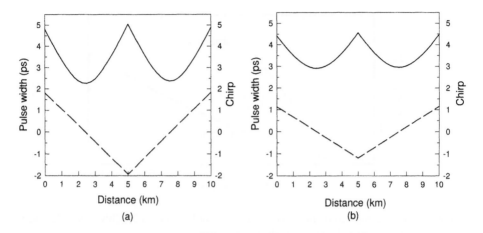

Figure 7.15: Variations of pulse width and chirp (dashed line) over one map period for DM solitons with the input energy (a) $E_0 = 0.1$ pJ and (b) E_0 close to E_c.

[166], solitons could be transmitted over 9000 km without using any in-line optical filters since the periodic use of DCFs reduced timing jitter by a factor of more than 3. A 1997 experiment focused on the transmission of DM solitons using dispersion maps such that solitons propagated most of the time in the normal-GVD regime [167]. This 10-Gb/s experiment transmitted signals over 28,000 km using a recirculating fiber loop consisting of 100 km of normal-GVD fiber and 8 km of anomalous-GVD fiber such that the average GVD was anomalous (about -0.1 ps^2/km). Periodic variations in the pulse width were also observed in such a fiber loop [168]. In a later experiment, the loop was modified to yield the average-GVD value of zero or a slightly positive value [169]. Stable transmission of 10-Gb/s solitons over 28,000 km was still observed. In all cases, experimental results were in excellent agreement with those of numerical simulations [170].

An important application of dispersion management consists of upgrading the existing terrestrial networks designed with standard fibers [171]–[173]. A 1997 experiment used fiber gratings for dispersion compensation and realized 10-Gb/s soliton transmission over 1000 km. Longer transmission distances were realized using a recirculating fiber loop [172] consisting of 102 km of standard fiber with anomalous GVD ($\beta_2 \approx -21$ ps^2/km) and 17.3 km of DCF with normal GVD ($\beta_2 \approx 160$ ps^2/km). The map strength S was quite large in this experiment when 30-ps (FWHM) pulses were launched into the loop. By 1999, 10-Gb/s DM solitons could be transmitted over 16,000 km of standard fiber when soliton interactions were minimized by choosing the location of amplifiers appropriately [85].

7.3.4 Timing Jitter

As discussed earlier in Section 7.1, noise added by optical amplifiers perturbs each optical pulse. It not only reduces the SNR but also shifts the pulse center. It turns out that such timing jitter limits the total transmission distance of any long-haul soliton

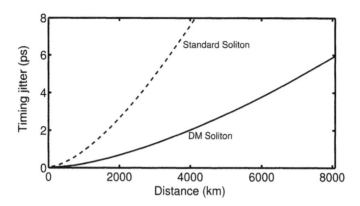

Figure 7.16: ASE-induced timing jitter as a function of length for a 20-Gb/s system designed with DM (solid curve) and standard (dashed line) solitons.

link. This limitation was first pointed out in 1986 in the context of standard constant-width solitons. It persists even for DM soliton systems, although jitter is reduced for them [174]–[186]. In all cases, the dominant source of timing jitter is related to changes in the soliton carrier frequency, occurring because of phase fluctuations induced by amplifier noise.

The moment method has been used to calculate the variance of fluctuations in the pulse position. In the case of DM solitons, the final result can be written as [183]

$$\sigma_t^2 = \frac{S_{ASE}T_m^2}{E_0}[N_A(1+C_0^2) + N_A(N_A-1)C_0d + \tfrac{1}{6}N_A(N_A-1)(2N_A-1)d^2], \quad (7.3.18)$$

where N_A is the number of amplifiers along the link. The dimensionless parameter d is defined as

$$d = \frac{1}{T_m^2}\int_0^{L_{\text{map}}} \beta_2(z)\,dz = \frac{\bar{\beta}_2 L_{\text{map}}}{T_0^2}, \quad (7.3.19)$$

where $\bar{\beta}_2$ is the average value of the dispersion parameter and T_m is the minimum pulse width over a map period L_{map}.

For a soliton system designed with $L_{\text{map}} = L_A$ and $N_A \gg 1$, jitter is dominated by the last term in Eq. (7.3.18) because of its N_A^3 dependence and is approximately given by

$$\frac{\sigma_t^2}{T_m^2} \approx \frac{S_{ASE}}{3E_0}N_A^3 d^2 = \frac{S_{ASE}L_T^3}{3E_0L_D^2L_A}, \quad (7.3.20)$$

where $L_D = T_m^2/|\bar{\beta}_2|$ and $N_A = L_T/L_A$ for a lightwave system with the total transmission distance L_T.

Because of the cubic dependence of σ_t^2 on the system length L_T, the timing jitter can become an appreciable fraction of the bit slot for long-haul systems, especially at bit rates exceeding 10 Gb/s for which the bit slot is shorter than 100 ps. Such jitter would lead to large power penalties if left uncontrolled. As discussed in Section 5.4.5, jitter should be less than 10% of the bit slot in practice. Figure 7.16 shows how timing jitter increases with L_T for a 20-Gb/s DM soliton system, designed using a dispersion map consisting of 10.5 km of anomalous-GVD fiber and 9.7 km of normal-GVD fiber [$D = \pm 4$ ps/(km-nm)]. Optical amplifiers with $n_{sp} = 1.3$ (or a noise figure of 4.1 dB) are placed every 80.8 km (four map periods) along the fiber link for compensating 0.2-dB/km losses. Variational equations were used to find the input pulse parameters for which solitons recover periodically after each map period ($T_0 = 6.87$ ps, $C_0 = 0.56$, and $E_0 = 0.4$ pJ). The nonlinear parameter γ was 1.7 W^{-1}/km.

An important question is whether the use of dispersion management is helpful or harmful from the standpoint of timing jitter. The timing jitter for standard solitons can also be found in a closed form using the moment method and is given by [183]

$$\sigma_t^2 = \frac{S_{ASE}T_0^2}{3E_s}[N_A + \tfrac{1}{6}N_A(N_A - 1)(2N_A - 1)d^2], \tag{7.3.21}$$

where we have used E_s for the input soliton energy to emphasize that it is different from the DM soliton energy E_0 used in Eq. (7.3.18). For a fair comparison of the DM and standard solitons, we consider an identical soliton system except that the dispersion map is replaced by a single fiber whose GVD is constant and equal to the average value $\bar{\beta}_2$. The soliton energy E_s can be found by using $E_0 = 2P_0T_0$ with $P_0 = |\bar{\beta}_2|/(\gamma T_0^2)$ and is given by

$$E_s = 2f_{LM}|\bar{\beta}_2|/(\gamma T_0), \tag{7.3.22}$$

where the factor f_{LM} is the enhancement factor resulting from loss management ($f_{LM} \approx$ 3.8 for a 16-dB gain). The dashed line in Figure 7.16 shows the timing jitter using Eqs. (7.3.21) and (7.3.22). A comparison of the two curves shows that the jitter is considerably smaller for DM solitons. The physical reason behind the jitter reduction is related to the enhanced energy of the DM solitons. From a practical standpoint, the reduced jitter of DM solitons permits much longer transmission distances as evident from Figure 7.16. Note that Eq. (7.3.21) also applies for DDFs because the GVD variations along the fiber can be included through the parameter d defined in Eq. (7.3.19).

For long-haul soliton systems, the number of amplifiers is large enough that the N_A^3 term dominates in Eq. (7.3.21), and the timing jitter for standard solitons is approximately given by [175]

$$\frac{\sigma_t^2}{T_0^2} = \frac{S_{ASE}L_T^3}{9E_sL_D^2L_A}. \tag{7.3.23}$$

Comparing Eqs. (7.3.20) and (7.3.23), we find that timing jitter is reduced for DM solitons because of their much larger pulse energies.

As the timing jitter ultimately limits the performance of soliton systems, it is essential to find a solution to the timing-jitter problem before the use of solitons can become practical. Several techniques were developed during the 1990s for controlling timing jitter [187]–[208]. The use of optical filters for controlling the timing jitter of solitons was proposed as early as 1991 [187]–[189]. This approach makes use of the fact that the ASE occurs over the entire amplifier bandwidth but the soliton spectrum occupies only a small fraction of it. The bandwidth of optical filters is chosen such that the soliton bit stream passes through the filter but most of the ASE is blocked. If an optical filter is placed after each amplifier, it improves the SNR because of the

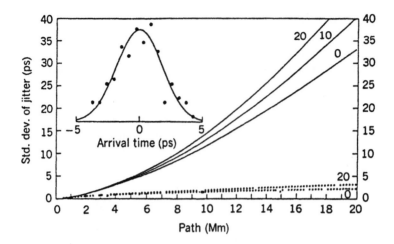

Figure 7.17: Timing jitter with (dotted curves) and without (solid curves) sliding-frequency filters at several bit rates as a function of distance. The inset shows a Gaussian fit to the numerically simulated jitter at 10,000 km for a 10-Gb/s system. (From Ref. [190]; ©1992 OSA.)

reduced ASE and reduces the timing jitter simultaneously. This was indeed found to be the case in a 1991 experiment [188] but the reduction in timing jitter was less than 50%.

The filter technique can be improved dramatically by allowing the center frequency of the successive optical filters to slide slowly along the link. Such sliding-frequency filters avoid the accumulation of ASE within the filter bandwidth and, at the same time, reduce the growth of timing jitter [190]. The physical mechanism behind the operation of such filters can be understood as follows. As the filter passband shifts, solitons shift their spectrum as well to minimize filter-induced losses. In contrast, the spectrum of ASE cannot change. The net result is that the ASE noise accumulated over a few amplifiers is filtered out later when the soliton spectrum has shifted by more than its own bandwidth.

Figure 7.17 shows the predicted reduction in the timing jitter for standard solitons. The bit-rate dependence of jitter is due to the contribution of acoustic waves; the $B = 0$ curves show the contribution of the Gordon–Haus jitter alone. Optical filters help in reducing both types of timing jitter and permit transmission of 10-Gb/s solitons over more than 20,000 km. In the absence of filters, timing jitter becomes so large that a 10-Gb/s soliton system cannot be operated beyond 8000 km. The inset in Figure 7.17 shows a Gaussian fit to the timing jitter of 10-Gb/s solitons at a distance of 10,000 km calculated by solving the NLS equation numerically after including the effects of both the ASE and sliding-frequency filters [190]. The timing-jitter distribution is approximately Gaussian with a standard deviation of about 1.76 ps. In the absence of filters, the jitter exceeds 10 ps under the same conditions.

Optical filters benefit a soliton system in many other ways. Their use reduces interaction between neighboring solitons [191]. The physical mechanism behind the reduced interaction is related to the change in the soliton phase at each filter. A rapid variation of the relative phase between neighboring solitons, occurring as a result of filtering, averages out the impact of soliton interaction by alternating the nature of the interaction force from attractive to repulsive. Optical filters also help in reducing the accumulation of dispersive waves [192]. The reason is easy to understand. As the soliton spectrum shifts with the filters, dispersive waves produced at earlier stages are blocked by filters together with the ASE. In spite of multiple benefits offered by sliding-frequency filters, they are rarely used in practice because of the the network-management issues.

Solitons can also be controlled in the time domain using the technique of *synchronous* amplitude modulation, implemented in practice using a LiNbO$_3$ modulator [193]. The technique works by introducing additional losses for those solitons that have shifted from their original position (center of the bit slot). The modulator forces solitons to move toward its transmission peak where the loss is minimum. As these peaks are uniformly spaced in time, timing jitter is reduced substantially.

The synchronous modulation technique can also be implemented by using a phase modulator [194]. One can understand the effect of periodic phase modulation by recalling that a frequency shift is associated with all time-dependent phase variations. Since a change in soliton frequency is equivalent to a change in the group velocity, phase modulation induces a temporal displacement. Synchronous phase modulation is implemented in such a way that the soliton experiences a frequency shift only if it moves away from the center of the bit slot, which confines it to its original position despite the timing jitter induced by ASE and other sources. Intensity and phase modulations can be combined to further improve system performance [195].

Synchronous modulation can be combined with optical filters to control solitons simultaneously in both the time and frequency domains. In fact, this combination permits arbitrarily long transmission distances [196]. The use of intensity modulators also permits a relatively large amplifier spacing by reducing the impact of dispersive waves. This property of modulators was exploited in 1995 to transmit a 20-Gb/s soliton train over 150,000 km with an amplifier spacing of 105 km [197]. Synchronous modulators also help in reducing the soliton interaction and in clamping the level of amplifier noise. The main drawback of modulators is that they require a clock signal that is synchronized with the original bit stream.

Several other techniques can be used for controlling timing jitter. One approach consists of inserting a fast saturable absorber periodically along the fiber link. Such a device absorbs low-intensity light, such as ASE and dispersive waves, but leaves the solitons intact by becoming transparent at high intensities. To be effective, it should respond at a time scale shorter than the soliton width. It is difficult to find an absorber that can respond at such short time scales. A nonlinear optical loop mirror (see Section 2.3.3) can act as a fast saturable absorber and reduce the timing jitter of solitons, while also stabilizing their amplitude [199]. Retiming of a soliton train can also be accomplished by taking advantage of XPM [200]. The technique overlaps the soliton data stream and another pulse train composed of only 1 bits (an optical clock) inside a fiber where XPM induces a nonlinear phase shift on each soliton in the signal bit stream. Such a phase modulation translates into a net frequency shift only when the soliton does not lie in the middle of the bit slot. Similar to the case of synchronous phase modulation, the direction of the frequency shift is such that the soliton is confined to the center of the bit slot. Other nonlinear effects such as SRS [201] and FWM can also be exploited for controlling the soliton parameters [202]. The technique of distributed amplification also helps in reducing timing jitter [186].

7.4   Pseudo-Linear Lightwave Systems

Pseudo-linear lightwave systems operate in the regime in which the local dispersion length is much shorter than the nonlinear length in all fiber sections of a dispersion-managed link. This approach is most suitable for systems operating at bit rates of 40 Gb/s or more and employing relatively short optical pulses that spread over multiple bits quickly as they propagate along the link. This spreading reduces the peak power and lowers the impact of SPM on each pulse. There are several ways one can design such systems. In one case, pulses spread throughout the link and are compressed back at the receiver end using a dispersion-compensating device. In another, pulses are spread even before the optical signal is launched into the fiber link using a DCF (precompensation) and they compress slowly within the fiber link, without requiring postcompensation.

As optical pulses spread considerably outside their assigned bit slot in all pseudo-linear systems, they overlap and interact with each other through the nonlinear term in the NLS equation. It turns out that the spreading of bits belonging to different WDM channels produces an averaging effect that reduces the *interchannel* nonlinear effects considerably [209]. However, at the same time, an enhanced nonlinear interaction among the 1 bits of the same channel produces new *intrachannel* nonlinear effects that limit the system performance, if left uncontrolled. Thus, pseudo-linear systems are far from being linear. The important question is whether pulse spreading helps to lower the overall impact of fiber nonlinearity and allows higher launched powers into the fiber link. The answer to this question turned out to be affirmative. In this section, we focus on the intrachannel nonlinear effects and study how they affect a pseudo-linear lightwave system.

7.4.1   Intrachannel Nonlinear Effects

The main limitation of pseudo-linear systems stems from the nonlinear interaction among the neighboring overlapping pulses. Starting in 1999, such intrachannel non-linear effects were studied extensively [210]–[224]. In a numerical approach, one solves the NLS equation (7.2.7) for a pseudo-random bit stream with the input

$$U(0,t) = \sum_{j=1}^{M} U_j(0, t - t_j), \qquad (7.4.1)$$

where $t_j = jT_b$, T_b is the duration of the bit slot, M is the total number of bits included in numerical simulations, and $U_m = 0$ if the mth pulse represents a 0 bit. In the case of 1 bits, U_m governs the shape of input pulses.

Although numerical simulations are essential for a realistic system design, considerable physical insight can be gained with a semianalytic approach that focuses on three neighboring pulses. If we write the total field as $U = U_1 + U_2 + U_3$ in Eq. (7.2.7), it reduces to the following set of three coupled NLS equations [209]:

$$i\frac{\partial U_1}{\partial z} - \frac{\beta_2}{2}\frac{\partial^2 U_1}{\partial t^2} + \gamma P_0 p(z)[(|U_1|^2 + 2|U_2|^2 + 2|U_3|^2)U_1 + U_2^2 U_3^*] = 0, \quad (7.4.2)$$

$$i\frac{\partial U_2}{\partial z} - \frac{\beta_2}{2}\frac{\partial^2 U_2}{\partial t^2} + \gamma P_0 p(z)[(|U_2|^2 + 2|U_1|^2 + 2|U_3|^2)U_2 + 2U_1 U_2^* U_3] = 0, \quad (7.4.3)$$

$$i\frac{\partial U_3}{\partial z} - \frac{\beta_2}{2}\frac{\partial^2 U_3}{\partial t^2} + \gamma P_0 p(z)[(|U_3|^2 + 2|U_1|^2 + 2|U_2|^2)U_3 + U_2^2 U_1^*] = 0. \quad (7.4.4)$$

The first nonlinear term corresponds to SPM. The next two terms result from XPM induced by the other two pulses. Since these terms represent XPM interaction between pulses belonging to the same channel, this phenomenon is referred to as *intrachannel XPM*. The last term is FWM-like (see Section 4.3) and is responsible for intrachannel FWM. Although it may seem odd at first sight that FWM can occur among pulses of the same channel, one should remember that the spectrum of each pulse has modulation sidebands located on both sides of the carrier frequency. If different sidebands of two or more overlapping pulses are present simultaneously in the same temporal window, they can interact through FWM and transfer energy among the interacting pulses. This phenomenon can also create new pulses in the time domain. Such pulses are referred to as a *shadow* pulse [210] or a *ghost* pulse [211]. They impact the system performance considerably, especially those created in the 0-bit time slots [219].

We can extend the preceding method to the case of more than three pulses. Assuming that Eq. (7.4.1) can be used at any distance z, the NLS equation (7.2.7) can be written as

$$\sum_{j=1}^{M}\left(i\frac{\partial U_j}{\partial z} - \frac{\beta_2}{2}\frac{\partial^2 U_j}{\partial t^2}\right) = -\gamma P_0 p(z)\sum_{j=1}^{M}\sum_{k=1}^{M}\sum_{l=1}^{M} U_j U_k^* U_l. \quad (7.4.5)$$

The triple sum on the right side includes all the nonlinear effects. SPM occurs when $j = k = l$. The terms responsible for XPM correspond to $j = k \neq l$ and $j \neq k = l$. The remaining terms lead to intrachannel FWM. Each nonlinear term in the triple sum on the right side of Eq. (7.4.5) provides its contribution in a temporal region near $t_j + t_l - t_k$, a relation analogous to the phase-matching condition among waves of different frequencies [209]. This relation can be used to identify all nonlinear terms that can contribute to a specific pulse. It is important to note that, whereas the total energy of all pulses remains constant during propagation, the energy of any individual pulse can change because of intrachannel FWM.

In the case of a single pulse surrounded by several zero bits on both sides, we can set $U_1 = U_3 = 0$ in Eqs. (7.4.2) through (7.4.4). The resulting equation for U_2 is identical to the original NLS equation (7.2.7). The SPM effects in this case have been studied in Section 7.3 through Eqs. (7.3.16) and (7.3.17), obtained with the help of the variational or the moment method. As was found there, the impact of SPM is reduced considerably for pseudo-linear systems because of the much lower peak power of the pulse. It is also reduced because of spectral breathing occurring as the pulse spectrum broadens and narrows from one fiber section to the next. However, the effects of intrachannel XPM and FWM are not negligible. Even though intrachannel

XPM affects only the phase of each pulse, this phase shift is time dependent and affects the carrier frequency of the pulse. As discussed later, the resulting frequency chirp leads to timing jitter through fiber dispersion [213].

The impact of intrachannel XPM and FWM on the performance of a pseudo-linear system depends on the choice of the dispersion map, among other things [209]. In general, optimization of a dispersion-managed system requires adjustment of many design parameters, such as launched power, amplifier spacing, and the location of DCFs [212]. In a 2000 experiment, a 40-Gb/s signal could be transmitted over transoceanic distances, in spite of its use of standard fibers, through a synchronous modulation technique [225]. In a 2002 experiment, distance could be increased to 10^6 km using synchronous modulation in combination with all-optical regeneration [226].

7.4.2   Intrachannel XPM

To understand how intrachannel XPM introduces timing jitter in a pseudo-linear system, consider two isolated 1 bits by setting $U_3 = 0$ in Eqs. (7.4.2) through (7.4.4). The optical field associated with each pulse satisfies an equation of the form

$$i\frac{\partial U_n}{\partial z} - \frac{\beta_2}{2}\frac{\partial^2 U_n}{\partial t^2} + \gamma P_0 p(z)(|U_n|^2 + 2|U_{3-n}|^2)U_n = 0, \qquad (7.4.6)$$

where $n = 1$ or 2. Clearly, the last term is due to XPM. If we ignore the effects of GVD for the moment, this term shows that, over a short distance Δz, the phase of each pulse is shifted nonlinearly by the other pulse by this amount:

$$\phi_n(z,t) = 2\gamma P_0 p(z)\Delta z|U_{3-n}(z,t)|^2. \qquad (7.4.7)$$

As this phase shift depends on pulse shape, it varies across the pulse and produces a frequency chirp

$$\delta\omega_n \equiv -\frac{\partial\phi_n}{\partial t} = -2\gamma P_0 p(z)\Delta z\frac{\partial}{\partial t}|U_{3-n}(z,t)|^2. \qquad (7.4.8)$$

This frequency shift is known as the XPM-induced chirp. Such a shift in the carrier frequency of the pulse translates into a shift in the pulse position through changes in the group velocity of the pulse. If all pulses were to shift in time by the same amount, this effect would be harmless. However, the time shift depends on the pattern of bits surrounding each pulse that varies from bit to bit depending on the data transmitted. As a result, pulses shift in their respective time slots by different amounts, a feature referred to as XPM-induced timing jitter. As will be seen later, XPM also introduces some amplitude fluctuations.

XPM-induced timing jitter depends on details of the dispersion map and can become quite large if care is not taken to suppress it. As an example of the degradation caused by such jitter [209], Figure 7.18 shows the results of numerical simulations when a 128-bit-long pseudo-random bit stream (with a 25-ps bit slot) was propagated through 80 km of dispersion-shifted fiber with $D = 4$ ps/(km-nm). Launched power was taken to be 18 dBm to enhance the XPM effects. The 5-ps Gaussian input pulses were first chirped by propagating them through a precompensation fiber with

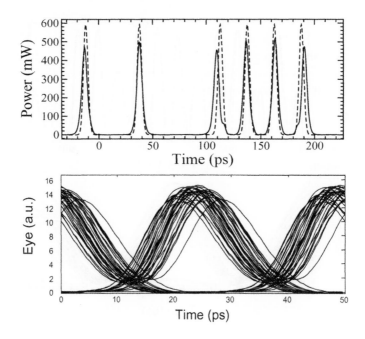

Figure 7.18: Bit stream and eye diagram at the end of a 80-km fiber with $D = 4$ ps/(km-nm). The dashed curve shows the input bit stream. (From Ref. [209]; ©2002 Academic Press.)

$DL = -17$ ps/nm. The output bit stream exhibits both the amplitude and timing jitters that produce severe degradation of the "eye" at the receiver. Numerical simulations, such as those shown in Figure 7.18, are performed with a pseudo-random sequence of 0 and 1 bits. The length of this sequence is typically kept below 512 bits to keep the computation time under control. It turns out that the results for some transmission lines depend on the sequence length and may not provide an accurate estimate until the number of bits is increased to beyond 10,000 [227].

7.4.3 Intrachannel FWM

In contrast to the case of intrachannel XPM, intrachannel FWM transfers energy from one pulse to neighboring pulses. In particular, it can create new pulses in bit slots that represent 0's and contain no pulse initially. Such FWM-generated pulses (called *ghost* or *shadow* pulses) are undesirable for any lightwave system because they can lead to additional errors if their amplitude becomes substantial [210]. Ghost pulses were observed as early as 1992 when a pair of ultrashort pulses, each stretched to 90 ps, was propagated through an optical fiber [228]. However, this phenomenon attracted attention only after 1999 when it was found to impact the performance of lightwave systems employing strong dispersion management [211].

As an example of the system degradation caused by intrachannel FWM [209], Figure 7.19 shows the results of numerical simulations for a 40-Gb/s system at the end of a 80-km-long standard fiber with $D = 17$ ps/(km-nm). The 5-ps Gaussian input

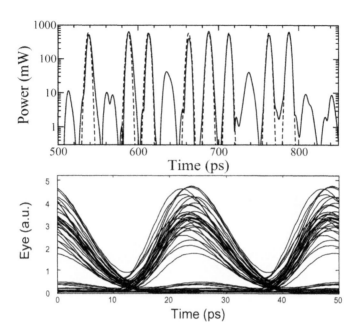

Figure 7.19: Bit stream and eye diagram at the end of a 80-km fiber with $D = 17$ ps/(km-nm). The dashed curve shows the input bit stream. (From Ref. [209]; ©2002 Academic Press.)

pulses were first chirped by propagating them through a precompensation fiber with $DL = -527$ ps/nm. Because of the rapid broadening of input pulses, timing jitter is reduced considerably. However, ghost pulses appear in all 0 slots, and they degrade the eye diagram considerably. Amplitude fluctuations seen in this figure also result from intrachannel FWM.

An analytic treatment of intrachannel FWM is more complex compared with the case of XPM. A perturbative approach has been used with considerable success to describe the impact of intrachannel nonlinearities [213], even though its accuracy decreases rapidly for large timing jitter. Its main advantages are that one does not need to assume a specific pulse shape and that it can be extended easily even to the case of a pseudo-random bit stream [216]–[218]. The main idea is to assume that the solution of the NLS equation (7.2.7) can be written in the form

$$U(z,t) = \sum_{j=1}^{M} U_j(z,t-t_j) + \sum_{j=1}^{M}\sum_{k=1}^{M}\sum_{l=1}^{M} \Delta U_{jkl}(z,t), \qquad (7.4.9)$$

where M the number of bits, U_j represents the amplitude of the jth bit located at $t = t_j$ initially, and ΔU_{jkl} is the perturbation created by the nonlinear term. The first term in Eq. (7.4.9) represents the zeroth-order solution obtained by neglecting the nonlinear term in the NLS equation ($\gamma = 0$). This solution can be obtained in an analytic form. The second term represents the contribution of all nonlinear effects. It can also be obtained in a closed form by employing first-order perturbation theory [213].

As mentioned earlier, intrachannel FWM also leads to amplitude fluctuations. Physically speaking, whenever the perturbation ΔU_{jkl} falls within the bit slots occupied by 1 bits, it beats with the amplitude of that bit. This beating modifies the amplitude of each 1 bit by an amount that depends not only on the pseudo-random bit pattern but also on the relative phases of neighboring pulses. In the case of a periodic dispersion map, energy fluctuations grow only linearly with length of the fiber link [218]. Moreover, they can be reduced considerably by adopting a distributed amplification scheme such that the average power does not vary much along the link.

Problems

7.1 Solve Eq. (7.1.1) for a short lumped amplifier ($\alpha = \beta_2 = \gamma = T_2 = 0$) and prove that the spectral density of ASE noise is given by $S_{sp} = (G - 1)n_{sp}h\nu_0$.

7.2 A 4000-km-long lightwave system is designed with 50 amplifiers spaced 80-km apart. How much will the ASE noise change if 40 amplifiers with 100-km spacing are employed. Assume that the fiber cable has a loss of 0.25 dB/km?

7.3 Explain the basic idea behind dispersion management. Prove that the input pulse shape is recovered at the end of a fiber link whose average GVD is zero when the nonlinear and third-order dispersive effects are negligible.

7.4 The Brillouin threshold of a 20-km-long fiber section is 5 mW. Describe two techniques that would increase it above the 10-mW level.

7.5 Why does the nonlinear phenomenon of SRS lead to crosstalk in a WDM system? How can the crosstalk be reduced in practice?

7.6 Calculate the SPM-induced phase shift at the end of a fiber of length L, neglecting GVD but including fiber losses. Estimate the power level for a π phase shift when $\alpha = 0.2$ dB/km, $\gamma = 2$ W^{-1}/km, and $L = 20$ km.

7.7 Explain why XPM-induced crosstalk is enhanced for certain values of channel spacings in a WDM system.

7.8 Solve Eq. (7.2.19) and prove that the FWM efficiency is indeed given by Eq. (7.2.21).

7.9 A 10-Gb/s soliton system is operating at 1.55 μm using fibers with a constant dispersion of $D = 2$ ps/(km-nm). The effective core area of the fiber is 50 μm^2. Calculate the peak power and the pulse energy required for fundamental solitons of 30-ps width (FWHM). Use $n_2 = 2.6 \times 10^{-20}$ m^2/W.

7.10 Verify by direct substitution that the soliton solution given in Eq. (7.3.9) satisfies the NLS equation.

7.11 Solve the NLS equation (7.3.1) numerically and plot the evolution of the third- and fifth-order solitons over one soliton period. Compare your results with those shown in Figure 7.11 and comment on the main differences.

7.12 Verify numerically by propagating a fundamental soliton over 100 dispersion lengths that the shape of the soliton does not change on propagation. Repeat the simulation using a Gaussian input pulse shape with the same peak power and explain the results.

7.13 A 10-Gb/s soliton lightwave system is designed with $T_0/T_b = 0.1$ to ensure well-separated solitons. Calculate pulse width, peak power, pulse energy, and the average power of the signal, assuming $\beta_2 = -1$ ps^2/km and $\gamma = 2$ W^{-1}/km.

7.14 Prove that the energy of standard solitons should be increased by the factor $G \ln G/(G-1)$ when fiber loss α is compensated periodically using optical amplifiers. Here, $G = \exp(\alpha L_A)$ is the amplifier gain and L_A is the spacing between two neighboring amplifiers.

7.15 A 10-Gb/s soliton communication system is designed with 50-km amplifier spacing. What should the peak power of the input pulse be to ensure that a fundamental soliton is maintained in an average sense in a fiber with 0.2 dB/km loss? Assume 50-ps pulse width (FWHM), $\beta_2 = -0.5$ ps^2/km, and $\gamma = 2$ W^{-1}/km. What is the average launched power for such a system?

7.16 Use the NLS equation (7.2.7) to prove that solitons remain unperturbed by fiber losses when fiber dispersion decreases exponentially as $\beta_2(z) = \beta_2(0) \exp(-\alpha z)$.

References

[1] R. H. Stolen, *Proc. IEEE* **68**, 1232 (1980).

[2] A. R. Chraplyvy, *J. Lightwave Technol.* **8**, 1548 (1990).

[3] F. Forghieri, R. W. Tkach, and A. R. Chraplyvy, in *Optical Fiber Telecommunications*, I. P. Kaminow and T. L. Koch, Eds., Vol. 3A (Academic Press, Boston, 1997), Chap. 8.

[4] I. P. Kaminow and T. Li, Eds., *Optical Fiber Telecommunications*, Vol. 4B (Academic Press, Boston, 2002), Chaps. 6 and 13.

[5] G. P. Agrawal, *Nonlinear Fiber Optics*, 4th ed. (Academic Press, Boston, 2007).

[6] G. P. Agrawal, *Fiber-Optic Communication Systems*, 3rd ed. (Wiley, Hoboken, NJ, 2002).

[7] G. Keiser, *Optical Fiber Communications*, 3rd ed. (McGraw-Hill, New York, 2000).

[8] G. P. Agrawal, *Lightwave Technology: Teleommunication Systems* (Wiley, Hoboken, NJ, 2005).

[9] P. C. Becker, N. A. Olsson, and J. R. Simpson, *Erbium-Doped Fiber Amplifiers: Fundamentals and Technology* (Academic Press, Boston, 1999).

[10] E. P. Ippen and R. H. Stolen, *Appl. Phys. Lett.* **21**, 539 (1972).

[11] R. G. Smith, *Appl. Opt.* **11**, 2489 (1972).

[12] D. Cotter, *Electron. Lett.* **18**, 495 (1982); *J. Opt. Commun.* **4**, 10 (1983).

[13] T. Sugie, *J. Lightwave Technol.* **9**, 1145 (1991); *IEEE Photon. Technol. Lett.* **5**, 102 (1992).

[14] X. P. Mao, R. W. Tkach, A. R. Chraplyvy, R. M. Jopson, and R. M. Derosier, *IEEE Photon. Technol. Lett.* **4**, 66 (1992).

[15] D. A. Fishman and J. A. Nagel, *J. Lightwave Technol.* **11**, 1721 (1993).

[16] T. Sugie, *Opt. Quantum Electron.* **27**, 643 (1995).

[17] D. Cotter, *Electron. Lett.* **18**, 504 (1982).

[18] R. G. Waarts and R. P. Braun, *Electron. Lett.* **21**, 1114 (1985).

[19] Y. Aoki, K. Tajima, and I. Mito, *J. Lightwave Technol.* **6**, 710 (1988).

[20] E. Lichtman, *Electron. Lett.* **27**, 759 (1991).

[21] A. Hirose, Y. Takushima, and T. Okoshi, *J. Opt. Commun.* **12**, 82 (1991).

[22] T. Sugie, *J. Lightwave Technol.* **9**, 1145 (1991); *IEEE Photon. Technol. Lett.* **5**, 102 (1993).

[23] M. O. van Deventer, J. J. G. M. van der Tol, and A. J. Boot, *IEEE Photon. Technol. Lett.* **6**, 291 (1994).

[24] S. Rae, I. Bennion, and M. J. Carswell, *Opt. Commun.* **123**, 611 (1996).

[25] A. Djupsjöbacka, G. Jacobsen, and B. Tromborg, *J. Lightwave Technol.* **18**, 416 (2000).

[26] D. Cotter, *Electron. Lett.* **18**, 638 (1982).

[27] M. Tsubokawa, S. Seikai, T. Nakashima, and N. Shibata, *Electron. Lett.* **22**, 473 (1986).

[28] N. Yoshizawa and T. Imai, *J. Lightwave Technol.* **11**, 1518 (1993).

[29] K. Shiraki, M. Ohashi, and M. Tateda, *Electron. Lett.* **31**, 668 (1995).

[30] K. Shiraki, M. Ohashi, and M. Tateda, *J. Lightwave Technol.* **14**, 50 (1996).

[31] M. M. Howerton, W. K. Burns, and G. K. Gopalakrishnan, *J. Lightwave Technol.* **14**, 417 (1996).

[32] J. Hansryd, F. Dross, M. Westlund, P. A. Andrekson, and S. N. Knudsen, *J. Lightwave Technol.* **19**, 1691 (2001).

[33] H. Lee and G. P. Agrawal, *Opt. Express* **11**, 3467 (2003).

[34] A. Kobyakov, S. Kumar, D. Q. Chowdhury, A. Boh Ruffin, M. Sauer, S. R. Bickham, and R. Mishra, *Opt. Express* **13**, 5338 (2005).

[35] R. H. Stolen, E. P. Ippen, and A. R. Tynes, *Appl. Phys. Lett.* **20**, 62 (1972).

[36] A. R. Chraplyvy, *Electron. Lett.* **20**, 58 (1984).

[37] A. M. Hill, D. Cotter, and I. Wright, *Electron. Lett.* **20**, 247 (1984).

[38] M. S. Kao and J. Wu, *J. Lightwave Technol.* **7**, 1290 (1989).

[39] A. R. Chraplyvy and R. W. Tkach, *IEEE Photon. Technol. Lett.* **5**, 666 (1993).

[40] F. Forghieri, R. W. Tkach, and A. R. Chraplyvy, *IEEE Photon. Technol. Lett.* **7**, 101 (1995).

[41] S. Tariq and J. C. Palais, *Fiber Integ. Opt.* **15**, 335 (1996).

[42] D. N. Christodoulides and R. B. Jander, *IEEE Photon. Technol. Lett.* **8**, 1722 (1996).

[43] J. Wang, X. Sun, and M. Zhang, *IEEE Photon. Technol. Lett.* **10**, 540 (1998).

[44] M. E. Marhic, F. S. Yang, and L. G. Kazovsky, *J. Opt. Soc. Am. B* **15**, 957 (1998).

[45] S. Bigo, S. Gauchard. A Bertaina, and J. P. Hamaide, *IEEE Photon. Technol. Lett.* **11**, 671 (1999).

[46] A. G. Grandpierre, D. N. Christodoulides, and J. Toulouse, *IEEE Photon. Technol. Lett.* **11**, 1271 (1999).

[47] K.-P. Ho, *J. Lightwave Technol.* **18**, 915 (2000).

[48] S. Norimatsu and T. Yamamoto, *J. Lightwave Technol.* **19**, 159 (2001).

[49] X, Zhou, and M. Birk, *J. Lightwave Technol.* **21**, 2194 (2003).

[50] T. Yamamoto and S. Norimatsu, *J. Lightwave Technol.* **21**, 2229 (2003).

[51] D. Anderson, *Phys. Rev. A* **27**, 3135 (1983).

[52] M. J. Potasek, G. P. Agrawal, and S. C. Pinault, *J. Opt. Soc. Am. B* **3**, 205 (1986).

[53] D. Marcuse, *J. Lightwave Technol.* **10**, 17 (1992).

[54] P. A. Bélanger and N. Bélanger, *Opt. Commun.* **117**, 56 (1995).

[55] N. Kikuchi and S. Sasaki, *J. Lightwave Technol.* **13**, 868 (1995).

[56] M. Florjanczyk and R. Tremblay, *J. Lightwave Technol.* **13**, 1801 (1995).

[57] Q. Yu and C. Fan, *J. Lightwave Technol.* **15**, 444 (1997).

[58] M. J. Potasek and G. P. Agrawal, *Electron. Lett.* **22**, 759 (1986).

[59] M. Suzuki, I. Morita, N. Edagawa, S. Yamamoto, H. Taga, and S. Akiba, *Electron. Lett.* **31**, 2027 (1995).

[60] S. Wabnitz, I. Uzunov, and F. Lederer, *IEEE Photon. Technol. Lett.* **8**, 1091 (1996).

[61] A. Naka, T. Matsuda, and S. Saito, *Electron. Lett.* **32**, 1694 (1996).

[62] J. P. Hamide, P. Emplit, and J. M. Gabriagues, *Electron. Lett.* **26**, 1452 (1990).

[63] D. Marcuse, *J. Lightwave Technol.* **9**, 356 (1991).

[64] M. Murakami and S. Saito, *IEEE Photon. Technol. Lett.* **4**, 1269 (1992).

[65] K. Kikuchi, *IEEE Photon. Technol. Lett.* **5**, 221 (1993).

[66] F. Matera, A. Mecozzi, M. Romagnoli, and M. Settembre, *Opt. Lett.* **18**, 1499 (1993).

[67] A. Mecozzi, *J. Opt. Soc. Am. B* **11**, 462 (1994).

[68] M. Yu, G. P. Agrawal, and C. J. McKinstrie, *J. Opt. Soc. Am. B* **12**, 1126 (1995).

[69] N. J. Smith and N. J. Doran, *Opt. Lett.* **21**, 570 (1996).

[70] C. Lorattanasane and K. Kikuchi, *IEEE J. Quantum Electron.* **33**, 1084 (1997).

[71] R. A. Saunders, B. A. Patel, and D. Garthe, *IEEE Photon. Technol. Lett.* **9**, 699 (1997).

[72] R. Q. Hui, M. O'Sullivan, A. Robinson, and M. Taylor, *J. Lightwave Technol.* **15**, 1071 (1997).

[73] E. Ciaramella and M. Tamburrini, *IEEE Photon. Technol. Lett.* **11**, 1608 (1999).

[74] A. R. Chraplyvy and J. Stone, *Electron. Lett.* **20**, 996 (1984).

[75] D. Marcuse, A. R. Chraplyvy, and R. W. Tkach, *J. Lightwave Technol.* **12**, 885 (1994).

[76] T. K. Chiang, N. Kagi, M. E. Marhic, and L. G. Kazovsky, *J. Lightwave Technol.* **14**, 249 (1996).

[77] M. Shtaif, M. Eiselt, R. W. Tkach, R. H. Stolen, and A. H. Gnauck, *IEEE Photon. Technol. Lett.* **10**, 1796 (1998).

[78] A. V. T. Cartaxo, *J. Lightwave Technol.* **17**, 178 (1999).

[79] L. Rapp, *J. Opt. Commun.* **20**, 29 (1999); *J. Opt. Commun.* **20**, 144 (1999).

[80] R. Hui, K. R. Demarest, and C. T. Allen, *J. Lightwave Technol.* **17**, 1018 (1999).

[81] S. Bigo, G. Bellotti, and M. W. Chbat, *IEEE Photon. Technol. Lett.* **11**, 605 (1999).

[82] M. Eiselt, M. Shtaif, and L. D. Garett, *IEEE Photon. Technol. Lett.* **11**, 748 (1999).

[83] L. E. Nelson, R. M. Jopson, A. H. Gnauck, and A. R. Chraplyvy, *IEEE Photon. Technol. Lett.* **11**, 907 (1999).

[84] M. Eiselt, M. Shtaif, R. W. Tkach, F. A. Flood, S. Ten, and D. Butler, *IEEE Photon. Technol. Lett.* **11**, 1575 (1999).

[85] G. J. Pendock, S. Y. Park, A. K. Srivastava, S. Radic, J. W. Sulhoff, C. L. Wolf, K. Kantor, and Y. Sun, *IEEE Photon. Technol. Lett.* **11**, 1578 (1999).

[86] M. Shtaif, M. Eiselt, and L. D. Garettt, *IEEE Photon. Technol. Lett.* **13**, 88 (2000).

[87] J. J. Yu and P. Jeppesen, *Opt. Commun.* **184**, 367 (2000).

[88] R. I. Killey, H. J. Thiele, V. Mikhailov, and P. Bayvel, *IEEE Photon. Technol. Lett.* **12**, 804 (2000).

[89] G. Bellotti, S. Bigo, P. Y. Cortes, S. Gauchard, and S. LaRochelle, *IEEE Photon. Technol. Lett.* **12**, 1403 (2000).

[90] S. Betti and M. Giaconi, *IEEE Photon. Technol. Lett.* **13**, 1304 (2001).

[91] H. J. Thiele, R. I. Killey, and P. Bayvel, *Opt. Fiber Technol.* **8**, 71 (2002).

[92] Q. Lin and G. P. Agrawal, *J. Lightwave Technol.* **22**, 977 (2004).

[93] R. S. Luís and A. V. T. Cartaxo, *J. Lightwave Technol.* **23**, 1503 (2005).

[94] H. Kim, *J. Lightwave Technol.* **21**, 1770 (2003).

[95] A. H. Gnauck and P. J. Winzer, *J. Lightwave Technol.* **23**, 115 (2005).

[96] K. P. Ho and H. C. Wang, *J. Lightwave Technol.* **24**, 396 (2006).

[97] N. Shibata, R. P. Braun, and R. G. Waarts, *IEEE J. Quantum Electron.* **23**, 1205 (1987).

[98] M. W. Maeda, W. B. Sessa, W. I. Way, A. Yi-Yan, L. Curtis, R. Spicer, and R. I. Laming, *J. Lightwave Technol.* **8**, 1402 (1990).

[99] K. Inoue, *J. Lightwave Technol.* **10**, 1553 (1992); **12**, 1023 (1994).

[100] A. R. Chraplyvy, A. H. Gnauck, R. W. Tkach, and R. M. Derosier, *IEEE Photon. Technol. Lett.* **5**, 1233 (1993).

[101] K. Inoue, K. Nakanishi, K. Oda, and H. Toba, *J. Lightwave Technol.* **12**, 1423 (1994).

[102] K. Inoue and H. Toba, *J. Lightwave Technol.* **13**, 88 (1995).

[103] F. Forghieri, R. W. Tkach, and A. R. Chraplyvy, *J. Lightwave Technol.* **13**, 889 (1995).

[104] A. Yu and M. J. O'Mahony, *IEE Proc.* **142**, 190 (1995).

[105] H. Taga, *J. Lightwave Technol.* **14**, 1287 (1996).

[106] W. Zeiler, F. Di Pasquale, P. Bayvel, and J. E. Midwinter, *J. Lightwave Technol.* **14**, 1933 (1996).

[107] J. S. Lee, D. H. Lee, and C. S. Park, *IEEE Photon. Technol. Lett.* **10**, 825 (1998).

[108] D. F. Grosz, C. Mazzali, S. Celaschi, A. Paradisi, and H. L. Fragnito, *IEEE Photon. Technol. Lett.* **11**, 379 (1999).

[109] M. Eiselt, *J. Lightwave Technol.* **17**, 2261 (1999).

[110] S. Song, C. T. Allen, K. R. Demarest, and R. Hui, *J. Lightwave Technol.* **17**, 2285 (1999).

[111] H. Suzuki, S. Ohteru, and N. Takachio, *IEEE Photon. Technol. Lett.* **11**, 1677 (1999).

[112] M. Nakajima, M. Ohashi, K. Shiraki, T. Horiguchi, K. Kurokawa, and Y. Miyajima, *J. Lightwave Technol.* **17**, 1814 (1999).

[113] M. Eiselt, *J. Lightwave Technol.* **17**, 2261 (1999).

[114] K.-D. Chang, G.-C. Yang, and W. C. Kwong, *J. Lightwave Technol.* **18**, 2113 (2000).

[115] M. Manna and E. A. Golovchenko, *IEEE Photon. Technol. Lett.* **13**, 929 (2002).

[116] S. Betti, M. Giaconi, and M. Nardini, *IEEE Photon. Technol. Lett.* **14**, 1079 (2003).

[117] M. Wu and W. I. Way, *J. Lightwave Technol.* **22**, 1483 (2004).

[118] I. Neokosmidis, T. Kamalakis, A. Chipouras, and T. Sphicopoulos, *J. Lightwave Technol.* **23**, 1137 (2005).

[119] E. Yamazaki, F. Inuzuka, K. Yonenaga, A. Takada, and M. Koga, *IEEE Photon. Technol. Lett.* **19**, 9 (2007).

[120] A. Hasegawa and M. Matsumoto, *Optical Solitons in Fibers* (Springer, New York, 2002).

[121] Y. S. Kivshar and G. P. Agrawal, *Optical Solitons* (Academic Press, Boston, 2003), Chap. 4.

[122] L. F. Mollenauer and J. P. Gordon, *Solitons in Optical Fibers* (Academic Press, Boston, 2007).

[123] M. J. Ablowitz and P. A. Clarkson, *Solitons, Nonlinear Evolution Equations, and Inverse Scattering* (Cambridge University Press, New York, 2003).

[124] P. G. Drazin and R. S. Johnson, *Solitons: an Introduction* (Cambridge University Press, New York, 2002).

[125] M. Remoissenet, *Waves Called Solitons: Concepts and Experiments* (Springer, New York, 2003).

[126] V. E. Zakharov and A. B. Shabat, *Sov. Phys. JETP* **34**, 62 (1972).

[127] A. Hasegawa and F. Tappert, *Appl. Phys. Lett.* **23**, 142 (1973).

[128] J. Satsuma and N. Yajima, *Prog. Theor. Phys.* **55**, 284 (1974).

[129] V. E. Zakharov and A. B. Shabat, *Sov. Phys. JETP* **37**, 823 (1973).

[130] A. Hasegawa and F. Tappert, *Appl. Phys. Lett.* **23**, 171 (1973).

[131] W. J. Tomlinson, R. J. Hawkins, A. M. Weiner, J. P. Heritage, and R. N. Thurston, *J. Opt. Soc. Am. B* **6**, 329 (1989).

[132] P. Emplit, M. Haelterman, and J. P. Hamaide, *Opt. Lett.* **18**, 1047 (1993).

[133] Y. S. Kivshar and B. Luther-Davies, *Phys. Rep.* **298**, 81 (1998).

[134] Y. Kodama and A. Hasegawa, *Opt. Lett.* **7**, 339 (1982); **8**, 342 (1983).

[135] Z. M. Liao, C. J. McKinstrie, and G. P. Agrawal, *J. Opt. Soc. Am. B* **17**, 514 (2000).

[136] A. Hasegawa, *Opt. Lett.* **8**, 650 (1983); *Appl. Opt.* **23**, 3302 (1984).

[137] L. F. Mollenauer, R. H. Stolen, and M. N. Islam, *Opt. Lett.* **10**, 229 (1985).

[138] L. F. Mollenauer and K. Smith, *Opt. Lett.* **13**, 675 (1988).

[139] L. F. Mollenauer, B. M. Nyman, M. J. Neubelt, G. Raybon, and S. G. Evangelides, *Electron. Lett.* **27**, 178 (1991).

[140] K. Tajima, *Opt. Lett.* **12**, 54 (1987).

[141] N. J. Smith, F. M. Knox, N. J. Doran, K. J. Blow, and I. Bennion, *Electron. Lett.* **32**, 54 (1996).

[142] M. Nakazawa, H. Kubota, and K. Tamura, *IEEE Photon. Technol. Lett.* **8**, 452 (1996).

[143] M. Nakazawa, H. Kubota, A. Sahara, and K. Tamura, *IEEE Photon. Technol. Lett.* **8**, 1088 (1996).

[144] A. B. Grudinin and I. A. Goncharenko, *Electron. Lett.* **32**, 1602 (1996).

[145] A. Berntson, N. J. Doran, W. Forysiak, and J. H. B. Nijhof, *Opt. Lett.* **23**, 900 (1998).

[146] J. N. Kutz, P. Holmes, S. G. Evangelides, and J. P. Gordon, *J. Opt. Soc. Am. B* **15**, 87 (1998).

[147] S. K. Turitsyn, I. Gabitov, E. W. Laedke, V. K. Mezentsev, S. L. Musher, E. G. Shapiro, T. Schafer, and K. H. Spatschek, *Opt. Commun.* **151**, 117 (1998).

[148] T. I. Lakoba and D. J. Kaup, *Phys. Rev. E* **58**, 6728 (1998).

[149] S. K. Turitsyn and E. G. Shapiro, *J. Opt. Soc. Am. B* **16**, 1321 (1999).

[150] I. R. Gabitov, E. G. Shapiro, and S. K. Turitsyn, *Phys. Rev. E* **55**, 3624 (1997).

[151] M. J. Ablowitz and G. Bioindini, *Opt. Lett.* **23**, 1668 (1998).

[152] C. Paré and P. A. Bélanger, *Opt. Lett.* **25**, 881 (2000).

[153] J. H. B. Nijhof, N. J. Doran, W. Forysiak, and F. M. Knox, *Electron. Lett.* **33**, 1726 (1997).

[154] V. S. Grigoryan and C. R. Menyuk, *Opt. Lett.* **23**, 609 (1998).

[155] J. N. Kutz and S. G. Evangelides, *Opt. Lett.* **23**, 685 (1998).

[156] Y. Chen and H. A. Haus, *Opt. Lett.* **23**, 1013 (1998).

[157] J. H. B. Nijhof, W. Forysiak, and N. J. Doran, *Opt. Lett.* **23**, 1674 (1998).

[158] S. K. Turitsyn, J. H. B. Nijhof, V. K. Mezentsev, and N. J. Doran, *Opt. Lett.* **24**, 1871 (1999).

[159] S. K. Turitsyn, M. P. Fedoruk, and A. Gornakova, *Opt. Lett.* **24**, 969 (1999).

[160] L. J. Richardson, W. Forysiak, and N. J. Doran, *IEEE Photon. Technol. Lett.* **13**, 209 (2001).

[161] E. Poutrina and G. P. Agrawal, *Opt. Commun.* **206**, 193 (2002).

[162] S. Waiyapot, S. K. Turitsyn, and V. K. Mezentsev, *J. Lightwave Technol.* **20**, 2220 (2002).

[163] C. Xie, L. F. Mollenauer, and N. Mamysheva, *J. Lightwave Technol.* **21**, 769 (2003).

[164] E. Poutrina and G. P. Agrawal, *J. Lightwave Technol.* **21**, 990 (2003).

[165] A. Naka, T. Matsuda, and S. Saito, *Electron. Lett.* **32**, 1694 (1996).

[166] I. Morita, M. Suzuki, N. Edagawa, S. Yamamoto, H. Taga, and S. Akiba, *IEEE Photon. Technol. Lett.* **8**, 1573 (1996).

[167] J. M. Jacob, E. A. Golovchenko, A. N. Pilipetskii, G. M. Carter, and C. R. Menyuk, *IEEE Photon. Technol. Lett.* **9**, 130 (1997).

[168] G. M. Carter and J. M. Jacob, *IEEE Photon. Technol. Lett.* **10**, 546 (1998).

[169] V. S. Grigoryan, R. M. Mu, G. M. Carter, and C. R. Menyuk, *IEEE Photon. Technol. Lett.* **10**, 45 (2000).

[170] R. M. Mu, C. R. Menyuk, G. M. Carter, and J. M. Jacob, *IEEE J. Sel. Topics Quantum Electron.* **6**, 248 (2000).

[171] A. B. Grudinin, M. Durkin, M. Isben, R. I. Laming, A. Schiffini, P. Franco, E. Grandi, and M. Romagnoli, *Electron. Lett.* **33**, 1572 (1997).

[172] F. Favre, D. Le Guen, and T. Georges, *J. Lightwave Technol.* **17**, 1032 (1999).

[173] M. Zitelli, F. Favre, D. Le Guen, and S. Del Burgo, *IEEE Photon. Technol. Lett.* **9**, 904 (1999).

[174] J. P. Gordon and H. A. Haus, *Opt. Lett.* **11**, 665 (1986).

[175] D. Marcuse, *J. Lightwave Technol.* **10**, 273 (1992).

[176] N. J. Smith, W. Foryisak, and N. J. Doran, *Electron. Lett.* **32**, 2085 (1996).

[177] G. M. Carter, J. M. Jacob, C. R. Menyuk, E. A. Golovchenko, A. N. Pilipetskii, *Opt. Lett.* **22**, 513 (1997).

[178] S. Kumar and F. Lederer, *Opt. Lett.* **22**, 1870 (1997).

[179] J. N. Kutz and P. K. A. Wai, *IEEE Photon. Technol. Lett.* **10**, 702 (1998).

[180] T. Okamawari, A. Maruta, and Y. Kodama, *Opt. Lett.* **23**, 694 (1998); *Opt. Commun.* **149**, 261 (1998).

[181] V. S. Grigoryan, C. R. Menyuk, and R. M. Mu, *J. Lightwave Technol.* **17**, 1347 (1999).

[182] M. F. S. Ferreira and S. C. V. Latas, *J. Lightwave Technol.* **19**, 332 (2001).

[183] J. Santhanam, C. J. McKinstrie, T. I. Lakoba, and G. P. Agrawal, *Opt. Lett.* **26**, 1131 (2001).

[184] C. J. McKinstrie, J. Santhanam, and G. P. Agrawal, *J. Opt. Soc. Am. B* **19**, 640 (2002).

[185] J. Santhanam and G. P. Agrawal, *IEEE J. Sel. Topics Quantum Electron.* **7**, 632 (2002).

[186] E. Poutrina and G. P. Agrawal, *IEEE Photon. Technol. Lett.* **14**, 39 (2002); *J. Lightwave Technol.* **20**, 762 (2002).

[187] A. Mecozzi, J. D. Moores, H. A. Haus, and Y. Lai, *Opt. Lett.* **16**, 1841 (1991).

[188] L. F. Mollenauer, M. J. Neubelt, M. Haner, E. Lichtman, S. G. Evangelides, and B. M. Nyman, *Electron. Lett.* **27**, 2055 (1991).

[189] Y. Kodama and A. Hasegawa, *Opt. Lett.* **17**, 31 (1992).

[190] L. F. Mollenauer, J. P. Gordon, and S. G. Evangelides, *Opt. Lett.* **17**, 1575 (1992).

[191] V. V. Afanasjev, *Opt. Lett.* **18**, 790 (1993).

[192] M. Romagnoli, S. Wabnitz, and M. Midrio, *Opt. Commun.* **104**, 293 (1994).

[193] M. Nakazawa, E. Yamada, H. Kubota, and K. Suzuki, *Electron. Lett.* **27**, 1270 (1991).

[194] N. J. Smith, W. J. Firth, K. J. Blow, and K. Smith, *Opt. Lett.* **19**, 16 (1994).

[195] S. Bigo, O. Audouin, and E. Desurvire, *Electron. Lett.* **31**, 2191 (1995).

[196] M. Nakazawa, K. Suzuki, E. Yamada, H. Kubota, Y. Kimura, and M. Takaya, *Electron. Lett.* **29**, 729 (1993).

[197] G. Aubin, E. Jeanny, T. Montalant, J. Moulu, F. Pirio, J.-B. Thomine, and F. Devaux, *Electron. Lett.* **31**, 1079 (1995).

[198] W. Forysiak, K. J. Blow. and N. J. Doran, *Electron. Lett.* **29**, 1225 (1993).

[199] M. Matsumoto, H. Ikeda, and A. Hasegawa, *Opt. Lett.* **19**, 183 (1994).

[200] T. Widdowson, D. J. Malyon, A. D. Ellis, K. Smith, and K. J. Blow, *Electron. Lett.* **30**, 990 (1994).

[201] S. Kumar and A. Hasegawa, *Opt. Lett.* **20**, 1856 (1995).

[202] V. S. Grigoryan, A. Hasegawa, and A. Maruta, *Opt. Lett.* **20**, 857 (1995).

[203] M. Matsumoto, *J. Opt. Soc. Am. B* **15**, 2831 (1998); *Opt. Lett.* **23**, 1901 (1998).

[204] S. K. Turitsyn and E. G. Shapiro, *J. Opt. Soc. Am. B* **16**, 1321 (1999).

[205] S. Waiyapot and M. Matsumoto, *IEEE Photon. Technol. Lett.* **11**, 1408 (1999).

[206] M. F. S. Ferreira and S. H. Sousa, *Electron. Lett.* **37**, 1184 (2001).

[207] M. Matsumoto, *Opt. Lett.* **23**, 1901 (2001).

[208] J. Santhanam and G. P. Agrawal, *J. Opt. Soc. Am. B* **20**, 284 (2003).

[209] R.-J. Essiambre, G. Raybon, and B. Mikkelsen, in *Optical Fiber Telecommunications*, Vol. 4B, I. P. Kaminow and T. Li, Eds. (Academic Press, Boston, 2002), Chap. 6.

[210] R.-J. Essiambre, B. Mikkelsen, and G. Raybon, *Electron. Lett.* **35**, 1576 (1999).

[211] P. V. Mamyshev and N. A. Mamysheva, *Opt. Lett.* **24**, 1454 (1999).

[212] M. Zitelli, F. Matera, and M. Settembre, *J. Lightwave Technol.* **17**, 2498 (1999).

[213] A. Mecozzi, C. B. Clausen, and M. Shtaif, *IEEE Photon. Technol. Lett.* **12**, 392 (2000).

[214] R. I. Killey, H. J. Thiele, V. Mikhailov, and P. Bayvel, *IEEE Photon. Technol. Lett.* **12**, 1624 (2000).

[215] J. Mårtensson, A. Berntson, M. Westlund, A. Danielsson, P. Johannisson, D. Anderson, and M. Lisak, *Opt. Lett.* **26**, 55 (2001).

[216] A. Mecozzi, C. B. Clausen, and M. Shtaif, *IEEE Photon. Technol. Lett.* **12**, 1633 (2000).

[217] S. Kumar, *IEEE Photon. Technol. Lett.* **13**, 800 (2001); S. Kumar, J. C. Mauro, S. Raghavan, and D. Q. Chowdhury, *IEEE J. Sel. Topics Quantum Electron.* **18**, 626 (2002).

[218] M. J. Ablowitz and T. Hirooka, *IEEE J. Sel. Topics Quantum Electron.* **18**, 603 (2002).

[219] P. Johannisson, D. Anderson, A. Berntson, and J. Mårtensson, *Opt. Lett.* **26**, 1227 (2001).

[220] P. Bayvel and R. I. Killey, in *Optical Fiber Telecommunications*, Vol. 4B, I. P. Kaminow and T. Li, Eds. (Academeic Press, Boston, 2002), Chap. 13.

[221] D. Duce, R. I. Killey, and P. Bayvel, *J. Lightwave Technol.* **22**, 1263 (2004).

[222] P. Minzioni and A. Schiffini, *Opt. Express* **13**, 8460 (2005).

[223] P. Johannisson, *J. Opt. Soc. Am. B* **24**, 729 (2007).

[224] D. Fonseca, A. V. T. Cartaxo, M. M. Monteiro, *J. Lightwave Technol.* **25**, 1447 (2007).

[225] A. Sahara, T. Inui, T. Komukai, H. Kubota, and M. Nakazawa, *J. Lightwave Technol.* **18**, 1364 (2000).

[226] G. Raybon, Y. Su, J. Leuthold, R.-J. Essiambre, T. Her, C. Joergensen, P. Steinvurzel, and K. D. K. Feder, in *Proc. Opt. Fiber Commun.* (OSA, Washington, DC, 2002), Paper FD-10.

[227] L. K. Wickham, R.-J. Essiambre, A. H. Gnauck, P. J. Winzer, and A. R. Chraplyvy, *IEEE Photon. Technol. Lett.* **16**, 1591 (2004).

[228] M. K. Jackson, G. R. Boyer, J. Paye, M. A. Franco, and A. Mysyrowicz, *Opt. Lett.* **17**, 1770 (1992).

Chapter 8

Optical Signal Processing

The preceding chapter focused mostly on the adverse impact of the nonlinear effects of fibers on the performance of lightwave systems. It turns out that the same nonlinear effects can be used to make useful all-optical devices capable of processing high-speed optical signals in a lightwave system. The use of stimulated Raman scattering (SRS) for making Raman amplifiers has been discussed in Section 4.7. Such amplifiers are often used for amplifying optical bit streams in modern long-haul wavelength-division multiplexing (WDM) systems. Similarly, four-wave mixing (FWM) can be used to make fiber-optic parametric amplifiers, as discussed in Chapter 10 of Ref. [1]. This section focuses on a variety of signal-processing devices that make use of the nonlinear effects such as self-phase modulation (SPM), cross-phase modulation (XPM), and FWM. Section 8.1 deals with wavelength converters that change the wavelength of a WDM channel without affecting its contents. The time-domain switches in which individual bits, or packets of bits, belonging to a specific channel are switched to a different port are described in Section 8.2. Applications of such switches for optical time-domain demultiplexing and format conversion are discussed in Section 8.3. Optical regenerators, devices that are used for reshaping and retiming of optical pulses belonging to an optical bit stream, are covered in Section 8.4.

8.1 Wavelength Conversion

Several nonlinear techniques for wavelength conversion make use of semiconductor optical amplifiers [2]–[5]. Wavelength converters can also be made using $LiNbO_3$ waveguides. This material exhibits a finite second-order susceptibility that can be used for difference-frequency generation, provided a phase-matching condition is satisfied. The technique of periodic poling is exploited for realizing quasi-phase matching, resulting in efficient wavelength conversion [6]–[8]. In this section we focus on fiber-based wavelength converters for which both the XPM and FWM phenomena have been successfully employed.

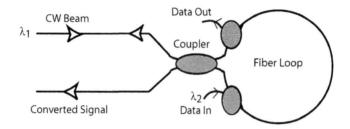

Figure 8.1: A wavelength converter making use of XPM-induced phase shifts inside a Sagnac interferometer to change the data wavelength from λ_2 to λ_1.

8.1.1   XPM-Based Wavelength Converters

The basic idea behind this technique has been discussed in Section 3.2.2 in the context of XPM-induced switching and is illustrated schematically in Figure 8.1. The data channel whose wavelength λ_2 needs to be changed is propagated inside a fiber of appropriate length together with a CW seed whose wavelength λ_1 is chosen to coincide with the desired wavelength of the converted signal. It acts as a pump and imposes an XPM-induced phase shift on the CW seed only in the time slots associated with the 1 bits. This phase shift is converted into amplitude modulation using an interferometer. In practice, a nonlinear optical loop mirror (NOLM) acting as a Sagnac interferometer is employed for this purpose [9]–[11]. The operation of an NOLM has been discussed in Section 3.2. The new feature in Figure 8.1 is that the data channel is launched such that it affects the CW seed in only one direction, resulting in a differential phase shift that can be used to copy the bit pattern of the data channel onto the transmitted λ_1 signal.

In a 1994 experiment, the wavelength of a 10-Gb/s data channel was shifted by 8 nm using 4.5-km of dispersion-shifted fiber inside an NOLM [9]. By the year 2000, this technique provided wavelength converters capable of operating at bit rates of up to 40 Gb/s [10]. The NOLM was made using 3 km of dispersion-shifted fiber with its zero-dispersion wavelength at 1555 nm, and it was used to shift the wavelength of a 1547-nm channel by as much as 20 nm. The on–off ratio between the maximum and minimum transmission states was measured to be 25 dB. The optical eye diagrams of the original and wavelength-converted signals indicated that individual pulses were almost unaffected during wavelength conversion.

One may ask what limits the extent of wavelength shift when an NOLM is used for wavelength conversion. The answer is related to different group velocities associated with the two fields interacting through XPM. The XPM-induced coupling between two optical fields is governed by the coupled nonlinear Schrödinger (NLS) equations of the form (see Section 7.4 of Ref. [1])

$$\frac{\partial A_1}{\partial z} + \frac{i\beta_{21}}{2}\frac{\partial^2 A_1}{\partial T^2} = i\gamma_1(|A_1|^2 + 2|A_2|^2)A_1, \qquad (8.1.1)$$

$$\frac{\partial A_2}{\partial z} + d_w\frac{\partial A_2}{\partial T} + \frac{i\beta_{22}}{2}\frac{\partial^2 A_2}{\partial T^2} = i\gamma_2(|A_2|^2 + 2|A_1|^2)A_2, \qquad (8.1.2)$$

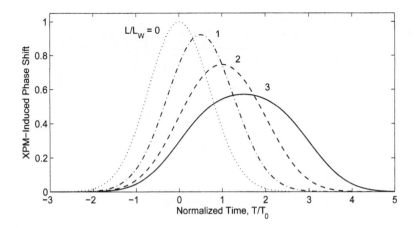

Figure 8.2: Temporal profile of the XPM-induced phase shift (normalized to ϕ_{max}) as a function of T/T_0 for several values of L/L_W.

where $T = t - z/v_{g1}$ is the time measured in a reference frame moving at the speed v_{g1} and

$$d_w = \frac{1}{v_{g1}} - \frac{1}{v_{g2}}, \qquad \gamma_j = \frac{2\pi n_2}{\lambda_j A_{\text{eff}}}. \tag{8.1.3}$$

The nonlinear parameters γ_j are different for the two fields because of their different wavelengths λ_j. The parameter d_w is a measure of group-velocity mismatch between the two optical fields. It affects the performance of a wavelength converter considerably.

One can simplify Eqs. (8.1.1) and (8.1.2) considerably for a typical experimental situation in which CW field A_1 is much weaker than the data channel A_2 because the nonlinear terms containing $|A_1|^2$ can then be ignored. Also, the dispersive effects governed by β_{21} and β_{22} can be neglected if fiber length is shorter than the dispersion length. As the pulse shape of data bits does not change under such conditions, the XPM-induced phase shift on the CW beam for a fiber of length L is given by

$$\phi_1(T) = 2\gamma_1 \int_0^L |A_2(0, T - d_w z)|^2 \, dz. \tag{8.1.4}$$

The XPM contribution changes along the fiber length because of the group-velocity mismatch. The total XPM contribution to the phase is obtained by integrating over the fiber length. If we assume that optical pulses in each bit slot have a Gaussian shape and use $|A_2(0,T)|^2 = P_0 \exp(-T^2/T_0^2)$, the XPM-induced phase shift becomes

$$\phi_1(T) = \sqrt{\pi} \gamma_1 P_0 L_W [\text{erf}(T/T_0) - \text{erf}(T/T_0 - L/L_W), \tag{8.1.5}$$

where $\text{erf}(x)$ stands for the error function and the walk-off length is defined as $L_W = T_0/d_w$. The integral in Eq. (8.1.4) can also be done for sech-shape pulses, and the result is of the form given in Eq. (3.2.10)

Figure 8.2 shows the XPM-induced phase shift ϕ_1, normalized to its maximum value, $\phi_{max} = 2\gamma_1 P_0 L$, as a function of T/T_0 for several values of L/L_W. The temporal

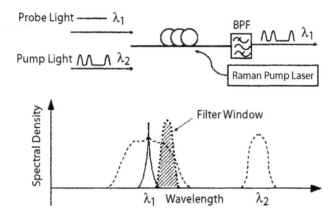

Figure 8.3: Schematic of an XPM-based wavelength converter; The bandpass filter (BPF) is offset to select only a portion of the XPM-broadened spectrum. (From Ref. [17]; ©2005 IEEE.)

phase profile mimics the pulse shape for small values of L/L_W, but its shape is distorted considerably as fiber length becomes larger than L_W. As one may expect, the maximum value of the phase shift is reduced by the walk-off effects, an undesirable feature as it increases the required peak power. However, even more detrimental from a practical standpoint is the broadening of the phase profile because it leads to broader pulses in the wavelength-converted channel. If pulses become so broad that neighboring pulses representing two 1 bits begin to overlap, the fidelity of the channel data is compromised.

To estimate the maximum wavelength shift, let us assume that L should be less than the walk-off length L_W. Noting that, for two channels separated in frequency by δv, the walk-off parameter d_w is related to fiber dispersion as $d_w = |\beta_2|(2\pi\delta v)$, the condition $L < L_W$ reduces to

$$\delta v < T_0/(2\pi\beta_2 L). \qquad (8.1.6)$$

The required fiber length is obtained from the requirement that $\phi_{\max} = 2\gamma_1 P_0 L$ must equal π for wavelength conversion to occur. For a dispersion shifted fiber, L should exceed 1 km if the input peak power P_0 is limited to near 1 W. Using $L = 2$ km and $T_0 = 20$ ps with $\beta_2 = 1$ ps^2/km, The frequency difference δv is close to 1.5 THz, a value that corresponds to a wavelength difference of 12 nm.

The required fiber length can be reduced considerably by employing highly nonlinear fibers for which γ exceeds 10 W^{-1}/km. In a 2001 experiment [11], wavelength conversion of 0.5-ps pulses was realized using am NOLM with just 50 m of a highly nonlinear fiber with $\gamma = 20.4$ W^{-1}/km. Because of such a short length of the fiber loop, the wavelength could be shifted by 26 nm even for such short pulses. The peak power of input pulses required for a π phase shift was close to 4 W in this experiment.

It is not essential to use an optical interferometer for XPM-based wavelength converters. As an alternative, one can launch the signal bit stream together with a CW probe beam into a fiber and pass the output of the fiber through a suitable optical filter [12]–[19]. Such a scheme is shown schematically in Figure 8.3. Signal pulses at the

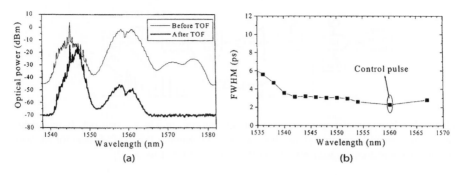

Figure 8.4: (a) Optical spectra measured before and after the tunable optical filter (TOF) for an XPM-based wavelength converter. (b) Pulse width as a function of converted wavelength. The width of the original data channel is marked as control pulse. (From Ref. [13]; ©2001 IEEE.)

wavelength λ_2 act as a pump and affect the phase and spectrum of the CW probe (at the desired wavelength λ_1) through XPM only in the time slots associated with the 1 bits. If the passband of the optical filter is shifted from λ_1 by a suitable amount, the output is a replica of the original bit stream at the new wavelength. Any optical filter with a bandwidth larger than that of the data channel (0.5 nm or so) can be used for this purpose, including a fiber grating or an optical interferometer. In a 2000 experiment [12], the wavelength of a 40-Gb/s signal was shifted by several nanometers through XPM inside a 10-km-long fiber. This experiment employed a 4-m loop made of polarization-maintaining fiber (PMF) that acted as a notch filter. The magnitude of wavelength shift was limited by the 10-km length of the fiber in which XPM occurred.

Much larger wavelength shifts have been realized by employing highly nonlinear fibers. In a 2001 experiment [13], wavelength conversion at a bit rate of 80 Gb/s was performed by using such a 1-km-long dispersion-shifted fiber for which γ was about 11 W^{-1}/km. The zero-dispersion wavelength of the fiber was 1552 nm with a relatively small dispersion slope near this wavelength. The 80-Gb/s data channel with the 1560-nm wavelength was first amplified to the 70-mW power level then coupled into the highly nonlinear fiber together with a CW seed at the new wavelength that was varied in the range of 1525–1554 nm. A tunable optical filter with a 1.5-nm bandwidth was used at the fiber output to produce the wavelength-converted channel. Figure 8.4(a) shows the optical spectra just before and after the optical filter when the CW-seed wavelength was 1545.6 nm. Before the filter, the spectrum shows multiple sidebands generated through XPM but the seed spectrum is dominated by the carrier. After the filter, the carrier has been suppressed relative to the sidebands, resulting in a wavelength-converted signal with a bit stream identical to that of the original channel. The pulse width of the converted signal is shown in Figure 8.4(b) as a function of the wavelength. As seen there, the width remains nearly unchanged over a wide bandwidth. The bit-error-rate measurements indicated negligible power penalty resulting from such a wavelength converter.

Wavelength conversion at a bit rate of 160 Gb/s was realized in a 2004 experiment in which a 0.5-km-long dispersion-shifted fiber was employed to produce the XPM-

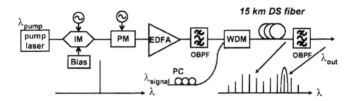

Figure 8.5: Experimental setup of a wavelength converter in which a sinusoidally modulated pump produces XPM-induced phase shifts on the data channel. IM and PM stand for intensity and phase modulators, respectively. (From Ref. [16]; ©2005 IEEE.)

induced phase shift [15]. This fiber was followed with a fiber grating that acted as a notch filter and two other optical filters with bandwidths of 5 and 4 nm whose pass bands were offset from the carrier frequency of the input signal by 160 GHz. It was observed that the wavelength converter maintained the phase of the input signal, a desirable property from the system standpoint. The performance of such wavelength converters improves further if the Raman gain of the fiber is utilized to enhance the XPM-induced phase shift by pumping the fiber in the backward direction, as indicated in Figure 8.3. In a 2005 experiment, the efficiency improved by 21 dB when a 1-km-long highly nonlinear fiber was pumped with 600-mW of pump power [17].

Fiber lengths required to impose sufficient XPM-induced phase shifts can be reduced to below 10 m if microstructured fibers with an ultrasmall effective mode area are employed. Indeed, only 5.8 m of a highly nonlinear fiber was needed in a 2003 experiment in which a fiber grating was used as a narrowband tunable filter [14]. The wavelength of the converted channel was tunable over a 15-nm bandwidth in the normal-dispersion region of the fiber. Such a large bandwidth was possible because of a reduction in the walk-off effects for short fibers. The use of normal dispersion eliminates the coherence degradation occurring in the case of anomalous dispersion because of the amplification of laser-intensity noise through modulation instability.

Even shorter fiber lengths are possible if one makes use of a nonsilica fiber with a relatively large value of the parameter n_2. In a 2006 experiment, only a 1-m-long piece of bismuth-oxide fiber was used to change the wavelength of a 10-Gb/s NRZ signal by as much as 15 nm [18]. The fiber exhibited normal dispersion of 330 ps^2/km with $\gamma \approx 1100$ W^{-1}/km at the 1550-nm wavelength. Because of its short length, the onset of SBS was not an issue as the SBS threshold was >1 W. For the same reason, the walk-off effects were negligible. The wavelength conversion in this experiment was based on the XPM-induced nonlinear polarization rotation that makes the fiber act as a Kerr shutter [1]. More specifically, the data channel causing the XPM and the CW beam are launched such that their linear states of polarizations are oriented at 45°. The XPM-induced phase shift changes the polarization of the CW beam only in time slots of 1 bits but leaves the polarization of 0 bits unchanged. A polarizer is used to select the wavelength-converted bits.

A new type of wavelength converter based on sinusoidal XPM was proposed in a 2005 study [16]. Figure 8.5 shows the experimental setup employed. In this scheme, a CW beam is first modulated sinusoidally at a frequency f_m higher than the bit rate of the data channel; phase modulation is used to suppress the onset of SBS. It is then amplified so that it acts as a pump that affects the data channel though XPM inside an optical fiber. Because of XPM, the spectrum of data channel develops AM sidebands separated in frequency by f_m such that each of the sidebands contains a faithful copy of the bit stream. An optical filter is used to select a specific sideband. If the nth sideband is filtered, the output is a data stream at a frequency shifted by nf_m. In the experiment, the wavelength of a 10-Gb/s signal was shifted by about 2 nm using $f_m = 25$ GHz. Larger wavelength shifts can be realized by using two CW lasers with a small wavelength spacing (0.5 nm or so) whose beating produces a sinusoidally modulated pump at a frequency $f_m > 50$ GHz without requiring an external modulator. This technique can, in principle, provide wavelength shifts of >20 nm.

A practical issue with most wavelength converters is related to the polarization-sensitive nature of XPM in optical fibers. It is well known that the XPM-induced nonlinear phase shift depends on the relative states of polarization of the pump and probe waves [1], and it acquires its maximum value when the two waves are copolarized. The polarization state of a data channel in a lightwave system is not fixed and can change with time in a virtually random fashion, and such polarization changes would manifest as fluctuations in the peak power of wavelength-converted pulses. It turns out that a polarization-insensitive wavelength converter can be realized by twisting the highly nonlinear fiber in which XPM occurs such that the fiber acquires a constant circular birefringence. In a 2006 experiment [19], this technique was employed to convert the wavelength of a 160-Gb/s data channel with only 0.7-dB polarization sensitivity. Optical fibers exhibiting circular birefringence should prove useful for a variety of nonlinear signal-processing applications [20].

8.1.2 FWM-Based Wavelength Converters

Although four-wave mixing inside a semiconductor optical amplifier is used often to make wavelength converters [2]–[4], an optical fiber has the distinct advantage that it can operate at high bit rates because of the ultrafast nature of the underlying nonlinearity. Such a wavelength converter requires parametric amplification of the signal bit stream inside an optical fiber (see Chapter 10 of [1]). The basic scheme is shown in Figure 8.6 schematically. If the frequency ω_s of a bit stream needs to be converted to ω_i, the signal is launched inside the fiber together with a CW pump whose frequency ω_p is exactly in the middle so that the energy conservation condition ($2\omega_p = \omega_s + \omega_i$) is satisfied for the four photons participating in the FWM process. Since two photons of the pump are converted into two new photons at frequencies ω_s and ω_i, amplification of the signal is accompanied by the generation of a new wave—the so-called idler wave. This wave mimics the bit pattern of the signal channel precisely because FWM occurs only during temporal slices allocated to 1 bits; no idler photons are produced during 0 bits because both the pump and signal photons must be present simultaneously for FWM to take place. As a result, if an optical filter is placed at the output end of the fiber that passes the idler but blocks the pump and signal, the output is a wavelength-converted replica of the original bit stream.

Although the use of fibers for wavelength conversion was investigated as early as 1992 [21]–[23], the technique matured only after the advent of highly nonlinear fibers,

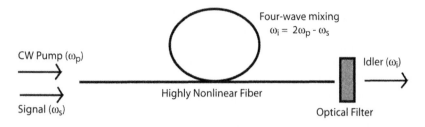

Figure 8.6: An FWM-based wavelength converter. The CW beam acts act as the pump and creates an idler at the desired wavelength in the form of a perfect replica of the signal bit stream. An optical filter passes the idler but blocks the pump and signal.

whose use reduced the length of the fiber in which FWM takes place [24]–[39]. To see why the fiber length plays an important role, we use the well-known theory of FWM developed for optical fibers [1]. When a single pump is used at power levels much higher than the signal, and depletion of the pump is ignored, the FWM process is governed by two coupled linear equations written in the Fourier domain as

$$\frac{dA_s}{dz} = 2i\gamma A_p^2 \exp(-i\kappa z)A_i^*, \qquad (8.1.7)$$

$$\frac{dA_i}{dz} = 2i\gamma A_p^2 \exp(-i\kappa z)A_s^*, \qquad (8.1.8)$$

where $A_s(\omega_s, z)$ and $A_i(\omega_i, z)$ represent the signal and idler fields, $A_p(\omega_p, 0)$ is the input pump field, and κ represents the total phase mismatch given by

$$\kappa = \beta(\omega_s) + \beta(\omega_i) - 2\beta(\omega_p) + 2\gamma P_0, \qquad (8.1.9)$$

where $\beta(\omega)$ is the propagation constant of the fiber mode at frequency ω and $P_0 \equiv |A_p|^2$ is the input pump power.

Equations (8.1.7) and (8.1.8) can be solved easily to study how the idler wave grows along the fiber length as a result of FWM. The conversion efficiency, defined as the ratio of output idler power at the end of a fiber of length L to the input signal power, is found to be [1]

$$\eta_c = |A_i(L)/A_s(0)|^2 = (\gamma P_0/g)^2 \sinh^2(gL), \qquad (8.1.10)$$

where the parametric gain g is given by

$$g(\delta) = [(\gamma P_0)^2 - \kappa^2/4)]^{1/2}, \qquad (8.1.11)$$

and $\delta = \omega_s - \omega_p$ is the frequency separation between the input signal and the pump. This solution shows that η_c can exceed 1 when the phase-matching condition is nearly satisfied. In fact, $\eta_c = \sinh^2(\gamma P_0 L) \gg 1$ when $\kappa = 0$ and $\gamma P_0 L > 1$. Thus, FWM-based wavelength converters can amplify a bit stream gain, while switching its wavelength from ω_s to ω_i. This is an extremely useful feature of such wavelength converters. Of course, the signal itself is also amplified. In essence, one has two amplified copies of the input bit stream at two different wavelengths. Such devices are often referred to as *fiber-optic parametric amplifiers* (FOPAs).

It follows from Eq. (8.1.10) and (8.1.11) that the conversion efficiency η_c depends on the phase mismatch κ and the fiber length L through the product κL. It turns out that the range of δ over which κL can be made small shrinks rapidly for long fibers. This feature can be seen more clearly by expanding $\beta(\omega_s)$ and $\beta(\omega_i)$ in a Taylor series around the pump frequency $\beta(\omega_p)$ to obtain

$$\kappa = \beta_2(\omega_p)\delta^2 + 2\gamma P_0, \tag{8.1.12}$$

where $\beta_2(\omega_p) \approx \beta_3(\omega_p - \omega_0)$ is the GVD parameter and β_3 is the third-order dispersion parameter at the zero-dispersion frequency ω_0 of the fiber. For a given value of δ, the phase mismatch κ can be made zero by choosing the pump wavelength in the anomalous-dispersion regime such that $\beta_2(\omega_p) = -2\gamma P_0/\delta^2$. However, if the signal wavelength deviates from this specific value of δ, the conversion efficiency η_c decreases at a rate that depends on the fiber length L. As a result, the bandwidth over which wavelength conversion can be realized for a specific pump wavelength is relatively narrow for long fibers (<10 nm for $L > 10$ km), but it can be increased to beyond 80 nm for fibers shorter than 100 m [25].

Experimental results on wavelength conversion are in agreement with this simple prediction of FWM theory. In the original 1992 experiment, the use of a 10-km-long dispersion-shifted fiber restricted the wavelength range to 8 nm or so [21]. By 1998, the use of a 720-m-long highly nonlinear fiber with a γ value of 10 W^{-1}/km permitted wavelength conversion over 40 nm with only 600 mW of pump powers [24]. The conversion efficiency varied with the signal wavelength but was as high as 28 dB, indicating that the wavelength-converted signal was amplified by up to a factor of 630 because of the FWM-induced parametric amplification. An added advantage of the wide bandwidth is that such a device can be used to convert wavelengths of multiple channels simultaneously. In a 2000 experiment [25], simultaneous wavelength conversion of 26 channels covering wavelengths ranging from 1570 to 1611 bandwidth was realized using a 100-m-long section of a highly nonlinear fiber with $\gamma = 13.8$ W^{-1}/km. The conversion efficiency was relatively low in this experiment (close to -19 dB) because the launched pump power was limited to 200 mW by the onset of SBS.

The onset of SBS becomes even more serious for wavelength converters that employ longer fibers and want to maintain a high conversion efficiency. The SBS threshold is around 5 mW for long fibers (>10 km) and increases to near 50 mW for fiber lengths of 1 km or so. Since FOPAs require pump power levels approaching 1 W, a suitable technique is needed that raises the threshold of SBS and suppresses it over the FOPA length. A common technique used in practice modulates the pump phase either at several fixed frequencies near 1 GHz [28], or over a broad frequency range using a pseudo-random bit pattern at a bit rate as high as 10 Gb/s [29]. This technique suppresses SBS by broadening the pump spectrum but it does not affect the FOPA gain much. In practice, dispersive effects within the fiber convert phase modulation into amplitude modulation of the pump. As a result, the signal-to-noise ratio of both the signal and the idler is reduced from the resulting undesirable power variations [37]. Pump-phase modulation also leads to broadening of the idler spectrum, making

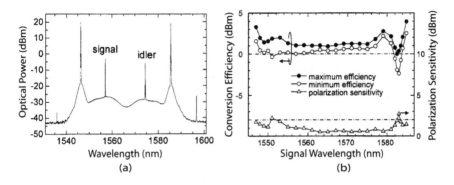

Figure 8.7: (a) Optical spectrum measured for a FWM-based wavelength converter. Two dominant peaks correspond to two orthogonally polarized pumps. (b) Conversion efficiency and its polarization sensitivity as a function of signal wavelength. (From Ref. [31]; ©2003 IEEE.)

it twice as broad as the pump spectrum. Such broadening of the idler is of concern for wavelength converters; it is avoided in practice using dual-pump FOPAs that also have other advantages.

The dual-pump configuration for FOPAs, first proposed in 1994 [23], can provide nearly uniform gain over a wide bandwidth, while allowing polarization-independent operation of the device. The suppression of idler broadening can be understood by noting that the complex amplitude A_i of the idler field resulting from the FWM process has the form $A_i \propto A_{p1}A_{p2}A_s^*$, where A_{p1} and A_{p2} are the pump amplitudes [1]. Clearly, the phase of the idler would vary with time if the two pumps are modulated either in phase or in a random manner. However, If the two pumps are modulated such that their phases are always equal but opposite in sign, the product $A_{p1}A_{p2}$ will not exhibit any modulation. As a result, even though the idler spectrum will be a mirror image of the signal spectrum, the bandwidth of the two spectra will be identical. A digital approach makes use of binary phase modulation so that the phase of both pumps is modulated in the same direction but takes only two discrete values, namely 0 and π. This approach works because the product $A_{p1}A_{p2}$ does not change under such a modulation scheme.

Another advantage of the dual-pump configuration is that the state of polarization (SOP) of the two pumps can be controlled to mitigate the dependence of the parametric gain on the signal SOP. This problem affects all FOPAs because the FWM process itself is strongly polarization-dependent. In the case of single-pump FOPAs, a polarization-diversity loop is sometimes employed. In this scheme, the two orthogonally polarized parts of the signal are amplified using a pump copolarzied with each of them, and then combined using a polarization beam splitter. In the case of dual-pump FOPAs, the problem can be solved simply by using orthogonally polarized pumps [29], although the FOPA gain is reduced considerably compared with the case of copolarized pumps.

Figure 8.7(a) shows the optical spectrum recorded at the output when a 1557-nm signal was launched inside a dual-pump wavelength converter [31]. The two pumps had power levels of 118 and 148 mW at wavelengths of 1585.5 and 1546.5 nm, respectively. The power was higher at the shorter wavelength to offset the Raman-induced

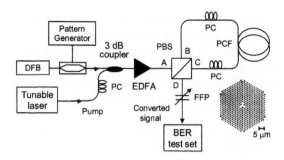

Figure 8.8: Schematic of a wavelength converter built using a 64-m-long photonic crystal fiber whose structure is shown in the inset. PC, PBS, FFP, and BER stand for polarization controller, polarization beam splitter, fiber Fabry–Perot, and bit-error rate, respectively. (From Ref. [34]; ©2005 IEEE.)

power transfer to the longer-wavelength pump. FWM occurred inside a 1-km-long highly nonlinear fiber ($\gamma = 18$ W^{-1}/km). The zero-dispersion wavelength of the fiber was 1566 nm with a dispersion slope of 0.027 ps/nm^2/km at this wavelength. The idler generated through FWM near 1570 nm had the same bit pattern as the signal. Its average power was also comparable to that of the signal, indicating nearly 100% efficiency for such a wavelength converter. In fact, as shown in Figure 8.7(b), high efficiency could be maintained over a bandwidth of 40 nm or so. The efficiency varied somewhat with the signal SOP but variations were below 2 dB over a 30-nm range. The wavelength of multiple channels can be converted simultaneously by such FOPAs [32]. It is also noteworthy that a single signal channel creates multiple idlers at different wavelengths that carry the same information as the signal, resulting in the so-called wavelength multicasting.

Wavelength converters have advanced considerably in recent years with the advent of photonic crystal and other microstructured fibers [33]–[36]. Figure 8.8 shows the schematic of an experiment in which a single pump was used with a polarization-diversity loop that contained a 64-m-long photonic crystal fiber (PCF) whose design is displayed as an inset [34]. This fiber had a triangular-shape core region with an effective mode area small enough that the nonlinear effects were enhanced. At the same time, it had a relatively constant dispersion over a 100-nm bandwidth centered near 1550 nm. These two features permit efficient FWM and provide wavelength conversion in spite of a relatively short length of the fiber.

A tunable semiconductor laser whose output was amplified to a power level close to 300 mW acted as the pump. It was not necessary to modulate the pump phase because laser's 1-GHz bandwidth was wide enough to suppress SBS over the short fiber length employed in this experiment. Both the pump and signal were separated into two parts with orthogonal SOPs that propagated in the opposite directions inside the fiber loop. The pump was polarized at 45° to ensure equal powers in both directions. A Fabry–Perot interferometer was used to filter the wavelength-converted signal. The conversion efficiency was about 1% because of a relatively low pump power that resulted in a low value of $\gamma P_0 L \approx 0.2$. However, this efficiency was constant over a

40-nm bandwidth. Wavelength conversion over this entire bandwidth was insensitive to the SOP of the 10-Gb/s signal and exhibited high fidelity, as confirmed through bit-error-rate measurements. In a later experiment, a similar device was used with success for converting the wavelength of a 40-Gb/s signal coded with the differential phase-shift keying (DPSK) format [35]. The highest conversion bandwidth of 100 nm was realized in another 2005 experiment in which the PCF length was reduced to only 20 m [36].

The shortest fiber length of only 40 cm for a wavelength converter was made possible with the advent of the bismuth-oxide fibers [38]. Such fibers exhibit a value of the Kerr nonlinearity n_2 that is 70 times larger than that of silica fibers. As a result, by reducing the core diameter to below 4 μm, the value of the nonlinear parameter γ can be increased to beyond 1000 W^{-1}/km. Such fibers exhibit FWM even when their length is below 1 m. Moreover, their SBS threshold is large enough that one does not have to modulate the pump phase. Further, their ends can be spliced to a standard silica fiber with <3-dB loss. Indeed, the wavelength of a 40-Gb/s signal could be converted with -16 dB conversion efficiency by launching about 1 W of CW pump power into a 40-cm-long bismuth-oxide fiber.

8.2   Ultrafast Optical Switching

Wavelength converters switch the entire bit stream at one wavelength to a different wavelength without affecting its temporal content. Some applications require switching of selected bits to a different port. An example is provided by packet switching in which a packet of tens or hundreds of bits should be selected from a bit stream. Another example is provided by optical time-domain demultiplexing in which a selected bit of a high-speed bit stream is sent to another port in a periodic fashion. Such applications require time-domain switches that can be turned on for a specific duration using an external control. This section focuses on ultrafast switches that make use of the XPM and FWM nonlinear phenomena inside optical fibers.

8.2.1   XPM-Based Sagnac-Loop Switches

The XPM-based switches make use of a Sagnac interferometer that converts phase modulation into amplitude modulation [40]–[46]. As discussed in Section 3.2.2, Sagnac interferometers employ a fiber loop and act as a perfect mirror when the 3-dB fiber coupler (used to form the loop) splits the input power equally (a coupler) across its two output ports. The reason for this behavior is that the linear as well as nonlinear phase shifts are the same in both directions in a symmetric Sagnac loop. To make a time-domain switch, this symmetry is broken by injecting a control signal into the Sagnac loop such that it propagates in only one direction and induces a nonlinear phase shift through XPM on only the copropagating wave (see Figure 8.1). Only a temporal slice of the input signal that overlaps with the control pulse and acquires an additional π phase shift is transmitted, while the rest of the input signal is reflected. As a result, such a scheme provides the ability to select different temporal slices of the signal, and

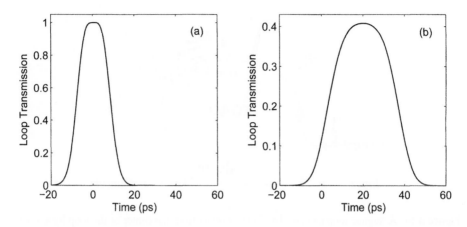

Figure 8.9: Switching window of a Sagnac-loop switch for (a) $T_0 = 10$ ps with $T_W = 2$ ps and (b) $T_0 = 10$ ps with $T_W = 40$ ps.

each slice can be quite short because of the femtosecond response time of the fiber nonlinearity.

Even though the response time of fiber nonlinearity allows switching at femtosecond time scales, several other physical mechanisms limit the switching time to >1 ps. For example, wavelengths of the control and signal pulses are often different in practice. Because of group-velocity dispersion, the signal and control pulses travel at different speeds, even when the two propagate in the same direction, and thus walk away from each other after some distance. The relative phase shift between the counterpropagating waves after one round trip inside the loop is given by Eq. (8.1.4). If we neglect the contribution of self-phase modulation, assuming that signal pulses are relatively weak.

In the case of a Gaussian-shape control pulse, the XPM-induced phase shift is given by Eq. (8.1.5). This equation can also be written as

$$\phi_1(T) = \frac{\gamma L E_c}{T_W} \left[\text{erf}\left(\frac{T}{T_0}\right) - \text{erf}\left(\frac{T - T_W}{T_0}\right) \right], \qquad (8.2.1)$$

where $E_c = \sqrt{\pi} P_0 T_0$ is the control-pulse energy and $T_W = d_w L$ is the total temporal shift induced by the walk-off effect in a loop of length L.

The Sagnac-loop transmissivity is obtained following the analysis in Section 3.2.1 and is given by

$$T_S = 1 - 2\rho(1 - \rho)[1 + \cos(\phi_1)] = \sin^2(\phi_1/2), \qquad (8.2.2)$$

where $\rho = \frac{1}{2}$ was used assuming a 3-dB fiber coupler. Figure 8.9 shows the "switching window" for a Sagnac loop for two combinations of T_0 and T_W when the loop length L and pulse energy E_c are large enough to produce a maximum phase shift of π. As T_W increases, peak transmissivity decreases and the switching window becomes wider because its width is set by the walk-off time rather than the control-pulse width.

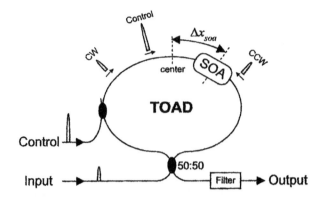

Figure 8.10: A Sagnac-loop switch. The SOA is offset from the center of the loop by a small but controlled amount. (From Ref. [51]; ©2003 Elsevier.)

The walk-off problem can be solved to a large extent by using a fiber whose zero-dispersion wavelength lies between the pump and signal wavelengths such that the two waves have nearly the same group velocity. It can be avoided completely by using a polarization-maintaining fiber and launching orthogonally polarized control pulses at the same wavelength as that of the signal [43]. A relatively small group-velocity mismatch still occurs because of polarization-mode dispersion, but its effects are negligible for typical loop lengths. Moreover, it can be used to advantage by constructing a Sagnac loop in which the slow and fast axes of the polarization-maintaining fiber are interchanged in a periodic fashion [44]. The parameter d_w in such a fiber changes from positive to negative in alternate sections in such a way that the control and signal pulses nearly overlap throughout the fiber. The XPM-induced phase shift is enhanced considerably with this configuration.

The walk-off problem can also be solved if a semiconductor optical amplifier (SOA), placed within the Sagnac loop, is used to impose the XPM-induced phase shifts [47]–[49]. The loop length in this case can be small (1 m or less) because the loop is used only to propagate an input signal into counterpropagating directions. SOAs produce a nonlinear phase shift when the refractive index of the active region changes in response to carrier density variations induced by gain saturation. The active nature of this process produces a relatively large nonlinear response, but the speed of such nonlinear changes is limited inherently by the carrier lifetime (typically >0.1 ns). It turns out that the speed limitation can be overcome by a clever trick [48]–[51]. The trick consists of placing the SOA such that it is shifted from the center of the loop by a small but precise distance. It is this shift that governs the temporal window over which switching occurs rather than the carrier lifetime. Figure 8.10 shows such a device schematically. It is referred to as a *semiconductor laser amplifier in a loop mirror* or SLALOM [48]. It is also called a *terahertz optical asymmetric demultiplexer* (TOAD) to emphasize the asymmetric placement of SOA and the potential of terahertz speeds [49].

In Figure 8.10, a 3-dB directional coupler splits the input pulse into two pulses propagating in the clockwise (CW) and counterclockwise (CCW) directions. The control pulse propagates only in the CW direction and is intense enough to saturate the SOA, which is offset from the loop center in the CW direction by an amount Δx_{soa}. In the absence of the control pulse, the signal pulse acquires the same phase shift in both directions (as it is too weak to saturate the SOA), and it is totally reflected by the Sagnac loop. The control pulse is timed such that it arrives at the SOA after the CCW signal pulse but before the CW pulse. As a result, the XPM-induced phase shift is produced on the CW pulse, but not on the CCW pulse. If this differential phase shift equals π, the signal pulse is transmitted, rather than being reflected. Neighboring signal pulses do not experience a large differential phase shift and are thus reflected. After a time interval comparable to the carrier lifetime, the gain recovers, and the process can be repeated as long as control pulses are separated by the gain-recovery time. The switching time of such a device is governed by the speed with which a short control pulse can saturate the SOA and change its refractive index.

The main advantage of using an SOA within the Sagnac loop is that it provides switching at relatively low energy levels of the control pulses. Typically, energy of control pulses should exceed 10% of the saturation energy for imposing a relative phase shift of π on counterpropagating signal pulses [52]. For typical values of the SOA parameters, such phase shifts can be realized at energy levels below 1 pJ. A drawback of using SOAs is that they have a relatively high noise figure and thus reduce the signal-to-noise ratio at the output because of spontaneous emission added to the signal during its amplification.

8.2.2 Polarization-Discriminating Switches

A Mach–Zennder interferometer (MZI) can be used in place of a Sagnac one but, as mentioned earlier, it suffers from stability problems when fibers are used to its two arms. The stability of a fiber-based MZI can be ensured by adopting a single-arm design in which the same physical path is shared by two orthogonally polarized components of the signal pulse. The basic idea was first demonstrated in 1995 and has been used since then in a number of configurations [53]–[60]. Such a switch is sometimes referred to as *polarization discriminating* as its operation is based on the state of polarization of the signal and control pulses. It is also called an *ultrafast nonlinear interferometric* (UNI) *switch*. It offers a number of advantages related to the single-arm nature of the underlying interferometer.

Figure 8.11 shows the design of a UNI switch. The signal pulse entering from the left is split into its orthogonally polarized components using a polarization-maintaining fiber (PMF) whose length L is chosen such that the two components are delayed by a fixed amount. The delay induced by the PMF depends on its birefringence and can be written as

$$\Delta t = |n_x - n_y| L / c. \qquad (8.2.3)$$

For typical values of the refractive-index difference ($\sim 10^{-4}$), a few meters of PMF can introduce relative delays of a few picoseconds. The two delayed signal pulses are then sent through a nonlinear medium, often an SOA but a highly nonlinear fiber can also be used for this purpose. The position of the control pulse is adjusted such

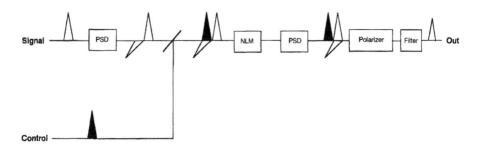

Figure 8.11: A polarization-discriminating switch: PSD and NLM stand for polarization-sensitive delay and nonlinear medium, respectively. (From Ref. [57]; ©1998 OSA.)

that it arrives in between the two components of the signal pulse. As a result, only the trailing component of the signal pulse acquires an additional XPM-induced phase shift. The two components of the signal are then synchronized by a second PMF whose fast and slow axes are reversed compared to the first PMF. A polarizer set at 45° forces interference between the two polarization components of the signal pulses and completes the XPM-induced switching. Control pulses are filtered at the output of the device. This filtering can be avoided by using a configuration in which signal pulses propagate in a direction opposite to that of the signal.

The main advantage of UNI geometry is its interferometric stability. Since all signals travel along the same path, they are exposed to identical environmentally induced fluctuations. As the output depends on the phase difference between the two polarization components of the signal pulse, the device is automatically immune to all environmental fluctuations. Indeed, UNI switches do not require any active stabilization in practice. For the same reason, any fluctuations in the linear refractive index of the nonlinear medium also cancel out and do not affect the device performance as long as such fluctuations occur over time scales longer than the signal-pulse width.

The use of folded geometry, as shown in Figure 8.12, can further improve the performance of UNI switches [58]. This configuration uses the same PMF for splitting and recombining signal pulses by placing the SOA inside a loop. Signal pulses are first polarized by the polarization controller PC-1 so that they leave the polarization beam splitter via port 3. The polarization controller PC-2 is used to align the polarization of the signal pulse at 45° with respect to the principal axes of the PMF, which splits it into two orthogonally polarized components and introduces a fixed relative delay between them. The XPM-induced phase shift on one of the components is induced by a polarization-insensitive SOA placed inside a loop; control pulses enter into this loop through a WDM coupler. A third polarization controller PC-3 placed after the SOA rotates the polarization such that the pulse pair is launched at −45° into the same PMF that was used to split the signal pulse. However, this PMF now reverses the delay and synchronizes the two polarization components.

As before, the two signal components recombine at the PBS. If both components encounter the same phase shifts in the loop containing the SOA, they leave the polarization beam splitter via port 1, and no signal is present at the output port 4. However, if a control pulse enters the SOA in between the two signal components, the trailing

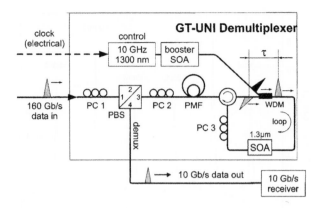

Figure 8.12: A gain-transparent (GT) UNI switch with SOA inside a loop. The same PMF splits and recombines signal pulses. (From Ref. [58]; ©2001 IEEE.)

component experiences an additional XPM-induced phase shift. This relative phase shift switches that specific signal pulse to the output port 4. The use of the same PMF for splitting and recombining the signal components ensures that any random phase shifts occurring in the PMF will not affect the switching quality as long as such shifts occur on a time scale longer than the round-trip time.

The device shown in Figure 8.12 also makes use of an additional improvement realized though a gain-transparent SOA. The main idea behind gain transparency is to use control pulses whose wavelength is tuned close to the gain peak while the wavelength of signal pulses falls outside the SOA-gain spectrum in the transparent regime below the bandgap of the active material [61]. Since the signal experiences virtually no gain (or loss), it does not suffer from the noise added by spontaneous emission. At the same time, control pulses saturate the gain and produce refractive-index changes at the signal wavelength. Other advantages offered by this scheme include low crosstalk, operation over a wide range of signal wavelengths, and negligible power variations among signal pulses at the device output.

8.2.3 FWM-Based Ultrafast Switches

Another nonlinear scheme for ultrafast switching makes use of FWM inside an optical fiber. It is similar to the scheme shown in Figure 8.6 in the context of wavelength conversion. In the switching context, the pump power is changed with time in a fashion dictated by the application. For example, if the objective is to select a packet of 40 bits from a 40-Gb/s signal, the pump is in the form of a 1-ns-wide nearly rectangular pulse. FWM creates the idler only over that 1-ns-wide temporal slice of the signal that overlaps with the pump, resulting in switching of the packet at the idler wavelength. An individual bit can be selected with this approach if the pump pulse lasts only over the duration of a single bit [62]. Moreover, multiple copies of the switched packet (or a specific bit) can be created at different wavelengths if the dual-pump configuration

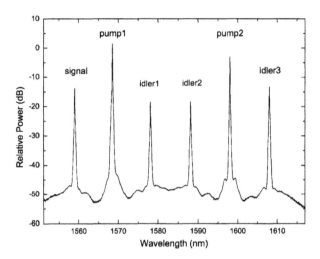

Figure 8.13: Optical spectrum recorded with 0.2-nm resolution when a 1559-nm signal is launched with two pumps inside a highly nonlinear fiber. (From Ref. [63]; ©2005 IEEE.)

is employed for FWM. This feature is referred to as wavelength multicasting and is useful for all-optical networks.

Packet switching as well as bit-level switching was realized at a 40-Gb/s bit rate in a 2005 experiment in which two pump beams were launched into a 520-m-long highly nonlinear fiber having its zero-dispersion wavelength at 1582 nm [63]. The two pumps (wavelengths 1568.5 and 1598 nm) were located on opposite sides of this wavelength. The average input power of the short-wavelength pump was chosen to be much larger (545 mW) compared with that of the other pump (155 mW) to offset the impact of SRS-induced power transfer. Also, the pump phases were modulated suitably to suppress the onset of SBS within the fiber. The shorter-wavelength pump was in the form of a CW beam, but the other pump could be modulated to produce pulses whose width varied from a single-bit duration of 25 ps to more than 1 ns. Figure 8.13 shows the optical spectrum recorded at the output end when a 1559-nm signal was launched into the fiber. Several different FWM processes occurring simultaneously produce three idlers at three distinct wavelengths, indicating the possibility of wavelength multicasting. The signal was amplified by 25.2 dB under the experimental conditions.

Figure 8.14 shows packet switching of a 40-Gb/s signal [part (a)], realized when the long-wavelength pump was in the form of 1-ns pulses [part (b)]. The two edges of pump pulses were aligned such that the pump pulse covered the time span of 10 signal bits precisely. As signal amplification occurred only over the pump-pulse duration, the amplified signal consists of the 10-bit packet [part (c)]. Moreover, a copy of this packet appears at the three idler wavelengths, one of which is shown in part (d). Such packet switching was realized with a low noise and high extinction ratio. A relatively high extinction ratio is related to the exponential dependence of the signal gain on the pump powers.

It is relatively easy to extend the FWM theory of Section 8.1.2 to the two-pump

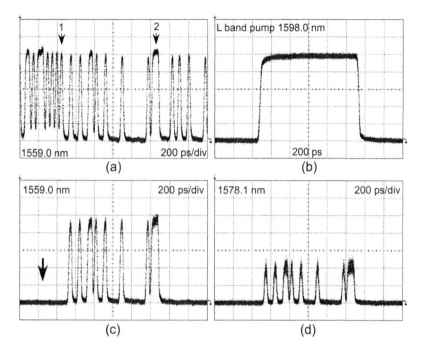

Figure 8.14: Temporal bit patterns showing packet switching: (a) bit stream of 1559-nm signal; (b) 1-nm-wide pump pulse; (c) amplified signal; (d) the idler at 1578.1 nm. Two arrows in (a) mark edges of the pump pulse. (From Ref. [63]; ©2005 IEEE.)

case [1] to obtain the following expression for the signal amplification factor:

$$G_s(t) = P_s(L)/P_s(0) = 1 + (2\gamma/g)^2 P_1 P_2(t) \sinh^2(gL), \quad (8.2.4)$$

where P_1 and P_2 are the pump powers. The parametric gain is still given by Eq. (8.1.11) provided $P_0 = 2(P_1 P_2)^{1/2}$. In the phase-matched region where $\kappa \approx 0$, the signal gain becomes

$$G_s(t) \approx \exp[4\gamma\sqrt{P_1 P_2(t)}L]/4. \quad (8.2.5)$$

The exponential dependence of the signal gain on the powers of the two pumps is helpful in practice because it helps mitigate the problems associated with the rise and fall times of pump pulses. When the peak value of $G_s(t)$ exceeds 100 (or 20 dB), the parametric response can be three to four times faster than the pump's rise or fall time. As a result, even a 25-ps rise/fall time of pump pulses was acceptable for switching a 40-Gb/s bit stream. The situation is even better in the single-pump case because Eq. (8.2.5) is replaced with

$$G_s(t) \approx \exp[2\gamma P(t)L]/4, \quad (8.2.6)$$

where $P(t)$ represents the power of the temporally modulated pump. The exponential dependence of the gain can be exploited for using a FOPA as a pulse compressor. In a 2005 experiment, 2-ps pulses were produced by amplifying 9-ps signal pulses with a single pump that was modulated at 40 GHz [64].

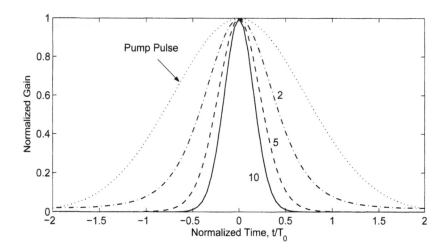

Figure 8.15: Switching window of a FWM-based switch for several values of $\gamma P_0 L$ in the case of Gaussian-shape pump pulses.

The time dependence of the parametric gain can also be used to convert a CW signal into a train of pulses whose width is shorter than that of the pump pulses [26]. In essence, a FOPA acts as a time-domain switch whose switching window is controlled by the width and peak power of pump pulses. As an example, assume that the pump is in the form of Gaussian pulses such that $P(t) = P_0 \exp(-t^2/T_0^2)$, where P_0 is the peak power and T_0 is the pulse width. Figure 8.15 shows the switching window by plotting $G_s(t)/G_0$ for several values of $\gamma P_0 L$, where G_0 is the maximum value occurring at the peak of the pump pulse. As seen there, the duration of switching window becomes a fraction of the pump-pulse width when the peak gain G_0 exceeds 20 dB.

8.3 Applications of Time-Domain Switching

Time-domain switching can be used for a number of applications ranging from channel demultiplexing to data-format conversion in lightwave systems [65]–[67]. Its use for demultiplexing a high-speed bit stream generated through optical time-domain multiplexing (OTDM) is indispensable because each optical pulse is only a few picosecond wide at aggregate bit rates exceeding 100 Gb/s. This section focuses on this application first and then considers several others.

8.3.1 Channel Demultiplexing

An OTDM signal consists of a high-speed bit stream that is composed of several channels, each operating at a lower bit rate and interwoven with others in a periodic fashion. If 10 channels, each operating at 40 Gb/s, are multiplexed in the time domain, every 10th bit of the composite 400-Gb/s bit stream belongs to the same channel. Demultiplexing a channel from such a high-speed OTDM signal requires optical switches

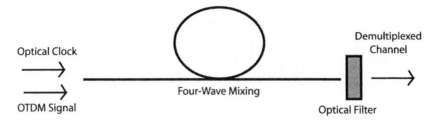

Figure 8.16: A FWM-based demultiplexing scheme. Optical clock acts as the pump and produces idler pulses only when one of signal pulses overlaps with a clock pulse.

that pick all the bits belonging to a specific channel and direct those bits to a different port. Such switches require control pulses at the single-channel bit rate (40 Gb/s in the preceding example) that is used to switch signal pulses selectively using a nonlinear phenomenon such as XPM or FWM. This control pulse train is referred to as an *optical clock*.

As shown schematically in Figure 8.16, one way to select individual pulses is to employ FWM inside an optical fiber. In this scheme, the optical clock is chosen to be at a wavelength that is a few nanometers away from the signal wavelength and it acts as a pump. Only when the signal and pump pulses are present simultaneously and overlap in the time domain, FWM creates an idler pulse at the new wavelength. An optical filter at this wavelength blocks the pump and signal pulses, resulting in an output bit stream that belongs to a specific channel. A different channel can be selected by adjusting the clock phase such that clock pulses overlap with the pulses belonging to another specific channel. This technique was first used in 1991 to demultiplex a 16-Gb/s bit stream [68].

In later experiments, the signal bit rate approached 1 THz, confirming that optical fibers can be used to switch pulses shorter than 1 ps. A polarization-maintaining fiber is often used as the nonlinear medium for FWM because of its ability to preserve the state of polarization despite environmental fluctuations. As early as 1996, demultiplexing of 10-Gb/s channels from a 500-Gb/s OTDM signal was demonstrated using clock pulses of 1-ps duration [69]. A distinct advantage of using FWM is that the demultiplexed channel is also amplified through parametric gain inside the same fiber [70].

A problem with FWM-based demultiplexer is related to the polarization sensitivity of the FWM process itself. In general, maximum parametric gain occurs when pump and signal are copolarized. If the state of polarization of a signal is not aligned with the pump and changes with time in an unpredictable manner, both the signal and idler power levels will fluctuate widely, resulting in poor performance. A polarization-diversity technique, in which the input signal is separated into two orthogonally polarized parts that are processed separately, can be used [71], but it adds considerable complexity. A simple scheme for solving the polarization problem was adopted in 2004. It consists of attaching a short piece of polarization-maintaining fiber to the input port of the highly nonlinear fiber used for FWM and using an optical phase-locked loop for locking clock pulses to the peak position of incoming signal pulses [72]. As shown in Figure 8.17, the control clock pulses are polarized at $45°$ with respect to

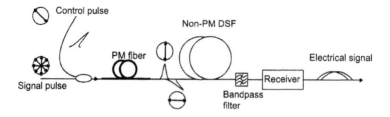

Figure 8.17: A polarization-insensitive FWM-based demultiplexing scheme; PM fiber and DSF stand for polarization-maintaining and dispersion-shifted fibers, respectively. (From Ref. [72]; ©2004 IEEE.)

the principal axes of the polarization-maintaining fiber that also splits and separates randomly polarized signal pulses into two orthogonally polarized parts. Since two separate FWM processes take place simultaneously within the same nonlinear fiber, in essence, polarization diversity is realized with this simple experimental arrangement. Such an approach was capable of demultiplexing a 160-Gb/s bit stream into 10-Gb/s individual channels with <0.5 dB polarization sensitivity.

In another approach to solving the polarization problem, the nonlinear fiber in which FWM occurs is itself made birefringent [73]. Moreover, it is divided into two equals sections in which the fast and slow axes are reversed. A single pump in the form of clock pulses, polarized at 45° with respect to the slow axis of the fiber, is launched together with the high-speed signal that needs to be demultiplexed. The orthogonally polarized components of the pump and signal interact through FWM and create the idler containing the demultiplexed channel. Even though the two polarization components separate from each other in the first section, they are brought back together in the second half of the fiber because of the reversal of the slow and fast axes in the second section. An optical filter at the end of the fiber blocks the pump and signal, resulting in the demultiplexed channel at the idler wavelength.

The nonlinear phenomenon of XPM can also be used for channel demultiplexing. Its use for this purpose was first demonstrated in 1990 using an NOLM [74], in a configuration similar to that shown in Figure 8.1 with a data channel entering at wavelength λ_1 and clock pulses at wavelength λ_2 acting as the pump. By 1998, such an

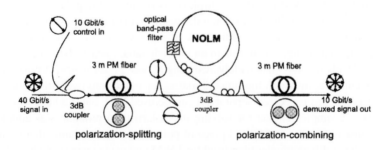

Figure 8.18: A polarization-insensitive NOLM-based demultiplexing scheme making use of two 3-m-long polarization-maintaining (PM) fibers. (From Ref. [80]; ©2002 IEEE.)

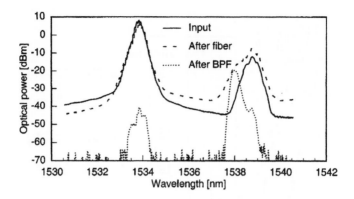

Figure 8.19: Optical spectra measured before (solid curve) and after (dashed curve) the fiber. The filtered spectrum is shown as a dotted curve. (From Ref. [81]; ©2001 IEEE.)

NOLM was employed for demultiplexing 10-Gb/s channels from a 640-Gb/s OTDM signal [75]. The walk-off effects in this experiment were minimized by using nine 50-m long sections of dispersion-flattened fibers with different group delay characteristics. Several techniques can be employed for polarization-insensitive operation of an NOLM [76]–[80], one of which is shown in Figure 8.18. The basic idea is similar to that behind the scheme in Figure 8.17. A short piece of polarization-maintaining fiber is used to split the pump and control pulses along its slow and fast axes. An optical bandpass filter centered at the signal wavelength is placed at one end of the NOLM so that it blocks the control pulse. However, this blocking occurs in one direction only after the pulse has passed through the loop and changed the phase of a specific signal pulse through XPM by π. As a result, data pulses belonging to the demultiplexed channel appear at the NOLM output, where a second polarization-maintaining fiber combines the two polarization components.

Similar to the case of wavelength conversion, it is not necessary to employ an NOLM for making use of XPM. A scheme is similar to that shown in Figure 8.3 in the context of wavelength conversion was used for demultiplexing in a 2001 experiment [81]. The only difference was that the role of probe at wavelength λ_1 was played by the OTDM data signal, while intense clock pulses at wavelength λ_2 played the role of the pump. The clock pulses shifted the spectrum through XPM of only those data pulses that overlap with them in the time domain. An optical filter was then used to select these pulses, resulting in a demultiplexed channel at the clock wavelength. This experiment used a 5-km-long fiber with its zero-dispersion wavelength at 1543 nm. The 14-ps control pulses at a repetition rate of 10 GHz had a wavelength of 1534 nm and were propagated with the 80-Gb/s OTDM signal at 1538.5 nm. Figure 8.19 shows the optical spectra before and after the fiber together with the filtered spectrum.

As discussed in Section 8.1.1, the group-velocity mismatch between the signal and control pulses plays a major role in XPM-based optical switching, and the switching window is set by the initial offset between the two. This mismatch can be reduced by locating the control and and signal pulses on the opposite sides of the zero-dispersion wavelength of the fiber. In addition, the use of a highly nonlinear fiber not only re-

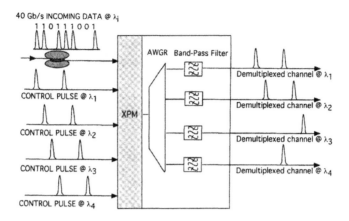

Figure 8.20: Simultaneous demultiplexing of four channels using four optical clocks at different wavelengths; AWGR stands for an arrayed-waveguide router. (From Ref. [82]; ©2002 IEEE.)

duces the required average power of control pulses but also helps with the problem of group-velocity mismatch as much shorter lengths are needed. An added benefit of this technique is that it can be used to demultiplex multiple channels simultaneously by simply employing multiple control pulses at different wavelengths. Figure 8.20 shows such a scheme schematically [82]. It was implemented in a 2002 experiment to demultiplex four 10-Gb/s channels from a 40-Gb/s composite bit stream through XPM inside a 500-m-long highly nonlinear fiber. Only a 100-m-long fiber was employed in another experiment to demultiplex 10-Gb/s channels from a 160-Gb/s bit stream [83].

Much smaller fiber lengths can be employed by using microstructured fibers or a nonsilica fiber made of a material with high values of n_2. Only a 1-m-long piece of bismuth-oxide fiber was needed in a 2005 experiment [84] because this fiber exhibited a value ~ 1100 W^{-1}/km for the nonlinear parameter γ. The train of 3.5-ps control pulses at a 10-GHz repetition rate was amplified to an average power level close to 0.4 W to ensure a high peak power ($P_0 > 10$ W) so that the value of $\gamma P_0 L$ exceeded 10 even for the 1-m-long fiber. This experiment employed the fiber as a Kerr shutter [1] and made use of the XPM-induced nonlinear birefringence that changed the state of polarization of selected signal pulses such that only they were transmitted through the polarizer placed at the output end of the fiber. Because the walk-off effects were negligible for the short fiber, the measured switching window was narrow enough (only 2.6-ps wide) to demultiplex a 160-Gb/s bit stream.

Polarization-independent operation can be realized with a variety of techniques such as polarization diversity or a spun fiber exhibiting circular birefringence [20]. A 30-m-long photonic crystal fiber exhibiting linear birefringence was employed in a 2006 experiment [85]. Clock pulses were polarized at 45° to the slow axis of the fiber so that their energy was divided equally between the slow and fast axes. The SOP of both the data and clock pulses evolved periodically with different beat lengths because of their different wavelengths. As a result, their relative SOP varied in a nearly random fashion. This feature resulted in an averaging of the XPM effect and produced

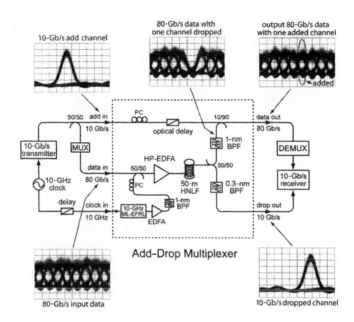

Figure 8.21: Experimental setup used to add and drop 10-Gb/s channels from a 80-Gb/s data stream. Eye patterns are also shown at five different locations. (From Ref. [87]; ©2005 IEEE.)

an output that was independent of the signal polarization. In a later experiment, a high-speed polarization scrambler was employed to randomize the SOP of 160-Gb/s data pulses but the SOP of the 10-Gb/s clock pulses was kept fixed [86]. The XPM-induced spectral broadening occurred inside a 2-m-long bismuth-oxide fiber. Because of polarization scrambling, the performance of such a demultiplexer exhibited little sensitivity to the input SOP of the data bit stream.

Add–drop multiplexers represent another key component for multichannel light-wave systems. Such devices have been developed for WDM systems and are finding commercial applications. One would need the same functionality for OTDM systems before they can reach a practical stage, and several semiconductor-based devices have been used for this purpose. It turns out that XPM-based demultiplexers can also function as add–drop multiplexers with minor design changes. Figure 8.21 shows the experimental setup used to add and drop 10-Gb/s channels from a 80-Gb/s data stream through XPM inside a 50-m-long highly nonlinear fiber together with five eye patterns [87]. In a later experiment, fiber length was reduced to 1 m by employing a bismuth-oxide fiber [88].

8.3.2 Data-Format Conversion

Optical communications systems employ a wide variety of formats for coding a bit stream. The NRZ format is often employed in WDM networks as it is most efficient spectrally. The use of RZ format, or one of its variants such as the carrier-suppressed RZ (CSRZ) format, becomes necessary at high bit rates, and it is the format of choice

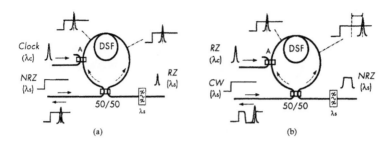

Figure 8.22: (a) NRZ-to-RZ and (b) RZ-to-NRZ conversion using an NOLM; DSF stands for dispersion-shifted fiber. (From Ref. [93]; ©1997 IEEE.)

for OTDM systems. More recently, the DPSK format has attracted attention because it reduces the extent of XPM-induced degradation. In a network environment, the conversion among the RZ, NRZ, CSRZ, and DPSK signals may become necessary. Several all-optical techniques for conversion between NRZ and RZ formats make use of the nonlinear effects occurring inside semiconductor lasers and amplifiers [89]–[92]. This subsection focuses on devices whose operation exploits nonlinear effects inside optical fibers.

It is easy to see from Figure 8.22 how XPM inside an NOLM can be used for conversion between the NRZ and RZ formats [93]. In the case of NRZ-to-RZ conversion, the phase of NRZ pulses is shifted inside the loop by launching an optical clock (a regular train of pulses at the bit rate) such that it propagates in one direction only (see Figure 8.1). In the case of RZ-to-NRZ conversion, the phase of a CW beam is shifted by the RZ data pulses propagating in one direction only. The main limitation is set by the walk-off effects that govern the switching window of the NOLM. The SOA-based NOLM shown in Figure 8.10 has been used to convert a RZ or NRZ bit stream into one with the CSRZ format [91].

Several other schemes have been developed in recent years for fiber-based format conversion [94]–[96]. In a 2005 experiment, the XPM-induced wavelength shift inside a nonlinear fiber was used for RZ-to-NRZ conversion [94]. The scheme is similar to that shown in Figure 8.3 (in the context of wavelength conversion), the only difference being that the optical filter is centered exactly at the wavelength of the CW probe. The RZ signal acts as the pump and modulates the phase of the CW probe. The resulting chirp shifts the wavelength of pulses representing 1 bits. The filter blocks these pulses but lets pass the 0 bits. The resulting bit stream is a polarity-reversed NRZ version of the original RZ signal.

A similar scheme can be adopted for NRZ-to-RZ conversion [95]. In this case, an optical clock acting as the pump is sent through the fiber together with the NRZ signal. The XPM interaction between the two broadens the signal spectrum. The optical filter is offset from the signal wavelength similar to the case of wavelength conversion. The output is an RZ version of the signal at the same wavelength. Such a scheme suffers from polarization sensitivity, because the nonlinear process of XPM itself is polarization dependent [1]. It can be made polarization insensitive by employing a polarization-diversity loop, as shown in Figure 8.23. The polarization of the clock

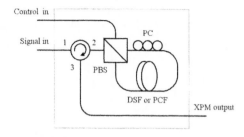

Figure 8.23: Experimental setup for NRZ-to-RZ conversion with a polarization-diversity loop whose output is directed toward an optical filter. (From Ref. [95]; ©2006 IEEE.)

(control) is oriented at $45°$ with respect to the principal axes of the polarizing beam splitter (PBS) so that its power is divided equally in two counterpropagating directions. The NRZ data (signal) with a random SOP is also divided into two orthogonally polarized parts. The same PBS combines the two parts. An optical circulator directs the output toward an optical filter whose passband is offset properly.

A scheme for RZ-to-NRZ conversion makes use of only SPM-induced spectral broadening inside a normally dispersive optical fiber [96]. The RZ pulses are chirped through SPM and undergo considerable broadening inside the fiber. If the fiber length is chosen such that this pulse broadening is large enough to fill the entire bit slot, the output is an NRZ version of the original bit stream.

8.3.3 All-Optical Sampling

Measurements of high-speed optical signals, such as a 40-Gb/s bit stream, require a temporal resolution of <10 ps. Although an electronic sampling oscilloscope with an ultrafast photodiode can provide a rise time of 8 ps, its use is limited for ODTM signals whose bandwidth can exceed 100 GHz. Several all-optical sampling techniques developed for solving this problem make use of XPM or FWM phenomenon within an optical fiber.

The FWM-based sampling technique was first implemented in 1991, but it required a 14-km-long dispersion-shifted fiber because of its relatively weak nonlinearity [97]. Soon after, FWM inside SOAs was employed for this purpose [98]–[100], and by 1999 this approach could be used at a bit rate of 160 Gb/s [101]. Its further development was fueled with the advent of highly nonlinear fibers [102]–[104], and it reached the commercial stage by 2006.

The underlying mechanism behind such an optical sampling technique can be understood [104] from Figure 8.24 and makes use of the periodic nature of a rapidly varying signal, say, at the bit rate B. The signal is launched inside a nonlinear fiber together with a high-power train of narrow intense pump pulses whose repetition rate f_s is chosen to satisfy $B = Mf_s + \Delta f$, where M is an integer and f_s sets the scanning rate. As seen in Figure 8.24, each successive pump pulse overlaps with a different part of the signal pulse. Because FWM can occur only when the signal and the pump are present simultaneously, the idler represents a sample of the signal whose duration is

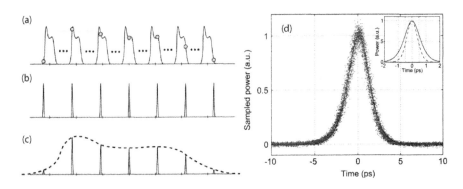

Figure 8.24: (a) Input signal and (b) a sampling train of short pump pulses through FWM create (c) a train of idler pulses. (d) An example of the sampled pulse (dots) and the fitted shape (solid line). The inset compares the pump pulse (solid line) with the idler pulse (dashed line). (From Ref. [104]; ©2005 IEEE.)

even shorter than the width of the pump pulse because of the exponential nature of the parametric gain. These samples occur at a relatively low rate (often <100 MHz) governed by the repetition rate of pump pulses and can be measured with a low-bandwidth detector. The resulting electrical signal represents a time-magnified copy of the original signal, and it can be used to reconstruct the shape of signal pulses and deduce their width.

Figure 8.24(d) shows, as an example, the reconstructed shape of 2.8-ps pulses emitted by a mode-locked fiber laser at a 10-GHz repetition rate (wavelength 1565 nm). The sampling pump pulses (wavelength 1541 nm) at a 100-MHz repetition rate had 1.3-ps widths and were launched with the signal into a 10-m-long highly nonlinear fiber. The resulting idler indicated that mode-locked pulses could be fitted well using a function $\text{sech}^2(at)$ with a FWHM of 2.95 ps. From the known value of 2.8-ps, the estimated temporal resolution of 0.93 ps is better than the 1.3-ps width of pump pulses [104]. This can be understood from the inset in Figure 8.24(d), which compares the pump pulse (solid line) with the idler pulse (dashed line). A shorter idler pulse is a consequence of the the fact that the the idler power depends on pump power exponentially under phase-matching conditions.

The XPM-based sampling technique was first demonstrated in 1991 using an NOLM as an optical switch [105]. Because of a relatively small value of the nonlinear parameter γ of conventional fibers, its use required high peak powers for loop lengths 100 m or less. This problem could be solved using the device shown in Figure 8.10, in which an SOA was used as the nonlinear element [105]. The use of highly nonlinear fibers in which γ is enhanced considerably provides an alternative solution. The experimental setup employed for this technique is similar to that shown in Figure 8.3 in the context of XPM-induced wavelength conversion. The main difference is that short pulses are employed at the wavelength λ_1 (in place of a CW beam), which shift the wavelength of the signal at the wavelength λ_2.

In a 2004 experiment, the use of XPM inside a 50-m-long highly nonlinear fiber allowed optical sampling at bit rates of up to 500 Gb/s [106]. A mode-locked fiber laser provided 1.1-ps-wide pump pulses, which were used to sample the signal. Each pump pulse shifted the wavelength of the signal through XPM over a sample whose duration was related to its width. An optical filter whose passband was offset from the signal wavelength by a suitable amount passed only these samples. The temporal resolution was estimated to be 0.7 ps for a 320-Gb/s signal.

8.4   Optical Regenerators

As discussed in Section 7.1, an optical bit stream used to transmit information in modern lightwave systems is degraded considerably during its propagation inside the fiber link. The peak power of optical pulses is reduced because of fiber losses, while their shape is distorted through a combination of the dispersive and nonlinear effects. Optical amplifiers can be used to manage fiber losses, but they degrade the bit stream even further by adding ASE noise that affects both the 0 and 1 bits and also shift pulse positions randomly (timing jitter).

An ideal optical regenerator is a device that transforms the degraded bit stream into its original form by performing three functions: reamplification, reshaping, and retiming. Such devices are often referred as *3R regenerators* to emphasize that they perform all three functions. With this terminology, optical amplifiers can be classified as 1R regenerators because they only reamplify the bit stream. Devices that perform the first two functions are called *2R regenerators*. Since 2R and 3R regenerators have to work at time scales shorter than the bit slot in order to carry out pulse reshaping and retiming, they must operate at time scales of 10 ps or less, depending on the bit rate of the optical signal. As nonlinear effects in optical fibers respond at femtosecond time scales, the use of highly nonlinear fibers has become quite common for such devices. Indeed, many techniques have been developed that make use of the three major nonlinear effects (SPM, XPM, and FWM) that were discussed in Section 7.2. This section focuses on the basic physics behind such devices together with their system applications.

8.4.1   SPM-Based Regenerators

Although the basic idea behind a SPM-based 2R regenerator was outlined in 1998 [107], it was only after 2003 that such devices were studied extensively [108]–[116]. Figure 8.25 shows the scheme employed schematically. The distorted noisy signal is first amplified by an EDFA before it is propagated through a highly nonlinear fiber, where its spectrum broadens considerably because of SPM-induced frequency chirping. It is subsequently passed through a bandpass filter, whose center wavelength is chosen judiciously, resulting in an output bit stream with much reduced noise and much improved pulse characteristics.

It may appear surprising at first sight that spectral filtering of a bit stream whose phase has been modified nonlinearly improves the signal in the time domain. However, it is easy to see why this scheme would remove noise from the 0 bits. As the noise power in 0 bits is relatively low, the spectrum does not broaden much during 0 bits. If the passband of the optical filter is offset enough from the peak of the input spectrum,

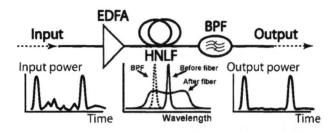

Figure 8.25: An SPM-based 2R regenerator (top) and its action on a bit stream (bottom). HNLF and BPF stand for a highly nonlinear fiber and a bandpass filter, respectively. (From Ref. [113]; ©2006 IEEE.)

this noise would be blocked by the filter. In practice, this offset is chosen such that pulses representing 1 bits pass though the filter without much distortion. The noise level of 1 bits is also reduced because a small change in the peak power does not affect the pulse spectrum significantly, resulting in a much cleaner output bit stream.

To understand the operation of SPM-based regenerators, one may employ the analysis given in Chapter 4 of Ref. [1]. If we neglect the dispersive effects within the highly nonlinear fiber, only the phase of the the optical filed is affected by SPM within the fiber such that

$$U(L,t) = U(0,t) \exp[i\gamma P_0 L_{\text{eff}} |U(0,t)|^2], \qquad (8.4.1)$$

where $L_{\text{eff}} = (1 - e^{-\alpha L})/\alpha$ is the effective length for a fiber of length L with the loss parameter α, P_0 is the peak power of pulses, and $U(0,t)$ represents bit pattern of the input bit stream. As an optical filter acts in the spectral domain, the optical field after the filter can be written as

$$U_f(t) = \mathscr{F}^{-1}\{H_f(\omega - \omega_f)\mathscr{F}[U(L,t)]\}, \qquad (8.4.2)$$

where $\mathscr{F}$ is the Fourier-transform operator and $H_f(\omega - \omega_f)$ is the transfer function of a filter offset from the carrier frequency of pulses by ω_f.

The performance of an SPM-based regenerator depends on three parameters; namely, the maximum nonlinear phase shift $\phi_{\text{NL}} \equiv \gamma P_0 L_{\text{eff}}$, the filter-passband offset ω_f, and the filter bandwidth $\delta\omega$, which must be large enough to accommodate the entire signal so that the width of optical pulses remains intact. This leaves only two design parameters whose optimum values were investigated in a 2005 study [112] using Gaussian-shape pulses and a Gaussian transfer function for the filter. In general, ϕ_{NL} should not be too large because, if the spectrum becomes too broad, filter-induced losses become too large. Its optimum value is close to $3\pi/2$ because the SPM-broadened spectrum then exhibits two peaks with a sharp dip at the original carrier frequency of the pulse [1]. Noting that $\phi_{\text{NL}} = L_{\text{eff}}/L_{\text{NL}}$, where L_{NL} is the nonlinear length, the optimum length L_{eff} is close to $5L_{\text{NL}}$. The optimum value of the filter offset in this case is found to be $\omega_f = 3/T_0$, where T_0 is the half-width of Gaussian pulses with the power profile $P(t) = P_0 \exp(-t^2/T_0^2)$.

Figure 8.26 shows a numerical example of the noise reduction provided by SPM-based 2R regenerators [112] in the case of $\phi_{\text{NL}} = 5$ and 2-ps-wide Gaussian pulses

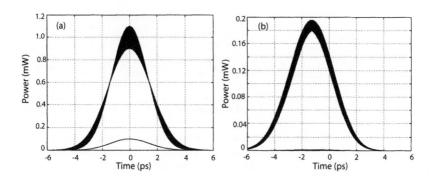

Figure 8.26: Numerically simulated pulse shapes at (a) input and (b) output ends of a SPM-based 2R regenerator designed with $\phi_{NL} = 5$. The noise pulse with 10% amplitude is almost completely blocked by the regenerator. (From Ref. [112]; ©2005 IEEE.)

(appropriate for a 160-Gb/s bit stream). Each input pulse could have up to 10% variations in its peak power (average value 1 mW) but its width was changed to keep the same pulse energy. At the output end, the noise power is reduced from 10% to 0.6% of the average peak power, and the amplitude of power variations is reduced from 10% to 4.6%. The reason behind a large reduction in noise power is related to almost complete blocking of noise pulses in 0-bit time slots. For example, a noise pulse with 0.1-mW peak power in Figure 8.26(a) is nearly blocked by the regenerator.

The preceding analysis holds as long as dispersive effects are negligible. At high bit rates, pulses become so short that such effects may not remain negligible. However, one must distinguish between the cases of normal and anomalous dispersion. The anomalous-GVD case was studied during the 1990s in the context of soliton-based systems after it was found that the use of an optical filter after each in-line amplifier helps to maintain solitons over thousands of kilometers [117]–[120]. In this case, SPM and GVD occur inside the transmission fiber itself. Soliton regenerators with a design similar to that shown in Figure 8.25 have also been considered [121], [122]. Their operating principle is, however, different from nonsoliton regenerators shown in Figure 8.25 because the optical filter is centered on the carrier frequency rather than at a shifted frequency.

In the case of normal GVD, the SPM-based regenerator is designed with a filter that is offset from the carrier frequency but it is important to include the dispersive effects. The optimum fiber length in this case depends on both the nonlinear length L_{NL} and the dispersion length $L_D = T_0^2/|\beta_2|$ and is found to be $L_{eff} = 2.4L_D/N$ [108], where the parameter N is defined as [1]

$$N^2 = \frac{L_D}{L_{NL}} = \frac{\gamma P_0 T_0^2}{|\beta_2|}. \tag{8.4.3}$$

It is therefore necessary to take into account the effects of dispersion whenever L_D and L_{NL} have comparable magnitudes.

Experimental optimization of an SPM-based regenerator at a bit rate of 40 Gb/s was carried out with the setup shown in Figure 8.27. Mode-locked 3-ps pulses were

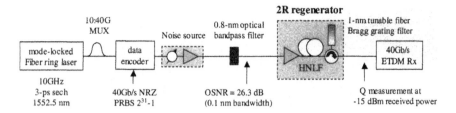

Figure 8.27: Experiment setup of an SPM-based 2R regenerator operating at 40 GB/s. HNLF, EDTM, and PRBS stand for highly nonlinear fiber, electrical time-domain depmultiplexing, and pseudorandom bit stream, respectively. (From Ref. [108]; ©2004 IEEE.)

intentionally broadened to 6.25 ps using a 0.8-nm bandpass filter. The 2R regenerator consisted of a high-power EDFA, a 2.5-km-long highly nonlinear fiber, and a 1-nm-bandwidth tunable filter in the form of a fiber grating. After fiber losses are taken into account, L_{eff} of the fiber is 2.1 km. For a 2-nm wavelength offset for the filter, the SNR of the regenerated bit stream (estimated by measuring the Q factor) was found to depend strongly on the parameter N and was maximum for $N = 12$. This experiment clearly showed that the power launched into the fiber depends on the fiber length and filter offset, and it must be optimized for such regenerators to work well.

The required fiber length can be reduced considerably by employing nonsilica fibers with large values of n_2. A 2.8-m-long piece of chalcogenide (As_2Se_3) fiber was employed in a 2005 experiment [111]. This fiber exhibited high normal dispersion near 1550 nm with $\beta_2 > 600$ ps^2/km. However, it turned out that this large value actually helped the device performance, rather than hindered it. The large value of the nonlinear parameter ($\gamma \approx 1200$ W^{-1}/km) reduced the required peak power to ~ 1 W, whereas large values of β_2 reduced the dispersion length L_D close to 18 m for 5.8-ps pulses employed in the experiment. The optimum fiber length under these conditions was close to 3 m. Figure 8.28 shows the impact of fiber dispersion on the SPM-broadened spectrum and the resulting changes in the transfer function of of the regenerator for a

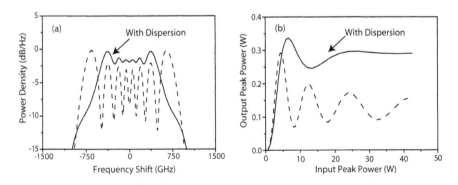

Figure 8.28: Effects of normal dispersion on (a) pulse spectrum and (b) power-transfer function for an SPM-based 2R regenerator made with a 2.8-m-long piece of chalcogenide fiber. Dashed curves show, for comparison, the dispersion-free case. (From Ref. [111]; ©2005 OSA.)

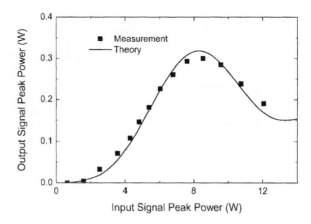

Figure 8.29: Measured and predicted power-transfer functions for an SPM-based regenerator made with a 1-m-long bismuth-oxide fiber. (From Ref. [115]; ©2006 IEEE.)

fixed position of the optical filter. Improvements in the transfer function result from the reduced amplitude of spectral oscillations, resulting in a relatively smooth spectrum. Even the presence of two-photon absorption in chalcogenide fibers, a normally undesirable phenomenon, helps in improving the device performance [114].

In a 2006 experiment, 1-m length of a fiber made with bismuth-oxide glass was employed, in combination with a tunable 1-nm bandpass filter, to build an SPM-based regenerator [115]. The center wavelength of the filter was offset by 1.7 nm from the carrier wavelength of the incoming 10-Gb/s bit stream. Losses were negligible (about 0.8 dB) for such a short fiber that also exhibited a normal dispersion of 330 ps^2/km at 1550 nm. The nonlinear parameter γ for this fiber was close to 1100 W^{-1}/km. Because of a high nonlinearity and normal dispersion, such a fiber performed well as a 2R regenerator when the peak power of input pulses was large enough (about 8 W) to induce significant spectral broadening. Figure 8.29 compares the measured power-transfer function with the theoretical prediction. A negligible output at low input powers and a relatively broad peak ensure that power fluctuations will be reduced considerably for both 0 and 1 bits.

The nonlinear phenomenon of XPM is also useful for optical regeneration [123]–[125]. The use of cascaded Mach–Zehnder interferometers for this purpose was proposed in one study with a semiconductor optical amplifier acting as the nonlinear element [123]. A fiber-based demultiplexing scheme was employed in a 2001 experiment in which the demultiplexed channel from an OTDM bit stream also exhibited regenerative properties with a reduced noise level [124]. The basic scheme is similar to that used for SPM-based regenerators (see Figure 8.25) except that the OTDM signal is launched together with control pulses in the form an optical clock. Inside the nonlinear fiber, the spectrum of all signal pulses is broadened through SPM, but pulses belonging to a specific channel experience additional XPM-induced spectral broadening resulting from their overlap with the control pulses. An optical filter, offset from the carrier frequency by an optimum amount, selects these pulses, while also improv-

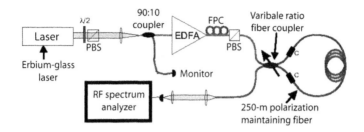

Figure 8.30: Experimental setup employed for a NOLM-based 2R regenerator; PBS and FPC stand for polarizing beam splitter and fiber polarization controller, respectively. (From Ref. [129]; ©2003 IEEE.)

ing their signal-to-noise ratio. Thus, such a scheme provides 2R regeneration together with demultiplexing.

A similar scheme was used in a 2003 experiment to convert the wavelength of a 80-Gb/s bit stream [125]. The new feature was that the 1-km-long highly nonlinear fiber was pumped backward to provide the Raman gain. The presence of Raman gain within the fiber providing XPM enhances the wavelength-conversion efficiency; a 21-dB enhancement was observed at a pump power of 600 mW. The resulting wavelength-converted signal exhibited considerable improvement in both the signal-to-noise ratio and extinction ratio because of the noise reduction provided by optical filtering in combination with the XPM-induced spectral broadening.

Any nonlinear device in which the SPM and XPM effects produce nonlinear power transfer characteristics can be used as a 2R regenerator. As discussed in Section 3.2, a Sagnac interferometer in the form of a NOLM is just such a nonlinear device. Indeed, it was used as early as 1992 for demonstrating optical regeneration [126]. In this experiment, the XPM-induced phase shift was employed to modify the NOLM transmission and to regenerate the bit stream. Soon after, such devices were analyzed [127] and employed for optical regeneration of pulses in soliton-based systems [93]. The use of a Kerr shutter where XPM is used to change the state of polarization provided regenerators operating at speeds of up to 40 Gb/s [128].

A highly asymmetric NOLM was employed in a 2003 experiment, and it reduced the signal noise by as much as 12 dB [129]. Figure 8.30 shows the experimental setup schematically. The NOLM was built using a fiber coupler whose splitting ratio could be varied to ensure that power levels in the counterpropagating directions differed substantially inside the Sagnac loop made with a 250-m-long polarization-maintaining fiber. For a splitting ratio of 90:10, the combination of SPM and XPM produced a relative phase shift in the two directions such that the power-transfer function of the NOLM exhibited a nearly flat region around 5 mW of input power, and the noise level was reduced considerably in this region. The optical SNR of a 40-Gb/s system could be improved by 3.9 dB with this approach [130].

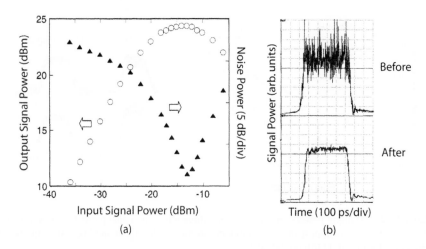

Figure 8.31: (a) Saturation of output signal power in a 2.5-km-long FOPA (circles) and the resulting drop in the noise power (triangles); (b) power fluctuations at the input and output ends. (From Ref. [134]; ©2002 IEEE.)

8.4.2 FWM-Based Regenerators

The nonlinear phenomenon of FWM attracted attention for 2R regeneration starting in 2000, and several experiments have demonstrated its use in practice [131]–[137]. As we have seen in Section 8.1.2, FWM converts a fiber into a parametric amplifier. Similar to any amplifier, the gain of a FOPA also saturates when signal power becomes large enough to saturate the amplifier [132]. Because of this gain saturation, fluctuations in the peak power of a pulse are reduced by a large factor. Figure 8.31(a) shows the improvement realized in the case of a FOPA made using a 2.5-km-long dispersion-shifted fiber and pumped close to the zero-dispersion wavelength with 500-ps pulses (peak power 1.26 W). The FOPA exhibited a gain of 45 dB at low signal powers, but the gain saturated when the output signal power approached 200 mW. Because of it, noise power of the signal was reduced by more than a factor of 20. This is also evident from the temporal patterns seen in part (b).

The simple theory of Section 8.1.2 cannot be used for describing gain saturation in FOPAs, because it assumes that the pump power remains nearly undepleted along the fiber. For a FOPA to be useful as a 2R regenerator, the signal powers become large enough that pump power is depleted significantly. Moreover, large power levels of the signal and idler initiate a cascaded FWM process by acting as the pump and creating multiple other waves [131]. All of these idlers act as a wavelength-shifted replica of the signal and exhibit much less noise compared with the signal. Experimental results for a single-pump FOPA agree with a theoretical model that takes into account pump depletion [133].

A dual-pump FOPA is often preferred in practice because it can provide nearly polarization-independent gain with a suitable choice of pump wavelengths and polarization states. As one may expect, such FOPAs can also be used as 2R regenerators [136]. Figure 8.32(a) shows the transfer functions for the signal and three idlers by

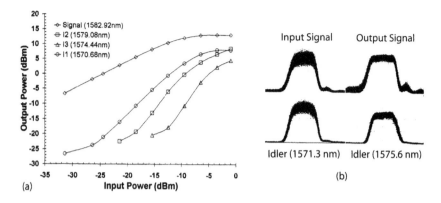

(a)

(b)

Figure 8.32: (a) Output powers for the signal and three idlers for a dual pump FOPA built using a 1-km-long highly nonlinear fiber; (b) temporal profiles of the signal and idlers showing noise reduction. (From Ref. [136]; ©2003 IEEE.)

plotting their output powers as a function of input signal power for a FOPA built using a 1-km-long highly nonlinear fiber (zero-dispersion wavelength at 1583 nm) and pumped using two lasers operating at 1567 and 1603.5 nm. The temporal profiles of the input and output signals and two idlers for a 10-Gb/s input bit stream are also shown. Signal noise is reduced considerably in the on state, but noise in the off state (0 bits) becomes worse because of the amplification provided by the FOPA. However, if one of the idlers is used, the noise in reduced even for 0 bits.

The performance of FWM-based regenerators can be improved further by cascading two FOPAs in series. In a 2006 experiment, the output of first FOPA was filtered with an optical filter to select a higher-order idler that acted as the pump for the second-stage FOPA [137]. A CW seed acted as the signal and created its corresponding idler. This idler had the same bit pattern as the signal launched at the input end of the first FOPA but with a much reduced noise level. Figure 8.33 shows the measured transfer functions after the first and second stages. A nearly step-function-like shape after the second stage indicates the extent of improvement possible with such a scheme.

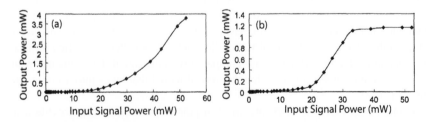

Figure 8.33: Measured transfer functions after the first (a) and second (b) stages for a two-stage FWM-based regenerator. (From Ref. [137]; ©2006 IEEE.)

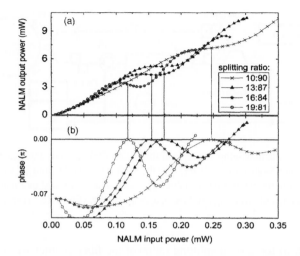

Figure 8.34: Measured (a) power and (b) phase transfer functions for several splitting ratios for a NOLM used for regenerating RZ-DPSK signals. (From Ref. [147]; ©2007 IEEE.)

8.4.3   Regeneration of DPSK Signals

So far we have considered all-optical regeneration of NRZ or RZ bit streams. In the case of the DPSK format, information is coded in the phase difference of two neighboring bits. Although the intensity of the electric field may remain constant, the RZ-DPSK format is often employed in practice by using identical pulses in all bit slots while coding their relative phases. Most of the schemes discussed so far cannot work for regeneration of DPSK signals because their operation is based on different power levels associated with the 0 and 1 bits. Indeed, several techniques discussed in this subsection have been developed for this purpose [138]–[147].

In a 2005 study, an NOLM similar to that shown in Figure 8.30 was employed with one crucial difference: an attenuator with different losses in the counterpropagating directions was inserted near one end of the fiber loop [140]. Such a device is similar to an optical isolator and can be fabricated using polarizers and a Faraday rotator. Transmission thorough such a device can be analyzed using the theory outlined in Section 3.2. Although much higher input powers are required, the power-transfer function exhibits a flat region around which the phase shift produced by the NOLM is also constant and relatively small. The experimental results were in agrement with these theoretical predictions [144].

A bidirectional EDFA (in place of a directional attenuator) at one end of the NOLM was used in a 2007 experiment to realize regeneration of RZ-DPSK signals [147]. The input signal was split asymmetrically at the fiber coupler such that each weaker subpulse was first amplified by the EDFA, while the stronger subpulse passed through it after traversing the Sagnac loop. As a result, the SPM-induced phase shift was much larger for the weaker subpulses. As the phase of output pulse is set by stronger subpulses, the NOLM does not distort the phase of outgoing pulses much. Figure 8.34 shows the measured power and phase characteristics for a 3-km-long fiber loop

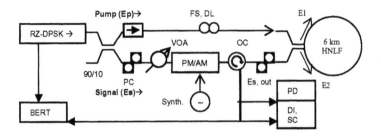

Figure 8.35: Experimental setup for regenerating RZ-DPSK signals through phase-sensitive amplification inside a Sagnac loop; BERT, FS-DL, VOA, OC, PD, DI, and SC stand for bit-error rate tester, fiber stretcher delay line, variable optical attenuator, optical circulator, photodiode, delay interferometer, and sampling oscilloscope, respectively. (From Ref. [145]; ©2006 OSA.)

($\gamma = 2.5$ W^{-1}/km) for several splitting ratios of the fiber coupler when the amplifier was pumped to provide a small-signal gain of 23 dB. As expected, the output power becomes nearly constant in a specific range of input powers, a feature that reduces noise. The important point is that the phase is nearly constant in this region, and amplitude noise of the signal can be suppressed without transferring it into phase jitter. At the same time, the relative phase shift between the 0 and 1 bits is so small ($<$ 0.07π) that it does not affect decoding of the DPSK bit stream. Indeed, measured bit-error rates for a 10-Gb/s DZ-DPSK bit stream were improved considerably with such a regenerator. The amplification can also be provided through the Raman gain by injecting pump light into the loop such that it propagates in one direction only [143].

The simplest kind of 2R regenerator, shown in Figure 8.25, can also be adopted for the RZ-DPSK format with suitable modifications. For example, the signal phase can be almost preserved over long distances if the nonlinear fiber provides anomalous dispersion and a saturable absorber is inserted before it [141]. In this case, the combination of soliton effects and narrowband filtering reduces amplitude noise and reshapes RZ pulses without affecting the signal phase significantly. A FWM-based approach can also be used by pumping the fiber near the zero-dispersion wavelength and increasing the signal power so that the parametric gain is saturated and multiple idlers are generated through cascaded FWM. However, one should set the optical filter such that it selects the signal and rejects all idlers to minimize degradation of the information contained in signal phase. An XPM-based scheme has also been proposed for regenerating DPSK signals [138].

The preceding schemes regenerate RZ pulses by reducing amplitude noise (while preserving their phases) but they do not reduce the phase noise. A novel approach accomplishes this task by making use of phase-sensitive amplification inside a Mach–Zender or Sagnac interferometer [139]. A 6-km-long Sagnac loop (or NOLM) was employed in a 2005 experiment [142] to realize $>$13 dB of phase-sensitive gain at a pump power of 100 mW. Phase noise was reduced enough to improve the bit-error rate of the regenerated DPSK signal by a factor of 100. In a later experiment, the same loop was used to reduce both the amplitude and phase noises by a relatively large factor [145].

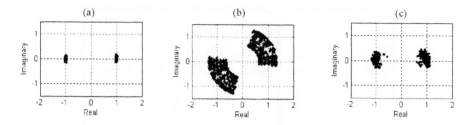

Figure 8.36: Constellation diagrams of the RZ-DPSK signal (a) before adding noise, (b) after adding noise, and (c) after phase-sensitive amplification. (From Ref. [145]; ©2006 OSA.)

Figure 8.35 shows the experimental setup employed for phase-sensitive amplification inside a Sagnac interferometer. The DPSK signal is first split into two parts using a 90:10 fiber coupler. The branch with 90% average power acts as a pump, while the low-power branch acts as the signal; a delay line in the pump branch ensures decorrelation between the two. Phase and amplitude noises are added to the signal before it enters the 6-km-long fiber loop, where a degenerate FWM process transfers power from the pump to the signal [139]. The extent of power transfer depends on the relative phase difference between the pump and signal. It is this feature that reduces phase noise at the NOLM output. Figure 8.36 shows the extent of improvement using constellation diagrams [145] in which the real and imaginary parts of the electric field are plotted at bit center for different bits of a DPSK signal. Both the amplitude and phase noises are reduced significantly after phase-sensitive amplification. A dual-pump FOPA can also be employed for this purpose provided signal frequency is located exactly in the middle of the two pumps so that it coincides with the idler frequency [146].

8.4.4   Optical 3R Regenerators

As mentioned earlier, a 3R regenerator should also perform the retiming function and reduce the timing jitter of a bit stream. An optical modulator was used during the 1990s for this purpose in the context of soliton systems [148], and its use is often necessary [149]. An electrical clock signal, extracted from the input data, drives the modulator and provides the timing information related to the center of each bit slot. An SPM-based 3R regenerator can be built by adding a modulator to the scheme shown in Figure 8.25. A schematic of such a device is shown in Figure 8.37. Numerical simulations

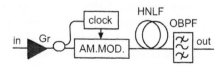

Figure 8.37: An SPM-based 3R regenerator; AM-MOD, HNLF, and OBPF stand for amplitude modulator, highly nonlinear fiber, and optical bandpass filter, respectively. (From Ref. [110]; ©2004 IEEE.)

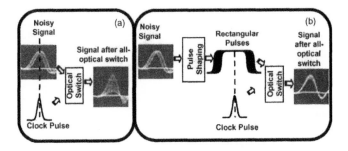

Figure 8.38: A retiming technique based on an NOLM acting as an optical switch. Two parts show switching (a) without and (b) with a sampled fiber grating used for pulse shaping. (From Ref. [154]; ©2006 IEEE.)

for a fiber link containing such 3R regenerators at periodic intervals indeed show a considerable reduction in timing jitter [110].

Several other schemes have been proposed for reducing timing jitter of a bit stream [150]–[156]. In one scheme, a single phase modulator in combination with a dispersive fiber is found to be effective in reducing timing jitter [150]. In another, an optical AND gate is used to correlate data pulses with clock pulses that have been chirped and broadened inside a dispersive fiber [151]. The combination of a dispersion-compensating fiber and a fiber grating is also found to be effective in suppressing timing jitter induced by the intrachannel XPM effects [152].

In an interesting scheme, a sampled fiber grating is used first to broaden and reshape data pulses into a nearly rectangular shape [154]. These pulses are then launched into an NOLM acting as an optical switch and driven by narrow clock pulses. Clock pulses shift the phase of each data pulse through XPM and direct only its central part to the output port, resulting in regenerated data with much reduced timing jitter. Figure 8.38 shows the basic idea schematically. In the absence of a fiber grating, such an optical switch does not reduce timing jitter much.

The simplest design of a 3R regenerator makes use of XPM inside a highly nonlinear fiber, followed with an optical filter. Figure 8.39 shows the configuration adopted in a 2005 experiment [153] together with the principle of operation. The clock pulses at the wavelength λ_2 are narrower than signal pulses and are delayed such that each of them continues overlapping with a signal pulse over the entire fiber length in spite of their different speeds. The optical filter is set at λ_2 with a bandwidth narrower than the clock spectrum. As the signal power increases in parts (b) to (d), the XPM-induced wavelength shift of clock pulses reduces their transmission, resulting in a power-transfer function, shown in Figure 8.39(e). The output of such a device is a wavelength-converted signal with a reversal of 1 and 0 bits. In the experiment, the 10-Gb/s signal at a wavelength of 1534 nm was launched inside a 750-m-long highly nonlinear fiber together with 2.9-ps clock pulses at 1552 nm at a 10-GHz repetition rate. The regeneration 10-Gb/s signal improved the bit-error rate significantly because of reduced noise level and timing jitter.

The XPM-based scheme shown in Figure 8.39 has been analyzed in detail theoretically [155]. It turns out that the improvement in BER after the regenerator occurs

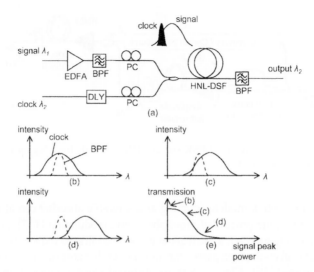

Figure 8.39: (a) Configuration of an XPM-based regenerator and its operating principle. As signal power increases from (b) to (d), the XPM-induced wavelength shift reduces transmission, resulting in the response shown in part (e). (From Ref. [153]; ©2005 IEEE.)

only if the power-transfer function of the regenerator is different for 0 and 1 bits. The scheme of Figure 8.39 exhibits this feature because the wavelength shift of the clock depends on the derivative of the signal power [1] as $\delta\omega = -2\gamma L_{\text{eff}}(dP/dt)$. Data bits representing a logical 1 shift the clock spectrum through XPM, and the filter blocks these clock bits. On the other hand, 0 bits containing only noise produce little spectral shift of clock pulses, which pass through the filter unchanged. Timing jitter is eliminated because clock pulses now represent the data with reversed polarity.

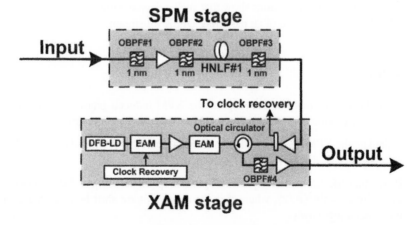

Figure 8.40: A 3R regenerator consisting of an SPM-based 2R regenerator followed with a cross-absorption modulation (XAM) stage. (From Ref. [159]; ©2006 IEEE.)

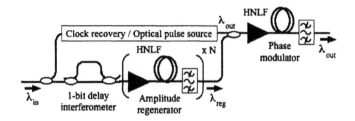

Figure 8.41: A 3R regenerator for DPSK signals; HNLF stands for a highly nonlinear fiber. (From Ref. [160]; ©2007 IEEE.)

An electro-absorption modulator acting as a saturable absorber can also eliminate timing jitter through the process of cross-absorption modulation [157]–[159]. In the scheme shown in Figure 8.40, a 2R regenerator is used first to reduce the noise level. The intense data pulses are then passed through a saturable absorber together with low-power clock pulses [159]. Clock pulses are absorbed when a logical 1 appears in the data stream but are transmitted otherwise. The resulting output is an inverted replica of the original bit stream with virtually no timing jitter.

The preceding schemes are appropriate for regenerating bit streams in which information is coded through on–off keying (OOK). As the use of phase-encoded signals is becoming common, 3R regenerators for them are needed. The design of a 3R regenerator for DPSK signals is shown in Figure 8.41. It adds a 1-bit delay interferometer in front of a an SPM-based 2R regenerator whose output is fed into a second fiber together with an optical clock at a different wavelength. Similar to the case of Figure 8.40, data is transferred to output clock pulses, which constitute the regenerated DPSK signal. The role of a delay interferometer is to convert the incoming DPSK signal into an OOK signal whose noise is reduced by the 2R amplitude regenerator. The resulting regenerated data stream is finally used to modulate the phase of clock pulses through XPM inside a highly nonlinear fiber. The last filter lets pass only the clock wavelength. Numerical simulations show that such a device reduces significantly both the amplitude and phase noise of an incoming DPSK bit stream.

Problems

8.1 Describe two techniques that exploit the XPM-induced phase shift for wavelength conversion of a WDM channel. Use diagrams as necessary.

8.2 Calculate the XPM-induced phase shift using Eqs. (8.1.4) when data pulses have a soliton shape; that is, $A_2(0, T) = \text{sech}(t/T_0)$.

8.3 Prove that the transmissivity of a Sagnac loop made using a 3-dB coupler is given by $T = \sin^2(\phi/2)$, where ϕ is the relative phase shift between the counterpropagating waves.

8.4 Solve Eqs. (8.1.7) and (8.1.8) describing the FWM process and prove that the wavelength-conversion efficiency is given by Eq. (8.1.10).

8.5 Prove that the phase-mismatch parameter κ, defined in Eq. (8.1.9), reduces to Eq. (8.1.12) when the pump wavelength is close to the zero-dispersion wavelength of the fiber.

8.6 Explain how a Sagnac loop can be used as a time-domain switch. Why is the switching window wider than the pump pulse when short pump pulses are employed?

8.7 Explain the physical mechanism that makes the switching window of a TOAD device to be much smaller than the carrier lifetime of the semiconductor optical amplifier placed inside the Sagnac loop.

8.8 Discuss how an ultrafast nonlinear interferometer works as a time-domain switch. How would you operate such a device in the gain-transparency mode?

8.9 Use the signal-gain expression in Eq. (8.2.6) to estimate the width of the switching window when a parametric amplifier is pumped using 5-ps (FWHM) Gaussian pulses with 5-W peak power. Assume $\gamma = 10 \text{ W}^{-1}$/km.

8.10 Describe two techniques that can be used to convert an NRZ bit stream into a RZ bit stream.

8.11 Explain how the nonlinear phenomenon of SPM can be used for regeneration of optical bit streams. Use diagrams as necessary.

8.12 Reproduce a figure similar to that shown in Figure 8.26 numerically by applying Eq. (8.4.2) to a set of 10-ps-wide (FWHM) noisy Gaussian pulses. Use $\phi_{NL} = 5$ and a frequency shift of 80 GHz for the optical filter.

References

[1] G. P. Agrawal, *Nonlinear Fiber Optics*, 4th ed. (Academic Press, Boston, 2007).

[2] G.-H. Duan, in *Semiconductor Lasers: Past, Present, and Future*, G. P. Agrawal, Ed. (AIP Press, Woodbury, NY, 1995), Chap. 10.

[3] T. Durhuus, B. Mikkelson, C. Joergensen, S. L. Danielsen, and K. E. Stubkjaer, *J. Lightwave Technol.* **14**, 942 (1996).

[4] S. J. B. Yoo, *J. Lightwave Technol.* **14**, 955 (1996).

[5] V. Lal, M. L. Mašanović, J. A. Summers, G. Fish, and D. J. Blumenthal, *IEEE J. Sel. Topics Quantum Electron.* **13**, 49 (2007).

[6] M. H. Chou, I. Brener, M. M. Fejer, E. E. Chaban, and S. B. Christman, *IEEE Photon. Technol. Lett.* **11**, 653 (1999).

[7] I. Cristiani, V. Degiorgio, L. Socci, F. Carbone, and M. Romagnoli, *IEEE Photon. Technol. Lett.* **14**, 669 (2002).

[8] Y. W. Lee, F. C. Fan, Y. C. Huang, B. Y. Gu, B. Z. Dong, and M. H. Chou, *Opt. Lett.* **27**, 2191 (2002).

[9] K. A. Rauschenbach, K. L. Hall, J. C. Livas, and G. Raybon, *IEEE Photon. Technol. Lett.* **6**, 1130 (1994).

[10] J. Yu, X. Zheng, C. Peucheret, A. T. Clausen, H. N. Poulsen, and P. Jeppesen, *J. Lightwave Technol.* **18**, 1001 (2000); *J. Lightwave Technol.* **18**, 1007 (2000).

[11] T. Sakamoto, F. Futami, K. Kikuchi, S. Takeda, Y. Sugaya, and S. Watanabe, *IEEE Photon. Technol. Lett.* **13**, 502 (2001).

[12] B. E. Olsson, P. Öhlén, L. Rau, and D. J. Blumenthal, *IEEE Photon. Technol. Lett.* **12**, 846 (2000).

[13] J. Yu and P. Jeppesen, *IEEE Photon. Technol. Lett.* **13**, 833 (2001).

[14] J. H. Lee, Z. Yusoff, W. Belardi, M. Ibsen, T. M Monro, and D. J. Richardson, *IEEE Photon. Technol. Lett.* **15**, 437 (2003).

[15] L. Rau, W. Wang, S. Camatel, H. Poulsen, and D. J. Blumenthal, *IEEE Photon. Technol. Lett.* **16**, 2520 (2004).

[16] W. Mao, P. A. Andrekson, and J. Toulouse, *IEEE Photon. Technol. Lett.* **17**, 420 (2005).

[17] W. Wang, H. N. Poulsen, L. Rau, H. F. Chou, J. E. Bowers, and D. J. Blumenthal, *J. Lightwave Technol.* **23**, 1105 (2005).

[18] J. H. Lee, T. Nagashima, T. Hasegawa, S. Ohara, N. Sugimoto, and K. Kikuchi, *IEEE Photon. Technol. Lett.* **18**, 298 (2006).

[19] T. Tanemura, J. H. Lee, D. Wang, K. Katoh, and K. Kikuchi, *Opt. Express* **14**, 1408 (2006).

[20] T. Tanemura and K. Kikuchi, *J. Lightwave Technol.* **24**, 4108 (2006).

[21] K. Inoue and H. Toba, *IEEE Photon. Technol. Lett.* **4**, 69 (1992).

[22] K. Inoue, T. Hasegawa, K. Oda, and H. Toba, *Electron. Lett.* **29**, 1708 (1993).

[23] K. Inoue, *J. Lightwave Technol.* **12**, 1916 (1994); *IEEE Photon. Technol. Lett.* **6**, 1451 (1993).

[24] G. A. Nowak, Y.-H. Kao, T. J. Xia, M. N. Islam, and D. Nolan, *Opt. Lett.* **23**, 936 (1998).

[25] O. Aso, S. Arai, T. Yagi, M. Tadakuma, Y. Suzuki and S. Namiki, *Electron. Lett.* **36**, 709 (2000).

[26] J. Hansryd, P. A. Andrekson, M. Westlund, J. Li, and P. O. Hedekvist, *IEEE J. Sel. Topics Quantum Electron.* **8**, 506 (2002).

[27] B. N. Islam and Ö Boyraz, *IEEE J. Sel. Topics Quantum Electron.* **8**, 527 (2002).

[28] K. K. Y. Wong, K. Shimizu, M. E. Marhic, K. Uesaka, G. Kalogerakis, and L. G. Kazovsky, *Opt. Lett.* **28**, 692 (2003).

[29] S. Radic and C. J. McKinstrie, *Opt. Fiber Technol.* **9**, 7 (2003).

[30] S. Radic, C. J. McKinstrie, R. M. Jopson, J. C. Centanni, Q. Lin, and G. P. Agrawal, *Electron. Lett.* **39**, 838 (2003).

[31] T. Tanemura and K. Kikuchi, *IEEE Photon. Technol. Lett.* **15**, 1573 (2003).

[32] Y. Wang, C. Yu, T. Luo, L. Yan, Z. Pan, and A. E. Willner, *J. Lightwave Technol.* **23**, 3331 (2005).

[33] J. H. Lee, W. Belardi, K. Furusawa, P. Petropoulos, Z. Yusoff, T. M. Monro, and D. J. Richardson, *IEEE Photon. Technol. Lett.* **15**, 440 (2003).

[34] K. K. Chow, C. Shu, C. Lin, and A. Bjarklev, *IEEE Photon. Technol. Lett.* **17**, 624 (2005).

[35] P. A. Andersen, T. Tokle, Y. Geng, C. Peucheret, and P. Jeppesen, *IEEE Photon. Technol. Lett.* **17**, 1908 (2005).

[36] A. Zhang and M. S. Demokan, *Opt. Lett.* **30**, 2375 (2005).

[37] F. Yaman, Q. Lin, S. Radic, and G. P. Agrawal, *IEEE Photon. Technol. Lett.* **17**, 2053 (2005).

[38] J. H. Lee, T. Nagashima, T. Hasegawa, S. Ohara, N. Sugimoto, and K. Kikuchi, *J. Lightwave Technol.* **24**, 22 (2006).

[39] G. Kalogerakis, M. E. Marhic, K. Uesaka, K. Shimizu, Member, K. K.-Y. Wong, and L. G. Kazovsky, *J. Lightwave Technol.* **24**, 3683 (2006).

[40] M. C. Farries and D. N. Payne, *Appl. Phys. Lett.* **55**, 25 (1989).

[41] K. J. Blow, N. J. Doran, B. K. Nayar, and B. P. Nelson, *Opt. Lett.* **15**, 248 (1990).

[42] M. Jinno and T. Matsumoto, *IEEE Photon. Technol. Lett.* **2**, 349 (1990); *Electron. Lett.* **27**, 75 (1991).

[43] H. Avramopoulos, P. M. W. French, M. C. Gabriel, H. H. Houh, N. A. Whitaker, and T. Morse, *IEEE Photon. Technol. Lett.* **3**, 235 (1991).

[44] J. D. Moores, K. Bergman, H. A. Haus, and E. P. Ippen, *Opt. Lett.* **16**, 138 (1991); *J. Opt. Soc. Am. B* **8**, 594 (1991).

[45] M. Jinno, *J. Lightwave Technol.* **10**, 1167 (1992); *Opt. Lett.* **18**, 726 (1993); *Opt. Lett.* **18**, 1409 (1993).

[46] H. Bülow and G. Veith, *Electron. Lett.* **29**, 588 (1993).

[47] A. W. O'Neil and R. P. Webb, *Electron. Lett.* **26**, 2008 (1990).

[48] M. Eiselt, *Electron. Lett.* **28**, 1505 (1992).

[49] J. P. Sokoloff, P. R. Prucnal, I. Glesk, and M. Kane, *IEEE Photon. Technol. Lett.* **5**, 787 (1993).

[50] M. Eiselt, W. Pieper, and H. G. Weber, *J. Lightwave Technol.* **13**, 2099 (1995).

[51] I. Glesk, B. C. Wang, L. Xu, V. Baby, and P. R. Prucnal, in *Progress in Optics*, Vol. 45, E. Wolf, Ed. (Elsevier, Amsterdam, 2003), Chap. 2.

[52] K. I. Kang, T. G. Chang, I. Glesk, and P. R. Prucnal, *Appl. Opt.* **35**, 417 (1996); *Appl. Opt.* **35**, 1485 (1996).

[53] K. Tajima, S. Nakamura, and Y. Sugimoto, *Appl. Phys. Lett.* **67**, 3709 (1995).

[54] N. S. Patel, K. L. Hall, and K. A. Rauschenbach, *Opt. Lett.* **21**, 1466 (1996).

[55] K. L. Hall and K. A. Rauschenbach, *Opt. Lett.* **23**, 1271 (1998).

[56] S. Nakamura, Y. Ueno, and K. Tajima, *IEEE Photon. Technol. Lett.* **10**, 1575 (1998).

[57] N. S. Patel, K. L. Hall, and K. A. Rauschenbach, *Appl. Opt.* **37**, 2831 (1998).

[58] C. Schubert, S. Diez, J. Berger, R. Ludwig, U. Feiste, H. G. Weber, G. Toptchiyski, K. Petermann, and V. Krajinovic, *IEEE Photon. Technol. Lett.* **13**, 475 (2001).

[59] B. S. Robinson, S. A. Hamilton, and E. P. Ippen, *IEEE Photon. Technol. Lett.* **14**, 206 (2002).

[60] C. Bintjas, K. Vlachos, N. Pleros, and H. Avramopoulos, *J. Lightwave Technol.* **21**, 2629 (2003).

[61] S. Diez, R. Ludwig, and H. G. Weber, *IEEE Photon. Technol. Lett.* **11**, 60 (1999).

[62] S. Radic, C. J. McKinstrie, R. M. Jopson, A. H. Gnauck, J. C. Centanni, and A. R. Chraplyvy, *IEEE Photon. Technol. Lett.* **16**, 852 (2004).

[63] Q. Lin, R. Jiang, C. F. Marki, C. J. McKinstrie, R. Jopson, J. Ford, G. P. Agrawal, and S. Radic, *IEEE Photon. Technol. Lett.* **17**, 2376 (2005).

[64] T. Torounidis, M. Westlund, H. Sunnerud, B. E. Olsson, and P. A. Andrekson, *IEEE Photon. Technol. Lett.* **17**, 312 (2005).

[65] P. Leclerc, B. Lavingne, and D. Chiaroni, in *Optical Fiber Telecommunications*, Vol. 4A, I. P. Kaminow and T. P. Lee, Eds. (Academic Press, Boston, 2002), Chap. 15.

[66] G. I. Papadimitriou, C. Papazoglou, and A. S. Pomportsis, *J. Lightwave Technol.* **21**, 384 (2003).

[67] K. Vlachos, N. Pleros, C. Bintjas, G. Theophilopoulos, and H. Avramopoulos, *J. Lightwave Technol.* **21**, 1857 (2003).

[68] P. A. Andrekson, N. A. Olsson, J. R. Simpson, T. Tanbun-Ek, R. A. Logan, and M. Haner, *Electron. Lett.* **27**, 695 (1991).

[69] T. Morioka, H. Takara, S. Kawanishi, T. Kitoh, and M. Saruwatari, *Electron. Lett.* **32**, 832 (1996).

[70] P. O. Hedekvist, M. Karlsson, and P. A. Andrekson, *J. Lightwave Technol.* **15**, 2051 (1997).

[71] T. Hasegawa, K. Inoue, and K. Oda, *IEEE Photon. Technol. Lett.* **5**, 947 (1993).

[72] T. Sakamoto, K. Seo, K. Taira, N. S. Moon, and K. Kikuchi, *IEEE Photon. Technol. Lett.* **16**, 563 (2004).

[73] F. Yaman, Q. Lin, and G. P. Agrawal, *IEEE Photon. Technol. Lett.* **18**, 2335 (2006).

[74] K. J. Blow, N. J. Doran, and P. B. Nelson, *Electron. Lett.* **26**, 962 (1990).

[75] T. Yamamoto, E. Yoshida, and M. Nakazawa, *Electron. Lett.* **34**, 1013 (1998).

[76] K. Uchiyama, T. Morioka, and M. Saruwatari, *Electron. Lett.* **31**, 1862 (1995).

[77] B. E. Olsson and P. A. Andrekson, *IEEE Photon. Technol. Lett.* **9**, 764 (1997).

[78] J. W. Lou, K. S. Jepsen, D. A. Nolan, S. H. Tarcza, W. J. Bouton, A. F. Evans, and M. N. Islam, *IEEE Photon. Technol. Lett.* **12**, 1701 (2000).

[79] H. C. Lim, T. Sakamoto, and K. Kikuchi, *IEEE Photon. Technol. Lett.* **12**, 1704 (2000).

[80] T. Sakamoto, H. C. Lim, and K. Kikuchi, *IEEE Photon. Technol. Lett.* **14**, 1737 (2002).

[81] B. E Olsson and D. J. Blumenthal, *IEEE Photon. Technol. Lett.* **13**, 875 (2001).

[82] L. Rau, W. Wang, B. E. Olsson, Y. Chiu, H. F. Chou, D. J. Blumenthal, and J. E. Bowers, *IEEE Photon. Technol. Lett.* **14**, 1725 (2002).

[83] J. Li, B. E. Olsson, M. A. Karlsson, and P. A. Andrekson, *IEEE Photon. Technol. Lett.* **15**, 1770 (2003).

[84] J. H. Lee, T. Tanemura, T. Nagashima, T. Hasegawa, S. Ohara, N. Sugimoto, and K. Kikuchi, *Opt. Lett.* **30**, 1267 (2005).

[85] A. S. Lenihan, R. Salem, T. E. Murphy, and G. M. Carter, *IEEE Photon. Technol. Lett.* **18**, 1329 (2006).

[86] R. Salem, A. S. Lenihan, G. M. Carter, and T. E. Murphy, *IEEE Photon. Technol. Lett.* **18**, 2254 (2006).

[87] J. Li, B. E. Olsson, M. A. Karlsson, and P. A. Andrekson, *J. Lightwave Technol.* **23**, 2654 (2005).

[88] J. H. Lee, K. Kikuchi, T. Nagashima, T. Hasegawa, S. Ohara, and N. Sugimoto, *Opt. Express* **13**, 6864 (2005).

[89] A. Reale, P. Lugli, and S. Betti, *IEEE J. Sel. Topics Quantum Electron.* **7**, 703 (2001).

[90] L. Xu, B. C.Wang, V. Baby, I. Glesk, and P. R. Prucnal, *IEEE Photon. Technol. Lett.* **15**, 308 (2003).

[91] W. Li, M. Chen, Y. Dong, and S. Xie, *IEEE Photon. Technol. Lett.* **16**, 203 (2004).

[92] C. G. Lee, Y. J. Kim, C. S. Park, H. J. Lee, and C. S. Park, *J. Lightwave Technol.* **23**, 834 (2005).

[93] S. Bigo, O. Leclerc, and E. Desurvire, *IEEE J. Sel. Topics Quantum Electron.* **3**, 1208 (1997).

[94] S. H. Lee, K. Chow, C. Shu, *Opt. Express* **13**, 1710 (2005).

[95] C. H. Kwok and C. Lin, *IEEE J. Sel. Topics Quantum Electron.* **12**, 451 (2006).

[96] S. H. Lee, K. Chow, C. Shu, *Opt. Commun.* **263**, 152 (2006).

[97] P. A. Andrekson, *Electron. Lett.* **27**, 1440 (1991).

[98] M. Jinno, J. B. Schlager, and D. L. Franzen, *Electron. Lett.* **30**, 1489 (1994).

[99] H. Kawaguchi and J. Inoue, *Proc. SPIE* **3283**, 477 (1998).

[100] S. Diez, C. Schmidt, D. Hoffmann, C. Bornholdt, B. Sartorius, H. G.Weber, L. Jiang, and A. Krotkus, *Appl. Phys. Lett.* **73**, 3821 (1998).

[101] S. Diez, R. Ludwig, C. Schmidt, U. Feiste, and H. G. Weber, *IEEE Photon. Technol. Lett.* **11**, 1402 (1999).

[102] J. Li, J. Hansryd, P. O. Hedekvist, P. A. Andrekson, and S. N. Knudsen, *IEEE Photon. Technol. Lett.* **13**, 987 (2001).

[103] M. Westlund, H. Sunnerud, B. E. Olsson, and P. A. Andrekson, *IEEE Photon. Technol. Lett.* **16**, 2108 (2004).

[104] M. Westlund, P. A. Andrekson, H. Sunnerud, J. Hansryd, and J. Li, *J. Lightwave Technol.* **23**, 2012 (2005).

[105] B. P. Nelson and N. J. Doran, *Electron. Lett.* **27**, 204 (1991).

[106] J. Li, M. Westlund, H. Sunnerud, B. E. Olsson, M. Karlsson, and P. A. Andrekson, *IEEE Photon. Technol. Lett.* **16**, 566 (2004).

[107] P. V. Mamyshev, *Proc. Eur. Conf. Opt. Commun.* **24**, 475 (1998).

[108] T. H. Her, G. Raybon, and C. Headley, *IEEE Photon. Technol. Lett.* **16**, 200 (2004).

[109] N. Yoshikane, I. Morita, T. Tsuritani, A. Agata, and N. Edagawa, *IEEE J. Sel. Topics Quantum Electron.* **10**, 412 (2004).

[110] M. Matsumoto, *J. Lightwave Technol.* **22**, 1472 (2004).

[111] L. B. Fu, M. Rochette, V. G. Ta'eed, D. J. Moss, and B. J. Eggleton, *Opt. Express* **13**, 7637 (2005).

[112] P. Johannisson and M. Karlsson, *IEEE Photon. Technol. Lett.* **17**, 2667 (2005).

[113] M. Rochette, L. Fu, V. Ta'eed, D. J. Moss, and B. J. Eggleton, *IEEE J. Sel. Topics Quantum Electron.* **12**, 736 (2006).

[114] M. R. E. Lamont, M. Rochette, D. J. Moss, and B. J. Eggleton, *IEEE Photon. Technol. Lett.* **18**, 1185 (2006).

[115] J. H. Lee, T. Nagashima, T. Hasegawa, S. Ohara, N. Sugimoto, Y.-G. Han, S. B. Lee, and K. Kikuchi, *IEEE Photon. Technol. Lett.* **18**, 1296 (2006).

[116] A. G. Striegler and B. Schmauss, *J. Lightwave Technol.* **24**, 2835 (2006).

[117] A. Hasegawa and Y. Kodama, *Solitons in Optical Communications* (Clarendon Press, Oxford, UK, 1995).

[118] E. Iannone, F. Matera, A. Mecozzi, and M. Settembre, *Nonlinear Optical Communication Networks* (Wiley, Hoboken, NJ, 1998).

[119] G. P. Agrawal, *Fiber-Optic Communication Systems*, 3rd ed. (Wiley, Hoboken, NJ, 2002).

[120] L. F. Mollenauer and J. P. Gordon, *Solitons in Optical Fibers* (Academic Press, Boston, 2007).

[121] P. Brindel, B. Dany, O. Leclerc and E. Desurvire, *Electron. Lett.* **35**, 480 (1999).

[122] M. Matsumoto and O. Leclerc, *Electron. Lett.* **38**, 576 (2002).

[123] M. H. Lee, J. M. Kang, and S. K. Han, *IEE Proc.* **148**, 189 (2001).

[124] J. Yu and P. Jeppesen, *J. Lightwave Technol.* **19**, 941 (2001).

[125] W. Wang, H. N. Poulsen, L. Rau, H. F. Chou, J. E. Bowers, D. J. Blumenthal, and L. Gruner-Nielsen, *IEEE Photon. Technol. Lett.* **15**, 1416 (2003).

[126] M. Jinno, *J. Lightwave Technol.* **12**, 1648 (1994).

[127] N. J. Smith and N. J. Doran, *J. Opt. Soc. Am. B* **12**, 1117 (1995).

[128] W. A. Pender, T. Widdowson, and A. D. Ellis, *Electron. Lett.* **32**, 567 (1996).

[129] M. Meissner, M. Rösch, B. Schmauss, and G. Leuchs, *IEEE Photon. Technol. Lett.* **15**, 1297 (2003).

[130] M. Meissner, K. Sponsel, K. Cvecek, A. Benz, S. Weisser, B. Schmauss, and G. Leuchs, *IEEE Photon. Technol. Lett.* **16**, 2105 (2004).

[131] E. Ciaramella and T. Stefano, *IEEE Photon. Technol. Lett.* **12**, 849 (2000).

[132] K. Inoue, *Electron. Lett.* **36**, 1016 (2000); *IEEE Photon. Technol. Lett.* **13**, 338 (2001).

[133] E. Ciaramella, F. Curti, and T. Stefano, *IEEE Photon. Technol. Lett.* **13**, 142 (2001).

[134] K. Inoue and T. Mukai, *J. Lightwave Technol.* **20**, 969 (2002).

[135] A. Bogris and D. Syvridis, *J. Lightwave Technol.* **21**, 1892 (2003).

[136] S. Radic, C. J. McKinstrie, R. M. Jopson, J. C. Centanni, and A. R. Chraplyvy, *IEEE Photon. Technol. Lett.* **15**, 957 (2003).

[137] S. Yamashita and M. Shahed, *IEEE Photon. Technol. Lett.* **18**, 1054 (2006).

[138] A. Striegler and B. Schmauss, *IEEE Photon. Technol. Lett.* **16**, 1083 (2004).

[139] K. Croussore, C. Kim and G. Li, *Opt. Lett.* **29**, 2357 (2004).

[140] A. G. Striegler, M. Meissner, K. Cvecek, K. Sponsel, G. Leuchs, and B. Schmauss, *IEEE Photon. Technol. Lett.* **17**, 639 (2005).

[141] M. Matsumoto, *IEEE Photon. Technol. Lett.* **17**, 1055 (2005); *J. Lightwave Technol.* **23**, 2696 (2005).

[142] K. Croussore, I. Kim, Y. Han, C. Kim, G. Li, and S. Radic, *Opt. Express* **13**, 3945 (2005).

[143] S. Boscolo, R. Bhamber, and S. K. Turitsyn *IEEE J. Quantum Electron.* **42**, 619 (2006).

[144] K. Cvecek, G. Onishchukov, K. Sponsel, A. G. Striegler, B. Schmauss, and G. Leuchs, *IEEE Photon. Technol. Lett.* **18**, 1801 (2006).

[145] K. Croussore, I. Kim, C. Kim, and G. Li, *Opt. Express* **14**, 2085 (2006).

[146] A. Bogris and D. Syvridis, *IEEE Photon. Technol. Lett.* **18**, 2144 (2006).

[147] K. Cvecek, K. Sponsel, G. Onishchukov, B. Schmauss, and G. Leuchs, *IEEE Photon. Technol. Lett.* **19**, 146 (2007).

[148] A. Sahara, T. Inui, T. Komukai, H. Kubota, and M. Nakazawa, *J. Lightwave Technol.* **18**, 1364 (2000).

[149] O. Leclerc, B. Lavigne, E. Balmefrezol, P. Brindel, L. Pierre, D. Rouvillain, F. Seguineau, *J. Lightwave Technol.* **21**, 2779 (2003).

[150] L. A. Jiang, M. E. Grein, H. A. Haus, and E. P. Ippen, *Opt. Lett.* **28**, 78 (2003).

[151] J. A. Harrison, K. J. Blow, and A. J. Poustie, *Opt. Commun.* **240**, 221 (2004).

[152] A. Striegler and B. Schmauss, *IEEE Photon. Technol. Lett.* **16**, 2574 (2004); *IEEE Photon. Technol. Lett.* **17**, 1310 (2005).

[153] J. Suzuki, T. Tanemura, K. Taira, Y. Ozeki, and K. Kikuchi, *IEEE Photon. Technol. Lett.* **17**, 423 (2005).

[154] F. Parmigiani, P. Petropoulos, M. Ibsen, and D. J. Richardson, *J. Lightwave Technol.* **24**, 357 (2006).

[155] M. Rochette, J. L. Blows, and B. J. Eggleton, *Opt. Express* **14**, 6414 (2006).

[156] Z. Zhu, M. Funabashi, Z. Pan, L. Paraschis, D. L. Harris, and S. J. B. Yoo, *J. Lightwave Technol.* **25**, 504 (2007).

[157] T. Otani, T. Miyazaki, and S. Yamamoto, *J. Lightwave Technol.* **20**, 195 (2002).

[158] H. Murai, M. Kagawa, H. Tsuji, and K. Fujii, *IEEE Photon. Technol. Lett.* **17**, 1965 (2005).

[159] M. Daikoku, N. Yoshikane, T. Otani, and H. Tanaka, *J. Lightwave Technol.* **24**, 1142 (2006).

[160] M. Matsumoto, *IEEE Photon. Technol. Lett.* **19**, 273 (2007).

Chapter 9

Highly Nonlinear Fibers

As we have seen in several preceding chapters of this book, new kinds of fibers, known as highly nonlinear fibers, are useful for a variety of applications. The parameter γ that governs most of the nonlinear phenomena in optical fibers is enhanced considerably in such fibers by confining the optical mode to a narrow central region. In practice, this is realized by introducing micrometer-size air holes within the cladding of a narrow-core fiber. Because of such structural modifications, these fibers are collectively referred to as *microstructured fibers*. This chapter focuses on several important applications of microstructured fibers. Their nonlinear and dispersive properties are discussed in Section 9.1. The next section describes how microstructured fibers can be used for tuning the wavelength of femtosecond pulses over a wide range by exploiting the nonlinear effects, such as stimulated Raman scattering (SRS) and four-wave mixing (FWM). The use of such fibers for generating a supercontinuum (an ultrabroadband spectrum) is discussed in Section 9.3. Since the bandwidth of output light in this case can cover a range of more than 500 nm, such a device is useful for several applications, discussed in Section 9.3, that fall in areas as diverse as biomedical imaging to frequency metrology. Section 9.4 is devoted to the applications of photonic-bandgap fibers whose hollow core is surrounded with a cladding containing a periodic array of air holes. The core can be filled with a gas or liquid to enhance the nonlinear effects.

9.1 Microstructured Fibers

The field of nonlinear fiber optics has been enriched considerably in recent years with the development of highly nonlinear fibers [1]. Microstructured fibers play an important role among such fibers. In contrast with the standard silica fibers, whose core is doped with the dopants, such as germania, to increase the refractive index inside the core, microstructured fibers typically use pure silica for both the core and cladding but embed multiple air holes within the cladding to lower its effective refractive index. For this reason, such fibers are also known as *holey fibers*.

Figure 9.1: Scanning electron micrographs showing the variety of microstructured fibers.

9.1.1   Design and Fabrication

Figure 9.1 shows several designs of microstructured fibers. In each case, the narrow silica core is surrounded by one or more rings of air holes that effectively lower the refractive index of the cladding region and thus help to confine the incident light to the core of the fiber. The size, shape, and the number of air holes vary for each design. In some cases, a narrow core is surrounded by a two-dimensional periodic array of air holes. Such fibers are referred to as the *photonic crystal fibers* (PCFs). In fact, such a fiber was first developed in 1996 in the form of a photonic-crystal cladding [2]. It was realized later that the periodic nature of air holes is not critical for silica-core fibers, as long as the cladding has multiple air holes that effectively reduce its refractive index compared with that of the central core. Since 1996, microstructured fibers have been developed extensively and have found a wide variety of applications [3]–[9].

The periodic nature of the air holes become important in the so-called *photonic bandgap* fibers in which optical modes are confined to the core by periodic variations of the refractive index within the cladding. The core of such fibers often contains air to which light is confined by the photonic bandgap [10]. Such true PCFs can act as a highly nonlinear medium if air is replaced with a suitable gas or liquid [5]. The applications of such fibers are considered in Section 9.4.

A common technique for fabricating microstructured fibers consists of first making a preform by stacking multiple capillary tubes of pure silica (diameter about 1 mm) in a hexagonal pattern around a solid silica rod [5]. The preform is then drawn into a fiber form using a standard fiber-drawing apparatus. A polymer coating is added on the outside to protect the resulting fiber. When viewed under a scanning electron microscope, such a fiber shows a two-dimensional pattern of air holes around the central region acting as a core. PCFs with a nonsilica core can be made with the same technique. The air channel is created by removing the central silica rod surrounding the capillary tubes before the preform is drawn into a fiber. This channel can later be filled with a gas or liquid that acts as the nonlinear medium. It is important to stress that microstructured fibers can be made relatively long (more than 1 km), while maintaining sufficient uniformity along their length. They are as easy to handle as conventional fibers because of a polymer coating added on top of the cladding. They can also be spliced to other kinds of fibers, although splice losses may exceed 1 dB, depending on the relative core sizes of two fibers.

Another technique used for fabricating microstructured fibers is known as the *extrusion technique* [5]. In this approach, the preform is produced by extruding material selectively from a solid glass rod of 1 to 2 cm diameter. More specifically, the molten glass rod is forced through a die containing the required pattern of holes. This technique allows one to draw fibers directly from any bulk material, whether crystalline or amorphous, and it is often used in practice with polymers or compound glasses. The structured preform with the desired pattern of holes is reduced in scale using a fiber-drawing tower in two steps. First, the outside diameter is reduced by a factor of 10 or so. The resulting "cane" is inserted into a glass tube whose size is then further reduced by a factor of more than 100.

A shortcoming of all microstructured fibers is that they exhibit higher losses than those associated with conventional fibers [11]. Typically, losses exceed 1000 dB/km when the core diameter is reduced to enhance the nonlinear parameter γ. The origin of such losses is related to the nature of mode confinement in such fibers. More specifically, both the core and the cladding are made of silica, and the mode confinement to the core is produced by air holes that are present in the cladding. Thus, the number and the size of air holes affect how the optical mode is guided inside such a waveguide. With a proper design, losses can be reduced to below 1 dB/km, if core diameter is made relatively large (>5 μm), but only at the expense of a reduced value of the nonlinear parameter [12]. In essence, a trade-off must be made between confinement losses and γ values. High values of γ require a narrow core and thus suffer from larger losses.

9.1.2 Nonlinear and Dispersive Properties

From the standpoint of application of microstructured fibers, what matters are their nonlinear and dispersive characteristics, rather than the details of how air holes are arranged around the core. Of course, the nonlinear parameter γ as well as the dispersion relation $\beta(\omega)$ depend on the actual details of air holes and can be controlled by changing the microstructure of a highly nonlinear fiber. For this reason, it is crucial to model the modal properties of a microstructured fiber as accurately as possible.

It is not easy to analyze the modes of a microstructured fiber because the refractive index of the cladding is far from being homogeneous and changes suddenly at the air–silica interface surrounding each air hole. A further complication is related to the noncircular shape of the core region, encountered commonly in such fibers and leading to birefringence effects. For this reason, several numerical techniques have been developed that make use of a variety of methods known under the names such as plane-wave expansion, multipole method, localized function approach, and finite-element method [13]–[26]. All of them are capable of solving Maxwell's equations with a realistic device geometry. The objective in each case is to find the propagation constant $\beta(\omega)$ and the effective mode area A_{eff} for various modes supported by such a fiber.

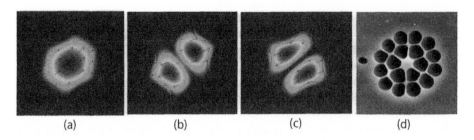

(a)   (b)   (c)   (d)

Figure 9.2: Spatial profiles associated with (a) the LP_{01} mode and (b) even and (c) odd LP_{11} modes of a microstructured fiber shown in part (d); arrows indicate direction of the electric field. (From Ref. [25]; ©2006 OSA.)

In the case of a PCF with a periodic array of air holes, the analysis is somewhat simpler because one can make use of the Bloch theorem, assuming an infinite-size cladding, and treat the core as a defect in an otherwise periodic structure. It turns out that the effective mode area A_{eff} is nearly independent of the number of hole rings for PCFs, even though confinement losses α_c depend strongly on this number [19]. These two parameters also depend on the ratios d/Λ and λ/Λ, where d is the air-hole diameter, Λ is the hole-to-hole spacing, and λ is the wavelength of light. The number of modes supported by a PCF also depends on the two ratios d/Λ and λ/Λ. No periodicity exists in the case of a microstructured fiber with only a few large-diameter holes. In this case, the multipole method is found to be quite useful as it can be applied to any cladding region with a finite number of holes of arbitrary refractive index [18]. Another general approach makes use of the finite-element method [19].

Several numerical methods were compared in a recent study for a specific microstructured fiber [25]. A microphotograph of this fiber is shown in Figure 9.2 together with the spatial profiles associated with the LP_{01} mode and the two LP_{11} modes (even and odd types). The effective refractive indices of these three modes are plotted in Figure 9.3(a) as a function of wavelength. These "dispersion curves" can be used to find various dispersion parameters at any specific frequency ω_0 using the definition $\beta_m = (d^m\beta/d\omega^m)_{\omega=\omega_0}$, where $\beta(\omega) = \bar{n}(\omega)\omega/c$ is the propagation constant of an optical mode. The first-order dispersion parameter β_1 is related to effective group index n_g as $\beta_1 = n_g/c$. The wavelength dependence of n_g is displayed in Figure 9.3(b).

Because of the numerical complexity associated with the modeling of microstructured fibers, several semianalytic models have been proposed that make use of the same concepts that are useful for standard optical fibers. The concept of an effective cladding index, n_{cl}, is quite useful in this context as it represents the extent to which the air holes reduce the refractive index of silica in the cladding region. One can even introduce an effective V parameter using the same relation used for standard fibers,

$$V_{\text{eff}} = (2\pi/\lambda)a_e(n_{co}^2 - n_{cl}^2)^{1/2}, \tag{9.1.1}$$

where a_e is the effective radius of the core region with the refractive index n_{co}. An inspection of fiber designs in Figure 9.1 shows that the choice of a_e is not obvious because, in contrast with the standard fibers, there is no real core–cladding boundary for microstructured fibers.

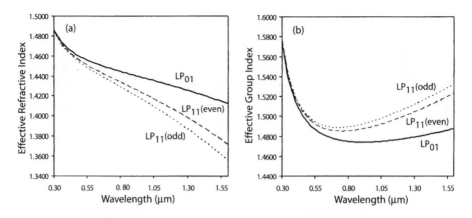

Figure 9.3: Wavelength dependence of (a) effective refractive index and (b) effective group index for the three modes shown in Figure 9.2. (From Ref. [25]; ©2006 OSA.)

In the case of a PCF with a periodic array of air holes, center-to-center spacing Λ between two neighboring air holes provides a natural length scale. The choice $a_e = \Lambda$ has been used in one study [20] to define V_{eff} in Eq. (9.1.1). However, the single-mode condition, $V_{\text{eff}} = \pi$, in this case is different from that found for conventional fibers ($V \approx 2.405$). In another study [23], the choice $a_e = \Lambda/\sqrt{3}$ was employed together with the single-mode condition $V_{\text{eff}} = 2.405$. More recent work indicates that effective core radius a_e should depend on the two parameters, d/Λ and λ/Λ, that characterize a PCF. In all cases, the number of modes supported by a PCF is found to depend on the size of air holes through the ratio d/Λ. An interesting property is that a PCF supports only the fundamental mode at all wavelengths if $d/\Lambda < 0.45$; such a fiber is referred to as the *endlessly single-mode* fiber [2].

The most relevant property of a fiber mode for nonlinear applications is the mode-field diameter, which sets the effective mode area A_{eff} of that mode. The nonlinear parameter, $\gamma = 2\pi n_2/(\lambda A_{\text{eff}})$, of a microstructured fiber can be enhanced compared with its values for conventional fibers by lowering the effective mode area. This area can be made quite small for such fibers by reducing the effective core radius to near 1 μm. Reducing the core size does not help in conventional fibers because the optical mode spreads deeper into the cladding. Such spreading does not occur in microstructured fibers because air holes keep the fundamental mode confined to the central core region. Because A_{eff} can become <1 μm^2 for properly designed fibers, γ is enhanced by a factor of 50 or more compared with the conventional fibers. Further enhancement can occur if silica glass is replaced with other glasses that make use of glassy materials such as bismuth oxide, lead silicates, or chalcogenides. As discussed in Chapter 8, bismuth-oxide fibers with values of $\gamma > 1000$ W^{-1}/km have been used for many signal-processing applications.

The dispersion parameters of microstructured fibers (β_2, β_3, etc.) are quite sensitive to the number, size, and shape of the air holes around the central core. In the case of a fiber design such as that shown in parts E and F of Figure 9.1, the core is surrounded mostly with air. As a result, a good understanding of the dispersion be-

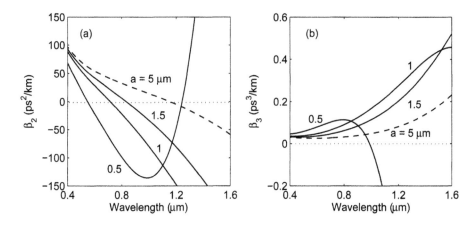

Figure 9.4: Wavelength dependence of β_2 and β_3 for a silica wire surrounded with air as its core radius a is reduced from 5 μm to 0.5 μm.

havior can be realized by considering a silica wire surrounded with air and varying the radius of this wire. Figure 9.4 shows the wavelength dependence of the dispersion parameters, β_2 and β_3, for a silica wire surrounded with air as its core radius a is varied from 0.5 to 5 μm. The last case (dashed curve) corresponds to the core size of conventional single-mode fibers. In this case, the material dispersion of silica glass dominates ($\beta_2 = 0$ near 1.25 μm) as the contribution of waveguide dispersion is relatively small. The zero-dispersion wavelength shifts in the vicinity of 0.8 μm when a is close to 1.5 μm because of a large waveguide contribution. A further qualitative change occurs when a is reduced to below 1 μm. As seen in Figure 9.4, the fiber develops two distinct zero-dispersion wavelengths for $a = 0.5$ μm. As a result, the third-order dispersion β_3 for such fibers becomes negative at wavelengths longer than 1 μm.

Similar features are found to occur for most microstructured fibers. The main point is that the dispersive properties of the fiber are quite sensitive to the exact value of the core radius in such fibers, and they can be further modified by changing the cladding design. In the case of PCFs, they can be characterized in terms of the same two parameters used earlier, namely, the air-hole diameter d and the hole-to-hole spacing Λ. Figure 9.5 shows the wavelength dependence of the dispersion parameter D, related to β_2 as $D = -(2\pi c/\lambda^2)\beta_2$, for $\Lambda = 1, 2, 2.5$, and 3 μm as the ratio d/Λ is varied [23]. For a relatively wide air-hole spacing ($\Lambda > 2$ μm), the PCF exhibits a single zero-dispersion wavelength in the spectral region extending from 0.5 to 1.8 μm that shifts toward shorter wavelengths as the air-hole diameter d is increased. For shorter values of Λ, D vanishes at two wavelengths that move closer as Λ becomes shorter and the ratio d/Λ decreases. Further tailoring of dispersion is possible by changing the size of holes closer to the core compared with the farther ones. The main advantage from the standpoint of applications is that the microstructured fibers offer so many design parameters that the dispersive properties of a fiber can be tailored to meet any application requirement.

A tapering technique is often employed for reducing the core diameter and air-hole

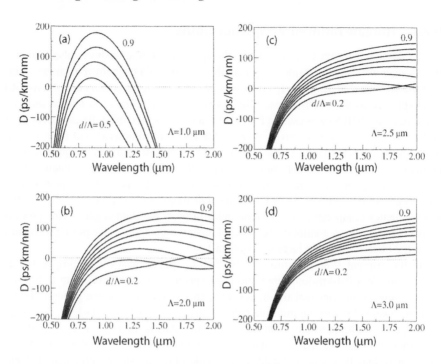

Figure 9.5: Calculated wavelength dependence of the dispersion parameter D for PCFs designed with $\Lambda = 1, 2, 2.5$, and 3 μm. The ratio d/Λ is varied from 0.5 to 0.9 in case (a) and from 0.2 to 0.9 in other cases. (From Ref. [23]; ©2005 IEEE.)

size of microstructured fibers after they have been manufactured [27]–[29]. The main advantage is that both the dispersive and nonlinear properties of a fiber can be altered with tapering, provided care is taken to ensure that air holes within the cladding do not collapse. In practice, the core size as well as the diameter of air holes can be reduced by a factor of 2 or so without collapsing the air holes. Even such small changes modify the characteristics of most microstructured fibers significantly because their dispersion is extremely sensitive to the size and spacing of air holes.

9.2 Wavelength Shifting and Tuning

An important application of highly nonlinear fibers consists of shifting the carrier wavelength of ultrashort pulses emitted from a mode-locked laser. The wavelength of such lasers can be tuned intrinsically over a spectral range set by the gain medium, but the tunic range is often limited to 50 nm or so for most lasers. (An exception occurs in the case of Ti:sapphire lasers, which are tunable over more than 200 nm.) In practice, it is desirable to find an extrinsic tuning mechanism that allows shifting the laser wavelength over a wide range. In the case of mode-locked pulses, several nonlinear effects occurring inside an optical fiber can be exploited for this purpose; two of them are discussed in this section.

9.2.1   Raman-Induced Frequency Shifts

As discussed in Section 12.1 of Ref. [1], the spectrum of an ultrashort pulse shifts toward longer wavelengths inside an optical fiber because of a phenomenon known as intrapulse Raman scattering. Such a shift was first noticed in the context of solitons [30] and explained as the self-frequency shift of solitons [31]. The term *Raman-induced frequency shift* (RIFS) was used in Ref. [1] because such spectral shifts can occur even in the normal-dispersion region of an optical fiber where solitons are not formed [32]. It was discovered during the 1990s that the RIFS of solitons can become quite large in highly nonlinear fibers. The physical origin of RIFS is related to the broadband nature of the Raman gain spectrum in silica fibers. Because the spectrum of femtosecond pulses can extend over more than 1 THz, its high-frequency components can pump the low-frequency components of the same pulse through the Raman gain. This process transfers pulse energy continuously toward longer wavelengths as the pulse propagates down the fiber, resulting in large spectral shifts.

In the original 1986 experiment [30], 560-fs optical pulses were propagated as solitons inside a 0.4-km-long fiber and their spectrum was shifted by up to 8 THz. It was found soon after that the fission of higher-order solitons can generate larger frequency shifts by forming multiple pulses known as Raman solitons [33]. Although the properties of Raman solitons attracted attention during the 1990s [34]–[38], it was only after 1999 that the RIFS mechanism was used for producing femtosecond pulses whose wavelengths could be tuned over a wide range by simply propagating them through microstructured or other types of narrow-core fibers [39]–[51].

Mathematically, the propagation of short optical pulses inside optical fibers is governed by the generalized nonlinear Schrödinger (NLS) equation [1]

$$\frac{\partial U}{\partial z} + \frac{i\beta_2}{2}\frac{\partial^2 U}{\partial T^2} - \frac{\beta_3}{6}\frac{\partial^3 U}{\partial T^3} = i\gamma P_0 e^{-\alpha z}\left(|U|^2 U + \frac{i}{\omega_0}\frac{\partial}{\partial T}(|U|^2 U) - T_R U\frac{\partial |U|^2}{\partial T}\right).$$

$$(9.2.1)$$

where the pulse amplitude $A(z,T)$ is related to $U(z,T)$ by the relation $A = U\sqrt{P_0}e^{-\alpha z}$, and P_0 is the peak power of the input pulse. The RIFS is governed by the last term containing the parameter T_R (related to the slope of the Raman gain spectrum) with a value close to 3 fs for silica fibers. As shown in Section 5.5 of Ref. [1], the moment method can be used to obtain the following expression for the RIFS at a distance z within the fiber [32]:

$$\Omega_p(z) = -\frac{8T_R}{15}\gamma P_0 T_p(0)\int_0^z \frac{e^{-\alpha z}}{T_p^3(z)}\,dz.$$

$$(9.2.2)$$

In deriving the preceding expression, the pulse was assumed to maintain its soliton profile, $|U(z,T)| = a_p\,\mathrm{sech}(T/T_p)$, but the peak amplitude a_p and the width T_p were allowed to evolve with z. The pulse width begins to change when pulse becomes chirped because of RIFS and fiber losses. Only when a fundamental soliton propagates inside a fiber with negligible losses and $\Omega_p \ll \omega_0$ can we assume that $T_p(z)$ remains close to its initial value T_0. Setting $\alpha = 0$ and $T_p(z) = T_0$ in Eq. (9.2.2), we obtain

$$\Omega_p(z) = -\frac{8T_R\gamma P_0}{15T_0^2}z \equiv -\frac{8T_R|\beta_2|}{15T_0^4}z,$$

$$(9.2.3)$$

where we used the condition that the soliton order $N = \gamma P_0 T_0^2 / |\beta_2| = 1$. The negative sign shows that the carrier frequency is reduced; that is, the soliton spectrum shifts toward longer wavelengths (the "red" side). The scaling of Ω_p with pulse width T_0 as T_0^{-4} was first found in 1986 using soliton perturbation theory [31]. It shows why the RIFS becomes important for only ultrashort pulses with widths ~ 1 ps or less. However, it should be kept in mind that such a dependence holds only over relatively short fiber lengths over which the soliton remains unchirped.

Equation (9.2.3) can also be written in terms of the nonlinear length, defined as $L_{NL} = 1/(\gamma P_0)$, by using $T_0^2 = |\beta_2| L_{NL}$ from the condition $N = 1$. If we introduce $\Delta \nu_R = \Omega_p/(2\pi)$, the RIFS grows along the fiber length linearly as

$$\Delta \nu_R(z) = -\frac{4 T_R z}{15 \pi |\beta_2| L_{NL}^2} = -\frac{4 T_R (\gamma P_0)^2 z}{15 \pi |\beta_2|}. \tag{9.2.4}$$

This equation shows that the RIFS scales quadratically with both the nonlinear parameter γ and the peak power P_0. Such a quadratic dependence of $\Delta \nu_R$ on the soliton peak power was seen in the original 1986 experiment [30].

It follows from Eqs. (9.2.3) and (9.2.4) that the RIFS can be made large by propagating shorter pulses with higher peak powers inside highly nonlinear fibers. As an example, if we use a highly nonlinear fiber with $\beta_2 = -30$ ps^2/km and $\gamma = 100$ W^{-1}/km, $L_{NL} = 10$ cm for 100-fs (FWHM) input pulses with $P_0 = 100$ W, and the spectral shift increases inside the fiber at a rate of about 1 THz/m. Under such conditions, pulse spectrum will shift by 50 THz over a 50-m-long fiber, provided that $N = 1$ can be maintained over this distance.

Indeed, such large values of RIFS were realized in a 2001 experiment [40] in which 200-fs pulses at an input wavelength of 1300 nm were launched into a 15-cm-long microstructured fiber with a 3-μm core diameter. The output spectrum revealed that most of the pulse energy was contained in a dominant spectral peak centered at 1550 nm. The spectral shift of 250 nm over 15-cm of fiber corresponds to a frequency shift rate of about 3 THz/cm, a value much larger than that expected from Eq. (9.2.4). This discrepancy can be resolved by noting that the soliton order N exceeded 1 under the experimental conditions and the input pulse propagated initially as a higher-order soliton inside the fiber.

It is well known that higher-order solitons split into multiple fundamental solitons because of perturbations induced by the third-order dispersion, self-steepening, and intrapulse Raman scattering [1]. For a noninteger value of N, the soliton order is the integer $\bar{N}$ closest to N. The fission of a higher-order soliton creates $\bar{N}$ fundamental solitons of different widths and peak powers. The inverse scattering method shows that their widths and peak powers are related to N as [35]

$$T_k = \frac{T_0}{2N + 1 - 2k}, \qquad P_k = \frac{(2N + 1 - 2k)^2}{N^2} P_0, \tag{9.2.5}$$

where $k = 1$ to $\bar{N}$. As an example, when $N = 2.1$, two fundamental solitons created through the fission process have widths of $T_0/3.2$ and $T_0/1.2$. The narrower soliton exhibits a much larger RIFS compared with the broader one because of the T_0^{-4} dependence seen in Eq. (9.2.3). For the $N = 2.1$ example, the RIFS is enhanced by a

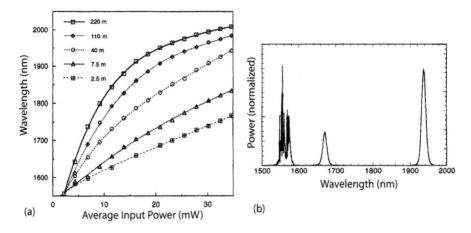

Figure 9.6: (a) Observed output wavelength as a function of average input power when 110-fs pulses at 1556 nm were launched inside fibers of different lengths. (b) Output spectrum in the case of 30-mW average power and 220-m-long fiber. (From Ref. [41]; ©2001 IEEE.)

factor of $(3.2)^4$ or about 105. Numerical simulations corresponding to the experimental situation of Ref. [40] reveal that fission occurs within 2 cm of propagation inside the fiber. The spectrum of the shortest soliton shifts rapidly toward the red side, as dictated by Eq. (9.2.3). The predictions of the generalized NLS equation (9.2.1) agree with the experimental data qualitatively.

The phenomenon of soliton fission can be used to tune the carrier wavelength of femtosecond pulses over a wide range by simply changing the peak power of such pulses [39]. In a 2001 experiment, the wavelength of a mode-locked fiber laser could be tuned from 1556 to 2033 nm by using polarization-maintaining fibers of lengths ranging from 2.5 to 220 m [41]. Figure 9.6 shows the experimentally measured values of the wavelength associated with the dominant Raman soliton as a function of the average power launched into the fiber (110-fs pulses at a 48-MHz repetition rate). An example of the output spectrum is also shown in the case of 30-mW average power launched inside a 220-m-long fiber. For such a long fiber, the spectral shift exceeded 470 nm at the 45 mW power level. These results show clearly the potential of RIFS for tuning the carrier wavelength of femtosecond pulses. The spectral peak located near 1950 nm corresponds to the shortest Raman soliton created during the fission process. The smaller peak located near 1680 nm belongs to the second Raman soliton, which is wider and thus experiences a much smaller wavelength shift.

In another experiment, the wavelength of 100-fs pulses, generated from a diode-pumped, Yb-doped, mode-locked fiber laser at a repetition rate of 108 MHz, was tuned from 1060 to 1320 nm using a 95-cm-long photonic crystal fiber [45]. The fiber had its zero-dispersion wavelength near 950 nm and exhibited anomalous dispersion at the input wavelength of 1060 nm ($\beta_2 \approx -20$ ps^2/km). As a result, pulses propagated in the form of higher-order solitons inside the fiber and underwent the fission process. Figure 9.7 shows the experimentally measured pulse spectra at average power levels raging from 0.5 to 30 mW. The formation of Raman solitons through the fission process

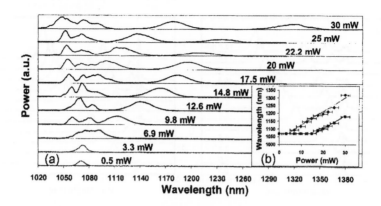

Figure 9.7: (a) Observed output spectra as a function of average input power when 100-fs pulses were launched inside a 95-cm-long photonic crystal fiber. The inset shows the wavelength of Raman solitons as a function of launched average power. (From Ref. [45]; ©2003 OSA.)

occurs at power levels above 7 mW, and the spectrum of the shortest Raman soliton shifts close to 1300 nm at a 30-mW power level. The inset shows how the wavelength of this soliton changes with the launched average power. The second soliton separates from the main pulse at power levels above 17 mW. In a 2005 experiment [48], an all-fiber source of femtosecond pulses, tunable from 1030 to 1330 nm, was realized by launching 2-ps pulses from a Yb-doped fiber laser into a 1.5-m-long microstructured fiber with a 2-μm-diameter core.

The experimental results seen in Figures 9.6 and 9.7 cannot be fully explained with Eq. (9.2.1) because it includes the Raman effect only approximately. For optical pulses shorter than 100 fs, one must employ the actual Raman response function and use Eq. (2.3.36) of Ref. [1] for a quantitative comparison of theory with the experiments. Using $A = U\sqrt{P_0}e^{-\alpha z}$ with $\alpha_1 = 0$, this equation can be written as

$$\frac{\partial U}{\partial z} + \sum_{m=2}^{M} i^{m-1}\frac{\beta_m}{m!}\frac{\partial^m U}{\partial T^m} = i\gamma P_0 e^{-\alpha z}$$
$$\times \left(1 + \frac{i}{\omega_0}\frac{\partial}{\partial T}\right)\left(U(z,T)\int_0^\infty R(t')|U(z,T-t')|^2 dt'\right), \qquad (9.2.6)$$

where M represents the order up to which dispersive effects are included. The nonlinear response function includes both the electronic (Kerr) and nuclear (Raman) contributions. Assuming that the electronic contribution is nearly instantaneous, $R(t)$ is of the form

$$R(t) = (1 - f_R)\delta(t) + f_R h_R(t), \qquad (9.2.7)$$

where f_R represents the fractional contribution of the delayed Raman response governed by $h_R(t)$; it has a value of about 0.18.

The form of the Raman response function h_R is set by vibrations of silica molecules induced by the optical field. It can be obtained numerically by using the actual Raman-gain curve. The following approximate analytic form of the Raman response function

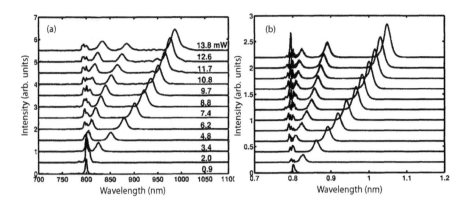

Figure 9.8: (a) Observed and (b) simulated output spectra as a function of average input power when 130-fs pulses at 800 nm are launched inside a 1.1-m-long microstructured fiber. (From Ref. [49]; ©2006 OSA.)

is also quite useful in practice [52]:

$$h_R(t) = \frac{\tau_1^2 + \tau_2^2}{\tau_1 \tau_2^2} \exp(-t/\tau_2) \sin(t/\tau_1) \tag{9.2.8}$$

with $\tau_1 = 12.2$ fs and $\tau_2 = 32$ fs. Figures 9.8 compares the theoretical predictions based on Eqs. (9.2.6)–(9.2.8) with the experimentally measured spectra [49]. In this experiment, 130-fs pulses, obtained from a Ti:sapphire laser operating near 800 nm at a 80-MHz repetition rate, were propagated through a 1.1-m-long microstructured fiber having a core diameter of only 1 μm. In numerical simulations, dispersion was included up to sixth order ($M = 6$), and parameters values were estimated as accurately as possible. Even though the agreement between the experiment and theory is quite good, the theory overestimates the RIFS by up to 20%. The reason is related to the use of an approximate form of the Raman response function given in Eq. (9.2.8). A different form of $h_R(t)$ has recently been proposed to include the anisotropic part of the Raman response [53], and it should be used for a more accurate description of the RIFS in optical fibers.

One may ask if much larger RIFS can be realized in practice by employing pulses shorter than 10 fs. To answer this question, pulses of 6-fs width were used in a 2005 experiment [50]. Figures 9.9(a) compares the input and output spectra when 6-fs pulses were transmitted through a 20-cm-long microstructured fiber. Part (b) shows numerical results based on Eqs. (9.2.6)–(9.2.8) and obtained using the experimentally measured intensity and chirp profiles of input pulses. It is important to note that the 6-fs input pulse has a spectrum that extends from 650 to 950 nm with a considerable internal structure. Although the output spectrum is wider and extends over 500 nm, it is hard to conclude that the RIFS has increased substantially for such short pulses. The dominant peak at the fiber output, located near 1060 nm, corresponds to a Raman soliton that has separated from the original pulses.

The output spectrum in Figure 9.9(a) shows a blue-shifted peak near 550 nm. The formation of such a peak occurs within 2-cm of fiber, as is apparent from numerical

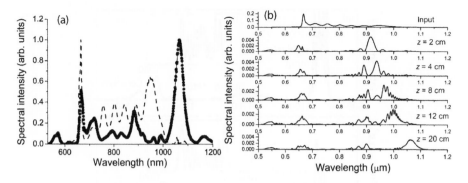

Figure 9.9: (a) Input (dashed) and output spectra when 6-fs pulses were launched inside a 20-cm-long microstructured fiber. (b) Numerically simulated evolution of pulse spectra as different distances along the fiber. (From Ref. [50]; ©2006 APS.)

simulations in part (b). The origin of this peak can be understood in terms of dispersive waves, or the so-called nonsolitonic Cherenkov radiation. It turns out that the Raman solitons, created through the fission process and perturbed by the third- and higher-order dispersion, lose energy by emitting dispersive waves. A specific dispersive wave, shifted from the soliton frequency by Ω_d at which its propagation constant (or phase velocity) matches that of the soliton, receives most of the shaded energy. This frequency shift is approximately given by [54]

$$\Omega_d \approx -\frac{3\beta_2}{\beta_3} + \frac{\gamma P_s \beta_3}{3\beta_2^2}. \tag{9.2.9}$$

where P_s is the peak power of the Raman soliton. When $\beta_2 < 0$ and $\beta_3 > 0$ at the soliton wavelength, the frequency shift is positive, and the Cherenkov radiation is emitted on the blue side of the input pulse spectrum. This was the case for the fiber used to obtain the experimental traces shown in Figure 9.9(a).

It follows from Eq. (9.2.9) that the Cherenkov radiation can be emitted toward longer wavelengths if $\beta_3 < 0$ for a fiber. This feature was used to advantage in a 2007 experiment to produce a wavelength shift larger than that was possible with the RIFS [55]. Negative values of β_3 were realized by propagating 200-fs pulses (from a fiber laser operating at 1064 nm) in a higher-order mode (LP$_{02}$ mode) of the fiber. A long-period grating was used to transfer the pulse energy to this higher-order mode of a 1-m-long fiber. The dispersion parameter β_2 for this mode was negative in the wavelength range of 908 to 1247 nm. As a result, β_3 was also negative near the zero-dispersion wavelength located at 1247 nm. Figure 9.10(a) shows the output spectrum when 3.31-nJ pulses were transmitted through the fiber; the spectrum of input pulses is shown as an inset. The output spectrum exhibits two well-defined peaks on the long-wavelength side of the input spectrum. The peak centered near 1200 nm corresponds to a Raman soliton that has undergone the RIFS. The second peak located near 1350 nm is due to the Cherenkov radiation; it contains about 20% of the input pulse energy. The numerically predicted spectrum, shown in Figure 9.10(b), is qualitatively similar to the experimental one and shows both of these peaks.

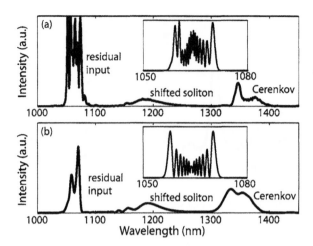

Figure 9.10: (a) Measured and (b) numerically simulated output spectra when 200-fs pulses were propagated into the LP_{02} mode of a 1-m-long fiber. In both cases, the inset shows the input spectrum. (From Ref. [55]; ©2007 OSA.)

The RIFS is useful for a variety of applications that benefit from a tunable optical source. In a 2006 study, RIFS inside a microstructured fiber was used for generating tunable Stokes pulses that were used for spectroscopy based on the process of coherent anti-Stokes Raman scattering [56]. The same laser pulse was used to provide the pump and Stokes pulses needed in this experiment. In another interesting application, the RIFS was used to create an all-optical tunable delay line [57]. The basic idea makes use of the fact that the group velocity of a pulse inside an optical fiber depends on its wavelength. As the carrier wavelength of a pulse shifts toward longer wavelengths through RIFS, the pulse slows down if it propagates in the anomalous-dispersion region of an optical fiber. The amount of delay depends on the RIFS and can be adjusted by changing the peak power of input pulses. In the experiment, 0.5-ps pulses could be delayed by up to 19.2 ps. Moreover, the pulse wavelength was converted back to its original value through SPM-induced spectral broadening inside a second highly nonlinear fiber and filtering it appropriately.

9.2.2 Four-Wave Mixing

Wavelength shifting and tuning can also be realized by exploiting the nonlinear phenomenon of FWM discussed in Chapter 10 of Ref. [1]. Highly nonlinear fibers were used for this purpose soon after their development [58]–[69]. As discussed in Section 8.1.2, one application uses FWM to make fiber-optic parametric amplifiers that can be employed as wavelength converters. By placing the highly nonlinear fiber inside an optical cavity, such an amplifier can be converted into a parametric oscillator. A 2.1-m section of a microstructured fiber was used in a 2002 experiment to realize a fiber-optic parametric oscillator that was tunable over a 40-nm bandwidth [58]. It was pumped with 630-fs pulses obtained from a mode-locked Ti:sapphire laser. The threshold for

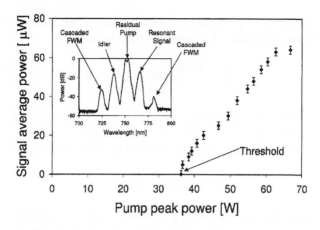

Figure 9.11: Measured average power of signal pulses as a function of the peak power of pump pulses. The inset shows the output spectrum at a pump wavelength of 751.8 nm. (From Ref. [58]; ©2002 OSA.)

the parametric oscillator was reached when pump pulses had a peak power of 34.4 W. Above this threshold, pump power was transferred to the signal and idler pulses whose frequencies satisfied the FWM condition $\omega_s + \omega_i = 2\omega_p$. Figure 9.11 shows the average power of signal pulses measured as a function of the peak power of pump pulses.

A fiber-optic parametric oscillator differs from other lasers inasmuch as it transfers pump power to the signal and idler simultaneously, and thus emits light at two wavelengths that fall on the opposite sides of the pump wavelength. In fact, the output spectrum may contain more than two peaks if the signal and idler become strong enough to act as pumps and generate new wavelengths through a process known as cascaded FWM. The inset in Figure 9.11 shows the output spectrum observed at a pump wavelength of 751.8 nm. It shows four peaks in addition to the residual pump peak. The signal and idler wavelengths change if the pump wavelength is tuned. The magnitude of change depends on the phase-matching condition. Up to six signal and six idler peaks were observed in a parametric oscillator in which a microstructured fiber was pumped continuously using more than 1.3 W of CW power [60]. However, the tuning range was relatively small in most experiments in which the pump experienced anomalous dispersion inside the fiber. It can be increased by using short fibers. In a recent study, a tuning range of more than 200 nm was realized using microstructured fibers of lengths ∼1 cm [61].

Under certain conditions, the tuning range of the signal and idler can exceed that of the pump by a large factor. To find the appropriate conditions, we should consider the phase-matching condition associated with the FWM process given in Eq. (8.1.9) in terms of the propagation constant $\beta(\omega)$ of the fiber mode at frequency ω. As seen in Section 9.1, the dispersion curve $\beta(\omega)$ can be quite different for photonic crystal and other microstructured fibers compared with the standard fibers. In particular, higher-order dispersive effects often become important for highly nonlinear fibers and should be considered. By expanding $\beta(\omega_s)$ and $\beta(\omega_i)$ in a Taylor series around the pump frequency ω_p in Eq. (8.1.9), we find that only the even-order dispersion terms affect the phase-matching condition, which can be written in the form

$$\sum_{m=2,4,\ldots}^{\infty} \frac{\beta_m(\omega_p)}{m!} \Omega_s^m + 2\gamma P_0 = 0, \tag{9.2.10}$$

where $\Omega_s = \omega_s - \omega_p$ is the shift of the signal frequency from the pump frequency ω_p.

In the anomalous-GVD regime ($\beta_2 < 0$), the $m = 2$ term dominates. If we neglect all higher-order terms, the frequency shift is given by the simple expression

$$\Omega_s = \sqrt{2\gamma P_0 / |\beta_2(\omega_p)|}. \tag{9.2.11}$$

It shows that the signal wavelength in a parametric oscillator depends on the pump power. This feature can be used to tune it over a limited range even when the pump wavelength is kept fixed. If pump wavelength is changed by $\delta\omega_p$, the signal frequency $\omega_s = \omega_p + \Omega_s$ changes with it by an amount that is different from $\delta\omega_p$. The reason is that Ω_s depends on $\beta_2(\omega_p)$, which varies with the pump wavelength. In practice, it is difficult to realize a wide tuning range if the pump wavelength lies in the anomalous-GVD regime of the fiber, unless the fiber length is reduced to ~ 1 cm [61].

The situation changes drastically if the pump wavelength lies in the normal-GVD regime of a fiber for which the fourth-order dispersion is negative [70]. If we neglect all other higher-order terms in Eq. (9.2.10), we obtain a fourth-order polynomial in Ω_s. Its roots can be easily analyzed to obtain the following expression for Ω_s:

$$\Omega_s^2 = \frac{6}{|\beta_4|} \left(\sqrt{\beta_2^2 + 2|\beta_4|\gamma P_0/3} + \beta_2 \right). \tag{9.2.12}$$

In the limit $|\beta_4|\gamma P_0 \ll \beta_2^2$, this shift is given by a simple expression $\Omega_s = (12\beta_2/|\beta_4|)^{1/2}$. For typical values of β_2 and β_4 occurring in microstructured fibers, the frequency shift can exceed 25 THz, indicating that FWM can amplify a signal that is more than 200 nm away from the pump wavelength. Even larger shifts are possible if both β_2 and β_4 are positive for a fiber but β_6 is negative.

To understand why the tuning range is enhanced in the normal dispersion region, we consider a specific case in which a microstructured fiber is pumped at 1060 nm and assume that the dispersion parameters of the fiber at this wavelength have values $\beta_2 = 2$ ps^2/km, $\beta_3 = 0.05$ ps^3/km, and $\beta_4 = -1 \times 10^{-4}$ ps^4/km. As the pump wavelength is changed in the vicinity of 1060 nm, β_2 changes approximately as $\beta_2(\omega_p + \delta) \approx \beta_2(\omega_p) + \beta_3\delta$, where δ is the change in pump frequency. Although β_4 also changes with δ, this change is relatively small, and we ignore it in this simple calculation. Figure 9.12 shows how the signal and idler wavelengths change with the pump wavelength. These wavelengths are quite sensitive to the magnitude of third-order dispersion. This sensitivity is shown in Figure 9.12 by changing β_3 from 0.05 to 0.1 ps^3/km. In this case, the qualitative change occurring after around 1075 nm is related to fact that β_2 becomes negative, and the pump travels in the anomalous-GVD regime of the fiber. The main point to note is that the signal and idler wavelengths can vary over a wide range exceeding 1000 nm, even though pump wavelength changes by only 40 nm.

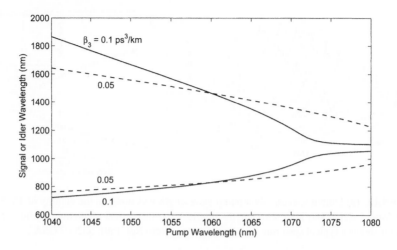

Figure 9.12: Changes in the signal and idler wavelengths occurring when pump wavelength is varied by 40 nm in the normal-dispersion region of a microstructured fiber.

Mathematically, this behavior is related to the appearance of fourth-order dispersion $|\beta_4|$ in the denominator of Eq. (9.2.12). Because of a relatively small value of $|\beta_4|$, small changes in β_2 produce large changes in Ω_s.

Several experiments have employed highly nonlinear fibers with $\beta_4 < 0$ to realize large wavelength shifts for the signal and idler pulses by pumping a fiber-optic parametric amplifier in the normal-dispersion region of the fiber [62]–[69]. In a 2003 experiment, 70-ps pump pulses at 647 nm, with peak powers of up to 1 kW, were launched inside a 1-m-long fiber placed inside the cavity of a parametric oscillator. In a 2005 experiment, pulses shorter than 0.5 ps and tunable over a 200-nm range were obtained using a ring cavity containing 65 cm of a PCF [67]. The laser was pumped synchronously with 1.3-ps pulses from a mode-locked ytterbium fiber laser capable of delivering up to 260 mW of average power.

The walk-off effects resulting from the group-velocity mismatch between the pump and signal (or idler) pulses limit the tuning range in practice. The pump and signal pulses should overlap over the entire fiber length to ensure transfer of energy from the pump to signal. The amount of walk-off delay for a fiber of length L is given by $T_w = d_w L$, where $d_w \equiv |\beta_1(\omega_p) - \beta_1(\omega_s)|$ depends on the group velocity of the two pulses as indicated in Eq. (8.1.3). Clearly, T_w should be less than the width of pump pulses. It was necessary to reduce the fiber length to 65 cm in the experiment reported in Ref. [67] to realize tuning over 200 nm, because pump pulses were only 1.3-ps wide. The tuning range of such parametric oscillators can be increased further by employing wider pump pulses. Indeed, the signal and idler wavelengths were separated by more than 400 nm in an experiment in which 20-ps pump pulses were launched inside a 1.3-m-long photonic crystal fiber [66].

Dispersion fluctuations along the fiber length also limit the tuning range of fiber-optic amplifiers and oscillators. Their impact was analyzed in the case of a dual-pump parametric amplifier in a 2004 study [71]. The case of a single pump propagating

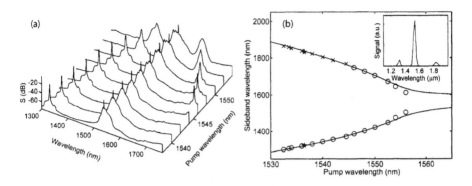

Figure 9.13: (a) Output spectra when pump wavelength was tuned in the vicinity of 1545 nm. (b) Measured signal and idler wavelengths (circles) and theoretical predictions (solid curves). The inset shows a typical spectrum on a linear scale. (From Ref. [69]; ©2007 OSA.)

in the normal-dispersion region of a fiber has also been considered [68]. The results show that the transfer of pump power to the signal and idler waves depends on the frequency difference between the pump and signal and becomes much less efficient as this difference increases beyond 40 THz.

The largest tuning range of a fiber-optic parametric oscillator was realized in a 2007 experiment [69] in which FWM occurred inside a 40-m-long dispersion-shifted fiber. The pump pulses used were relatively wide (>1 ns) and were obtained by modulating the output of a CW semiconductor laser at a repetition rate of ~1 MHz with a Mach–Zehnder modulator. As a result, the walk-off effects over the fiber length were negligible even for a signal-pump frequency difference of 50 THz (about 400 nm in the 1.55-μm spectral region). Figure 9.13(a) shows the spectra at the device output obtained when the wavelength of 50-W pump pulses was tuned by 24 nm in the vicinity of 1545 nm. The measured signal and idler wavelengths (circles) are compared with the theoretical prediction in Figure 9.13(b). Wavelengths longer than 1760 nm (crosses) could not be measured by an optical spectrum analyzer and were estimated from the FWM condition $\omega_s + \omega_i = 2\omega_p$. The combined tuning range on both sides of the pump wavelength exceeded 450 nm in this experiment. Such a widely tunable, all-fiber, parametric oscillator should find a multitude of applications as it does not require a bulky mode-locked laser.

9.3 Supercontinuum Generation

The temporal as well as spectral evolution of optical pulses, launched inside a highly nonlinear fiber, is affected not only by a multitude of nonlinear effects, such as SPM, XPM, FWM, and SRS, but also by the dispersive properties of the fiber. All of these nonlinear processes are capable of generating new frequencies within the pulse spectrum. For sufficiently intense pulses, the pulse spectrum becomes so broad that it may extend over a frequency range exceeding 100 THz. Such extreme spectral broadening is referred to as *supercontinuum generation*, a phenomenon first observed around 1970 in solid and gaseous nonlinear media [72]–[74].

In the context of optical fibers, supercontinuum was first observed in 1976 by launching Q-switched pulses (widths ~10 ns) from a dye laser into a 20-m-long fiber with a core diameter of 7 μm [75]. The output spectrum spread as much as over 180 nm when more than 1 kW of peak power was injected into the fiber. A 1-km-long single-mode fiber was used in a 1987 experiment [33] in which 830-fs pulses with 530-W peak power produced a 200-nm-wide spectrum. During the 1990s, supercontinuum attracted attention for telecommunication applications, but the required bandwidth was less than 100 nm.

The situation changed completely with the advent of microstructured fibers. In a 2000 experiment, 100-fs pulses at 790 nm were launched into a 75-cm section of a microstructured fiber exhibiting anomalous dispersion [76]. The observed spectrum at the fiber output extended over a bandwidth extending from 400 to 1500 nm. Since then, much progress has been made in generating such wide spectra using pulses ranging from a few femtosecond to a few nanoseconds [1]; even continuous-wave (CW) light can be used to produce a supercontinuum. In the case of femtosecond pulses, the formation of supercontinuum can be understood in terms of the fission of higher-order solitons. The basic physics has already been discussed in Section 9.2.1 in the context of intrapulse Raman scattering that creates multiple solitons whose spectra shift toward longer wavelengths with further propagation. When input pulses are intense enough that the soliton order exceeds 10 or so, the spectra of different Raman solitons overlap to form a supercontinuum. Spectral broadening toward shorter wavelengths occurs through the Cherenkov radiation emitted by solitons. Since the physics behind supercontinuum generation has been discussed in a recent review [77] and in Chapter 12 of Ref. [1], this section focuses on the applications of supercontinuum sources.

9.3.1 Multichannel Telecommunication Sources

The technique of wavelength-division multiplexing (WDM) requires a laser with a narrow spectral bandwidth, such as a distributed feedback (DFB) semiconductor laser, for each channel. The wavelengths or the carrier frequencies of two neighboring channels should be different by an amount known as the *channel spacing*, with a value typically falling in the range of 25 to 100 GHz. The laser wavelengths are chosen in practice to match precisely a frequency grid standardized by the International Telecommunication Union (ITU).

The use of individual optical transmitters containing a fixed-wavelength DFB laser for each channel becomes impractical when the number of channels becomes large. This is usually the case for dense WDM systems. A unique approach to solving this problem was developed during the 1990s [78]–[81]. It creates a broadband spectrum by exploiting the technique of supercontinuum generation in optical fibers, which is then sliced spectrally using an optical filter with equally spaced spectral bands. Especially designed highly nonlinear fibers were soon developed for enhancing the spectral bandwidth and employed for system demonstration of supercontinuum-based WDM sources [82]–[87]. By 2003, this approach resulted in multichannel WDM transmitters that were capable of providing more than 1000 channels on the ITU grid [88]–[90].

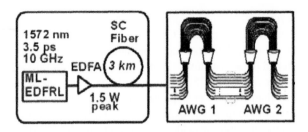

Figure 9.14: A fiber-based WDM source based on spectral slicing of a supercontinuum (SC). ML-EDFRL and AWG stand for a mode-locked erbium-doped fiber-ring laser and an arrayed-waveguide grating, respectively. (From Ref. [81]; ©1996 IEE.)

In an earlier implementation of the basic idea, picosecond pulses from a gain-switched semiconductor laser were first amplified using an EDFA and then broadened spectrally inside a 4.9-km-long standard single-mode fiber to a bandwidth as large as 80 nm through supercontinuum generation [79]. The output of this fiber was passed through a birefringent periodic optical filter that acted as a demultiplexer and created 40 channels, separated by 1.2 nm, in the so-called C band covering the wavelength range from 1525 to 1575 nm. A harmonically mode-locked fiber laser (see Section 5.3.4), operating at a repetition rate of 10 GHz, was used in a 1996 experiment to demonstrate 1-Tb/s lightwave system with a single supercontinuum WDM source [81]. Figure 9.14 shows the design of such a source schematically. This and later experiments employed arrayed-waveguide grating (AWG) filters that are fabricated with the silica-on-silicon technology and are capable of producing a large number of WDM channels with a channel spacing of 1 nm or less.

In a 2000 experiment, the supercontinuum-based technique was used to produce 1000 channels with 12.5-GHz channel spacing [85]. By 2003, this technique resulted in a transmitter that delivered 50-GHz spaced optical carriers (on the ITU grid) over a spectral range from 1425–1675 nm covering the S, C, and L bands [88]. The channel spacing was reduced to only 6.25 GHz in a 2005 experiment that demonstrated transmission of more than 1000 channels at a bit rate of 2.67 Gb/s over a field-installed fiber link [89]. This experiment did not employ a mode-locked laser. Rather, the output of a continuously operating semiconductor laser was phase-modulated at 6.25 GHz. Figure 9.15 shows the design of such a WDM source schematically [90]. Dispersion inside a standard single-mode fiber was used to convert phase modulation into amplitude modulation, resulting in a 6.25-GHz pulse train. This pulse train was amplified and launched inside a polarization-maintaining, dispersion-decreasing fiber that was designed with a convex dispersion profile to produce a 80-nm-wide supercontinuum. Even though channels powers were not uniform over the entire spectral range, the data could be transmitted with success over 126 km of fiber length.

9.3.2   Nonlinear Spectroscopy

Extreme spectral broadening of femtosecond optical pulses inside a microstructured fiber is useful for any application requiring a broadband source. An obvious example

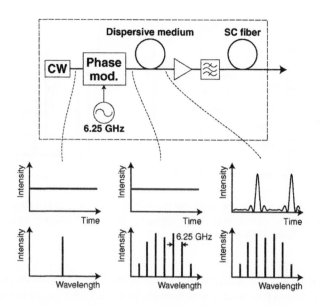

Figure 9.15: A 1000-channel WDM transmitter. Temporal and spectral traces are shown at three locations indicated by dashed lines. SC stands for supercontinuum. (From Ref. [90]; ©2006 IEEE.)

is provided by spectroscopy. Indeed, fiber-based supercontinuum sources have found a variety of applications related to pump-probe spectroscopy [91], coherent Raman spectroscopy [92], near-field optical microscopy [93], and other forms of coherent nonlinear spectroscopy. These techniques are useful for identifying unknown molecular species as well as for imaging biological samples. A femtosecond spectrometer was built as early as 2002 by using the supercontinuum as a wideband probe that interacted through cross-phase modulation with an ultrashort pump pulse inside a sample [91]. A similar scheme was later used to make ultrafast measurements of carrier dynamics in semiconductor lasers through pump-probe spectroscopy [94].

A common technique used for nonlinear imaging makes use of coherent anti-Stokes Raman scattering (CARS) microscopy [95]–[103]. It works by focusing the pump and Stokes pulses onto a sample, as shown schematically in Figure 9.16. Both pulses are obtained from the same mode-locked laser, but the probe pulses are sent through a microstructured fiber to broaden their spectra through supercontinuum generation [95]. Whenever their frequency difference $\omega_p - \omega_s$ coincides nearly with a vibrational resonance of a specific molecule, a blue-shifted anti-Stokes signal at the frequency $2\omega_p - \omega_s$ is generated through a FWM-like process. Because of the broadband nature of the probe spectrum, different molecular species emit radiation at different wavelengths through the CARS process and are thus easily distinguishable in the resulting image of the sample.

In a 2003 experiment, the CARS technique was employed by sending 50-fs pulses, obtained from a mode-locked Ti:sapphire laser, through a 4-cm-long PCF, resulting in a spectrum, shown in Figure 9.17(a), that extended from 625 to 900 nm. The same

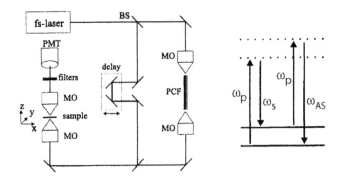

Figure 9.16: A CARS microscope (left) and the energy levels involved (right). BS, MO, and PMT stand for a beam splitter, microscope objective, and photomultiplier tube, respectively. (From Ref. [95]; ©2003 OSA.)

50-fs input pulses with the spectrum shown in the inset were used as the pump pulses. The pump and Stokes pulses were sent through the CARS microscope to image a glass plate containing a layer of polystyrene beads of 4.8-μm diameter. The resulting CARS signal and the image displayed in Figure 9.17(b) show clearly that a resolution of less than 5 μm could be realized in this experiment.

The CARS microscopy is used extensively for imaging biological samples. The use of optical fibers for generating a supercontinuum makes this approach quite practical. In a 2005 experiment, the CARS technique was employed to image a living yeast cell [98]. Because of an ultrabroadband spectrum associated with the Stokes pulses, multiple vibrational resonances were detected simultaneously. It was found later that, in addition to the CARS signal, multiple electronic states can also be excited through the two-photon absorption process, resulting in a two-photon fluorescence signal [100]. This combination of the two nonlinear signals allows one to image living cells with a

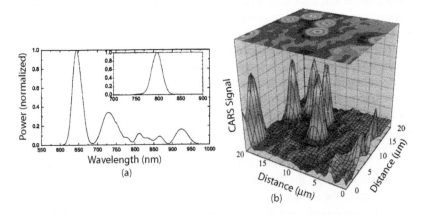

Figure 9.17: (a) Supercontinuum produced with a 4-cm-long PCF and (b) the image of polystyrene beads of 4.8-μm diameter formed using it. (From Ref. [95]; ©2003 OSA.)

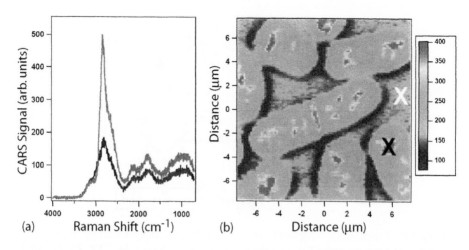

(a) Raman Shift (cm^{-1})   **(b)**   Distance (μm)

Figure 9.18: (a) CARS spectra of a living yeast cell (light curve) and surrounding water (dark curve). (b) CARS image for a Raman shift of 2840 cm^{-1}. Black and white crosses mark the location of the two spectra. (From Ref. [100]; ©2006 OSA.)

clear view of its internal structure. Figure 9.18(b) shows the image of a living yeast cell when its nucleus is labeled by a green fluorescent protein. The image is able to resolve the organelles such as mitochondria, septum, and nucleus. The CARS spectra at the two locations marked by crosses are displayed in part (a). The dominant peak corresponding to a Raman shift of 2840 cm^{-1} is due to the stretching vibrational mode of the CH$_2$ molecule.

In 2002, a single-beam CARS technique making use of the coherent-control concept was introduced [104]. It was later adopted for fiber-based supercontinuum sources [101]. The basic idea behind this scheme is shown in Figure 9.19. In general, the CARS process requires up to three separate beams at frequencies ω_p, ω_s, and ω_{pr} to generate the CARS signal at the frequency $\omega_p - \omega_s + \omega_{pr}$. Only the pump and Stokes beams are needed when the pump plays the role of the probe at the frequency ω_{pr}. A single beam with a broad spectrum can be employed if the pump and Stokes frequencies participating in the CARS process are extracted from the same beam. In this case, a collinear configuration can be used with success. The single beam is focused onto the sample (through a microscope objective), where the CARS signal is generated. A bandpass filter inserted after the sample lets pass only the CARS signal and allows one to form an image of the sample.

The pulse-shaper section in Figure 9.19 is a critical part of such a single-beam CARS microscope. Here, a grating is used to separate spatially different wavelengths within the supercontinuum. A spatial light modulator then modifies phases associated with various spectral components. A knife edge is used to block the blue wing of the supercontinuum so that it does not interfere with the CARS signal generated inside the sample. Remaining spectral components are them combined back using a second grating to form compressed pulses as short as 20 fs. Spectral phase modulation introduced by the spatial light modulator allows one to shape the supercontinuum actively

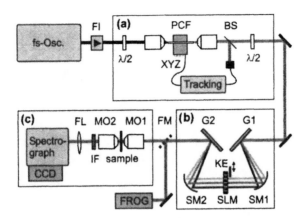

Figure 9.19: Experimental setup employed for single-beam CARS microscopy. It consists of (a) supercontinuum generator, (b) pulse shaper, and (c) image former. FI, XYZ, BS, FL, MO1, FM, SLM, SM1, KE, and G1 stand for a Faraday isolator, piezo stage, beam splitter, focusing lens, microscope objective, flipping mirror, spatial light modulator, spherical mirror, knife edge, and grating, respectively. (From Ref. [101]; ©2006 OSA.)

to any desired form. In particular, it can be used to form two pulses at two different wavelengths that act as the pump and Stokes pulses within the sample [102].

9.3.3 Optical Coherence Tomography

Optical coherence tomography (OCT) is a linear imaging technique in which the short coherence time of a broadband source is used to improve the image resolution [105]–[107]. It is capable of providing high-resolution images of biological tissues even in vivo. Superluminescent diodes, used initially for OCT imaging, have a spectral bandwidth around 40 nm. As this translates into a resolution in the range of 10 to 15 μm, such sources are unable to identify subcellular structure, and even individual cells in some cases. The use of mode-locked femtosecond pulses improves the OCT resolution considerably in view of their wide bandwidth and a short coherence time. A resolution of 3.7 μm was realized in a 1995 experiment by using a mode-locked Ti:sapphire laser [108]. By 1999, this approach could provide in vivo images of biological samples with a resolution close to 1 μm [109] using 5-fs pulses whose spectral bandwidth exceeded 250 nm.

Figure 9.20 shows the experimental setup for a high-resolution OCT system making use of femtosecond pulses [109]. It is essentially a Michelson interferometer, in which the biological sample acts as a mirror in one arm. The sample is scanned in the transverse dimensions using a XY translation stage. Imaging occurs because the sample reflectivity is different depending on the spot where the light is focused. The interference pattern forms only when the length of two arms is matched to within the coherence length, $l_c = c/(n_s \Delta v)$, where n_s is the refractive index of the sample and Δv is the source bandwidth. A longitudinal resolution of 1 μm requires Δv to be close to 200 THz. Even a 100-fs pulse has a spectral bandwidth under 10 THz. Indeed, it was

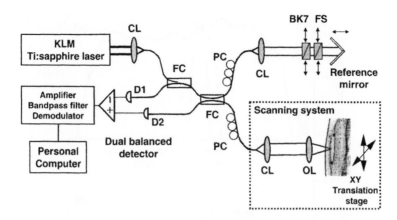

Figure 9.20: An OCT system with a mode-locked Ti:sapphire laser acting as a broadband source. KLM, CL, D1, FC, PC, OL, BK7, and FS stand for Kerr-lens mode locking, coupling lens, detector, fiber coupler, polarization controller, objective lens, glass prism, and fused silica respectively. (From Ref. [109]; ©1999 OSA.)

necessary to employ a Ti:sapphire laser emitting pulses shorter than two optical cycles (<6 fs) to realize a longitudinal resolution close to 1 μm in the 1999 experiment.

The use of ultrashort pulses has a fundamental limitation resulting from chromatic dispersion. Resolution of an OCT system is degraded if there is a significant dispersion mismatch between the reference and the sample arms. For this reason, BK7 prisms and a variable-thickness fused-silica plate were placed in the reference arm in Figure 9.20 to match the dispersion in the two interferometer arms [109]. The thickness of the silica plate was adjusted to compensate for fiber-length differences, and the prisms were used to offset higher-order dispersive effects. A balanced detection scheme was employed to minimize the impact of intensity fluctuations. With these additions, the OCT system shown in Figure 9.20 provided in vivo subcellular images with a ultrahigh resolution. It should be stressed, however, that such a system is not really practical for field applications.

The situation changed around 2000 with the advent of highly nonlinear fibers that could generate a supercontinuum whose bandwidth exceeded 200 THz by using 100-fs or even wider pulses [1]. The system setup shown in Figure 9.20 changes only slightly as one only needs to insert a short piece of microstructured fiber after the mode-locked laser. Indeed, this approach was adopted soon after the formation of supercontinuum was demonstrated in laboratory experiments [110]–[119]. In the initial 2001 experiment, 100-fs pulses obtained from a mode-locked Ti:sapphire laser were coupled into a 1-m-long microstructured fiber [110]. Even though the resulting supercontinuum extended from 400 to 1600 nm, an interference filter, centered at 1300 nm, was used to select a 370-nm-wide spectral slice. Its use for OCT provided a longitudinal resolution of about 2 μm.

By 2002, an OCT resolution of about 0.5 μm was demonstrated in an experiment that employed sub-10-fs pulses at 800 nm to generate a supercontinuum extending from 550 to 950 nm inside a 6-mm-long PCF [111]. The use of a bulky Ti:sapphire

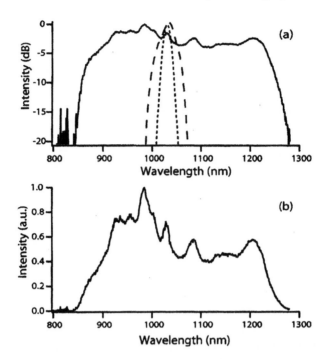

Figure 9.21: Supercontinuum plotted (a) on a logarithmic scale and (b) on a linear scale. The dashed and dotted curves show the spectra of the laser pulse and that of the amplified pulse, respectively. (From Ref. [116]; ©2005 OSA.)

laser can be avoided if a mode-locked fiber laser is employed as a source of femtosecond pulses. As early as 2003, a compact light source was used for OCT imaging, in which a pulsed erbium-doped fiber laser was combined with a microstructured fiber to produce a supercontinuum extending from 1100 to 1800 nm [113]. Its use resulted in OCT images with a longitudinal resolution of about 1.4 μm. In a 2004 experiment, an erbium-doped fiber laser, mode-locked passively and operating near 1.55-μm, was used to create a supercontinuum extending from 1.4 to 1.7 μm, but the resolution was limited to around 5.5 μm [115]. In a later experiment, a mode-locked, Yb-doped fiber laser operating near 1050 nm was employed [116]. Figure 9.21 shows the supercontinuum generated when pulses, first amplified using an Yb-doped fiber amplifier, were launched into a 2-m-long PCF.

As the spectral bandwidth of the supercontinuum in Figure 9.21 exceeds 300 nm, a longitudinal resolution close to 1 μm is expected. However, a nonuniform intensity profile and several other factors kept the resolution to around 1.5 μm inside a biological tissue. Figure 9.22 shows three OCT images acquired by such a fiber-based system [116]. The top image represents an area of size 2 mm × 0.4 mm on a bovine bone. The middle image corresponds to a same-size area on onion skin. The bottom image represents an area of size 1 mm × 0.4 mm near the eye of an in vivo African tadpole. These results indicate that a portable fiber-based OCT system can be built. Such a biomedical instrument would be extremely useful in a clinical setting.

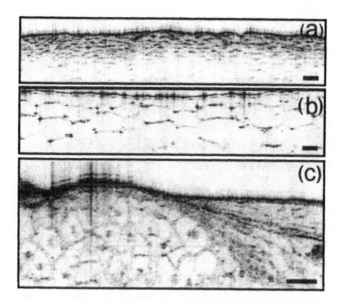

Figure 9.22: Images of biological samples acquired with an OCT system built using a Yb-doped fiber laser: (a) in vitro bovine bone, (b) onion skin, and (c) in vivo African tadpole. The scale bar is 100 μm long in each case. (From Ref. [116]; ©2005 OSA.)

Microstructured fibers offer another benefit for OCT applications as they allow dispersion tailoring. As discussed in Section 9.1, such fibers may even exhibit two closely spaced zero-dispersion wavelengths between which the group-velocity dispersion parameter β_2 has small negative values (anomalous GVD). Such a suitably designed fiber can create two broadband spectral bands centered in the spectral regions near 800 and 1300 nm when pumped around 1060 nm. Figure 9.23 shows such a supercontinuum generated by launching 85-fs pulses from a Nd:glass laser into a 1-m-long PCF [117]. The two spectral bands were generated simultaneously by pumping the fiber with 78-mW average power. The peak centered at 800 nm had a spectral width (full width at half maximum) of 116 nm. The second peak centered at 1300 nm had a spectral width of 156 nm. Their bandwidth was large enough that the same source could be used to create OCT images in two different spectral regions with a resolution in tissue of under 5 μm at 800 nm and under 3 μm at 1300 nm.

An undesirable feature of supercontinuum generated using microstructured fibers is that, depending on the input conditions, it can be quite noisy and may exhibit considerable internal structure. OCT applications require a broadband source with a relatively smooth spectrum. Considerable attention has been paid to understanding the launch conditions under which a smoother spectrum can be created [1]. In general, the spectrum is much smoother when input pulses experience normal dispersion ($\beta_2 > 0$) inside the fiber, although its bandwidth is also reduced in this case. If pulse propagation in the anomalous-dispersion regime is necessary to increase the bandwidth, shorter pump pulses typically create smoother spectra [77]. One can also play with the state of polarization of pump pulses if the fiber exhibits birefringence. All of these techniques

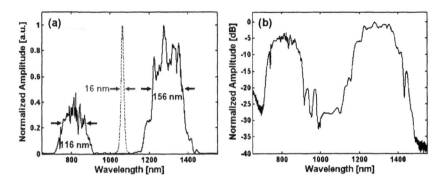

Figure 9.23: Supercontinuum displayed on (a) linear and (b) logarithmic scales. It was generated by launching 85-fs pulses from a Nd:glass laser into a 1-m-long PCF designed to exhibit two closely spaced zero-dispersion wavelengths around 1060 nm. (From Ref. [117]; ©2006 OSA.)

were employed in one study [118] aimed at optimizing a dual-band supercontinuum similar to that shown in Figure 9.23.

In a 2007 experiment, a special 0.8-m-long PCF with a core size of 2.3 μm and exhibiting no zero-dispersion wavelength was employed [119]. Figure 9.24 shows the wavelength dependence of the dispersion parameter D together with the fiber structure. The fiber exhibits normal dispersion ($\beta_2 > 0$), with the minimum value of β_2 occurring around a wavelength of 1 μm. When 130-fs pulses, obtained from a Nd:glass laser operating near 1060 nm, were launched into this fiber, a relatively smooth supercontinuum extending from 800 to 1300 nm was observed at the fiber output. Its use for OCT imaging resulted in a longitudinal resolution close to 2.8 μm in air (and < 2.5 μm inside biological tissue). The main point to note is that the use of microstructured fibers provides considerable flexibility in optimizing the performance of OCT systems by tailoring their dispersion characteristics.

9.3.4 Optical Frequency Metrology

A somewhat unexpected application of the supercontinuum is related to the field of frequency metrology, a field dedicated to developing techniques for making precise measurements of frequencies [120]–[122]. The need for such measurements stems from the fact that the fundamental unit of time, or the *time standard*, is defined as the duration of 9,192,631,770 cycles of microwave radiation corresponding to the transition between two hyperfine levels of the ground state of cesium. The so-called international atomic time is a time scale based on a statistical average of a large number of cesium-based atomic clocks.

It has been recognized for some time that the use of an optical frequency standard, based on a suitable atomic transition, would improve the accuracy of the time standard, and several atomic transitions have been proposed [120]. However, before any new frequency standard can be adopted, one should be able to measure the optical frequency as precisely as possible. The use of frequency combs is common for

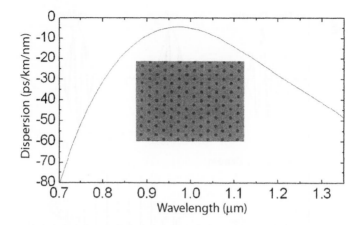

Figure 9.24: Wavelength dependence of the dispersion parameter D for a PCF whose structure is shown in the inset. Note that the second-order dispersion does not vanish at any wavelength for this fiber. (From Ref. [119]; ©2007 OSA.)

this purpose [122]. A frequency comb consists of a large number of equally spaced spectral lines and thus acts as a ruler for measuring frequencies.

Considerable efforts was directed in the 1990s toward developing frequency combs with a variety of techniques [123]–[129]. Even the use of self-phase modulation (SPM) inside optical fibers was investigated for extending the frequency range of an optical comb. In a 1998 study, SPM-induced spectral broadening allowed a frequency comb to span a 30-THz range [127]. When the output of a CW semiconductor laser was modulated at 6.06 GHz using a LiNbO$_3$ modulator, the resulting pulse train exhibited a frequency comb that extended to about 7 THz. By transmitting this pulse train through a 1-km-long dispersion-flattened fiber, SPM increased the range of frequency comb to 50 THz [128]. One may regard this effort as the first application of supercontinuum to the field of optical frequency metrology.

A mode-locked laser provides an excellent example of a frequency comb because the spectrum of mode-locked pulses is naturally in the form of equally spaced spectral lines [129]. The spacing among these lines equals the repetition rate of pulses that depends on the cavity length and can be controlled precisely. This rate is $\sim$100 MHz for most mode-locked lasers but can be increased toward 1 GHz by shortening the cavity length. The spectral range of the frequency comb is inversely related to the temporal width of mode-locked pulse. It the case of femtosecond pulses, the spectral range can easily exceed 1 THz. In particular, as we have seen earlier, Ti:sapphire lasers can be designed to emit pulses shorter than 10 fs, resulting in frequency combs that extend over a range as large as 100 THz. Indeed, such femtosecond frequency combs have attracted considerable attention in recent years [130]–[134].

However, one feature of such frequency combs requires careful calibration before they can be used. Figure 9.25 shows a mode-locked pulse train in the time domain and its spectrum in the frequency domain [132]. As seen there, the electric field associated with a pulse train peaks periodically, and it vanishes between two successive pulses

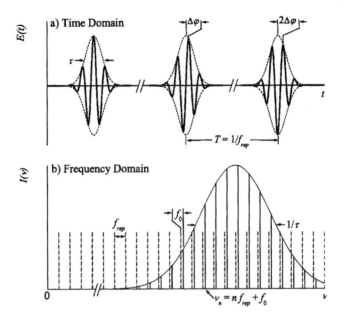

Figure 9.25: (a) Electric field of a mode-locked pulse train as a function of time. Dashed curve shows the pulse envelope. (b) Pulse-train spectrum in the frequency domain. Vertical dashed lines show an ideal frequency comb with $v_n = n f_{rep}$. (From Ref. [132]; ©2003 APS.)

separated by a time interval $T = 1/f_{rep}$. Here, the repetition frequency is given by $f_{rep} = c/L_{opt}$ and L_{opt} represents the optical path length during one round trip inside the laser cavity. However, the peak of the electric field does not coincide with the peak of the pulse envelope. Rather, it shifts at a constant rate from one pulse to next. This so-called carrier-envelope phase mismatch $\Delta\phi$ occurs because the phase and group velocities inside the cavity of a mode-locked lasers are not the same.

The pulse-train spectrum in Figure 9.25(b) shows a consequence of the carrier-envelope phase mismatch in the spectral domain. An ideal frequency comb (vertical dashed lines) should have its nth spectral line at the frequency $v_n = n f_{rep}$. The laser spectrum (solid lines) exhibits spectral lines shifted from this value by a constant amount f_0 that is related to $\Delta\phi$ as [132]

$$f_0 = f_{rep}\Delta\phi/(2\pi). \tag{9.3.1}$$

As a result of this carrier-envelope frequency offset, frequencies associated with a mode-locked pulse trains are given by the relation

$$f_n = n f_{rep} + f_0. \tag{9.3.2}$$

A practical consequence of the offset frequency f_0 is that such a frequency comb can be used for metrological applications only if f_0 can be measured precisely. Moreover, the carrier-envelope phase $\Delta\phi$ should be stabilized to ensure that f_0 does not change with time in a random fashion.

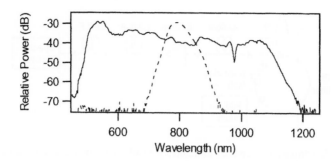

Figure 9.26: Supercontinuum generated inside a 10-cm-long microstructured fiber (solid curve) when 10-fs pulses with the input spectrum shown dashed are launched inside it. (From Ref. [135]; ©2000 APS.)

An appropriate technique for measuring f_0 makes use of second-harmonic generation inside a nonlinear crystal [132]. It assumes that the frequency comb is wide enough to span an octave in frequency; that is, the ratio of the highest and lowest frequencies exceeds a factor of 2. If we double a relatively low frequency f_n using a nonlinear crystal and mix it coherently with the frequency f_{2n} using a heterodyne detector, the resulting microwave signal will have the frequency

$$2f_n - f_{2n} = 2(nf_{\text{rep}} + f_0) - (2nf_{\text{rep}} + f_0) = f_0; \qquad (9.3.3)$$

that is, the beat frequency equals the offset frequency f_0 precisely. Such a scheme is called *self-referencing*, as it uses two frequencies from the same comb and does not require an external reference frequency. A CW laser with a well-stabilized operating frequency can also be used to measure f_0.

The only problem in using the preceding scheme is that the spectrum of a mode-locked laser cannot be an octave wide unless the pulse width is reduced to below a single cycle. This is not easy in practice, although mode-locked Ti:sapphire lasers can be designed to emit pulses shorter than two cycles. This is where supercontinuum generation inside optical fibers comes to the rescue. As we have seen earlier, a microstructured fiber can easily broaden the spectrum of a mode-locked laser that covers more than an octave. Indeed, soon after the 2000 experiment [76], in which a 75-cm section of a microstructured fiber broadened the spectrum of 100-fs input pulses to cover a bandwidth extending from 400 to 1500 nm, such a fiber was used for making measurements of the offset frequency f_0 [135]–[137].

In an initial 2000 experiment, 10-fs pulses from a Ti:sapphire laser, operating near 800 nm at a repetition rate 100 MHz were launched into a 10-cm-long piece of a microstructured fiber [135]. Figure 9.26 compares the output supercontinuum with the input spectrum of mode-locked pulses at a 40-mW average power level. As seen there, even such a short piece of fiber broadens the spectrum enough to cover more than one octave. In this experiment, the frequency of an iodine-stabilized Nd:YAG laser was measured directly in terms of the microwave frequency that controlled the comb spacing. The same femtosecond comb was used to measure optical frequencies of two other lasers operating at 633 and 778 nm with an accuracy of 3 kHz. These measure-

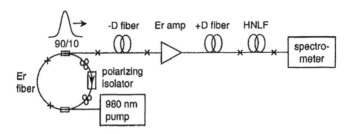

Figure 9.27: Experimental setup used for realizing an all-fiber source of octave-spanning super-continuum. Er amp and HNLF stand for a fiber amplifier and a highly nonlinear fiber, respectively. (From Ref. [140]; ©2003 OSA.)

ments indicated an upper bound of 3×10^{-12} for the uniformity of comb frequencies. In another experiment, a 8-cm-long PCF was used to create the supercontinuum and to realize a frequency chain linking a 10-MHz reference frequency to the optical region [137]. The upper limit for the measurement uncertainty was less than 10^{-15} in this experiment. Clearly, the use of nonlinear effects inside microstructured fibers has revolutionized the field of optical frequency metrology.

More recently, the attention has focused on developing frequency combs that employ a mode-locked fiber laser [138]–[152]. Since fiber lasers are commonly pumped with semiconductor lasers, such a system is much more compact and portable than the one based on a Ti:sapphire laser. Erbium-doped fiber lasers operating near 1550 nm are employed most commonly for the purpose of frequency metrology [139]. Figure 9.27 shows the setup employed in a 2003 experiment for realizing an all-fiber source of octave-spanning supercontinuum [140]. The erbium-doped fiber laser was passively mode-locked at a repetition rate of 33 MHz. It generated pulses centered at 1550 nm that were <200 fs wide, but their average power was limited to around 7 mW.

It was necessary to boost the average laser power by using an erbium-doped fiber amplifier (EDFA) in combination with the chirped-pulse amplification technique [140]. The EDFA was sandwiched between two fibers, marked as $-D$ (normal GVD) and $+D$ (anomalous GVD) fibers in Figure 9.27. The average power of the mode-locked pulse train could be boosted to up to 50 mW with this approach. Such power levels were high enough to create an octave-spanning supercontinuum inside a 6-m-long conventional fiber (no microstructure) designed with $\gamma = 8.5$ W^{-1}/km. It was necessary to vary dispersion along the fiber length by combining four pieces (each 1.5 m long) whose dispersion at 1550 nm was 3.8, 2.2, 0, and -2 ps/(km-nm). Figure 9.28 compares the output spectra when laser pulses were launched into the opposite ends of such a hybrid fiber. The middle trace shows the supercontinuum generated when fiber dispersion was 3.8 ps/(km-nm) along the entire length of a 10-m-long fiber.

A figure-eight fiber laser was employed in a 2004 experiment [141]. It provided 130-fs mode-locked pulses at a 50-MHz repetition rate with energies around 50 pJ, resulting in a 3-mW average power. As before, their power level was increased to near 100 mW with the chirped-pulse amplification and their width was compressed down to below 70 fs. The resulting pulse train was launched into a 23-cm-long dispersion-flattened fiber to create an octave-spanning supercontinuum. The resulting frequency

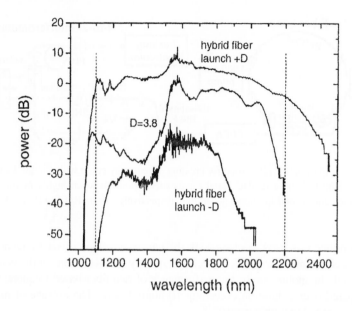

Figure 9.28: Octave-spanning supercontinuum generated inside a 6-m-long dispersion-tailored hybrid fiber (top trace) and a 10-m-long constant-dispersion fiber (middle trace). The bottom trace shows the case when pulses are launched into the hybrid fiber in the opposite direction. Dashed lines mark the boundaries of an octave. (From Ref. [140]; ©2003 OSA.)

comb was characterized with the self-referencing technique discussed earlier. The offset frequency f_0 (around 15 MHz) could be measured with an uncertainty of 10 mHz.

The repetition rate of the fiber laser was made tunable from 49.3 to 50.1 MHz in another experiment in which a tunable delay line was inserted inside the ring cavity of an erbium-doped fiber laser [142]. Figure 9.29 shows the setup employed for creating a supercontinuum and using it to measure f_0 with the self-referencing technique. The laser produced 210-fs soliton pulses that were amplified to an average power of 60 mW and compressed down to below 90 fs. The octave-spanning supercontinuum was generated by injecting these pulses into a 40-cm-long highly nonlinear fiber (without any microstructure) exhibiting $\gamma = 10.6$ W^{-1}/km. The carrier-envelope offset frequency was measured by mixing 1030-nm radiation with its second harmonic of 2060 nm inside an interferometer. Figure 9.30 shows (a) the supercontinuum and (b) the power spectrum exhibiting the two sidebands around the repetition frequency f_r that result from the offset frequency f_0.

A turn-key, all-fiber system for frequency metrology was demonstrated in 2004 by phase-locking the repetition rate and the carrier-envelope offset frequency to a hydrogen maser whose frequency was calibrated with a cesium atomic clock [143]. With this setup, a fully phase-locked optical frequency comb with well-defined absolute frequencies was obtained. It was used to measure optical frequencies of two lasers operating at wavelengths of 1064 and 1542 nm with a relative accuracy of 2×10^{-14}. In a different implementation, mode-locked pulses from a fiber laser were amplified by two parallel amplifiers, seeded by the same master oscillator [144]. The output

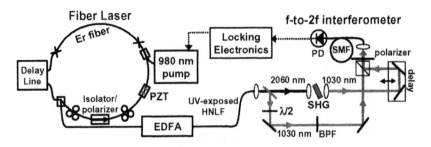

Figure 9.29: Experimental setup used for creating a frequency comb with a mode-locked fiber laser. PZT, HNLF, BPF, and SHG stand for piezoelectric transducer, highly nonlinear fiber, bandpass filter, and second-harmonic generation, respectively. (From Ref. [142]; ©2004 OSA.)

of one amplifier was used to phase lock the carrier-envelope offset frequency. The output of second amplifier enabled precise frequency measurements in the visible and near-infrared. In another study, the performance of two fiber-based frequency combs was compared over a duration exceeding 10 hours [145]. The average of measured frequencies agreed to within 6×10^{-16}.

An optical element that is not easy to integrate within a fiber-based turn-key system is the nonlinear crystal used for generating the second harmonic, required for measuring the carrier-envelope offset frequency. This problem was solved in a 2005 experiment [146] with the setup shown in Figure 9.31. It employed a periodically poled LiNbO$_3$ waveguide (with a poling period of 26.45 μm) that was designed to double the frequency associated with the 2128-nm radiation. The mode-locked, erbium-doped fiber laser employed a Fabry–Perot cavity formed by a saturable-absorber mirror and a fiber Bragg grating.

A problem common to all fiber-laser-based frequency combs is the frequency jitter induced by intensity noise of the semiconductor laser used for pumping the fiber laser [147]. Pump-laser noise affects both the repetition rate f_r and the carrier-envelope

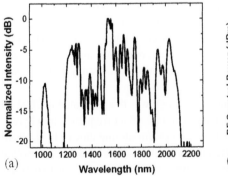
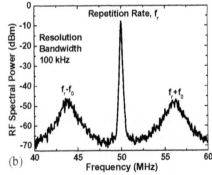

Figure 9.30: (a) Octave-spanning supercontinuum and (b) observed power spectrum. Two sidebands around f_r result from the offset frequency f_0. (From Ref. [142]; ©2004 OSA.)

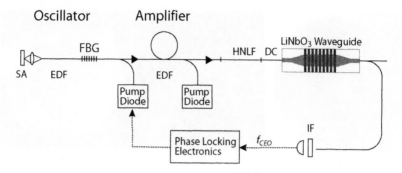

Figure 9.31: Experimental setup for an integrated fiber-based system. SA, EDF, FBG, HNLF, DC, and IF stand for saturable absorber, erbium-doped fiber, fiber Bragg grating, highly nonlinear fiber, dispersion compensation fiber, and interference filter, respectively. (From Ref. [146]; ©2005 OSA.)

offset frequency f_0 and manifests as breathing-mode motion of the comb about the central frequency of the laser output. It results in a substantial spectral broadening of the individual lines of the comb that becomes particularly large in the comb wings. This is the reason behind the relatively wide sidebands seen in Figure 9.30(b). In a 2006 experiment, the bandwidth of these sidebands was reduced from 250 kHz to below 1 Hz by reducing intensity noise of the pump laser through a feedback loop [148]. With such an active stabilization scheme, the noise of a fiber-based frequency comb was low enough to make frequency measurements with a relative accuracy of 5.7×10^{-15} over an averaging duration of more than 25 hours [149]. Moreover, the stability of the system allowed continuous measurements for more than one week. The line width of individual frequencies in fiber-laser-based combs has been reduced to below 1 Hz, corresponding to a timing jitter of around 1 fs for mode-locked pulses [150]. This value is comparable to that realized with Ti:sapphire lasers.

9.4 Photonic Bandgap Fibers

As mentioned earlier in Section 9.1.1, photonic bandgap fibers contain a hollow core (normally air filled) that is surrounded by a silica cladding in which periodic variations of the refractive index are introduced through a regular array of air holes [5]–[7]. Because of a lower refractive index of the core, compared with the cladding, total internal reflection cannot occur, resulting in the absence of traditional waveguide modes. However, light of certain frequencies can still be confined to the core through the so-called photonic bandgap mechanism. In this picture, a two-dimensional periodic array of air holes creates photonic frequency bands, analogous to the electronic energy bands of crystals. In particular, light cannot leak out of the core if its frequency falls within the photonic bandgap between the two allowed frequency bands. Such fibers were developed during the 1990s [10]. They have attracted considerable attention in recent years and have found a number of applications [153]–[159].

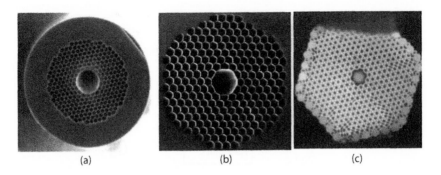

(a)   (b)   (c)

Figure 9.32: Scanning electron micrographs of three hollow-core PCFs. (From Ref. [7]; ©2006 IEEE.)

9.4.1 Properties of Hollow-Core PCFs

Figure 9.32 shows structural details in micrographs of three hollow-core PCFs. A nearly hexagonal geometry is a natural consequence of the capillary-stacking method used to fabricate such fibers [5]. The diameter of the central core can vary from ~ 1 to more than 20 μm, depending on the application. Moreover, the core can remain empty (air filled) or it may be filled with a suitable gas, again depending on the application. In this section we focus on the properties of the optical modes supported by hollow-core PCFs, with emphasis on their dispersive and nonlinear characteristics.

To understand the guiding mechanism in photonic bandgap fibers, it is useful to consider first a perfectly periodic structure consisting of a hexagonal array of circular air rods ($n = 1$) in a fused-silica background ($n \approx 1.45$). The important parameter is the center-to-center spacing Λ between two air holes. The size of air holes sets the air-filling fraction. Modes of such a structure can be analyzed in term of a propagation diagram shown in Figure 9.33 for a 45% air-filling fraction [7]. Three slanted lines divide the possible values of the propagation constant β at a fixed frequency ω into four regions. Light can propagate everywhere (air, holy cladding, and silica) in region 1, only in the cladding and silica in region 2, only in silica in region 3, and nowhere in region 4. Four fingerlike black regions represent photonic bandgaps. If the optical frequency lies within such a bandgap (point B), the light of that frequency cannot escape from the core; that is, it remains confined to the core in spite of the lower refractive index of the air compared with the surrounding cladding. More than one such mode may exist, but all of them have a finite bandwidth set by the photonic bandgaps of such fibers. The number of such modes depends on the size of hollow core, and no mode may exist unless the size is increased above a certain value. Often, seven capillary tubes are omitted from the center to create a large-enough hollow core. Core sizes of 20 μm or more are realized by removing 19 capillaries from the center.

From the standpoint of applications, loss and dispersion characteristics of the modes guided inside the hollow core of a PCF are most relevant. These can be calculated numerically for any specific structure by using one of the techniques discussed in Section 9.1. They can also be measured experimentally. As an example, Figure 9.34 shows wavelength dependence of the loss and dispersion parameters for the PCF whose mi-

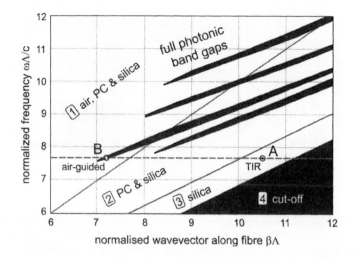

Figure 9.33: Propagation diagram for a PCF designed with 45% air-filling fraction. No light can propagate in black regions. PC and TIR stand for photonic cladding and total internal reflection, respectively. Four propagation regimes are discussed in the text. (From Ref. [7]; ©2006 IEEE.)

crostructure and near-field distribution at a wavelength of 848 nm are shown in the inset [7]. Losses are quite large in comparison with standard fibers but are acceptable for short fiber lengths ($\sim$1 m). The reason behind such large losses is related to the scattering of light taking place at glass–air interfaces. Losses can be decreased by reducing the fraction of mode power residing outside the hollow core of the fiber. Values as low as 1.2 dB/km has been realized for some hollow-core PCFs [157].

Dispersion characteristics of hollow-core PCFs are similar to those occurring for solid-core PCFs, even though the optical mode propagates mostly through the air. The reason is related to the fact that the waveguide contribution dominates in both cases. The air core (diameter about 6 μm) is slightly elliptical for the PCF used for Figure 9.34. As a result, dispersion is different for light polarized along the two principle axes of the fiber. The zero-dispersion wavelengths are also different for the two polarization modes and occur around 830 nm.

The value of the nonlinear parameter γ is the relevant quantity from a practical perspective. Even if the optical mode is confined mostly to the hollow core, a part of it still resides in the silica glass. Thus, the effective value of γ is given by

$$\gamma = \frac{2\pi}{\lambda A_{\text{eff}}} [\Gamma n_2(\text{air}) + (1 - \Gamma) n_2(\text{silica})], \tag{9.4.1}$$

where Γ is the mode-confinement factor and A_{eff} is the effective mode area. The values of n_2 in air and silica are given by $n_2(\text{air}) = 2.9 \times 10^{-23} \text{m}^2/\text{W}$ and $n_2(\text{silica}) = 2.6 \times 10^{-20} \text{m}^2/\text{W}$; that is, n_2 is smaller by three orders of magnitude in air compared with silica. Thus, the nonlinear parameter γ for a hollow-core PCF can be thousands of times smaller compared with solid-core microstructured fibers. For example, whereas γ can exceed 100 W^{-1}/km for PCFs designed with a narrow solid core, values as small as 0.023 W^{-1}/km have been realized for hollow-core PCFs [160].

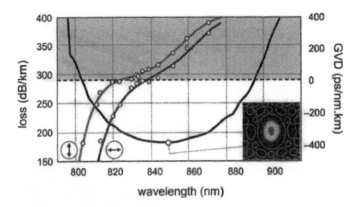

Figure 9.34: Measured wavelength dependence of the loss (black curve) and dispersion parameters (gray curves) for a PCF whose microstructure and near-field distribution are shown in the inset. Dispersion is anomalous in the shaded region. (From Ref. [7]; ©2006 IEEE.)

The applications of hollow-core PCFs fall in two categories. In one category, the weak nonlinearity of air is exploited to transport high-energy pulses to a target. Propagation of such pulses directly in air suffers from diffraction spreading, whereas the pulse can remain confined within the hollow core of a PCF over long distances. In the other category, the core of the fiber is filled with a gas or liquid with a high value of n_2 to enhance the nonlinear effects. The following subsections consider these two types of applications separately.

9.4.2 Applications of Air-Core PCFs

PCFs with an air-filled core should be able to transport ultrashort pulses with high peak powers. Since the dispersive effects are not negligible in such fibers, femtosecond pulses begins to broaden when the fiber length exceeds the dispersion length. Using $L_D = T_0^2 / |\beta_2|$, this length is only ~ 10 cm for 100-fs pulses if we assume $|\beta_2| \sim 10$ ps^2/km. As discussed earlier, the nonlinear effects are weaker but not negligible in such fibers. Thus, a simple way to keep the width of a pulse intact is to increase its peak power P_0 enough that the pulse propagates as a fundamental soliton. The necessary condition for formation of the soliton is $L_D = L_{NL}$, where the nonlinear length is defined as $L_{NL} = 1/(\gamma P_0)$. The required peak power is thus given by

$$P_0 = |\beta_2|/(\gamma T_0^2). \qquad (9.4.2)$$

It follows that P_0 can be increased considerably by increasing β_2 and reducing the pulse width T_0. As an example, if $|\beta_2| = 100$ ps^2/km and $\gamma = 0.03$ W^{-1}/km, 100-fs-wide "sech" pulses ($T_0 = 57$ fs) should be launched with a 1-MW peak power for propagating them as a soliton.

The formation of optical solitons in a PCF with the 12.7-μm-diameter air core was observed in a 2003 experiment in which 110-fs pulses, centered at a wavelength of 1470 nm, were launched into a 3-m-long fiber [161]. Pulse broadening was observed

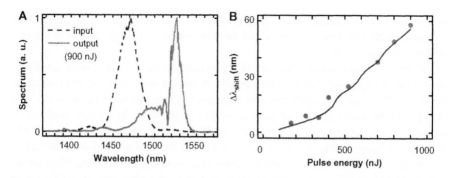

Figure 9.35: (A) Input and output spectra of pulses launched with 900-nJ energy inside a hollow-core PCF. (B) Experimental (circles) and theoretical (solid line) values of Raman-induced wavelength shift in air as a function of pulse energy. (From Ref. [161]; ©2003 AAAS.)

at low pulse energies, but it became relatively small at pulse energies around 450 nJ. The peak power of such pulses exceeded 2 MW. Even higher-energy pulses could be transmitted, but they exhibited considerable Raman-induced spectral shift as expected from the discussion in Section 9.2.1. As an example, Figure 9.35(A) shows the input and output spectra of pulses launched with 900-nJ energy. The central wavelength of each pulse shifts by nearly 60 nm because of Raman scattering in air. Part (B) shows how this wavelength shift depends on pulse energy. Experimental results (circles) are consistent with the prediction of the generalized NLS equation (solid line). Similar results have been obtained for solitons formed at a wavelength near 800 nm [160].

Much higher pulse energies can be transmitted through hollow-core PCFs for wider pulses with widths of 10 ps or more. Dispersion-induced pulse broadening is negligible for such pulses because dispersion length exceeds 10 km, whereas fiber length is typically a small fraction of it. Although soliton formation does not occur for wide pulses, the nonlinear effects in air still manifest as SPM-induced spectral broadening [162]. The extent of this broadening is governed by the maximum nonlinear phase shift

$$\phi_{\max} = \gamma P_0 L_{\text{eff}} = L_{\text{eff}}/L_{\text{NL}}, \qquad (9.4.3)$$

where $L_{\text{eff}} = (1 - e^{-\alpha L})/\alpha$ is the effective length for a fiber of length L with losses α. If spectral broadening is not of concern for a specific application, the peak power of input pulses can be increased substantially.

One may ask what limits the maximum pulse energy that can be transmitted through a hollow-core PCF. The answer is provided by the finite bandwidth associated with such fibers. In essence a hollow-core PCF acts as an optical filter. When the spectral width of pulses exceeds this bandwidth, the energy lying outside its passband is lost as it does not remain confined to the air core. Because of SPM, the fiber acts as a nonlinear energy limiter in the sense that the energy of output pulses saturates to a nearly constant value after input energy exceeds a certain value [162]. This limiting value is ~1 μJ for picosecond pulses, but it can be increased to close to 1 mJ for nanosecond pulses. In a 2004 experiment, 65-ns-wide Q-switched pulses, obtained from a Nd:YAG laser operating at a 15-kHz repetition rate, could be transmitted through a 20-m-long hollow-core PCF at energy levels as high as 0.38 mJ [163]. In a later experiment, pulses energies exceeding 0.5 mJ could be transported in a single spatial mode of a hollow-core fiber [164].

Many application require delivery of high-energy pulses to a relatively small area on a suitable target and therefore may benefit from hollow-core PCFs. Micromachining of metals is one such application as it requires cutting and welding of metals with high precision. Laser ignition of gases is another application. In a recent study, 0.55-mJ pulses were delivered using a hollow-core PCF and focused to form a spark [165]. Another application of high-energy pulses occurs in the field of dentistry. Laser ablation of dental tissues was performed in a 2004 experiment by transporting 40-ps pulses with 2-mJ energy (at a wavelength of 1060 nm) through a photonic-crystal fiber with a 14-μm-diameter air core [166].

A potential application of hollow-core PCFs is for transporting digital signals in fiber-optic WDM networks. Such networks are often limited by the nonlinear effects occurring inside the silica core of conventional fibers. The use of PCFs with air cores would reduce these nonlinear effects to a negligible level. However, this application cannot become a reality until losses of PCFs are reduced to a level below 1 dB/km, and considerable effort has been directed in that direction [167]–[170]. Although PCFs with a solid silica core have already been employed for system experiments at 10 Gb/s [171], hollow-core PCFs still need improvement before their use becomes practical.

In an interesting application, micron-sized dielectric particles were laser-guided through a 15-cm-long hollow-core PCF [172]. The 5-μm polystyrene spheres were first levitated in air using the radiation pressure applied by a CW laser (power 80 mW). They were then forced through the air core (diameter 20 μm) of a PCF by using 514-nm light from an argon-ion laser operating at 514 nm. The speed of guided particles was close to 1 cm/s in this application.

9.4.3 PCFs with Fluid-Filled Cores

PCFs whose hollow core is filled with a suitable fluid (in a gaseous or liquid form) that acts as a nonlinear medium have a wide variety of applications. In a 2002 experiment, a 1-m-long hollow-core PCF was filled with hydrogen gas under pressure and used for observing SRS by launching 6-ns pulses at a wavelength of 532 nm [173]. The Raman threshold for Stokes generation [1] was reduced considerably because the PCF acted as a 1-m-long cell that confined both the hydrogen gas and the pump light. The output spectrum revealed the creation of a blue-shifted peak though CARS when the energy of pump pulses exceeded 3.4 μJ. This experiment indicated clearly the potential of PCFs as gas cells that allow nonlinear interaction over their entire length and thus permit observation of nonlinear phenomena at lower power levels. When the pump light is focused onto a traditional gas cell, a narrow spot size can be maintained only over a diffraction length (or the Rayleigh range), defined as $L_{\text{diff}} = \pi w_0^2 / \lambda$, where w_0 is the radius of the focused spot. As an example, this length is only 74 μm for $w_0 = 5$ μm and $\lambda = 1.06$ μm.

Since 2002, many nonlinear effects have been observed in fluid-filled PCFs [174]–[184]. Even the air inside the core can act as a nonlinear medium if the peak intensity of input pulses is large enough. This was shown clearly in a 2003 experiment in which

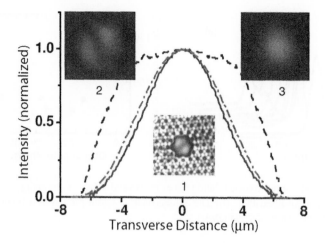

Figure 9.36: Output intensity profiles for input pulses with an energy of 0.5 μJ (dashed curve) and 4 μJ (solid curve); the dot-dashed curve shows theoretical prediction. Three insets show images of the output beam at input energies of 0.1, 0.5, and 4 μJ. (From Ref. [176]; ©2004 OSA.)

the third harmonic was generated through a FWM process by launching the pump and its second harmonic simultaneously inside a 9-cm-long PCF with air-filled core [174]. The diameter of the air core was relatively large (about 13 μm). Even then, energies of 30-ps pulses required for FWM to occur exceeded 30 μJ. In a later experiment, 30-fs pulses with energies of up to 10 μJ (obtained from a Ti:sapphire laser) were launched into a 6-cm-long PCF whose 14-μm-diameter core filled with argon, nitrogen, or atmospheric air [176]. Figure 9.36 shows that the width as well as the intensity profile of the output beam changed considerably when pulse energy was increased from 0.5 to 4 μJ. These results can be understood as a self-channeling of pulses inside a nonlinear waveguide created by the Kerr nonlinearity of gases.

Fluid-filled PCFs can also be used for enhancing the resonant effects associated with specific atomic or molecular transitions. This was demonstrated in a 2005 experiment in which coherent three-level spectroscopy of acetylene molecules was used to observe the electromagnetically induced transparency and slow-light effects [178]. A 1.33-m-long PCF with a 12-μm-diameter was filled with acetylene at pressure levels ranging from 10 to 200 mTorr. Two external-cavity semiconductor lasers operating at wavelengths near 1517 and 1535 nm were tuned close to two molecular transitions forming a Λ-type three-level system. As expected, a narrow transparency window opened up in the presence of the 1535-nm control beam. Changes in the refractive index associated with this window led to slowing down of a 19-ns-wide probe pulse. Transparency exceeded 70% in another experiment that explored other transitions [179].

In a 2005 experiment, 120-fs input pulses were compressed down to 50 fs inside a 24-cm-long PCF whose core was filled with Xe to a pressure of 4.5 atm [180]. The compression mechanism made use of the soliton-effect compression technique

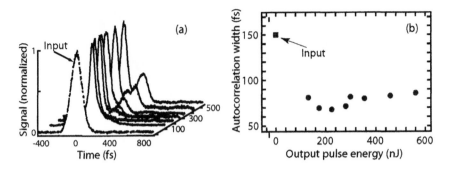

Figure 9.37: (a) Measured autocorrelation traces for several values of the energy of input pulses launched inside a Xe-filled PCF. (b) The autocorrelation width plotted as a function of pulse energy. (From Ref. [180]; ©2005 OSA.)

discussed in Section 6.3. The energy of input pulses was varied from low values to 550 nJ, and pulse compression was observed at energy levels exceeding 100 nJ that were large enough to excite a higher-order soliton inside the Xe gas. Figure 9.37(a) shows the autocorrelation traces measured for several values of pulse energy. Changes in the width of the main autocorrelation peak with increasing pulse energies are shown in part (b). These results indicate pulse compression by more than a factor of 2 at energy levels close to 200 nJ. The second peak in the autocorrelation traces suggests splitting of higher-order solitons. The extent of pulse compression was limited in this experiment by a relatively large third-order dispersion common in PCFs.

The use of fluid-filled PCFs for SRS has continued to attract attention [181]–[184]. In a 2005 study, nine Stokes components were observed covering a wavelength range extending from 950 to 1200 nm, when Q-switched pulses with energies ∼100 nJ were launched inside a 11-m-long PCF whose core filled with hydrogen gas [182]. Moreover, a clear transition was observed from the transient to steady-state regime for input pulses as long as 14 ns. More recently, ethanol has been used as a Raman-active medium inside the core of a PCF [183]. Ethanol exhibits a Raman gain of 5 cm/GW with a relatively large Raman shift of about 88 THz. In this experiment, the 11-μm-diameter core of a 2.8-m-long PCF was filled with ethanol. When the fiber was pumped with 0.5-ns pulses at 532 nm, the output spectrum exhibited two Stokes lines at wavelengths of 630 and 772 nm. Figure 9.38 shows the output peak power at the pump and the two Stokes wavelengths as a function of the peak power associated with input pump pulses. The Raman threshold for the first and second Stokes lines is reached at input power levels of 500 and 900 W, respectively.

In a 2007 experiment [184], the use of ethanol inside the core of a PCF permitted efficient generation of 630-nm light from 532-nm pulses in a single spatial mode with high output-beam quality. The PCF was designed such that the second Stokes line at 772 nm was outside of its transmission bandwidth. As a result, the peak power of input pulses could be increased up to 2 kW, with more than 350-W appearing at the Stokes wavelength. These results show that hollow-core PCFs can act as a nonlinear cell that can be filled with a variety of fluids. Theoretical considerations indicate that,

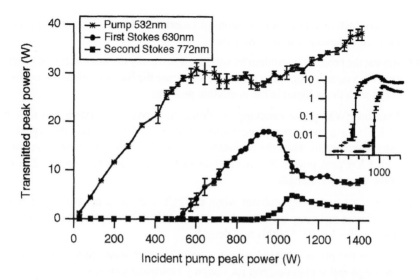

Figure 9.38: Output peak powers at the pump wavelength and the two Stokes wavelengths measured as a function of the peak power of input pump pulses. The inset shows the Stokes powers on a log scale to identify the Raman threshold. (From Ref. [183]; ©2006 OSA.)

with a proper design, the nonlinear parameter γ can exceed 5 W^{-1}/m for liquids such as carbon disulfide and nitrobenzene [185]. The SPM-induced nonlinear phase shift, $\phi_{NL} = \gamma P_0 L$, can exceed 50 for such a 10-m-long fiber even at a relatively low 1-W power level. The application of such fibers can range from a simple wavelength shifter to nonlinear pulse compressors. They can also be used for supercontinuum generation and the related applications discussed in Section 9.3.

Problems

9.1 Explain how a microstructured fiber guides light inside its core. Sketch two designs for such fibers.

9.2 Explain why large values of γ cannot be realized by reducing the core diameter to below 1 μm and using a conventional design of silica fibers containing a germania-doped core?

9.3 Solve the eigenvalue equation given in Section 2.2.1 of Ref. [1] for a thin silica wire with air cladding. Use the Sellmeier equation on page 6 of Ref. [1] and reproduce the dispersion curves given in Figure 9.4.

9.4 What is meant by intrapulse Raman scattering? Why does this effect shift the wavelength of femtosecond pulses toward longer wavelengths.

9.5 Solve the generalized NLS equation (9.2.1) with the split-step Fourier method for a 1-m-long PCF with $\beta_2 = -20$ ps^2/km, $\beta_3 = 0.1$ ps^3/km, and $\gamma = 20$ W^{-1}/km. Neglect fiber losses and assume $T_R = 3$ fs. Plot output pulse shapes and spectra when 100-fs sech-shape pulses at 1060 nm are launched with peak powers such that they excite solitons of orders $N = 1$ to 5. Estimate the Raman-induced wavelength shift in each case.

9.6 Repeat the preceding problem by solving numerically the more accurate generalized NLS equation (9.2.6) with $M = 3$. Compare the Raman-induced wavelength shifts with those found in the preceding problem.

9.7 Use the FWM phase-matching condition to derive the relation given in Eq. (9.2.10) for a single-pump parametric oscillator. Use it to find the signal and idler wavelengths when the oscillator is pumped at 1060 nm using 10-ps pulses with 1-kW peak power. Assume that the PCF exhibits a dispersion of -2 ps^2/km at the pump wavelength.

9.8 Repeat the preceding problem when the PCF exhibits a normal dispersion of 2 ps^2/km at the pump wavelength and the fourth-order dispersion parameter has a value of -1×10^{-4} at this wavelength.

9.9 Explain the physics behind coherent anti-Stokes Raman scattering. How can one employ this phenomenon for imaging biological samples.

9.10 Explain the operating principle behind optical coherence tomography. How wide should the source spectrum be to realize a longitudinal resolution of 1 μm for imaging a biological sample with an average refractive index of 1.3?

9.11 Describe the self-referencing technique used in frequency metrology for measuring the carrier-envelope offset frequency. Sketch the design of a fiber-based system based on this technique.

9.12 Explain how a photonic bandgap fiber can guide light inside its core in spite of a lower refractive index of the core compared with that of the cladding.

References

[1] G. P. Agrawal, *Nonlinear Fiber Optics*, 4th ed. (Academic Press, Boston, 2007).

[2] J. C. Knight, T. A. Birks, P. St. J. Russell, and D. M. Atkin, *Opt. Lett.* **21**, 1547 (1996).

[3] J. Broeng, D. Mogilevstev, S. B. Barkou, and A. Bjarklev, *Opt. Fiber Technol.* **5**, 305 (1999).

[4] B. J. Eggleton, C. Kerbage, P. S. Westbrook, R. S. Windeler, and A. Hale, *Opt. Express* **9**, 698 (2001).

[5] P. St. J. Russell, *Science* **299**, 358 (2003).

[6] A. Bjarklev, J. Broeng, and A. S. Bjarklev, *Photonic Crystal Fibres* (Kluwer Academic, Boston, 2003).

[7] P. St. J. Russell, *J. Lightwave Technol.* **24**, 4729 (2006).

[8] F. Zolla, G. Renversez, A. Nicolet, and B. Kuhlmey, S. Guenneau, and D. Felbacq, *Foundations of Photonic Crystal Fibres* (Imperial College Press, London, 2005).

[9] F. Poli, A. Cucinotta, and S. Selleri, *Photonic Crystal Fibers: Properties and Applications* (Springer, New York, 2007).

[10] R. F. Cregan, B. J. Mangan, J. C. Knight, T. A. Birks, P. St. J. Russell, P. J. Roberts, and D. C. Allan, *Science* **285**, 1537 (1999).

[11] V. Finazzi, T. M. Monro, and D. J. Richardson, *J. Opt. Soc. Am. B* **20**, 1427 (2003).

[12] K. Nakajima, J. Zhou, K. Tajima, K. Kurokawa, C. Fukai, and I. Sankawa, *J. Lightwave Technol.* **17**, 7 (2005).

[13] D. Mogilevtsev, T. A. Birks, and P. St. J. Russell, *J. Lightwave Technol.* **17**, 2078 (1999).

[14] A. Ferrando, E. Silvestre, J. J. Miret, P. Andrés, and M. V. Andrés, *Opt. Lett.* **24**, 276 (1999).

[15] T. M. Monro, D. J. Richardson, N. G. R. Broderick, and P. J. Bennett, *J. Lightwave Technol.* **18**, 50 (2000).

[16] F. Brechet, J. Marcou, D. Pagnoux, and P. Roy, *Opt. Fiber Technol.* **6**, 181 (2000).

[17] Z. Zhu and T. G. Brown, *Opt. Express* **8**, 547 (2001); *Opt. Commun.* **206**, 333 (2002).

[18] T. P. White, B. T. Kuhlmey, R. C. McPhedran, D. Maystre, G. Renversez, C. Martijn de Sterke, and L. C. Botten, *J. Opt. Soc. Am. B* **19**, 2322 (2002).

[19] M. Koshiba and K. Saitoh, *Opt. Express* **11**, 1746 (2003).

[20] M. D. Nielsen, N. A. Mortensen, J. R. Folkenberg, and A. Bjarklev, *Opt. Lett.* **28**, 2309 (2003).

[21] H. Uranus and H. Hoekstra, *Opt. Express* **12**, 2795 (2004).

[22] N. A. Mortensen, *Opt. Lett.* **30**, 1455 (2005).

[23] K. Saitoh and M. Koshiba, *J. Lightwave Technol.* **23**, 3580 (2005); *Opt. Express* **13**, 267 (2005).

[24] P. Kowalczyk, M. Wiktor, and M. Mrozowski, *Opt. Express* **13**, 10349 (2005).

[25] M. Szpulak, W. Urbanczyk, E. Serebryannikov, A. Zheltikov, A. Hochman, Y. Leviatan, R. Kotynski, and K. Panajotov, *Opt. Express* **14**, 5699 (2006).

[26] H. Li, A. Mafi, A. Schlzgen, L. Li, V. L. Temyanko, N. Peyghambarian, and J. V. Moloney, *J. Lightwave Technol.* **25**, 1224 (2007).

[27] E. C. Mägi, P. Steinvurzel, and B. J. Eggleton, *Opt. Express* **12**, 776 (2004).

[28] H. C. Nguyen, B. T. Kuhlmey, E. C. Mägi, M. J. Steel, P. Domachuk, C. L. Smith, and B. J. Eggleton, *Appl. Phys. B* **81**, 377 (2005).

[29] W. J. Wadsworth, A. Witkowska, S. G. Leon-Saval, and T. A. Birks, *Opt. Express* **13**, 6541 (2005).

[30] F. M. Mitschke and L. F. Mollenauer, *Opt. Lett.* **11**, 659 (1986).

[31] J. P. Gordon, *Opt. Lett.* **11**, 662 (1986).

[32] J. Santhanam and G. P. Agrawal, *Opt. Commun.* **222**, 413 (2003).

[33] P. Beaud, W. Hodel, B. Zysset, and H. P. Weber, *IEEE J. Quantum Electron.* **23**, 1938 (1987).

[34] G. P. Agrawal, *Opt. Lett.* **16**, 226 (1991).

[35] J. K. Lucek and K. J. Blow, *Phys. Rev. A* **45**, 666 (1992).

[36] T. Sugawa, K. Kurokawa, H. Kubota, and M. Nakazawa, *Electron. Lett.* **30**, 1963 (1994).

[37] C. Hedaley and G. P. Agrawal, *J. Opt. Soc. Am. B* **13**, 2170 (1996).

[38] M. Golles, I. M. Uzunov, and F. Lederer, *Phys. Lett. A* **231**, 195 (1997).

[39] N. Nishizawa and T. Goto, *IEEE Photon. Technol. Lett.* **11**, 325 (1999).

[40] X. Liu, C. Xu, W. H. Knox, J. K. Chandalia, B. J. Eggleton, S. G. Kosinski, and R. S. Windeler, *Opt. Lett.* **26**, 358 (2001).

[41] N. Nishizawa and T. Goto, *IEEE J. Sel. Topics Quantum Electron.* **7**, 518 (2001).

[42] D. T. Reid, I. G. Cormack, W. J. Wadsworth, J. C. Knight, and P. St. J. Russell, *J. Mod. Opt.* **49**, 757 (2002).

[43] N. Nishizawa, Y. Ito, and T. Goto, *IEEE Photon. Technol. Lett.* **14**, 986 (2002).

[44] D. A. Chestnut and J. R. Taylor, *Opt. Lett.* **28**, 2512 (2003).

[45] F. Druon, N. Sanner, G. Lucas-Leclin, P. Georges, K. P. Hansen, and A. Petersson, *Appl. Opt.* **42**, 6768 (2003).

[46] M. Kato, K. Fujiura, and T. Kurihara, *Appl. Opt.* **43**, 5481 (2004).

[47] K. S. Abedin and F. Kubota, *IEEE J. Sel. Topics Quantum Electron.* **10**, 1203 (2004).

[48] H. Lim, J. Buckley, A. Chong, and F. W. Wise, *Electron. Lett.* **40**, 1523 (2005).

[49] M. G. Banaee and J. F. Young, *J. Opt. Soc. Am. B* **23**, 1484 (2006).

[50] N. Ishii, C. Y. Teisset, S. Köhler, E. E. Serebryannikov, T. Fuji, T. Metzger, F. Krausz, A. Baltuska, and A. M. Zheltikov, *Phys. Rev. E* **74**, 036617 (2006).

[51] J. van Howe, J. H. Lee, S. Zhou, F. Wise, C. Xu, S. Ramachandran, S. Ghalmi, and M. F. Yan, *Opt. Lett.* **32**, 340 (2007).

[52] K. J. Blow and D. Wood, *IEEE J. Quantum Electron.* **25**, 2665 (1989).

[53] Q. Lin and G. P. Agrawal, *Opt. Lett.* **31**, 3086 (2006).

[54] N. Akhmediev and M. Karlsson, *Phys. Rev. A* **51**, 2602 (1995).

[55] J. H. Lee, J. van Howe, C. Xu, S. Ramachandran, S. Ghalmi, and M. F. Yan, *Opt. Lett.* **32**, 1053 (2007).

[56] E. R. Andresen, V. Birkedal, J. Thøgersen, and S. R. Keiding, *Opt. Lett.* **31**, 1328 (2006).

[57] S. Oda and A. Maruta, *Opt. Express* **14**, 7895 (2006).

[58] J. E. Sharping, M. Fiorentino, P. Kumar, and R. S. Windeler, *Opt. Lett.* **27**, 1675 (2002).

[59] M. E. Marhic, K. K. Y. Wong, L. G. Kazovsky, and T. E. Tsai, *Opt. Lett.* **27**, 1439 (2002).

[60] C. J. S. de Matos, J. R. Taylor, and K. P. Hansen, *Opt. Lett.* **29**, 983 (2004).

[61] J. E. Sharping, M. A. Foster, A. L. Gaeta, J. Lasri, O. Lyngnes, and K. Vogel, *Opt. Express* **15**, 1474 (2007).

[62] S. Pitois and G. Millot, *Opt. Commun.* **226**, 415 (2003).

[63] J. D. Harvey, R. Leonhardt, S. Coen, G. K. L. Wong, J. C. Knight, W. J. Wadsworth, and P. St. J. Russell, *Opt. Lett.* **28**, 2225 (2003).

[64] W. Wadsworth, N. Joly, J. Knight, T. Birks, F. Biancalana, and P. Russell, *Opt. Express* **12**, 299 (2004).

[65] T. V. Andersen, K. M. Hilligse, C. K. Nielsen, J. Thgersen, K. P. Hansen, S. R. Keiding, and J. J. Larsen, *Opt. Express* **12**, 4113 (2004).

[66] A. Y. H. Chen, G. K. L. Wong, S. G. Murdoch, R. Leonhardt, J. D. Harvey, J. C. Knight, W. J. Wadsworth, and P. St. J. Russell, *Opt. Lett.* **30**, 762 (2005).

[67] Y. Deng, Q. Lin, F. Lu, G. P. Agrawal, and W. H. Knox, *Opt. Lett.* **30**, 1234 (2005).

[68] J. S. Y. Chen, S. G. Murdoch, R. Leonhardt, and J. D. Harvey, *Opt. Express* **14**, 9491 (2006).

[69] G. K. L. Wong, S. G. Murdoch, R. Leonhardt, J. D. Harvey, and V. Marie, *Opt. Express* **15**, 2947 (2007).

[70] M. Yu, C. J. McKinstrie, and G. P. Agrawal, *Phys. Rev. E* **52**, 1072 (1995).

[71] F. Yaman, Q. Lin, S. Radic, and G. P. Agrawal, *IEEE Photon. Technol. Lett.* **16**, 1292 (2004).

[72] F. Shimizu, *Phys. Rev. Lett.* **19**, 1097 (1967).

[73] R. R. Alfano and S. L. Shapiro, *Phys. Rev. Lett.* **24**, 584 (1970).

[74] Q. Z. Wang, P. P. Ho, and R. R. Alfano, in *The Supercontinuum Laser Source*, 2nd ed., R. R. Alfano, Ed. (Springer, New York, 2005), Chap. 2.

[75] C. Lin and R. H. Stolen, *Appl. Phys. Lett.* **28**, 216 (1976).

[76] J. K. Ranka, R. S. Windeler, and A. J. Stentz, *Opt. Lett.* **25**, 25 (2000).

[77] J. M. Dudley, G. Genty, and S. Coen, *Rev. Mod. Phys.* **78**, 1135 (2006).

[78] T. Morioka, K. Mori, and M. Saruwatari, *Electron. Lett.* **29**, 862 (1993).

[79] T. Morioka, K. Mori, S. Kawanishi, and M. Saruwatari, *IEEE Photon. Technol. Lett.* **6**, 365 (1994).

[80] T. Morioka, K. Uchiyama, S. Kawanishi, S. Suzuki, and M. Saruwatari, *Electron. Lett.* **31**, 1064 (1995).

[81] T. Morioka, H. Takara, S. Kawanishi, O. Kamatani, K. Takiguchi, K. Uchiyama, M. Saruwatari, H. Takahashi, M. Yamada, T. Kanamori and H. Ono, *Electron. Lett.* **32**, 906 (1996).

[82] T. Okuno, M. Onishi, T. Kashiwada, S. Ishikawa, and M. Nishimura, *IEEE J. Sel. Topics Quantum Electron.* **5**, 1385 (1999).

[83] L. Boivin, S. Taccheo, C. R. Doerr, L. W. Stulz, R. Monnard, W. Lin, and W. C. Fang, *IEEE Photon. Technol. Lett.* **12**, 1695 (2000).

[84] Ö. Boyraz, J. Kim, M. N. Islam, F. Coppinger, and B. Jalali, *J. Lightwave Technol.* **18**, 2167 (2000).

[85] H. Takara, T. Ohara, K. Mori, K. Sato, E. Yamada, Y. Inoue, T. Shibata, M. Abe, T. Morioka, and K. I. Sato, *Electron. Lett.* **36**, 2089 (2000).

[86] F. Futami and K. Kikuchi, *IEEE Photon. Technol. Lett.* **13**, 73 (2001).

[87] Ö. Boyraz and M. N. Islam, *J. Lightwave Technol.* **20**, 1493 (2002).

[88] K. Mori, K. Sato, H. Takara, and T. Ohara, *Electron. Lett.* **39**, 544 (2003).

[89] H. Takara, T. Ohara, T. Yamamoto, H. Masuda, M. Abe, H. Takahashi, and T. Morioka, *Electron. Lett.* **41**, 270 (2005).

[90] T. Ohara, H. Takara, T. Yamamoto, H. Masuda, T. Morioka, M. Abe, and H. Takahashi, *J. Lightwave Technol.* **24**, 2311 (2006).

[91] V. Nagarajan, E. Johnson, P. Schellenberg, W. Parson, and R. Windeler, *Rev. Sci. Instrum.* **73**, 4145 (2002).

[92] H. Kano and H. Hamaguchi, *Opt. Lett.* **28**, 2360 (2003).

[93] T. Nagahara, K, Imura, and H. Okamotoa, *Rev. Sci. Instrum.* **75**, 4528 (2004).

[94] M. Punke, F. Hoos, C. Karnutsch, U. Lemmer, N. Linder, and K. Streubel, *Opt. Lett.* **31**, 1157 (2006).

[95] H. N. Paulsen, K. M. Hilligse, J. Thgersen, S. R. Keiding, and J. Larsen, *Opt. Lett.* **28**, 1123 (2003).

[96] H. Kano and H. Hamaguchi, *Appl. Phys. Lett.* **85**, 4298 (2004); **86**, 12113 (2005).

[97] S. O. Konorov, D. A. Akimov, E. E. Serebryannikov, A. A. Ivanov, M. V. Alfimov, and A. M. Zheltikov, *Phys. Rev. E* **70**, 057601 (2004).

[98] H. Kano and H. Hamaguchi, *Opt. Express* **13**, 1322 (2005).

[99] R. Shimada, H. Kano, and H. Hamaguchi, *Opt. Lett.* **31**, 320 (2006).

[100] H. Kano and H. Hamaguchi, *Opt. Express* **14**, 2798 (2006).

[101] B. von Vacano, W. Wohlleben, and M. Motzkus, *Opt. Lett.* **31**, 413 (2006).

[102] B. von Vacanoa and M. Motzkus, *Opt. Commun.* **264**, 488 (2006).

[103] K. Shi, P. Li, and Z. Liu, *Appl. Phys. Lett.* **90**, 141116 (2007).

[104] N. Dudovich, D. Oron, and Y. Silberberg, *Nature* **418**, 512 (2002).

[105] D. Huang, E. Swanson, C. P. Lin, J. S. Schuman, W. G. Stinson, W. Chang, M. R. Hee, T. Flotte, K. Gregory, C. A. Puliafito, and J. G. Fujimoto, *Science* **254**, 1178 (1991).

[106] J. G. Fujimoto, M. E. Brezinski, G. J. Tearney, S. A. Boppart, B. E. Bouma, M. R. Hee, J. F. Southern, and E. A. Swanson, *Nature Med.* **1**, 970 (1995).

[107] M. E. Brezinski, *Optical Coherence Tomography: Principles and Applications* (Academic Press, Boston, 2006).

[108] B. Bouma, G. J. Tearney, S. A. Boppart, and M. R. Hee, M. E. Brezinski, and J. G. Fujimoto, *Opt. Lett.* **20**, 1486 (1995).

[109] W. Drexler, U. Morgner, F. X. Kärtner, C. Pitris, S. A. Boppart, X. D. Li, E. P. Ippen, and J. G. Fujimoto, *Opt. Lett.* **24**, 1221 (1999).

[110] I. Hartl, X. D. Li, C. Chudoba, R. K. Hganta, T. H. Ko, J. G. Fujimoto, J. K. Ranka, and R. S. Windeler, *Opt. Lett.* **26**, 608 (2001).

[111] B. Povazay, K. Bizheva, A. Unterhuber, B. Hermann, H. Sattmann, A. F. Fercher, W. Drexler, A. Apolonski, W. J. Wadsworth, J. C. Knight, P. St. J. Russell, M. Vetterlein, and E. Scherzer, *Opt. Lett.* **27**, 1800 (2002).

[112] Y. Wang, Y. Zhao, J. S. Nelson, Z. Chen, and R. S. Windeler, *Opt. Lett.* **28**, 182 (2003).

[113] K. Bizheva, B. Povazay, B. Hermann, H. Sattmann, W. Drexler, M. Mei, R. Holzwarth, T. Hoelzenbein, V. Wacheck, and H. Pehamberger *Opt. Lett.* **28**, 707 (2003).

[114] S. Bourquin, A. D. Aguirre, I. Hartl, P. Hsiung, T. H. Ko, J. G. Fujimoto, T. A. Birks, W. J. Wadsworth, U. Bunting, and D. Kopf, *Opt. Express* **11**, 3290 (2003).

[115] N. Nishizawa, Y. Chen, P. Hsiung, E. P. Ippen, and J. G. Fujimoto, *Opt. Lett.* **29**, 2846 (2004).

[116] H. Lim, Y. Jiang, Y. Wang, Y. C. Huang, Z. Chen, and F. W. Wise, *Opt. Lett.* **30**, 1171 (2005).

[117] A. D. Aguirre, N. Nishizawa, J. G. Fujimoto, W. Seitz, M. Lederer, and D. Kopf, *Opt. Express* **14**, 1145 (2006).

[118] H. Wang and A. M. Rollins, *Appl. Opt.* **46**, 1787 (2007).

[119] H. Wang, C. P. Fleming, and A. M. Rollins, *Opt. Express* **15**, 3095 (2007).

[120] L. Hollberg, C. W. Oates, E. A. Curtis, E. N. Ivanov, S. A. Diddams, T. Udem, H. G. Robinson, J. C. Bergquist, R. J. Rafac, W. M. Itano, R. E. Drullinger, and D. J. Wineland, *IEEE J. Quantum Electron.* **37**, 1502 (2001).

[121] S. A. Diddams, J. C. Bergquist, S. R. Jefferts, C. W. Oates, *Science* **306**, 1318 (2002).

[122] S. Cundiff and J. Ye, Eds., *Femtosecond Optical Frequency Comb: Principle, Operation and Applications* (Springer, New York, 2004).

[123] M. Kourogi, T. Enami, and M. Ohtsu, *IEEE Photon. Technol. Lett.* **6**, 214 (1994).

[124] A. S. Bell, G. M. Macfarlane, E. Riis, and A. I. Ferguson, *Opt. Lett.* **20**, 1435 (1995).

[125] M. Kourogi, B. Widiyatmoko, and M. Ohtsu, *IEEE Photon. Technol. Lett.* **8**, 560 (1996).

[126] T. Udem, M. Kourogi, J. Reichert, and T. W. Hänsch, *Opt. Lett.* **23**, 1387 (1998).

[127] K. Imai, M. Kourogi, and M. Ohtsu, *IEEE J. Quantum Electron.* **34**, 54 (1998).

[128] K. Imai, B. Widiyatmoko, M. Kourogi, and M. Ohtsu, *IEEE J. Quantum Electron.* **35**, 559 (1999).

[129] T. Udem, J. Reichert, R. Holzwarth, and T. W. Hänsch, *Opt. Lett.* **24**, 881 (1995).

[130] R. Holzwarth, M. Zimmermann, T. Udem, and T.W. Hänsch, *IEEE J. Quantum Electron.* **37**, 1493 (2001).

[131] T. Udem, R. Holzwarth, and T. W. Hänsch, *Nature* **416**, 233 (2002).

[132] S. T. Cundiff and Jun Ye, *Rev. Mod. Phys.* **75**, 325 (2003).

[133] L. S. Ma, L. Robertsson, S. Picard, M. Zucco, S. Bi, S. Wu, and R. S. Windeler, *Opt. Lett.* **29**, 641 (2004).

[134] L. S. Ma, Z. Bi, A. Bartels, K. Kim, L. Robertsson, M. Zucco, R. S. Windeler, G. Wilpers, C. Oates, L. Hollberg, and S. A. Diddams, *IEEE J. Quantum Electron.* **43**, 139 (2007).

[135] S. A. Diddams, D. J. Jones, J. Ye, S. T. Cundiff, J. L. Hall, J. K. Ranka, R. S. Windeler, R. Holzwarth, T. Udem, and T.W. Hänsch, *Phys. Rev. Lett.* **84**, 5102 (2000).

[136] D. J. Jones, S. A. Diddams, J. K. Ranka, A. Stentz, R. S. Windeler, J. L. Hall, and S. T. Cundiff, *Science* **288**, 635 (2000).

[137] R. Holzwarth, T. Udem, T. W. Hänsch, J. C. Knight, W. J. Wadsworth, and P. St. J. Russell, *Phys. Rev. Lett.* **85**, 2264 (2000).

[138] J. Rauschenberger, T. Fortier, D. Jones, J. Ye, and S. Cundiff, *Opt. Express* **10**, 1404 (2002).

[139] F. Tauser, A. Leitenstorfer, and W. Zinth, *Opt. Express* **11**, 594 (2003).

[140] J. W. Nicholson, M. F. Yan, P. Wisk, J. Fleming, F. DiMarcello, E. Monberg, A. Yablon, C. Jørgensen, and T. Veng, *Opt. Lett.* **28**, 643 (2003).

[141] B. R. Washburn, S. A. Diddams, N. R. Newbury, J. W. Nicholson, M. F. Yan, and C. G. Jørgensen, *Opt. Lett.* **29**, 250 (2004).

[142] B. R. Washburn, R. Fox, N. Newbury, J. Nicholson, K. Feder, P. Westbrook, and C. Jørgensen, *Opt. Express* **12**, 4999 (2004).

[143] T. R. Schibli, K. Minoshima, F.-L. Hong, H. Inaba, A. Onae, H. Matsumoto, I. Hartl, and M. E. Ferman, *Opt. Lett.* **29**, 2467 (2004).

[144] F. Adler, K. Moutzouris, A. Leitenstorfer, H. Schnatz, B. Lipphardt, G. Grosche, and F. Tauser, *Opt. Express* **12**, 5872 (2004).

[145] P. Kubina, P. Adel, F. Adler, G. Grosche, T. Hänsch, R. Holzwarth, A. Leitenstorfer, B. Lipphardt, and H. Schnatz, *Opt. Express* **13**, 904 (2005).

[146] I. Hartl, G. Imeshev, M. Fermann, C. Langrock, and M. Fejer, *Opt. Express* **13**, 6490 (2005).

[147] E. Benkler, H. Telle, A. Zach, and F. Tauser, *Opt. Express* **13**, 5662 (2005).

[148] J. J. McFerran, W. C. Swann, B. R. Washburn, and N. R. Newbury, *Opt. Lett.* **31**, 1997 (2006).

[149] H. Inaba, Y. Daimon, F.-L. Hong, A. Onae, K. Minoshima, T. R. Schibli, H. Matsumoto, M. Hirano, T. Okuno, M. Onishi, and M. Nakazawa, *Opt. Express* **14**, 5223 (2006).

[150] W. C. Swann, J. J. McFerran, I. Coddington, N. R. Newbury, I. Hartl, M. E. Fermann, P. S. Westbrook, J. W. Nicholson, K. S. Feder, C. Langrock, and M. M. Fejer, *Opt. Lett.* **31**, 1997 (2006).

[151] J.-L. Peng and R.-H. Shu, *Opt. Express* **15**, 4485 (2007).

[152] N. R. Newbury and W. C. Swann, *J. Opt. Soc. Am. B* **24**, 1756 (2007).

[153] C. M. Smith, N. Venkataraman, M. T. Gallagher, D. Müller, J. A. West, N. F. Borrelli, D. C. Allan, and K. W. Koch, *Nature*, **424**, 657 (2003).

[154] T. P. Hansen, J. Broeng, C. Jakobsen, G. Vienne, H. R. Simonsen, M. D. Nielsen, P. M. W. Skovgaard, J. R. Folkenberg, and A. Bjarklev, *J. Lightwave Technol.* **22**, 11 (2004).

[155] G. Humbert, J. C. Knight, G. Bouwmans, P. St. J. Russell, D. P. Williams, P. J. Roberts, and B. J. Mangan, *Opt. Express* **12**, 1477 (2004).

[156] J. A. West, C. M. Smith, N. F. Borrelli, D. C. Allen, and K. W. Koch, *Opt. Express* **12**, 1485 (2004).

[157] P. J. Roberts, F. Couny, H. Sabert, B. J. Mangan, D. P. Williams, L. Farr, M. W. Mason, A. Tomlinson, T. A. Birks, J. C. Knight, and P. St. J. Russell, *Opt. Express* **13**, 236 (2005).

[158] L. Vincetti, F. Poli, and S. Selleri, *IEEE Photon. Technol. Lett.* **18**, 508 (2006).

[159] F. Benabid, *Philo. Trans. Royal Soc. A*, **364**, 3439 (2006).

[160] F. Luan, J. C. Knight, P. St. J. Russell, S. Campbell, D. Xiao, D. T. Reid, B. J. Mangan, D. P. Williams, and P. J. Roberts, *Opt. Express* **12**, 835 (2004).

[161] D. G. Ouzounov, F. R. Ahmad, D. Muller, N. Venkataramen, M. T. Gallagher, M. G. Thomas, J. Silcox, K. W. Koch, and A. L. Gaeta, *Science*, **301**, 1702 (2003).

[162] S. O. Konorov, D. A. Sidorov-Biryukov, I. Bugar, D. Chorvat, Jr., D. Chorvat, E. E. Serebryannikov, M. J. Bloemer, M. Scalora, R. B. Miles, and A. M. Zheltikov, *Phys. Rev. A* **70**, 023807 (2004).

[163] J. Shephard, J. Jones, D. Hand, G. Bouwmans, J. Knight, P. Russell, and B. Mangan, *Opt. Express* **12**, 717 (2004).

[164] J. D. Shephard, F. Couny, P. St. J. Russell, J. D. C. Jones, J. C. Knight, and D. P. Hand, *Appl. Opt.* **44**, 4582 (2005).

[165] S. Joshi, A. P. Yalin, and A. Galvanauskas, *Appl. Opt.* **46**, 4057 (2007).

[166] S. O. Konorov, V. P. Mitrokhin, A. B. Fedotov, D. A. Sidorov-Biryukov, V. I. Beloglazov, N. B. Skibina, A. V. Shcherbakov, E. Wintner, M. Scalora, and A. M. Zheltikov, *Appl. Opt.* **43**, 2251 (2004).

[167] P. J. Roberts, D. P. Williams, B. J. Mangan, H. Sabert, F. Couny, W. J. Wadsworth, T. A. Birks, J. C. Knight, and P. St. J. Russell, *Opt. Express* **13**, 8277 (2005).

[168] N. J. Florous, K. Saitoh, T. Murao, and M. Koshiba, *Opt. Express* **14**, 4861 (2006).

[169] P. J. Roberts, D. P. Williams, H. Sabert, B. J. Mangan, D. M. Bird, T. A. Birks, J. C. Knight, and P. St. J. Russell, *Opt. Express* **14**, 7329 (2006).

[170] T. Murao, K. Saitoh, N. J. Florous, and M. Koshiba, *Opt. Express* **15**, 4268 (2007).

[171] K. Kurokawa, K. Tajima, K. Tsujikawa, K. Nakajima, T. Matsui, I. Sankawa, and T. Haibara, *J. Lightwave Technol.* **24**, 32 (2006).

[172] F. Benabid, J. C. Knight, and P. St. J. Russell, *Opt. Express* **10**, 1195 (2002).

[173] F. Benabid, J. C. Knight, G. Antonopoulos, and P. St. J. Russell, *Science*, **298**, 399 (2002).

[174] S. O. Konorov, A. B. Fedotov, and A. M. Zheltikov, *Opt. Lett.* **28**, 1448 (2003).

[175] F. Benabid, G. Bouwmans, J. C. Knight, P. St. J. Russell, and F. Couny, *Phys. Rev. Lett.* **93** 123903 (2004).

[176] S. O. Konorov, A. M. Zheltikov, P. Zhou, A. P. Tarasevitch, and D. von der Linde, *Opt. Lett.* **29**, 1521 (2004).

[177] S. O. Konorov, A. B. Fedotov, A. M. Zheltikov, and R. B. Miles, *J. Opt. Soc. Am. B* **22**, 2049 (2005).

[178] S. Ghosh, J. E. Sharping, D. G. Ouzounov, and A. L. Gaeta, *Phys. Rev. Lett.* **94**, 093902 (2005).

[179] F. Benabid, P. S. Light, F. Couny, and P. St. J. Russell, *Opt. Express* **13**, 5694 (2005).

[180] D. Ouzounov, C. Hensley, A. Gaeta, N. Venkateraman, M. Gallagher, and K. Koch, *Opt. Express* **13**, 6153 (2005).

[181] F. Benabid, F. Couny, J. C. Knight, T. A. Birks, and P. St. J. Russell, *Nature* **434**, 488 (2005).

[182] F. Benabid, G. Antonopoulos, J. C. Knight, and P. St. J. Russell, *Phys. Rev. Lett.* **95**, 213903 (2005).

[183] S. Yiou, P. Delaye, A. Rouvie, J. Chinaud, R. Frey, G. Roosen, P. Viale, S. Février, P. Roy, J.-L. Auguste, and J.-M. Blondy, *Opt. Express* **13**, 4786 (2005).

[184] S. Lebrun, P.e Delaye, R. Frey, and G. Roosen, *Opt. Lett.* **32**, 337 (2007).

[185] R. Zhang, J. Teipel, and H. Giessen, *Opt. Express* **14**, 6800 (2006).

Chapter 10

Quantum Applications

It may appear surprising at first that the nonlinear effects in optical fibers have found applications for observing the quantum phenomena. Quantum effects in optics often manifest at low intensity levels, because the relative strength of quantum fluctuations scales inversely with the number of photons. In contrast, fiber nonlinearity is relatively weak and its use requires high intensity levels. Nevertheless, it was realized during the 1980s that optical fibers would be useful for observing quantum phenomena such as squeezing because they can confine light in the form of a single spatial mode over long distances. More recently, optical fibers have attracted attention for generating photon pairs that are correlated in the quantum sense, and therefore they can be employed for applications such as quantum cryptography that require entangled photons. This chapter covers quantum applications in which the nonlinear effects such as self-and cross-phase modulations (SPM and XPM) and four-wave mixing (FWM) are exploited. Section 10.1 provides the mathematical apparatus needed for including quantum noise while treating propagation of the optical pulses inside optical fibers. Section 10.2 focuses on several types of noise squeezing that can be realized using optical fibers. Section 10.3 deals with quantum nondemolition schemes. The important topic of quantum entanglement is covered in Section 10.4 with emphasis on the use of fibers for generating polarization-entangled photon pairs. Continuous-variable entanglement is also covered in this section. Section 10.5 focuses on a specific application of such photon pairs in the emerging field of quantum cryptography.

10.1 Quantum Theory of Pulse Propagation

As we have seen in earlier chapters, most nonlinear effects inside optical fibers are governed by the nonlinear Schrödinger (NLS) equation. Therefore, the first task is to find a quantum version of this equation in which the complex field amplitude $A(z,t)$ is replaced with an operator. Although this is not an easy task because of the presence of the nonlinear term $|A|^2A$, such an equation has been derived using several approaches [1]–[4].

10.1.1   Quantum Nonlinear Schrödinger Equation

In one approach [2], the classical NLS equation (3.1.11) is first written in the frequency domain using the Fourier transform

$$\tilde{A}(z,\omega) = \int_{-\infty}^{\infty} \frac{A(z,t)}{\sqrt{\hbar\omega_0}} \exp(i\omega t)\, dt, \tag{10.1.1}$$

where $\hbar\omega_0$ represents the photon energy at the carrier frequency ω_0 of the pulse. As will be apparent soon, the use of this factor helps us to write various commutation relations in a standard form. In the frequency domain, the NLS equation involves a convolution because of the nonlinear term and is given by

$$\frac{\partial \tilde{A}}{\partial z} = \frac{i\beta_2}{2}\omega^2 \tilde{A}(z,\omega) + \frac{i\kappa}{4\pi^2} \int\!\!\int_{-\infty}^{\infty} \tilde{A}^*(z,\omega_1)\tilde{A}(z,\omega_2)\tilde{A}(z,\omega+\omega_1-\omega_2)\, d\omega_1\, d\omega_2, \tag{10.1.2}$$

where $\kappa = \gamma\hbar\omega_0$ plays the role of the nonlinear parameter.

To quantize Eq. (10.1.2), we replace the classical field amplitude $\tilde{A}(z,\omega)$ with the operator $\hat{A}(z,\omega)$ such that it satisfies the commutation relations

$$[\hat{A}(z,\omega),\, \hat{A}^\dagger(z,\omega')] = \delta(\omega-\omega'), \tag{10.1.3}$$

$$[\hat{A}(z,\omega),\, \hat{A}(z,\omega')] = 0, \tag{10.1.4}$$

$$[\hat{A}^\dagger(z,\omega),\, \hat{A}^\dagger(z,\omega')] = 0, \tag{10.1.5}$$

In the quantized version of Eq. (10.1.2), $\tilde{A}(z,\omega)$ is replaced with the operator $\hat{A}(z,\omega)$. Taking the inverse the Fourier transform of this equation, we obtain the following quantum NLS equation [2]:

$$i\frac{\partial \hat{A}}{\partial z} - \frac{\beta_2}{2}\frac{\partial^2 \hat{A}}{\partial t^2} + \kappa \hat{A}^\dagger \hat{A}\hat{A} = 0, \tag{10.1.6}$$

where the operator $\hat{A}(z,t)$ satisfies the commutation relations

$$[\hat{A}(z,t),\, \hat{A}^\dagger(z,t')] = \delta(t-t'), \tag{10.1.7}$$

$$[\hat{A}(z,t),\, \hat{A}(z,t')] = 0, \tag{10.1.8}$$

$$[\hat{A}^\dagger(z,t),\, \hat{A}^\dagger(z,t')] = 0. \tag{10.1.9}$$

Physically, the operator $\hat{A}^\dagger \hat{A}$ represents the photon flux (number of photons per unit time), and the total number of photons within the pulse is obtained by integrating this operator over time.

Quantum problems deal with a Hamiltonian $\hat{H}$ that describes evolution of the quantum state $|\psi\rangle$ of a system through the standard Schrödinger equation [5],

$$i\hbar\frac{\partial |\psi\rangle}{\partial z} = \hat{H}|\psi\rangle. \tag{10.1.10}$$

It turns out that Eq. (10.1.6) can be written in the form of an Heisenberg equation of motion,

$$i\hbar\frac{\partial \hat{A}}{\partial z} = [\hat{A}, \hat{H}], \tag{10.1.11}$$

where the so-called Hamiltonian $\hat{H}$ is given by

$$\hat{H} = -\frac{\hbar \beta_2}{2} \int_{-\infty}^{\infty} \frac{\partial \hat{A}^{\dagger}}{\partial t} \frac{\partial \hat{A}}{\partial t} dt - \frac{\hbar \kappa}{2} \int_{-\infty}^{\infty} \hat{A}^{\dagger 2}(z,t) \hat{A}^2(z,t) dt. \tag{10.1.12}$$

A problem with the quantization of Eq. (10.1.2) is related to the fact that the classical NLS equation in the fiber context describes evolution of an optical field along the fiber length [6], whereas quantum mechanics deals with the evolution in time. Because of this, the Hamiltonian in Eq. (10.1.12) does not have units of energy. If necessary, one can introduce a temporal evolution variable, $\tau = z/v_g$, through a linear transformation [2]. In this chapter, we continue to use z but think of it as a temporal variable. It should be noted that we have used the variable t in this section in place of the reduced time T used in earlier chapters (to avoid conflict with T used to denote absolute temperature). This variable describes pulse shape and should not be confused with the evolution temporal variable τ that governs the time taken by the pulse to arrive at a certain distance z within the fiber.

10.1.2 Quantum Theory of Self-Phase Modulation

Although the dispersive effects governed by β_2 in Eq. (10.1.6) are often important and must be included for solitons, considerable insight is gained by first ignoring them and focusing on the simplest case of pure SPM [7]–[13]. Setting $\beta_2 = 0$ in Eq. (10.1.6), the SPM effects are governed by

$$\frac{\partial \hat{A}}{\partial z} = i\kappa \hat{A}^{\dagger}(z,t) \hat{A}(z,t) \hat{A}(z,t). \tag{10.1.13}$$

This equation can be solved immediately by noting that SPM affects only the phase of light and does not change the photon flux governed by the operator $\hat{A}^{\dagger}\hat{A}$. Mathematically, it is easy to verify that $\hat{A}^{\dagger}\hat{A}$ is preserved during propagation inside a fiber (in the absence of losses) because this operator commutes with the Hamiltonian given in Eq. (10.1.12). The solution of Eq. (10.1.13) is thus given by

$$\hat{A}(z,t) = \exp[i\kappa z \hat{A}^{\dagger}(0,t) \hat{A}(0,t)] \hat{A}(0,t). \tag{10.1.14}$$

To study how SPM affects an optical field initially in a coherent state $|\alpha(t)\rangle$ [5], we need to calculate the average value,

$$\bar{A}(z,t) = \langle \alpha(t) | \hat{A}(z,t) | \alpha(t) \rangle, \tag{10.1.15}$$

using $\hat{A}(0,t)|\alpha(t)\rangle = \alpha(t)|\alpha(t)\rangle$, where $\alpha(t)$ is the input field amplitude. This average is not well defined for an instantaneously responding nonlinear medium, as has been noted in several studies [9]–[13]. In one approach [11], one considers a small time interval τ and writes the exponential operator in Eq. (10.1.14) in a normal-ordered form as

$$\exp\left[\frac{i\kappa z}{\tau} \int_{t}^{t+\tau} \hat{A}^{\dagger}(0,t') \hat{A}(0,t') dt'\right] =\, :\exp\left[(e^{i\kappa z/\tau} - 1) \int_{t}^{t+\tau} \hat{A}^{\dagger}(0,t') \hat{A}(0,t') dt'\right]:, \tag{10.1.16}$$

where the two colons around an operator indicate normal ordering [5]. Using this form, the average in Eq. (10.1.14) becomes

$$\bar{A}(z,t) = \lim_{\tau \to 0} \exp\left[(e^{i\kappa z/\tau} - 1)\int_t^{t+\tau} |\alpha(t')|^2 dt'\right]\alpha(t). \tag{10.1.17}$$

This limit is not well defined and leads to an erroneous conclusion that the input field remains unchanged during propagation [11].

The physical origin of the difficulty lies in the observation that quantum fluctuations in an instantaneously responding nonlinear medium can occur over an infinite bandwidth. Fortunately, there is a simple solution to this problem if we note that any realistic medium responds on a finite time scale [9]. Such a finite response time can be included by replacing Eq. (10.1.13) with

$$\frac{\partial \hat{A}}{\partial z} = i\kappa\left[\int_{-\infty}^t dt' \, R(t-t')\hat{A}^\dagger(z,t')\hat{A}(z,t')\right]\hat{A}(z,t) + i\hat{\Gamma}_R\hat{A}(z,t). \tag{10.1.18}$$

where $R(t)$ is the response function associated with the fiber nonlinearity, normalized such that $\int_0^\infty R(t)\,dt = 1$ and $\hat{\Gamma}_R(z,t)$ is a noise operator that includes any fluctuations associated with this nonlinear response [14]. In the case of optical fibers, Raman scattering involving optical photons provides a delayed response, and the noise is governed by the phonon population.

Ignoring the noise term in Eq. (10.1.18) for the moment (it is considered later in this section), we can integrate this equation as before. In the normal-ordered form, the result is

$$\hat{A}(z,t) =: \exp\left[\int_{-\infty}^t dt' \left(e^{i\kappa z R(t-t')} - 1\right)\hat{A}^\dagger(0,t')\hat{A}(0,t')\right] : \hat{A}(0,t). \tag{10.1.19}$$

The average indicated in Eq. (10.1.14) can now be performed without difficulty by just replacing $\hat{A}(0,t)$ with $\alpha(t)$. The important point is that the result is different from that obtained in the classical case. To see under what conditions it reduces to the classical result, we employ a simple model for the nonlinear response function and assume that $R(t) = 1/\tau_0$ in the range $0 < t < \tau_0$ and zero otherwise [9]. Here, τ_0 is the response time of electrons with a value ~ 1 fs.

Assuming that the input photon flux $|\alpha(t)|^2$ does not change much over this tiny interval, the integration in Eq. (10.1.19) can be carried out analytically to obtain

$$\bar{A}(z,t) = \exp[(e^{i\phi_v} - 1)\tau_0|\alpha(t)|^2]\alpha(t), \tag{10.1.20}$$

where

$$\phi_v = \kappa z/\tau_0 = \gamma z(\hbar\omega_0/\tau_0) \tag{10.1.21}$$

is the nonlinear phase shift induced by the vacuum fluctuations over a bandwidth $1/\tau_0$. For a standard fiber of 1-km-length, its magnitude is $\phi_v \approx 2 \times 10^{-4}$ if we use $\gamma = 1.6$ W^{-1}/km and $\hbar\omega_0 = 0.8$ eV for the photon energy in the wavelength region near 1550 nm, together with $\tau_0 = 1$ fs. Notice that $\phi_v \to \infty$ in the limit $\tau_0 \to 0$, indicating once again the need to include the finite response time of the nonlinear medium.

It is easy to show that the quantum result in Eq. (10.1.20) is quite close to the classical theory of SPM. Because of a relatively small value of ϕ_v, we can replace $e^{i\phi_v} - 1$ with $i\phi_v$ in this equation, resulting in

$$\bar{A}(z,t) = \exp[i\kappa z|\alpha(t)|^2]\alpha(t) = \exp[i\gamma z P(t)]\alpha(t), \qquad (10.1.22)$$

where $P(t) = \hbar\omega_0|\alpha(t)|^2$ is the optical power. As expected, the classical phase shift, $\phi_c(t) = \gamma z P(t)$ depends linearly on the optical power. As an example, its value for a standard fiber of 1-km-length is $\phi_c = 1.6$ at the pulse center for a pulse with a peak power of 1 W, using again $\gamma = 1.6$ W^{-1}/km. Even though $\phi_v \ll \phi_c$ under typical experimental conditions, quantum noise has considerable implications for the evolution of optical pulses in the presence of SPM inside an optical fiber [9]. In particular, it leads to squeezing of noise, a phenomenon discussed in Section 10.2.

10.1.3 Generalized NLS Equation

As we have seen, a proper quantum description of pulse propagation inside optical fibers must include the finite response time associated with the third-order nonlinearity. In the case of optical fibers, the Raman contribution to the third-order nonlinearity is known to be important, even in a classical context, as it leads to Raman-induced spectral shifts for ultrashort optical pulses [6]. In the quantum case, spontaneous Raman scattering acts as an additional noise source, and it must be included for a realistic description [14]–[16]. A more complete quantum theory describing evolution of pulses inside optical fibers has been developed in recent years and leads to the following generalized NLS equation [3]:

$$\frac{\partial \hat{A}}{\partial z} = \frac{i\beta_2}{2}\frac{\partial^2 \hat{A}}{\partial t^2} + i\kappa \int_{-\infty}^{t} dt' R(t-t')\hat{A}^{\dagger}(z,t')\hat{A}(z,t')\hat{A}(z,t) + i\hat{\Gamma}_R(z,t)\hat{A}(z,t), \quad (10.1.23)$$

where the nonlinear response function $R(t)$ includes both the nearly instantaneous electronic response and the delayed Raman response though the expression [6]

$$R(t) = (1 - f_R)h_e(t) + f_R h_R(t). \qquad (10.1.24)$$

Here f_R represents the fractional contribution of the delayed Raman response to nonlinear polarization. The functional form of the Raman response function $h_R(t)$ is set by vibrations of silica molecules induced by the optical field that may last for more than 100 fs in the case of optical fibers. In contrast, the electronic response, governed by $h_e(t)$, decays down to zero on a time scale ~ 1 fs because electrons respond so fast that they can follow optical oscillations.

Equation (10.1.23) is similar to its classical counterpart, used often in dealing with the nonlinear phenomena in optical fibers [6]. Indeed, if the last term is ignored and the operator $\hat{A}$ is replaced with A, we recover the classical NLS equation. In the quantum description, the last term is needed to ensure that the commutation relation in Eq. (10.1.7) is satisfied at any distance z along the fiber. Even though the average value of the noise operator $\hat{\Gamma}_R$ vanishes, the correlation function of its Fourier transform $\hat{\Gamma}_R(z,\omega)$ must satisfy [3]

$$\langle \hat{\Gamma}_R^{\dagger}(z',\omega')\hat{\Gamma}_R(z,\omega)\rangle = g_R(|\omega|)\bar{N}_p^{-1}[n_{\text{th}}(|\omega|) + \Theta(-\omega)]\delta(z-z')\delta(\omega-\omega'), \quad (10.1.25)$$

where $g_R(\omega)$ is the Raman gain spectrum, $\bar{N}_p$ is the average number of photons within the pulse, and $n_{th}(\omega) = [\exp(\hbar\omega/k_BT) - 1]^{-1}$ is the Bose distribution of phonons at a finite absolute temperature T. The phonon population appears because spontaneous Raman scattering depends on it. The step function $\Theta(-\omega)$ in Eq. (10.1.25) equals 1 for Stokes photons ($\omega < 0$) but is zero for anti-Stokes ones ($\omega > 0$) to ensure that no up-shifted photons are created in the absence of thermal phonons.

Equation (10.1.23), although quite general, does not include the power loss (or gain) that may occur within the fiber. This can be included by adding another noise operator $\hat{\Gamma}(z,t)$ on the right side of this equation with a correlation function that depends on the magnitude of the loss parameter α [3]. It is important to note that, whereas fluctuations associated with linear losses appear as an additive noise, those associated with spontaneous Raman scattering add multiplicative noise. Although Eq. (10.1.23) is often necessary for ultrashort pulses, it can be simplified considerably for pulses broader than 1 ps. The Raman response function $h_R(t)$ for such pulses can be approximated by a delta function, and Eq. (10.1.23) reduces to the much simpler quantum NLS equation given in Eq. (10.1.6), provided we ignore the noise term as well. The dispersion term in this equation leads to pulse broadening that removes the infinite-bandwidth problem noted earlier in the case of pure SPM [10]. As a result, Eq. (10.1.6) provides a valid quantum description of pulses propagating inside optical fibers.

10.1.4 Quantum Solitons

The classical NLS equation has specific analytic solutions, known as *optical solitons*, that have proved quite beneficial in understanding nonlinear phenomena inside optical fibers [6]. In particular, a fundamental soliton is formed when the pulse and fiber parameters satisfy the condition

$$N^2 = \gamma P_0 T_0^2 / |\beta_2| = \kappa F_0 T_0^2 / |\beta_2| = 1, \qquad (10.1.26)$$

where $F_0 = P_0/(\hbar\omega_0)$ is the photon flux and N is the soliton order. Fundamental solitons are characterized with an input field $A(0,t) = \sqrt{P_0} \, \text{sech}(t/T_0)$ and have the property that they maintain their shape and width as they propagate inside the fiber. A question of considerable interest is whether the quantum NLS equation (10.1.6) also has solutions that are analogous to classical solitons. The answer turned out to be positive. Such solutions are referred to as *quantum solitons* and have properties that are quite distinct from their classical counterparts [2].

The classical NLS equation is solved analytically with the inverse scattering method [6]. It turns out that this method can also be used for solving the quantum NLS equation [17], and it was used in a 1989 study to discuss the properties of quantum solitons [18]. In another approach [2], the time-dependent Hartree approximation was used to solve the Schrödinger equation (10.1.10) for the wave function $|\psi\rangle$ with the Hamiltonian given in Eq. (10.1.12). The phase-space techniques have also been used to convert the quantum NLS equation into a Fokker–Planck equation or into a set of stochastic Langevin equations [19]. In the so-called positive-P representation, the quantum NLS equation is replaced with a classical stochastic NLS equation in which the quantum effects are included through a noise source [1].

To understand how quantum noise affects a soliton, it is helpful to consider the case of a classical fundamental soliton ($N = 1$). Such a soliton is characterized by four parameters and has the general form [18]

$$A_c(z,t) = (|\beta_2|/\kappa)^{1/2}\eta \, \text{sech}[\eta(t - t_s + |\beta_2|\xi z)] \exp[i|\beta_2|(\eta^2 - \xi^2)z/2 - i\xi(t - t_s) + i\phi], \tag{10.1.27}$$

where η governs the width (and amplitude) of the soliton, ξ is a frequency shift related to its momentum, t_s represents its initial position, and ϕ is a constant phase. In the quantum case, these four parameters should be treated as operators. The photon-number operator, defined as

$$\hat{N}_p = \int_{-\infty}^{\infty} \hat{A}^\dagger(z,t)\hat{A}(z,t)\, dt, \tag{10.1.28}$$

is related to the operator η by the relation $\hat{N}_p = (2|\beta_2|/\kappa)\hat{\eta}$.

It turns out that $\hat{N}_p$ and $\hat{\phi}$ form a pair of conjugate variables [18] that satisfy the commutation relation $[\hat{N}_p, \hat{\phi}] = 1$. The operators $\hat{t}_s$ and $\hat{\xi}$ form the second pair of conjugate variables. If a quantum soliton is in an eigenstate $|N\rangle$ of the photon-number operator, its amplitude and width are uniquely determined but its phase is completely random. Similarly, if that soliton is in an eigenstate $|\xi\rangle$ of the "momentum" operator, its central position t_s cannot be precise, in accordance with the Heisenberg uncertainty principle. Of course, in general, the soliton does not have to be in these eigenstates. In that case, one can expand the soliton wave function into a linear superposition of different $|N, \xi\rangle$ states. The four soliton parameters in this case fluctuate around their average values. One consequence of such fluctuations is that, in contrast with a classical soliton, a quantum soliton broadens slowly as it propagates down the fiber. Of course, this broadening depends on the average number of photons contained inside the optical pulse propagating as a soliton. The distance over which soliton width doubles scales with $\langle \hat{N}_p \rangle$ as $z_d \sim \langle \hat{N}_p \rangle L_D$, where L_D is the dispersion length. When $L_D > 10$ km, this distance exceeds 10^5 km for a soliton containing only 10^4 photons.

A perturbative approach was developed in 1990 for analyzing quantum solitons [20]. It is based on the observation that perturbations induced on the soliton by the quantum noise are relatively small. One can thus write the general solution of the quantum NLS equation (10.1.6) in the form

$$\hat{A}(z,t) = A_c(z,t) + \hat{a}(z,t), \tag{10.1.29}$$

where $A_c(z,t)$ is the classical solution given in Eq. (10.1.27) and $\hat{a}(z,t)$ is a small perturbation induced by quantum noise. Because of its smallness, we can linearize Eq. (10.1.6) in terms of $\hat{a}(z,t)$ and obtain

$$i\frac{\partial \hat{a}}{\partial z} = \frac{i\beta_2}{2}\frac{\partial^2 \hat{a}}{\partial t^2} - \kappa[|A_c(z,t)|^2 \hat{a}^\dagger(z,t) + 2|A_c(z,t)|^2 \hat{a}(z,t)]. \tag{10.1.30}$$

This operator equation can be solved easily because of its linear nature. Moreover, the solution can be expanded as [20]

$$\hat{a}(z,t) = \frac{\partial A_c}{\partial \eta}\Delta\hat{\eta} + \frac{\partial A_c}{\partial \phi}\Delta\hat{\phi} + \frac{\partial A_c}{\partial \xi}\Delta\hat{\xi} + \frac{\partial A_c}{\partial t_s}\Delta\hat{t}_s + \Delta\hat{a}(z,t), \tag{10.1.31}$$

where $\Delta\hat{\eta}$, $\Delta\hat{\phi}$, $\Delta\hat{\xi}$, and $\Delta\hat{t}_s$ represent quantum fluctuations in the four soliton parameters and $\Delta\hat{a}(z,t)$ is the radiation (dispersive waves) emitted by a soliton because of these fluctuations. Perturbation theory shows that fluctuations in the soliton parameters evolve with z as

$$\Delta\hat{\eta}(z) = \Delta\hat{\eta}_0, \qquad \Delta\hat{\phi}(z) = \Delta\hat{\phi}_0 + 2\eta\Delta\hat{\eta}_0|\beta_2|z, \qquad (10.1.32)$$

$$\Delta\hat{\xi}(z) = \Delta\hat{\xi}_0, \qquad \Delta\hat{t}_s(z) = \Delta\hat{t}_{s0} + |\beta_2|\Delta\hat{\xi}_0 z, \qquad (10.1.33)$$

where the subscript 0 denotes the initial value at $z = 0$. Fluctuations in the number of photons are obtained from the relation $\Delta\hat{N}_p = (2|\beta_2|/\kappa)\Delta\hat{\eta}$. Note that, even though fluctuations in the photon number and soliton momentum (or frequency ξ) do not change during propagation, fluctuations in both the phase and the position of the soliton increase continuously.

The initial values of the variance of fluctuations in the soliton parameter can also be calculated by performing averages with respect to the vacuum state. For example, one can show that the variance $\langle(\Delta\hat{N}_p)^2\rangle$ equals the average number of photons $\langle\hat{N}_p\rangle$. This is expected as fluctuations in $\hat{N}_p$ follow the Poisson distribution [5]. The initial value of the variance of phase fluctuations, found to be $\langle(\Delta\hat{\phi}_0)^2\rangle = 0.6075/\bar{N}$, is more than twice that expected for a minimum uncertainty state [20]. Changes in phase fluctuations with propagation are related to noise squeezing, a topic we turn to next.

10.2 Squeezing of Quantum Noise

Optical fibers were first used in 1985 for observing quantum effects in the context of a phenomenon known as *squeezing* [21]. Since then, quantum squeezing has been studied in several contexts [22]–[26]. It refers to the process of generating the special states of an electromagnetic field for which noise fluctuations in some frequency range are reduced below the shot-noise level that is imposed by quantum mechanics in the absence of squeezing.

10.2.1 Physics behind Quadrature Squeezing

Several kinds of squeezing are possible depending on the physical parameter that exhibits noise below the shot-noise level. For example, photon-number squeezing occurs if the variance of the photon-number operator is less than the shot-noise level. The squeezing that attracted the attention initially is referred to as the *quadrature squeezing*. To understand its meaning, note first that a monochromatic optical field, written in the phasor notation as $E(t) = A_0\exp(i\phi - i\omega_0 t)$ at a specific frequency ω_0, is characterized in terms of its amplitude A_0 and phase ϕ. The real and imaginary parts of $E(t)$ correspond to two field quadratures, both of which fluctuate around their mean values because of quantum noise.

When the optical field is quantized, the complex amplitude $A = A_0 e^{i\phi}$ is related to the expectation value of the annihilation operator $\hat{a}$. The two quadrature components $\hat{X}$ and $\hat{Y}$ can then be introduced as

$$\hat{X} = \hat{a}^\dagger e^{i\theta} + \hat{a}e^{-i\theta}, \qquad \hat{Y} = i(\hat{a}^\dagger e^{i\theta} - \hat{a}e^{-i\theta}), \qquad (10.2.1)$$

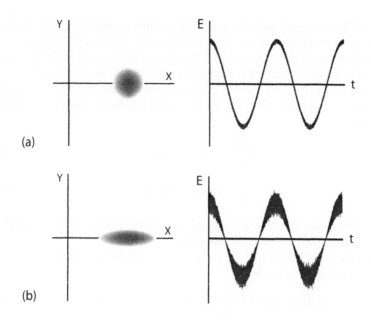

Figure 10.1: (a) Coherent and (b) squeezed quantum states. Noise distribution along the two quadratures is shown together with the time dependence of the optical field.

where θ is an arbitrary angle. Using the commutation relation $[\hat{a}, \hat{a}^\dagger] = 1$, it is easy to show that $[\hat{X}, \hat{Y}] = 2i$ for any value of θ. For any quantum state $|\psi\rangle$, the Heisenberg uncertainty principle imposes the constraint $\sigma_X^2 \sigma_Y^2 \geq 1$, where the noise variance is calculated using $\sigma_X^2 = \langle \psi | (\hat{X} - \bar{X})^2 | \psi \rangle$ and $\bar{X} = \langle \psi | X | \psi \rangle$ is the mean value. In the case of a coherent state $\sigma_X = \sigma_Y$.

Figure 10.1(a) shows a coherent state schematically with the same amount of noise in its two quadratures, resulting in a circular shape of fluctuations around the mean value. It turns out that the nonlinear effects inside a medium can produce a squeezed state similar to that shown in part (b). Such states have the property that quantum fluctuations in one quadrature are reduced below those of a coherent state. Of course, this reduction in noise can occur only if the noise in the other quadrature increases enough that the Heisenberg uncertainty relation remains satisfied. In the case of Figure 10.1(b), fluctuations decrease in the Y quadrature but are enhanced in the X quadrature, resulting in an elongated noise ellipse. A phase-sensitive detection scheme (homodyne or heterodyne) is employed to observe the noise reduction in one quadrature.

10.2.2   FWM-Induced Quadrature Squeezing

A nonlinear process is required to transform a coherent state into a squeezed state. The use of FWM for squeezing was suggested during the 1980s, and a detailed theory was developed in a 1985 paper [21]. The basic scheme is similar to that shown in Figure 8.6 to the extent that a pump beam initiates the FWM process inside an optical fiber, but no signal is launched at the input end. Rather, vacuum fluctuations provide the initial seed for the signal and idler fields. Such a process may be called *spontaneous FWM* to differentiate it from its stimulated version. Squeezing occurs because the noise components at the signal and idler frequencies are coupled through the fiber nonlinearity. Mathematically, it is governed by two equations, similar to those in Eqs. (8.1.7) and (8.1.8). The only difference is that A_s and A_i should be treated as operators. Introducing $\hat{a}_j = \hat{A}_j \exp(-i\delta z/2)$ as the new variables, these equations can be written in the form

$$\frac{d\hat{a}_s}{dz} = \frac{i\delta}{2}\hat{a}_s + 2i\gamma A_p^2 \hat{a}_i^\dagger, \tag{10.2.2}$$

$$\frac{d\hat{a}_i}{dz} = \frac{i\delta}{2}\hat{a}_i + 2i\gamma A_p^2 \hat{a}_s^\dagger, \tag{10.2.3}$$

Vacuum fluctuations are included through the commutation relation, $[\hat{a}_j, \hat{a}_k^\dagger] = \delta_{jk}$ (where $j,k = s$ or i) satisfied by the operators $\hat{a}_s$ and $\hat{a}_i$.

Equations (10.2.2) and (10.2.3) can be solved easily because of their linear nature, and the solution is given by [6]

$$\hat{a}_s(z) = \hat{a}_s(0)[\cosh(gz) + (i\delta/2g)\sinh(gz)] + i(\gamma/g)A_p^2\hat{a}_i^\dagger(0)\sinh(gz), \tag{10.2.4}$$

$$\hat{a}_i(z) = \hat{a}_i(0)[\cosh(gz) + (i\delta/2g)\sinh(gz)] + i(\gamma/g)A_p^2\hat{a}_s^\dagger(0)\sinh(gz), \tag{10.2.5}$$

where the parametric gain g is defined as $g = (|\gamma|^2|A_p|^4 - \delta^2/4)^{1/2}$. This solution reduces to that given in Ref. [21] in the case of perfect phase matching ($\delta = 0$). Note that the signal amplitude at a distance z inside the fiber evolves as a linear combination of $\hat{a}_s(0)$ and $\hat{a}_i^\dagger(0)$. It is this feature that is responsible for squeezing. The total field at the output end of a fiber of length L is given by

$$\hat{A}_t(t) = A_p(L) + \hat{a}_s(L)\exp(-i\Omega t) + \hat{a}_i(L)\exp(i\Omega t). \tag{10.2.6}$$

where $\Omega \equiv \omega_s - \omega_p$ is the signal detuning from the pump frequency ω_p.

From a physical standpoint, squeezing can be understood as deamplification of signal and idler waves for certain values of the relative phase between them. A phase-sensitive detection scheme is employed, and the phase of the local oscillator is adjusted to change the relative phase. In practice, the pump itself is used as a local oscillator with an adjustable phase θ. Its beating with the signal and idler fields at a photodetector generates an electric current whose noise power varies with both Ω and θ. In particular, at any Ω, noise power becomes minimum for a specific value of θ.

Figure 10.2 shows this minimum value measured as a function of $\Omega/(2\pi)$ at the output end of a 114-m-long optical fiber in the original 1986 experiment in which a 647-nm CW beam was used as a pump [27]. It was necessary to cool the fiber to liquid-helium temperature to overcome the noise produced by spontaneous guided acoustic-wave Brillouin scattering [28]. Cooling also reduced the threshold of stimulated Brillouin scattering (SBS), which was suppressed by modulating the pump beam at a frequency much larger than the bandwidth of Brillouin-gain spectrum. Thermal Brillouin scattering from guided acoustic waves was still the most limiting factor in the experiment, and it limited both the frequency range and the amount of noise squeezing. The two peaks in Figure 10.2 are due to this scattering. Squeezing occurs in the two

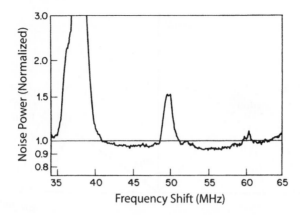

Figure 10.2: Noise power (normalized to the shot-noise level) as a function of detuning from pump under minimum-noise conditions. Reduced noise around 45 and 55 MHz is a manifestation of FWM-induced squeezing in optical fibers. (From Ref. [27]; ©1986 APS.)

spectral bands located around 45 and 55 MHz, where noise power is reduced below the shot noise. We can define the squeezing ratio R_s as the measured noise power relative the shot-noise power. Maximum squeezing in Figure 10.2 occurs near 55 MHz, where $R_s = 0.87$ or -0.6 dB on the decibel scale. A value close to 1 dB was realized later through four-mode squeezing using a dual-pump configuration [29], but the FWM approach is limited in practice because of Brillouin scattering.

10.2.3 SPM-Induced Quadrature Squeezing

A different approach to squeezing makes use of the fact that the SPM-induced phase shift inside an optical fiber can also squeeze the noise associated with an intense input beam [24]. This feature can be understood from Figure 10.3, showing how the noise is affected by SPM as the input beam of average power P_0 propagates through the fiber. The noise initially is the same in the two quadratures of a coherent state, resulting in the circular shape. However, the nonlinear phase shift of $\phi_{NL} = \gamma P_0 L$ at the output of a fiber of length L distorts this circle into an ellipse, because SPM converts power fluctuations into phase fluctuations. This can be seen by noting that a change in P_0 by ΔP introduces an additional change in phase by $\Delta\phi = (\gamma L)\Delta P$. As a result, the phase angle ϕ_{NL} in Figure 10.3 is larger for the upper circle compared with the lower one, resulting in an elliptical shape for noise fluctuations. This kind of squeezing is different from that shown in Figure 10.2, because the input field itself is squeezed rather than a vacuum state (squeezing occurs for $\Omega = 0$).

We can use the solution given in Eq. (10.1.19) to calculate the amount of noise squeezing produced by SPM. Assuming again that the nonlinear response function $R(t) = 1/\tau_0$ in the range $0 < t < \tau_0$ and zero otherwise [9] with $\tau_0 \sim 1$ fs, the solution at the output end of a fiber of length L is given by

$$\hat{A}_o(t) =: \exp\left[(e^{i\phi_\nu - 1})\tau_0\hat{A}^\dagger(0,t)\hat{A}(0,t)\right] : \hat{A}(0,t), \qquad (10.2.7)$$

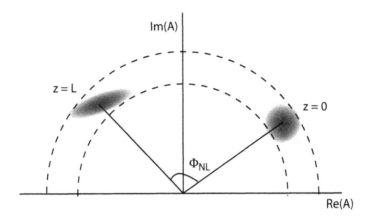

Figure 10.3: Squeezing of a coherent state through SPM-induced phase shift ϕ_{NL}. Shaded circle between two dashed lines, showing the range of input fluctuations, is transformed by SPM into an elongated ellipse.

where $\phi_v = \kappa L/\tau_0$, as defined in Eq. (10.1.21). One of the quadrature components is then obtained using the relation

$$\hat{X} = \hat{A}_o^\dagger e^{i\theta} + \hat{A}_o e^{-i\theta},\qquad(10.2.8)$$

where the phase θ is controlled by a local oscillator. The noise spectrum of this component is given by [12]

$$S(\omega,\theta) = \frac{1}{T}\int_{-T/2}^{T/2}dt\int_{-T/2}^{T/2}dt'[\langle\hat{X}(t)\hat{X}(t')\rangle - \langle\hat{X}(t)\rangle\langle\hat{X}(t')\rangle]\exp[i\omega(t-t')],\quad(10.2.9)$$

where T is the integration time of the spectrum analyzer.

The noise spectrum obtained using Eq. (10.2.7)–(10.2.9) is found to be independent of ω in the range $0 < \omega < 2\pi/\tau_0$. It also does not depend on the integration time T. The resulting expression is still quite complicated [9]. It can be simplified considerably if we make use of the fact that $\phi_v \ll 1$ and keep terms only up to first order in ϕ_v. The result is [12]

$$S(\theta) = 1 + 2\phi_c^2 - 4\phi_c^3\phi_v/3 - \phi_c(2\phi_c + \phi_v - 8\phi_c^2\phi_v/3)\cos\Theta - \phi_c(2 - 5\phi_c\phi_v)\sin\Theta,$$
$$(10.2.10)$$

where $\Theta = 2\theta + 2(\phi_c - \phi_0)$, ϕ_0 is the phase of the input field, and $\phi_c = \gamma L P_0$ is the classical value of the SPM-induced phase shift for an input CW beam with power P_0. The noise spectrum is normalized to the standard quantum limit such that $S = 1$ in the absence of the input beam ($\phi_c = 0$).

The θ-dependence of the noise spectrum indicates that the quantum noise may be above or below the standard quantum limit, depending on the value of θ. Figure 10.4(a) shows this behavior for two values of ϕ_c, using $\phi_v = 2 \times 10^{-4}$ and $\phi_0 = 0$. The maximum and minimum values of S can be deduced by setting $dS/d\Theta = 0$ in Eq. (10.2.10), finding the angle Θ at which they occur, and substituting this value back in

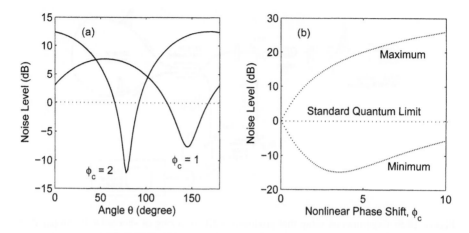

Figure 10.4: (a) Noise variance in one quadrature as a function of the local-oscillator phase θ for several values of the SPM-induced phase shift ϕ_c. (b) Maximum and minimum values of noise variances plotted as a function of ϕ_c. In both cases, $\phi_v = 2 \times 10^{-4}$.

Eq. (10.2.7). The result is given by [12]

$$S_\pm = 1 + 2\phi_c^2 - 4\phi_c^3 \phi_v/3 \pm 2\phi_c(1 + \phi_c^2 - 4\phi_c\phi_v - 8\phi_c^3 \phi_v/3)^{1/2}, \qquad (10.2.11)$$

Figure 10.4(b) shows how the maximum and minimum values of S vary with the magnitude of the SPM-induced phase shift ϕ_c. In practice, ϕ_c can be varied by changing the input power P_0 for a fixed fiber length L. The main point to note is that the quantum noise can be reduced, in theory, by more than 10 dB in one quadrature, but only at the expense of a large increase in the other quadrature.

The preceding analysis ignores the effects of fiber dispersion that must be included when input field is in the form of a short pulse. In the case of anomalous dispersion, the pulse can propagate as a fundamental soliton when its peak power is chosen to ensure that $N = 1$ in Eq. (10.1.26). The linearized theory of quantum fluctuations discussed in Section 10.1.4 has been used to analyze squeezing for fundamental solitons [20]. A numerical approach has also been used for this purpose [30]. The results show that squeezing of quantum noise by more than 10 dB can be realized when the classical nonlinear phase shift ϕ_v exceeds 2 at the peak of the pulse.

It is difficult to observe the SPM-induced squeezing associated with an intense beam or pulse because of a large background. However, this kind of squeezing can be converted into vacuum squeezing by using a Mach–Zehnder or Sagnac interferometer [31]–[36]. The reason is that such devices have two input and two outport ports. When an intense optical beam is launched through one of the input ports, the vacuum noise enters from the other one and interferes with it coherently. One can employ the quantum NLS equation to analyze this situation and find the operators that govern the output from the two ports [31]. The analysis is simpler when the dispersive effects are negligible. In the case of a balanced Sagnac interferometer built using a 3-dB coupler, the intense beam is reflected but the transmitted quantum state represents squeezed

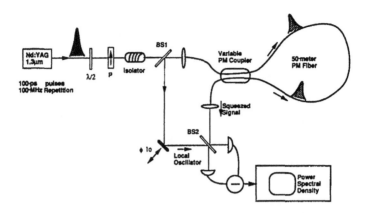

Figure 10.5: Experimental setup that produced 5-dB squeezing of shot noise by propagating 100-ps pulses inside a 50-m-long Sagnac loop. (From Ref. [33]; ©1991 OSA.)

vacuum. As shown schematically in Figure 10.5, one can employ the reflected beam as a local oscillator if its phase can be adjusted. The predicted squeezing of shot noise depends on the magnitude of nonlinear phase shift $\phi_{NL} = \gamma P_0 L$ and can exceed 10 dB under ideal conditions. It is limited to near 7 dB for Gaussian pulses because the nonlinear phase shift is not uniform across the pulse.

In practice, many effects, such as losses, fiber dispersion, and detection efficiency, limit further the magnitude of observed squeezing. The dispersive effects can be used to advantage to form solitons if the wavelength of the input pulse falls in the anomalous-GVD regime. A nonlinear optical-loop mirror (NOLM) acting as a Sagnac interferometer was first employed in a 1991 experiment in which 0.2-ps pulses were propagated inside a 5-m-long fiber loop as solitons. Squeezing by 1.1 dB was observed at room temperature, and its magnitude could be increased to 1.7 dB at the liquid-nitrogen temperature [32]. In another experiment, noise reduction by more than 5 dB could be observed at room temperature by propagating 100-ps pulses whose wavelength coincided nearly with the zero-dispersion wavelength of the fiber [33]. Figure 10.5 shows schematically the experimental setup employed. The Sagnac loop employed 50-m of polarization-maintaining fiber (PMF) in combination with a PMF coupler that ensured 50:50 splitting to within 0.2%.

The XPM inside optical fibers can also be used for squeezing a vacuum state using an intense optical beam. Such a scheme does not even require an interferometer. Figure 10.6 shows the basic idea schematically [36]. The input optical pulse, acting as a pump, is polarized linearly along a principal axis of an isotropic fiber. It affects, through XPM, the quantum noise associated with the vacuum state that is polarized orthogonal to the pump pulse. The XPM-induced phase shift changes the noise circle centered at the origin into an ellipse, similar to the way shown in Figure 10.3. As a result, the vacuum state is squeezed at the output end of the fiber. The attainable squeezing is limited in practice because an ideal isotropic fiber does not exist. Any residual birefringence of the fiber produces slightly different phase shifts as well as group velocities for the light polarized along the two principal axes and is harmful

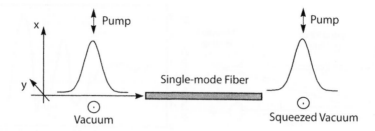

Figure 10.6: Squeezed vacuum produced through an XPM-induced phase shift inside a fiber using an orthogonally polarized pump pulse. (From Ref. [36]; ©1998 OSA.)

for squeezing. In a 1998 experiment, 150-fs pulses with nanojoule energy levels were launched inside a standard fiber whose length of 20 cm was so short that these effects were relatively small. The short fiber also helped in reducing the guided acoustic-wave Brillouin scattering. As a result, the shot noise was found to be squeezed by as much as 3 dB. This value corresponds to a squeezing level of 5 dB when detection efficiency is taken into account.

10.2.4   SPM-Induced Amplitude Squeezing

Amplitude squeezing, also known as *photon-number squeezing*, refers to a reduction in intensity fluctuations associated with an optical signal below the shot-noise level [37]. This type of squeezing does not require a local oscillator (for homodyne or heterodyne detection) and can be observed with direct detection. A practical advantage of direct detection is that frequency chirping and phase fluctuations associated with the optical signal do not affect the photocurrent noise.

The nonlinear phenomena of SPM and XPM affect only the optical phase and thus leave amplitude fluctuations of an optical signal unchanged. However, it was discovered in a 1996 experiment that amplitude squeezing can be realized by propagating optical pulses in the anomalous-GVD regime of an optical fiber and passing them through an optical filter of suitable bandwidth [37]. Figure 10.7 shows the experimental setup in which 2.7-ps pulses were launched inside a 1.5-km-long optical fiber with energies larger than that of a fundamental soliton so that the parameter N governing the soliton order exceeded 1. Pulses narrowed and their spectrum broadened under such conditions. The grating–slit combination acted as an optical filter.

The soliton evolution is quite sensitive to the average value E_p of the input pulse energy, which sets the initial value of the parameter N. As a result, one would expect amplitude squeezing to depend on E_p strongly. As shown in Figure 10.7, this was indeed the case in the 1996 experiment [37]. The amplitude noise of the filtered light was reduced to below the shot-noise level only in a certain range of pulse energies. The maximum observed reduction was 2.3 dB when the filter bandwidth was optimized suitably. Accounting for losses, this value corresponds to 3.7-dB amplitude squeezing. The quantum NLS equation (10.1.6) has been employed to study the extent of amplitude squeezing that is possible with this technique [37]–[41].

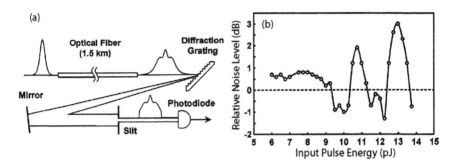

Figure 10.7: (a) Experimental setup used and (b) observed amplitude squeezing as a function of input pulse energy when 2.7-ps pulses are launched inside a 1.5-km-long optical fiber. (From Ref. [37]; ©1996 APS.)

Why does spectral filtering of optical pulses produce amplitude squeezing? The answer is related to how solitons evolve inside an optical fiber [6]. The soliton order N for a pulse of width T_0 and peak power P_0 is defined by the relation $N^2 = \gamma P_0 T_0^2/|\beta_2|$. A fundamental soliton is formed when the input pulse is launched such that $N = 1$, or $P_0 = |\beta_2|/(\gamma T_0^2)$. If N exceeds 1 because of a larger peak power, the pulse compresses and its spectrum broadens inside the optical fiber. The bandpass filter transmits the central part of this spectrum but blocks the spectral wings, in essence introducing spectral losses that depend on the peak power P_0. Because of quantum noise, P_0 fluctuates for any pulse around its average value. If the pulse spectrum becomes broader for positive power fluctuations, the filter-induced loss is larger than that occurring for negative fluctuations. The net result is that the amplitude noise is reduced below the standard quantum limit (SQL) set by the shot noise.

Another way to understand the origin of filter-induced amplitude squeezing is to ask how the SPM-induced squeezing discussed earlier affects an optical soliton. The answer turns out to be that the SPM produces spectral correlations within the pulse. Such quantum correlations were calculated, as well as measured, in a 1998 study [42]. Figure 10.8 shows the extent of spectral correlation after 130-fs solitons were propagated through a 2.7-m-long fiber. The gray shading of each square represents the value of the correlation coefficient between the corresponding spectral components, lighter shades reflecting a negative correlation. The data were obtained by splitting the soliton spectrum into 15 wavelength intervals. If n_i represents the number of photons in the ith wavelength slot, the measured correlation coefficient corresponds to

$$C_{ij} = \langle \Delta n_i \Delta n_j \rangle/(\sigma_i \sigma_j), \tag{10.2.12}$$

where Δn_i is a fluctuation in n_i and $\sigma_i^2 = \langle (\Delta n_i)^2 \rangle$ is its variance. The results show that the correlation coefficient could be negative or positive depending on the spectral region. If an optical filter removes selectively a part of the spectrum exhibiting positive correlations, the resulting pulse exhibits amplitude squeezing. Such spectral correlations are enhanced further for higher-order solitons, and they can be used for a variety of applications [43]–[45].

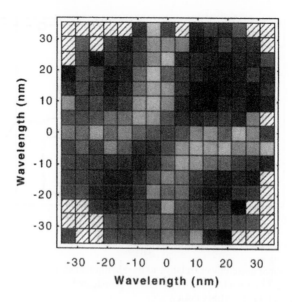

Figure 10.8: Map of intensity correlations across the soliton spectrum. The gray scale of each square represents the value of the correlation coefficient in the range of −1 to 1; lighter and darker shades indicate negative and positive correlations, respectively. Hatched squares correspond to correlation data with large error bars. (From Ref. [42]; ©1998 APS.)

Mathematically, one should solve the quantum NLS equation (10.1.6) to study the extent of amplitude squeezing realized by filtering the spectrum of a soliton. In one study, this equation was transformed into a classical equation by employing the positive-P representation [39]. In another, a perturbation approach was employed first to linearize the NLS equation, which could be solved using a standard technique [40]. The later approach provides an analytic expression for the variance of photon-number fluctuations when the spectral passband of the filter is approximated by a parabola. However, the amount of squeezing is quite sensitive to the shape of the passband, and most squeezing occurs when it has a rectangular shape with sharp edges (the so-called brick-wall filter). In this case, amplitude noise for fundamental solitons ($N = 1$) can be reduced by 6.5 dB at a distance of about three soliton periods. The noise begins to increase beyond that distance and displays an oscillatory behavior as a function of fiber length. A more careful analysis included the impact of spontaneous Raman scattering, which acts as a noise source at any finite temperature [38]. This kind of noise reduces the magnitude of amplitude squeezing from the value predicted by Eq. (10.1.6). However, squeezing can be enhanced by using more energetic pulses for which $N > 1$, provided fiber length is chosen appropriately.

The 1996 experiment, whose results are shown in Figure 10.7, used 2.7-ps pulses and required a 1.5-km-long fiber because the soliton period z_0 equals $(\pi/2)L_D$, and the dispersion length L_D scales with the pulse width as $L_D = T_0^2/|\beta_2|$. Losses associated with such a long fiber are detrimental to squeezing; they can be reduced by using shorter fibers if shorter pulses are employed. Several experiments have employed pulse

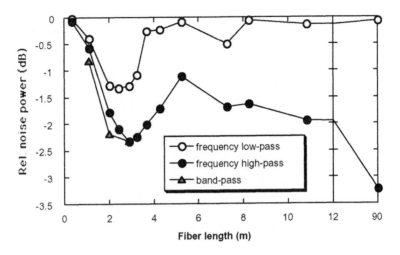

Figure 10.9: Measured noise power (relative to the shot noise) as a function of fiber length for three types of optical filters when 135-fs pulses are propagated as solitons inside a 90-m-long fiber fiber. (From Ref. [47]; ©1998 OSA.)

widths ~0.1 ps together with fiber lengths ranging from 3 to 90 m [46]–[59]. Amplitude squeezing by 3.8 dB was observed in a 1998 experiment by propagating 135-fs solitons with $N = 1.3$ inside a 90-m-long fiber [47]. This length corresponded to more than 100 soliton periods for such short pulses. Indeed, the spectrum of output pulses shifted considerably toward longer wavelengths (by 40 nm) because of intrapulse Raman scattering [6]. This spectral shift could be reduced by using a much shorter fiber. Figure 10.9 shows the measured noise power (relative to the shot noise) as a function of fiber length for fundamental solitons ($N = 1$) using three kinds of filters. Although squeezing occurred for fiber lengths ranging from 1 to 90 m, its magnitude varied with length and also with the type of filter employed. The first noise minimum occurred close to three soliton periods (magnitude 2.3 dB), as expected from theory. However, larger squeezing (close to 3.2 dB) was measured for a 90-m-long fiber.

An interesting question is whether the soliton effects are essential for amplitude squeezing. If they are not, one should be able to observe squeezing even in the normal-dispersion region of an optical fiber. Of course, a practical problem in this case is that short pulses spread rapidly inside the fiber, resulting in a reduced peak power and weakening of SPM. This problem forces one to employ a relatively short fiber. In one experiment, 29-fs pulses at a wavelength of 809 nm were launched inside a fiber only 2 m long [48]. As the dispersion length was only 6 mm for such short pulses, pulses spread rapidly. The nonlinear length could also be reduced to below 1 cm by increasing pulse energies close to 0.3 nJ. Under such conditions, the effective fiber length over which squeezing occurred was only about 2 cm. Nevertheless, amplitude noise could be squeezed by up to 1.2 dB by optimizing the spectral response of a band-pass filter. This observation established that the SPM-induced spectral broadening of optical pulses is the main ingredient; the use of solitons is helpful because it suppresses pulse broadening that occurs in the case of normal dispersion.

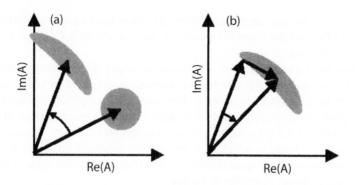

Figure 10.10: Amplitude squeezing realized using a fiber interferometer. (a) SPM-induced deformation the noise circle into an ellipse; (b) rotation of this ellipse by the field entering from the other port. (From Ref. [52]; ©2001 APS.)

Several experiments have employed interferometric or other nonlinear techniques for amplitude squeezing [49]–[56]. It turns out that an unbalanced Mach–Zehnder interferometer can squeeze the amplitude noise if the light experiences SPM-induced nonlinear phase shift in both of its arms [57]. The same can occur inside an unbalanced Sagnac interferometer in which two arms correspond to propagation of light in opposite directions. A linear configuration can also be employed when the orthogonally polarized modes of an optical fiber mimic the two arms of a Mach–Zehnder interferometer. All of these schemes have been deployed for realizing amplitude squeezing.

To understand the origin of amplitude squeezing, recall from Figure 10.3 that SPM inside an optical fiber normally produces quadrature squeezing by turning the noise circle into an ellipse. The interferometer rotates this ellipse, in the fashion shown in Figure 10.10, by adding coherently a small portion of the original field with a $\pi/2$ phase shift. Because of this rotation, the combined field at the output exhibits less amplitude noise than the input field itself. In a 1998 experiment, amplitude noise was reduced by up to 5.7 dB when 182-fs pulses were propagated as solitons inside a 3.5-m-long NOLM acting as a Sagnac interferometer [50]. The polarization-maintaining fiber exhibited anomalous dispersion with $\beta_2 = -19$ ps^2/km at the operating wavelength of 1550 nm, resulting in a dispersion length of about 54 cm. The Sagnac loop was unbalanced by using a 82:18 beam splitter so that optical pulses were considerably more intense in one direction compared with the other one. The extent of squeezing depends on the pulse energy that sets the soliton order N. Theoretically, squeezing can exceed 10 dB with a proper optimization of the pulse energy and fiber length [24]. As before, the use of solitons is not essential because noise squeezing by 2.5 dB was observed even when pulses were propagated in the normal-GVD regime of a fiber loop [51].

The Mach–Zehnder configuration was first deployed in a 2001 experiment in which orthogonally polarized 180-fs pulses of different energies were propagated along the principal axes of a polarization-maintaining optical fiber then combined together using a polarization-beam splitter [52]. Amplitude noise was observed to be squeezed by up to 4.4 dB in the case of 82% detection efficiency (or by 6.3 dB when corrected for linear losses). Microstructured fibers (of length 1 m or less) were employed in a later experiment [55], but the observed squeezing was limited to 2.7 dB (or 4 dB after correcting for losses).

Another experiment employed a balanced nonlinear Sagnac interferometer that also acted as a phase-sensitive fiber-optic parametric amplifier in which the pump and signal fields propagated in opposite directions [58]. Amplitude squeezing in this case was produced through parametric deamplification of noise, and its magnitude was only 0.6 dB, or 1.4 dB after correcting for detection efficiency. This experiment also employed a scheme in which two orthogonally polarized pulses were launched simultaneously inside the Sagnac interferometer to cancel the noise induced by guided acoustic-wave Brillouin scattering in the fiber.

As discussed in Chapter 9, microstructured fibers are often used to generate a supercontinuum by propagating femtosecond pulses inside them. In the temporal domain, fission of a higher-order soliton splits the input pulse into several fundamental solitons whose spectrum shifts rapidly toward the red side because of intrapulse Raman scattering [6]. As one may expect, amplitude squeezing can be produced by filtering the supercontinuum appropriately. In a 2005 experiment, femtosecond pulses at a wavelength of 810 nm were launched inside a 30-cm-long microstructured fiber [59]. The output spectrum extended over 300 nm for pulses of energies close to 100 pJ. Amplitude squeezing was observed for both low- and high-pass filtering with a maximum value of 4.6 dB that corresponds to 10.3-dB squeezing when corrected for detection losses. Extensive measurements and numerical simulations [60] indicate that the Raman-shifted fundamental solitons produced though soliton fission ($N > 2$) exhibit larger squeezing compared with those obtained using pulses with N close to 1. Indeed, maximum squeezing was observed with an optical filter that preserved the soliton whose spectrum was red-shifted most.

10.2.5 Polarization Squeezing

A new kind of squeezing, known as *polarization squeezing*, has received considerable attention in recent years [61]–[70]. The basic idea, first proposed in 1993, consists of employing the polarization dependence of the third-order nonlinearity to reduce noise below the standard quantum limit in one of the Stokes parameters that characterize the state of polarization (SOP) of an optical field [61]. It is common to represent the SOP of an optical field by the Stokes vector whose components are related to the two polarization components A_x and A_y of the optical field as

$$S_1 = |A_x|^2 - |A_y|^2, \quad S_2 = A_x^* A_y + A_y^* A_x, \quad S_3 = i(A_y^* A_x - A_x^* A_y). \quad (10.2.13)$$

It is easy to verify that the magnitude of this vector is related to the total intensity $S_0 = |A_x|^2 + |A_y|^2$ and remains constant even if the SOP of the optical field changes. It is therefore common to represent the SOP by the tip of the Stokes vector on the Poincaré sphere as shown in Figure 10.11(a). In the quantum case shown in part (b), the thickness of the sphere represent intensity fluctuations, whereas the spherical ball centered on the tip of the Stokes vector represents fluctuations in the three components of the Stokes vector. Polarization squeezing occurs when this ball is deformed such that the noise in one of the components is reduced below the standard quantum limit.

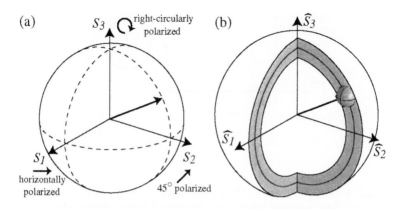

Figure 10.11: (a) Classical and (b) quantum Stokes vectors on a Poincaré sphere. The spherical ball centered on the tip of the Stokes vector represents fluctuations in the three components of the Stokes vector. (From Ref. [65]; ©2002 APS.)

In a 2001 experiment, polarization squeezing was realized by mixing coherently an intense optical beam with a squeezed vacuum state that was polarized orthogonal to the SOP associated with this beam [63]. In another experiment, polarization squeezing was produced by mixing two quadrature-squeezed beams [65]. The spherical ball on the Poincaré sphere turned into a cigarlike or pancakelike ellipsoid depending on the squeezed quadrature of the two beams. This experiment employed two CW beams that were squeezed using LiNbO$_3$ crystals acting as optical parametric amplifiers. More than 3 dB of squeezing in three of the four Stokes parameters could be realized simultaneously.

Optical fibers were first employed for polarization squeezing in a 2002 experiment [67], and by 2005 their use produced squeezing of up to 5.1 dB [68]. Figure 10.12 shows the experimental setup for the later experiment, in which 130-fs pulses were first split into two orthogonally polarized pulses with an adjustable delay between the two. They were then launched into a 13.3-m polarization-maintaining fiber such that their SOP was along the two principle axes of the fiber. Both pulses were amplitude-squeezed through SPM inside the fiber. The output pulses were combined using a polarizing beam splitter and aligned to overlap temporally by adjusting the initial tunable delay between them. A relative phase of $\pi/2$ between them ensured that the output beam was circularly polarized. The average values of the two Stokes parameters, $\hat{S}_1$ and $\hat{S}_2$, vanishes in this case. Fluctuations associated with the quantity $\hat{S}_\theta = \cos(\theta)\hat{S}_1 + \sin(\theta)\hat{S}_2$ were measured as a function of θ by rotating a half-wave plate and using two photodetectors; squeezing by up to 5.1 dB was observed under appropriate experimental conditions. This value corresponds to a maximum squeezing of 8.8 dB when corrected for losses and detection efficiency.

A quantum theory of polarization squeezing has been developed to understand how it depends on various input parameters, such as the pulse energy [69]. The quantum NLS equation (10.1.23) was solved in a recent study by employing the phase-space methods [70]. It was used to predict the magnitude of polarization squeezing, and the

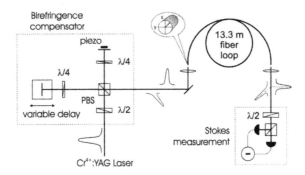

Figure 10.12: Experimental setup used for polarization squeezing inside a 13.3-m-long fiber loop. PBS stands for a polarizing beam splitter. (From Ref. [68]; ©2005 OSA.)

results were in good agreement with the experimental data obtained with the setup shown in Figure 10.12.

10.3   Quantum Nondemolition Schemes

A fundamental limitation imposed by quantum mechanics is that any measurement of a physical quantity affects the quantum state of the system—unless the system is in one of the eigenstates of that physical variable—and thus demolishes the original state of the system. This property is behind the Heisenberg uncertainty relation. It is also behind the so-called no-cloning theorem, which states that a general quantum state cannot be duplicated perfectly because any attempt to duplicate it changes the original state [71].

A quantum nondemolition (QND) scheme tries to overcome this fundamental limitation by ensuring that the act of measuring a physical variable changes the quantum state such that the variable being measured is not affected by that measurement [72]. This requirement is satisfied if the operator $\hat{O}$ associated with that physical variable (in the interaction picture) evolves such that $[\hat{O}(t), \hat{O}(t')] = 0$ for $t' > t$. Variables that evolve in this manner are called *QND observables* [23]. If the operator $\hat{O}$ commutes with the Hamiltonian describing the coupling between that variable and the measuring system, the measurement is referred to as a *back-action evading measurement*. In the case of optical fields, it turns out that the nonlinear effects inside optical fibers can be employed for a variety of QND measurements [73]–[79].

10.3.1   QND Measurements through Soliton Collisions

A simple example of a QND observable in the case of optical fibers is the number of photons associated with an optical pulse. Classically, as long as fiber losses remain negligible, neither the dispersive nor the nonlinear effects inside the fiber change the number of photons contained within a pulse, even though its amplitude, phase, and spectrum may evolve considerably. This feature also holds for the quantum NLS

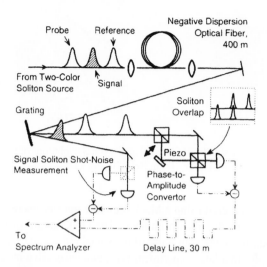

Figure 10.13: Experimental setup used for QND measurements of the photon number by measuring the phase shift induced on a probe soliton during its collision with the signal soliton. The reference pulse is needed for the phase measurement. (From Ref. [75]; ©1992 APS.)

equation (10.1.6) as the photon-number operator $\hat{N}_p$ commutes with the Hamiltonian in Eq. (10.1.12).

Similar to the classical case, the quantum NLS equation permits propagation of pulses as fundamental solitons when the two pulse parameters, T_0 and P_0, satisfy the relation $N = 1$, or $\gamma P_0 T_0^2 / |\beta_2| = 1$. However, as seen in Section 9.1.4, in contrast with the classical case, soliton parameters such as its energy, phase, position, and momentum (related to soliton frequency) fluctuate because of quantum noise [18]. Nevertheless, the photon-number operator $\hat{N}_p$, defined as in Eq. (10.1.28), is a QND-observable variable, and it allows measurements of the photon number without disturbing its value. Of course, such a measurement would affect the conjugate variable, the optical phase.

It was suggested in 1989 that collision of a soliton with a probe soliton at a different wavelength can be used for a QND measurement of the photon number [18]. Since the two solitons travel at different speeds, they can be arranged to collide in the middle of an optical fiber. It is well known that such a collision leaves the two solitons intact, except for shifting their phases and temporal positions [6]. The collision-induced phase shift of the probe soliton depends on the number of photons contained in the signal pulse on which the QND measurement is being performed. This phase shift was measured interferometrically in a 1992 experiment [75].

Figure 10.13 shows the experimental setup used for this experiment. As seen there, a reference pulse was launched inside a 400-m-long fiber in addition to the signal and probe pulses [75]. The identical 3.6-ps-wide probe and references pulses were separated by 30 ps. The 2.6-ps-wide signal pulse was situated in the middle, and its wavelength differed from other two pulses by 5.7 nm. At the fiber output, a grating was used to separate the signal pulse from the other two pulses, whose phases were different because of the signal-induced phase shift on the probe pulse. This phase

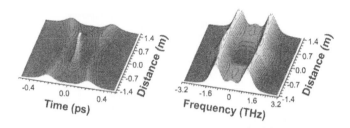

Figure 10.14: Numerically simulated temporal and spectral evolutions of two fundamental solitons colliding in the middle of a fiber. (From Ref. [79]; ©2002 APS.)

difference could be measured using a Mach–Zehnder interferometer, thus allowing an indirect measurement of the photon number. The QND nature of this measurement was verified by measuring the shot noise of the signal pulse and correlating it with the phase measurements. In a later experiment [78], two successive measurements of the photon number were performed to verify the QND nature of such measurements. The first measurement collapses the quantum state toward an eigenstate of the system but does not affect the second measurement of the photon number.

Several factors affect the accuracy of such QND measurements. An obvious source of noise is the SPM phenomenon that affects the phase of the probe pulse as it propagates through the fiber [76]. SPM converts amplitude fluctuations into phase noise that affects the measurement of any collision-induced phase shift. Another potential noise source is Brillouin scattering through guided acoustic waves. Both of these noise sources can be mitigated by making use of a phase-shifting cavity [77]. The basic idea was first used in a squeezing experiment [27]; it makes use of the fact that noise measurements are often performed at a frequency shifted from the carrier frequency of the signal. Because of this frequency shift, an optical cavity introduces a relative phase shift between the mean field and its fluctuations.

10.3.2 QND Measurements through Spectral Filtering

Spectral filtering, a technique discussed earlier in the context of amplitude squeezing, has also been used for performing QND measurements [79]. In this approach, rather then measuring the collision-induced phase shift with an interferometer, spectral changes induced by the collision are quantified using an optical filter. Figure 10.14 shows the temporal and spectral evolutions of two fundamental solitons, colliding in the middle of a fiber, by solving the NLS equation numerically. As seen there, spectra of both solitons shift as they collide. Such a spectral shift can be used to make a QND measurement of the photon number.

Figure 10.15 shows the experimental setup employed in a 2002 experiment to measure the collision-induced spectral changes [79]. One difference compared with the scheme shown in Figure 10.13 is that no reference pulse is needed. For this reason, only the signal and probe pulses (each 200 fs wide), separated temporally as well as spectrally by appropriate amounts, were launched inside a polarization-maintaining fiber. The initial pulse separation and the 6.3 m length of the fiber were chosen care-

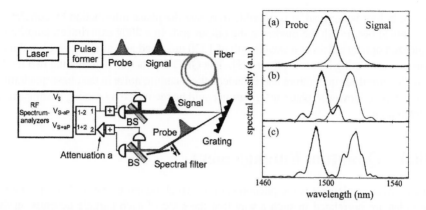

Figure 10.15: Experimental setup used for QND measurements of the photon number by measuring the spectral shift induced on a probe soliton during its collision with the signal soliton. The spectra of the signal and probe pulses are shown (a) before and (c) after collision; part (b) shows these spectra when each pulse propagates alone. (From Ref. [79]; ©2002 APS.)

fully to ensure that the two solitons collide near the output end of the fiber. As a result of this collision, their spectra were found to be shifted by 5 nm or so though the XPM effects.

The pulse spectra of the signal and probe pulses are displayed in Figure 10.15 at the input end (a) and at the output end (c) of the fiber; part (b) shows the spectra when each pulse propagates alone. In the experiment, a grating was used to separate the signal and probe pulses. A knife edge in the probe path acted as a low-pass spectral filter. Two balanced detectors were used to measure amplitude fluctuations for the signal and probe pulses. It was found that there was a strong negative correlation in the amplitude fluctuations associated with the two pulses. Physically speaking, an increase in the photon number of the signal enhances the XPM-induced spectral shift, resulting in increased losses for the spectrally filtered probe pulse. A detailed analysis of the experimental data showed that the condition required for a QND measurement was satisfied in this experiment.

The QND phenomenon is related to back-action-evading measurements in which back action of the measuring apparatus affects only the complementary variable of the variable being measured [80]–[83]. Such measurements can be used for quantum erasing. The idea behind quantum erasing can be understood as follows. In a Young's double-slit experiment, interference fringes are destroyed, if it is possible to tell through which slit a photon passes through. However, if the "which-slit" information is erased even after photons have already passed through the slits, the interference pattern can be revived.

Recently, continuous-variable quantum erasing has attracted considerable attention [84]–[86]. Consider an optical beam characterized by two conjugate quadrature variables, namely, its amplitude and phase. If this beam is combined with a quantum marker that provides a back-action-evading measurement of its amplitude, excess noise is produced in the phase quadrature. However, if the amplitude information is erased before being read, it is possible to revive the phase information by correlating the optical beam with the marker at the output end. In a 2004 experiment, amplitude-squeezed optical pulses at a wavelength of 1530 nm, used as a marker, were combined with a signal using a beam splitter [85]. The QND interaction between the two modified the amplitude quadrature and enhanced the quantum noise in the phase quadrature of the signal. However, phase information could be recovered by erasing the amplitude information.

10.4   Quantum Entanglement

Quantum entanglement is the phenomenon in which quantum states of two (or more) particles are correlated in such a way that the state of each particle depends on the other one, even when the two particles are separated by a large distance [87]–[89]. As a result, measurements performed on one particle appear to instantaneously affect the other particle entangled with it. This appears to contradict the special relativity at first sight, but quantum entanglement does not enable the transmission of classical information faster than the speed of light. Quantum entanglement has attracted considerable attention in recent years because it has applications in the emerging fields such as quantum computing, quantum teleportation, and quantum cryptography [90]–[93].

In the context of optics, one is interested in the entanglement of two or more photons. Entangled photons were used in a 1972 experiment to demonstrate violation of Bell inequalities that characterize the presence of entanglement [94]. Since then, optical beams and pulses have been used in a wide variety of quantum experiments [89]–[93]. This section focuses on some of these experiments making use of the nonlinear effects inside optical fibers.

10.4.1   Photon-Pair Generation

Before entangled photons can be used for any application, one needs an optical source capable of generating them. Historically, entangled photon pairs were first generated through the process of spontaneous parametric down-conversion inside a nonlinear crystal using its second-order susceptibility [95]–[98]. This nonlinear process splits the energy $\hbar\omega_p$ of a pump photon into two photons whose frequencies satisfy the energy-conservation relation $\omega_s + \omega_i = \omega_p$. The important point is that two photons in each pair are created simultaneously [96]. Thus, if a photon is detected in a certain time slot, the existence of a second photon is guaranteed in that slot, even if the two photons have traveled far apart.

The use of spontaneous FWM inside optical fibers for creating entangled photon pairs was analyzed in a 2001 study [99], and the FWM phenomenon has been employed in several experiments [100]–[109]. The main advantage of this scheme is that photon pairs are created in a single spatial mode directly inside a single-mode fiber. The nonlinear phenomenon of FWM in optical fibers makes use of the third-order susceptibility in which two pump photons transfer their energy to two new photons whose frequencies satisfy the energy-conservation relation $\omega_s + \omega_i = 2\omega_p$. It is also possible to employ a dual-pumping scheme in which two pump photons involved in the FWM process are not identical and originate from two different lasers [6]. In both cases, the two photons in each signal–idler photon pair are correlated in the quantum sense as they are created simultaneously.

Equations (10.2.2) and (10.2.3) given earlier in the context of squeezing apply to the simultaneous generation of photon pairs as well. In particular, the analytic solution given in Eqs. (10.2.4) and (10.2.5) shows how the signal and idler field operators evolve inside the fiber. In the case of photon-pair generation, stimulated FWM is avoided by keeping the input pump power low enough that the condition $gz \ll 1$ over the entire fiber length L. Typically, $gL = 0.1$ or less. The average number of photons at the end of the fiber in the signal and idler fields is found by using

$$\bar{N}_s = \langle \hat{a}_s^\dagger(L)\hat{a}_s(L) \rangle, \qquad \bar{N}_i = \langle \hat{a}_i^\dagger(L)\hat{a}_i(L) \rangle. \tag{10.4.1}$$

Because neither the signal nor idler field is present at the input end of the fiber, the average in Eq. (10.4.1) is performed with respect to the vacuum state. The degree of quantum correlation between the signal and idler photons is characterized by a parameter C_q defined as

$$C_q = \frac{\langle \hat{a}_i^\dagger(L)\hat{a}_s^\dagger(L)\hat{a}_s(L)\hat{a}_i(L) \rangle}{\bar{N}_s \bar{N}_i} - 1. \tag{10.4.2}$$

This quantity is measured experimentally by taking the ratio of true and accidental coincidence rates.

Experimentally, FWM was first used in a 2001 experiment to show that signal and idler beams produced through FWM were correlated in the quantum sense [100]. Soon after, the same technique was used to generate photon pairs [101]. It was found that spontaneous Raman scattering (SpRS) reduced the value of parameter C_q considerably by increasing the accidental coincidence rate. This degradation depends on frequency difference Δv between the pump and signal and is largest when Δv is close to 13.2 THz, the value where the Raman gain is maximum. However, because of the broadband nature of the Raman gain, significant degradation occurs even when Δv is near 5 THz.

In a 2004 experiment, the impact of SpRS was reduced by using $\Delta v = 1.25$ THz [102]. Figure 10.16 shows the experimental setup employed. FWM occurred inside a nonlinear fiber-loop mirror (or Sagnac interferometer), containing 300 m of dispersion-shifted fiber (DSF). Its use helps in reflecting most of the unused pump light (in the form of 5-ps pulses) back toward the input port. Any pump photons leaking through the output port are further rejected by the two gratings used to direct the signal and idler photons toward two InGaAs avalanche photodiodes (APDs) operating in the gated-Geiger mode. The use of polarizers before the APDs helps reducing the impact of SpRS by rejecting the orthogonally polarized photon pairs. With these precautions the value of C_q was close to 10 in this experiment [102].

The phenomenon of SpRS creates Stokes and anti-Stokes photons whose number depends on the pump power and the frequency difference Δv. It also depends on the phonon population that varies with temperature as

$$n(\Delta v) = [\exp(h\Delta v/k_B T_f) - 1]^{-1}, \tag{10.4.3}$$

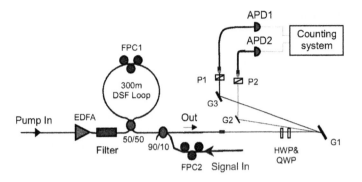

Figure 10.16: Experimental setup used for generating photon pairs through FWM inside a fiber. FPC, HWP, QWP, G, and P stand for fiber polarization controller, half-wave plate, quarter-wave plate, grating, and polarizer, respectively. (From Ref. [102]; ©2004 OSA.)

where k_B is the Boltzmann constant and T_f is the fiber temperature (in Kelvins). Thus, a simple way to reduce the impact of SpRS is to cool the fiber to below 80 K using liquid air. This was the approach adopted in a 2005 experiment [103]. Figure 10.17 shows the experimental setup employed. The CW radiation at a wavelength of 1551 nm was first modulated to create 100-ps pulses at a 100-MHz repetition rate. The pulses were amplified and filtered to reduce the noise added by the EDFA before they were launched into a 500-m-long DSF that was cooled to liquid-nitrogen temperature. The zero-dispersion wavelength of the DSF nearly coincided with the laser wavelength to ensure phase matching for the FWM process.

Photon pairs generated through FWM were passed through a polarizer and a fiber grating that reflected the residual pump backward. The output was passed through an arrayed waveguide grating that sent the signal and idler photons, separated in frequency by 400 GHz, to different output ports. Two band-pass filters (BPF) selected the signal and idler photons within their bandwidth and rejected any residual pump photons. Two APDs operating in the gated-Geiger mode acted as photon counters. They were used to measure both the photon flux and the coincidence rate for the signal

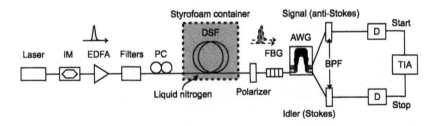

Figure 10.17: Experimental setup for generating photon pairs through FWM inside a fiber. IM, PC, FBG, AWG, and TIA stand for intensity modulator, polarization controller, fiber Bragg grating, arrayed waveguide grating, and time-interval analyzer, respectively. (From Ref. [103]; ©2005 OSA.)

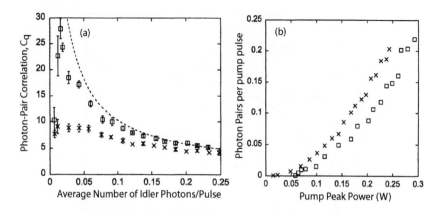

Figure 10.18: (a) Photon-pair correlation as a function of idler flux and (b) average number of photon pairs per pulse as a function of peak power of pump pulses. In each case, squares and crosses show the data for cooled and uncooled fibers. The dashed line shows the ideal case without the SpRS degradation. (From Ref. [103]; ©2005 OSA.)

and idler photons. Figure 10.18(a) compares the measured value of C_q as a function of the idler flux at 77 K with that obtained without cooling the fiber. Values close to 30 could be realized at low pump powers by cooling the fiber. The average number of photon pairs per pump pulse is shown in Figure 10.18(b) as a function of the peak power of pump pulses for cooled and uncooled fibers. It is reduced slightly in the case of a cooled fiber because of a 0.9-dB increase in the fiber loss [103]. Nevertheless, at the 100-MHz repetition rate of pump pulses, several millions of photon pairs per second could be generated in this experiment.

Several experiments have employed photonic-crystal or microstructured fibers for photon-pair generation [104]–[109]. The enhanced intensities realized within a narrow core of such fibers allow one to reduce the fiber length to below 10 m. However, the zero-dispersion wavelength of such fibers also shifts near 800 nm, forcing one to use a pump laser operating in this spectral region. In a 2004 experiment, a 5.8-m-long microstructured fiber was pumped at 749 nm (close to its zero-dispersion wavelength) using 3-ps pulses, and gratings were used to select photon pairs with wavelengths at 736 and 761 nm. Accidental counts resulting from SpRS limited the value of C_q to relatively low values. In another experiment, a 1.8-m-long microstructured fiber was pumped using 4-ps pulses at 737.5 nm, producing correlated photon pairs at a rate of 37.6 kHz with wavelengths of 688.5 and 798.8 nm [109]. Although SpRS was still the limiting factor, measured values of C_q were as large as 10.

The extent of SpRS-induced degradation on photon-pair correlation has been analyzed in detail [110]–[112]. As mentioned earlier, cooling of the fiber helps to reduce the impact of SpRS by reducing the phonon population. The use of a birefringent fiber can phase-match a FWM process that creates photon pairs polarized orthogonal to the pump. Since SpRS generates photons that are mostly copolarized with the pump, its impact on photon-pair correlation is reduced considerably [111]. Figure 10.19 compares the predicted value of C_q for the FWM configurations in which photons pairs are

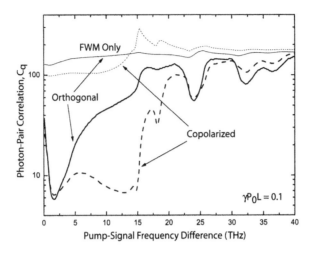

Figure 10.19: Predicted pair correlation as a function of Δv when photon pairs are polarized parallel (dashed curve) or orthogonal (solid curve) to the pump. In both cases, C_q values are also shown when FWM alone creates photon pairs. (From Ref. [112]; ©2007 APS.)

polarized either parallel or orthogonal to the pump [112]. Considerable improvement occurs in the frequency range of 5 to 15 THz, and C_q values close to 50 are possible in the orthogonal case, even when Δv is near the Raman-gain peak. The theory predicts $C_q > 100$ if the SpRS phenomenon does not affect the FWM-created photon pairs and the pump power is kept low enough to ensure $\gamma P_0 L = 0.1$.

Microstructured fibers offer another mechanism for avoiding SpRS. If the pump wavelength falls in the normal-dispersion regime of the fiber, the dispersive properties of such fibers can phase-match a FWM process such that the signal and idler frequencies differ from the pump frequency by more than 25 THz [6]. As the Raman gain peaks for a frequency difference Δv close to 13 THz and nearly vanishes for $\Delta v > 25$ THz, the impact of SpRS should become negligible in such a situation. This was the approach adopted in a 2005 experiment [105] in which a photonic-crystal fiber with its zero-dispersion wavelength at 1065 nm was pumped at 1047 nm on the normal-dispersion side, resulting in correlated photon pairs at wavelengths of 839 and 1392 nm, with Δv exceeding 40 THz. In another experiment, a microstructured fiber with its zero-dispersion wavelength at 715 nm was pumped using 4-ps pulses at 708.4 nm, resulting in the generation of up to 10^7 photon pairs/s at wavelengths of 587 and 897 nm [108]. No SpRS-induced degradation was observed, and the measured value of C_q was close to 40 at relatively low values of pump powers.

As mentioned earlier, two pumps at different wavelengths can also be used for FWM inside optical fibers [6]. In this case, the frequencies of the four photons involved satisfy the energy-conservation condition $\omega_s + \omega_i = \omega_l + \omega_h$, where ω_l and ω_h are the pump frequencies. Of course, FWM occurs only if the phase-matching condition (related to momentum conservation) is also satisfied. In the case of photon-pair generation, spontaneous FWM occurs at the specific signal and idler frequencies for which this condition holds.

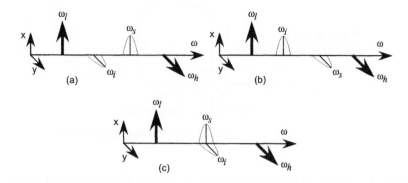

Figure 10.20: Three FWM configurations involving two orthogonally polarized pumps at different frequencies. Polarization states or frequencies of the signal and idler photons are different in each case. (From Ref. [112]; ©2007 APS.)

There are several situations in which a dual-pump configuration is advantageous. For example, when a single pump is employed, the signal and idler frequencies lie on opposite sides of the pump, resulting in signal and idler photons (forming a photon pair) that are distinguishable on the basis of their frequencies. Some applications may require that the two photons forming an entangled pair have the same frequency. This is easily realized in the case of two pumps if phase-matching condition requires that the signal and idler frequencies lie exactly in the middle of the two pumps; that is, $\omega_s = \omega_i = (\omega_l + \omega_h)/2$. This configuration was employed in a 2005 experiment [106]. The two pumps were separated by only 4 nm and had wavelengths of 833 and 837 nm. When they were launched inside a 1.5-m-long microstructured fiber, the signal and idler photons at the 835-nm wavelength exhibited a coincidence counting rate that was eight times larger than that of accidental coincidences ($C_q = 8$) at a peak pump power of 250 mW.

The use of two pumps allows one to create signal and idler photons that differ in their state of polarization. Figure 10.20 shows three FWM configurations in which the two pumps are polarized orthogonally. In each case, the signal and idler photons are also orthogonally polarized, as dictated by the conservation of angular momentum. In case (c), the two photons in each pair are degenerate in frequency but distinguishable on the basis of their polarization. This situation is desirable in when photon pairs exhibiting polarization entanglement are needed and is discussed next.

10.4.2 Polarization Entanglement

Many quantum applications require photon pairs that exhibit polarization entanglement, and several techniques based on parametric down-conversion inside a nonlinear crystal have been developed [113]–[115]. Some of them have been employed for creating polarization entanglement between photons of each signal–idler pair produced through FWM inside an optical fiber, resulting in an all-fiber source of polarization-entangled photons [116]–[122].

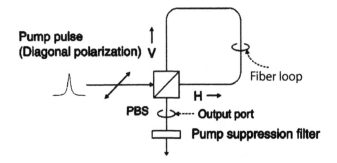

Figure 10.21: A FWM configuration suitable for creating polarization-entangled photons. V and H denote verical and horizontal SOPs, respectively. (From Ref. [116]; ©2004 APS.)

When a single pump is employed, FWM inside an optical fiber generates signal and idler photons, forming a pair, such that they have the same SOP as the pump (or orthogonal to the pump). To create polarization entanglement, two photons should have orthogonal SOPs. In a relatively simple polarization-diversity scheme [116], a polarizing beam splitter (PBS) is employed to form a fiber loop as shown schematically in Figure 10.21. The input pump pulse is polarized linearly at 45° such that the orthogonally polarized pump pulses enter the fiber from opposite directions. Each pump pulse creates signal and idler photons that are copolarized with it and copropagte in the its own direction. At the fiber output, the two sets of photon pairs are mixed by the PBS to create polarization-entangled photons.

To see more clearly how this scheme creates polarization entanglement, we denote the SOP of two pump pulses as being horizontal (H) or vertical (V). The quantum states of photon pairs created by these two pulses can then be denoted as $|H\rangle_s|H\rangle_i$ and $|V\rangle_s|V\rangle_i$. When these states enter the PBS, as shown in Figure 10.21, the output from the PBS is a superposition of the two product states, resulting in a state that is polarization entangled. The fiber loop in this case does not function as a Sagnac interferometer but it provides two fundamental functions needed for creating polarization entanglement generation. It separates the path of the two pumps by propagating them in opposite directions and also helps in removing the path distinguishability between the two product states. This configuration has another advantage. It sets the relative phase between $|H\rangle_s|H\rangle_i$ and $|V\rangle_s|V\rangle_i$ to be 0 or π, the two values that yield a maximally entangled state. Indeed, coincidence fringes with >90% visibility were observed in the experiment together with the violation of a Bell inequality by seven standard deviations [116]. Moreover, quantum correlation between the photon pair was preserved even after the two photons had been separated by 20 km of optical fiber.

A different technique, shown schematically in Figure 10.22, was used to create polarization entanglement in a 2005 experiment [117]. The pump pulse was still separated into its H and V components, but these two were propagated along the same direction inside the fiber after introducing a time delay T_d between them. FWM inside the fiber creates photon pairs in the states $|H\rangle_s|H\rangle_i$ and $|V\rangle_s|V\rangle_i$ that are separated in time by T_d. This distinguishing time delay between the orthogonally polarized photon pairs is then removed by passing them through a birefringent fiber of suitable

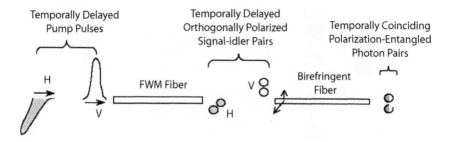

Figure 10.22: A technique used for creating polarization-entangled photons. PM stands for a polarization-maintaining fiber. (From Ref. [117]; ©2005 APS.)

length. This is possible because such a fiber propagates orthogonally polarized pulses at slightly different speeds. If v_{gx} and v_{gy} denote the group velocities along the slow and fast axes of the fiber, respectively, the fiber length should satisfy the condition $L_f = (v_{gx}^{-1} - v_{gy}^{-1})T_d$. When such a fiber is employed, the resulting temporal overlapping of the two states creates the polarization-entangled state, $|H\rangle_s|H\rangle_i + e^{i\phi}|V\rangle_s|V\rangle_i$, where ϕ is the relative phase difference between the two product states.

The scheme shown in Figure 10.22 was implemented in a 2005 experiment with 30-ps delay between the two orthogonally polarized pump pulses [117]. The relative phase ϕ between the two sets of photon pairs was adjusted to create the following four Bell states:

$$|\Psi^\pm\rangle = \frac{1}{\sqrt{2}}\left(|H\rangle_s|H\rangle_i \pm |V\rangle_s|V\rangle_i\right), \qquad |\Phi^\pm\rangle = \frac{1}{\sqrt{2}}\left(|H\rangle_s|V\rangle_i \pm |V\rangle_s|H\rangle_i\right). \quad (10.4.4)$$

The coincidence fringe visibility was >90% in this experiment, and a violation of a Bell inequality up to 10 standard deviations was observed. In a later experiment, the polarization-entangled photons could be stored for a duration of 125 μs using a 25-km-long fiber spool [118]. Moreover, even when such photons were separated by 50 km, the measured two-photon fringe visibility was close to 86%, indicating near perfect preservation of polarization entanglement over such a long distance.

Several recent experiments have yielded further improvement in the quality of entangled photons. In a 2006 experiment, the polarization-diversity scheme was implemented using 5-ps pump pulses that were divided into its H and V components and then launched in opposite directions into a 30-m-long dispersion-shifted fiber [119]. The frequency difference between the pump and signal was kept relatively small (about 0.5 THz) and the fiber was cooled to 77 K to minimize the impact of SpRS. As a result, the ratio of coincidence to accidental-coincidence counts exceeded 100 in this experiment, resulting in two-photon interference visibility of more than 98%. The same technique was used in a later experiment to produce polarization entanglement between photons of nearly the same frequency by pumping the fiber at two wavelengths that were 10-nm apart [120]. In both experiments, the pump, signal, and idler wavelengths were in the spectral region near 1550 nm. The fibers exhibit minimum loss in this telecommunication spectral band. As a result, entangled photons can be separated over considerable distances without affecting the entanglement quality too much.

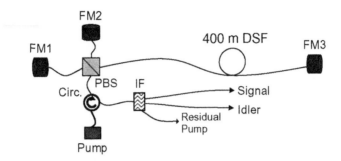

Figure 10.23: A setup used for creating polarization-entangled photons. Faraday mirrors (FM) are used to counteract undesirable polarization changes inside fibers. IF and DSF stand for interference filter and dispersion-shifted fiber, respectively. (From Ref. [122]; ©2006 OSA.)

An integrable fiber-based source of polarization-entangled photon pairs has also been realized in this spectral band [121].

As mentioned earlier, polarization-entangled photons can also be generated during a single pass through an optical fiber, provided spontaneous FWM is induced by two orthogonally polarized pumps at different wavelengths. In this case, signal and idler photons, forming a pair, are automatically created with orthogonal SOPs, as shown schematically in Figure 10.20. These two photons can also be made degenerate in frequency with an appropriate phase-matching condition. The three schemes shown in Figure 10.20 have been analyzed in a recent study [112]. Polarization entanglement was quantified by the violation of a Bell inequality, $|S| \leq 2$, where S is the parameter introduced by Clauser et al. [123] and measured experimentally by changing four polarization angles. The results show that $|S|$ can exceed 2 for a wide range of pulse and fiber parameters as long as accidental coincidences are minimized by controlling SpRS.

A problem with using fibers for polarization entanglement is that the SOP of the signal and idler photons can change in a random fashion along the fiber because of its residual birefringence. This problem can be solved with the scheme shown in Figure 10.23, allowing development of alignment-free all-fiber sources of polarization-entangled photon pairs [122]. A 400-m-long piece of dispersion-shifted fiber was used for FWM in a linear configuration. Three Faraday mirrors were employed to create polarization entanglement by injecting into the fiber two orthogonally polarized pump pulses with a relative time delay between them. A Faraday mirror is a nonreciprocal optical element that reflects incident light with an SOP orthogonal to the input SOP. Its use allows one to employ regular fibers that do not maintain SOP of pulses propagating inside them, as long as these pulses make a round trip inside the fiber. Any polarization changes occurring in one direction are automatically compensated during the round trip. The two pump pulses create photon pairs in states $|H\rangle_s|H\rangle_i$ and $|V\rangle_s|V\rangle_i$ that are separated in time. However, this time distinguishability is removed when pump pulses travel back in the same fiber with their SOPs reversed. Up to 92% visibility of two-photon interference fringes was observed in this experiment without cooling the fiber.

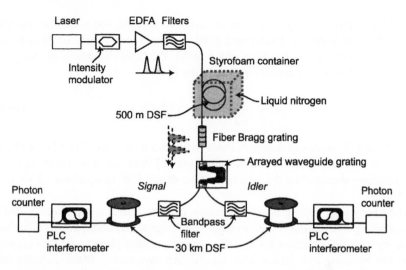

Figure 10.24: Experimental setup employed for creating and distributing time-bin-entangled photons. PLC and DSF stand for planar lightwave circuit and dispersion-shifted fiber, respectively. (From Ref. [127]; ©2006 OSA.)

10.4.3   Time-Bin Entanglement

Entangled photon pairs are useful for a variety of applications including quantum cryptography, quantum computing, and quantum teleportation [90]–[93]. However, it is not enough to create such pairs: one should also be able to distribute them over long distances without destroying their entanglement. In practice, optical fiber networks are likely to be used for this purpose. Since such networks operate near 1550 nm, the source should be able to generate photon pairs in this spectral region. This is already the case for sources making use of the nonlinear effects in fibers. However, a problem arises when polarization entanglement is employed. The transmission of polarization-entangled photons over optical fibers suffers from polarization-mode dispersion, a phenomenon that causes decoherence and limits the transmission distance.

Time-bin entanglement has been proposed to overcome this problem [124]. In this scheme, each photon pair can occupy two time slots, and the entanglement results from a superposition of these two possibilities. Time-bin entanglement of the signal and idler photons has been employed in several experiments to show that the photons can be distributed them over tens of kilometers of optical fibers without degrading the entanglement severely [125]–[127]. A fiber-based source of entangled photons was first used in a 2005 experiment to distribute entangled photons over 20 km of standard fiber without deterioration in the extent of quantum correlation [126]. The number of accidental coincidences was relatively large in this experiment because of the the SpRS effects. To solve this problem, the fiber was cooled to liquid-air temperature in a later experiment.

Figure 10.24 shows the experimental setup employed [127]. The LiNbO3 modulator is used to produce double pulses from the CW light at a wavelength near 1551, each of them 100-ps wide and the two separated by 1 ns. They are amplified, filtered, then launched into a 500-m-long dispersion-shifted fiber, spooled, and placed in a container filled with liquid nitrogen. Each pump pulse generates its own photon pairs through spontaneous FWM inside the fiber. As a result, the time-bin entangled state at the output of the fiber is a superposition state in the form

$$|\Psi\rangle = \frac{1}{\sqrt{2}}\left(|1\rangle_s|1\rangle_i \pm |2\rangle_s|2\rangle_i\right), \qquad (10.4.5)$$

where $|1\rangle$ and $|2\rangle$ represent quantum states in the two time slots. The relative phase ϕ depends on the pump phase ($\phi = 2\phi_p$ in the case of FWM) and does not change much over durations shorter than the coherence time of the CW laser (about 10 μs in the experiment).

The fiber Bragg grating is used in Figure 10.24 to reflect pump photons, and the arrayed waveguide grating is used to separate the signal and idler photons differing in frequency by 0.8 THz (about 6.4 nm). The signal and idler photons are then separated by 60 km by using two 30-km fiber spools. An asymmetric Mach–Zehnder interferometer, formed using silica waveguides and fabricated with the planar lightwave technology, was used in this 2006 experiment to remove the time distinguishability of two sets of photon pairs [127]. The path lengths in two arms of this interferometer differed by a precise amount that delayed one set of photon pairs by 1 ns, thus overlapping them in time with the other set of photon pairs. Two APDs operating in the gated-Geiger mode were used to measure the coincidence rates. Cooling of the fiber improved considerably the visibility of two-photon interference fringes. Even without removing accidental coincidences, visibility exceeded 75% after 60 km of optical fiber.

10.4.4   Continuous-Variable Entanglement

Polarization and time-bin entanglement represent examples in which quantum variables with two discrete possible values (two polarization states or two distinct time slots) are used to form an entangled state. Continuous-variable entanglement refers to the situation in which a continuous variable, such as the amplitude or phase of light, is employed for entanglement. This type of entanglement was first realized with optical fibers in a 2001 experiment [128], and it has attracted considerable attention since then [129]–[134].

The basic idea behind the scheme, used often in practice [133], is shown schematically in Figure 10.25. Two amplitude-squeezed pulses are first generated through SPM inside a fiber with the technique discussed in Section 10.3. If they are then forced to interfere at a beam splitter after introducing a relative phase θ between them, the transmitted and reflected pulses exhibit continuous-variable entanglement, sometimes referred to as the *Einstein–Podolsky–Rosen entanglement* [135]. The entanglement results from superposition of two squeezed states. This type of entanglement is quite different from the photon-pair entanglement, because each pulse may contain a large number of photons.

The idea displayed in Figure 10.25 was first implemented in a 2001 experiment [128] with the schematic shown in Figure 10.26. Two amplitude-squeezed beams

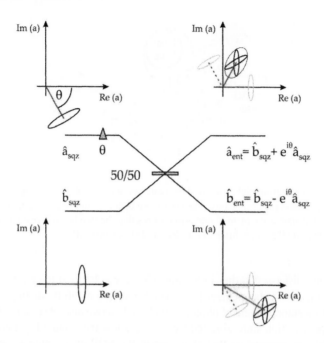

Figure 10.25: Scheme showing how continuous-variable entanglement can be produced by interfering two amplitude-squeezed beams at a 50:50 beam splitter. Maximum entanglement occurs for $\theta = \pi/2$. (From Ref. [133]; ©2006 APS.)

were generated inside an asymmetric Sagnac loop containing 8-m-long polarization-maintaining fiber. Input pulses were 130 fs wide and were polarized at 45° with respect to the slow and fast axes of the fiber. A 90:10 beam spitter launched 90% of the input power in the clockwise direction such that both polarization modes were excited equally and their amplitude noise squeezed by the same amount through SPM inside the fiber. A polarizing beam splitter was used to separate the two squeezed beams. A half-wave plate flipped the polarization of one beam before the beams were forced to interfere at the 50:50 beam splitter.

The continuous-variable entanglement created by the scheme shown in Figure 10.25 is of a different nature than the polarization or time-bin entanglement of photon pairs discussed earlier. To understand in what sense the two beams are entangled, one has to measure the spectrum of intensity fluctuations associated with each beam using a balanced detector. In the 2001 experiment [128], each amplitude-squeezed beam, before arriving at the 50:50 beam splitter, exhibited a noise level below the shot noise in a specific range of input pulse energies. However, no such noise reduction was observed for the two individual beams after the beam splitter. It was only when both beams were detected simultaneously and their photocurrents were added that the noise dipped below the shot-noise level. Amplitudes of the two beams were entangled in this sense.

The idea that superposition of two squeezed states creates an entangled state can be used to create other types of continuous-variable entanglement. For example, polariza-

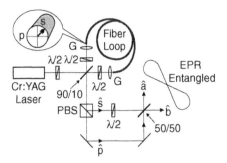

Figure 10.26: The experimental setup used to create continuous-variable entanglement by interfering two amplitude-squeezed beams. The polarizing beam splitter (PBS) separates the s and p polarizations. The half-wave plate in one arm ensures that the two beams are copolarized when they arrive at the 50:50 beam splitter. (From Ref. [128]; ©2001 APS.)

tion squeezing, discussed earlier in Section 10.3.2, can be used to create continuous-variable polarization entanglement [64], which must be carefully distinguished from the polarization entanglement of photon pairs based on two discrete polarization states. Recall that polarization squeezing reduces noise below the standard quantum limit in one of the Stokes parameters that characterize the SOP of an intense optical pulse containing a large number of photons. When two such pulses interfere at a 50:50 beam splitter, after introducing a $\pi/2$ shift on one of them, the noise fluctuations in the Stokes parameters become entangled in a way analogous to the case of amplitude-squeezed pulses discussed earlier.

Several other factors must be considered while dealing with entangled pulses. In any practical implementation, it is important to verify (or witness) that the pulses are indeed entangled without affecting the quality of entanglement. Homodyne detection or single-photon counters, often used when dealing with entangled photon pairs, are not suitable for bright entangled pulses containing a large number of photons on average. For this reason, several interferometric techniques have been developed for verifying continuous-variable entanglement [133]. In one scheme, phase measurements are performed on each beam by using an asymmetric Mach–Zehnder interferometer.

In a different scheme, shown schematically in Figure 10.27, entanglement is verified by combining the two entangled pulses on a beam splitter after introducing a phase shift ϕ in one of the paths. In the notation of Figure 10.27, the beam splitter creates superposition of the two entangled pulses, and its two outputs are governed by the operators

$$\hat{c} = \frac{1}{\sqrt{2}} \left(\hat{a}_{\text{ent}} - e^{i\phi} \hat{b}_{\text{ent}} \right), \qquad \hat{d} = \frac{1}{\sqrt{2}} \left(\hat{a}_{\text{ent}} + e^{i\phi} \hat{b}_{\text{ent}} \right). \tag{10.4.6}$$

Two photodetectors measure the photon numbers $\hat{c}^{\dagger}\hat{c}$ and $\hat{d}^{\dagger}\hat{d}$. The sum and difference of fluctuations in these two photocurrents provide the correlation signal for the amplitude and phase quadratures of the entangled beams and serve to witness the presence of entanglement.

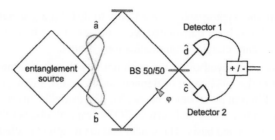

Figure 10.27: Setup employed for verifying continuous-variable entanglement of two intense pulses. BS stands for beam splitter. (From Ref. [133]; ©2006 APS.)

Another issue is related to the teleportation of an entangled state requiring entanglement swapping [136]–[138]. In a 2003 experiment related to continuous-variable entanglement swapping [131], a highly correlated four-partite entangled state was generated using two sources, each providing two entangled pulses whose amplitude noises were entangled using the experimental scheme shown in Figure 10.26. It was found experimentally that the amplitudes of the four optical pulses were quantum-correlated and exhibited noise that was 3-dB below the shot noise level, indicating the presence of four-pulse entanglement. Potential applications of entangled pulses include quantum key distribution required for quantum cryptography, a topic we turn to next.

10.5   Quantum Cryptography

Among the applications of quantum phenomena in optics, the field of quantum cryptography has advanced considerably in recent years [139]–[143]. In addition to a 2002 review [90], several recent articles have summarized the progress realized in this field [144]–[146]. In fact, the development has been so rapid that quantum cryptography has reached the commercial stage [147]. This section describes briefly the main ideas behind quantum cryptography with emphasis on the role of photon-pair sources.

Cryptography is an old art used to communicate a secret message between two parties. Its use has become much more common with the advent of the Internet. A commonly used technique is based on an asymmetrical public-key cryptography system. It relies on the difficulty in factoring a very large integer formed by multiplying two large prime numbers. In principle, such a system is not completely secure and may be compromised in future with the advent of quantum computers .

Quantum cryptography employs quantum mechanics to guarantee secure communication. The basic idea is to develop a scheme that allows two parties to share a "quantum key" in such a fashion that the presence of an eavesdropper trying to obtain the key is revealed. Since such a quantum key must be transmitted over a classical channel (often a fiber-optic channel), the quantum effects must be employed to guarantee its secure transmission. The resulting scheme is referred to as *quantum key distribution*. It makes use of the fact that any measurement of a quantum system produces disturbances that can be detected by the intended recipient, thus revealing the presence of an eavesdropper. Quantum entanglement is often employed for this purpose. It should be stressed that quantum cryptography is used only to distribute a secret key. However, once a "secret key" has been transmitted with success, it can be used to encrypt any message over a standard classical communication channel.

In one scheme, employed often for quantum key distribution and known as the *BB84 protocol* after the 1984 paper of Bennett and Brassard, single photons are transmitted using two different polarization bases [90]. For example, in one basis, bits 0 and 1 may correspond to horizontal and vertical SOPs, respectively, but the two SOPs are rotated by 45° in the second basis. The sender and the recipient (called Alice and Bob, traditionally) choose these two bases for each bit randomly. After Bob has measured all the photons, they share the basis information over a public classical channel, and they discard the bits for which Bob used a different basis. The remaining half of the bits (on average) constitute the shared key.

Another scheme, proposed in 1991 by Ekert [148], makes use of a source of photon pairs exhibiting polarization entanglement. These photons are distributed such that Alice and Bob each end up with one photon from each pair. They measure the SOPs of incoming photons using two randomly switched, nonorthogonal bases, similar to the case of the BB84 protocol. The resulting random string of bits is then used to construct a shared key. Any attempt at eavesdropping weakens the pair correlation in such a way that Alice and Bob can detect the attempt.

The BB84 scheme has been used extensively because it does not require a source of entangled photons. Often, optical pulses are attenuated to the extent that each pulse carries less than one photon on average. This technique was used in 2002 to send single photons through a fiber running under Lake Geneva [90]. However, even when the average number of photons per pulse is optimized for a single-photon source, the probability of two photons in a single pulse is not negligible. Such systems are therefore vulnerable to the so-called photon-number-splitting attack [149]. The use of decoy pulses can solve this problem to a large extent [150].

Another approach makes use of a photon-pair source. Since the detection of one of the photons forming a pair guarantees the existence of a single photon in that time slot, this information can be used to "herald" single photons to the receiver, resulting in a secure distribution of the quantum key. In a 2007 experiment, quantum key distribution over 40 km of optical fiber was realized using such a pulse-heralded, single-photon source [143]. Entangled photons were used as early as 2000 for demonstrating quantum key distribution [151]–[153]. By 2004, polarization-entangled photons were employed for this purpose over a 1.45-km-long fiber link installed in the Vienna sewer network [139].

Recent experiments have increased the fiber length over which a quantum key can be distributed securely to beyond 100 km [154]–[156]. The main problem from a practical standpoint is related to fiber losses that decrease the signal strength at the receiver. In contrast with the classical communication systems, optical amplifiers or regenerators cannot be employed for compensating fiber losses because they affect the same quantum properties on which the security of a quantum key is based. Therefore, the only solution is to minimize fiber losses by operating in the wavelength region near 1550 nm and to employ as sensitive single-photon detectors as possible.

In a 2006 experiment, the use of ultra-low-noise, transition-edge sensor, single-photon detectors made it possible to distribute a quantum key over 148.7-km of fiber by using weak coherent pulses with a mean photon number of 0.1 per pulse [154]. However, this distance was reduced to 67.5 km to ensure security against photon-number-splitting attacks. In a later experiment, secure key distribution that was immune to photon-number-splitting attacks was realized over 107 km of optical fiber by employing decoy-state protocol [155]. Further progress was made in a 2007 experiment that demonstrated quantum key distribution over 200 km of fiber by using a technique known as differential phase-shift keying [156]. In this modulation format, the information is coded in the phase difference between two neighboring pulses. The use of such coding makes the system robust to photon-number-splitting attacks [157]. A Mach–Zehnder interferometer with a 1-bit delay between its two arms converts phase information into amplitude variations. Two single-photon detectors, based on super-conducting nanowires, were employed for constructing the quantum key securely even after 40 dB of losses over a 200-km-long fiber link.

The important question is whether the nonlinear effects within a fiber have played any role in these experiments on quantum cryptography. The answer is negative so far. Even the experiments making use of photon-pair sources have employed spontaneous parametric down-conversion inside a crystal or planar waveguide. The situation may change in the future. As discussed in Section 10.4, FWM inside optical fiber has been used to realize photon pairs that are polarization-entangled. Such a source of entangled photon pairs operates in the wavelength region near 1550 nm and can launch these photons into a fiber link with minimum coupling losses. These advantages point to the potential applications of nonlinear fiber optics for quantum information processing.

Problems

10.1 Verify that the quantum NLS equation is recovered from the Heisenberg equation of motion with the Hamiltonian given in Eq. (10.1.12).

10.2 Prove by using Eq. (10.1.13) that self-phase modulation does not affect the photon flux operator $\hat{A}^{\dagger}\hat{A}$ along the fiber.

10.3 Prove that the normal-ordered form of the operator $\exp(c\hat{a}^{\dagger}\hat{a})$ is given by : $\exp[(e^c - 1)\hat{a}^{\dagger}\hat{a}]$:, where c is a constant and the operator $\hat{a}$ satisfies the commutation relation $[\hat{a}, \hat{a}^{\dagger}] = 1$.

10.4 Prove that the variances of the two quadrature components $\hat{X}$ and $\hat{Y}$, defined as in Eq. (10.2.1), are equal ($\sigma_X = \sigma_Y$) in the case of a coherent state. Also provide their numerical values.

10.5 Solve the FWM equations (10.2.2) and (10.2.3) and verify that the solution is indeed given by Eqs. (10.2.4) and (10.2.5).

10.6 Derive Eq. (10.2.10) from Eq. (10.2.9). You may consult Ref. [12].

10.7 Explain how cross-phase modulation inside a birefringent fiber can be exploited to produce a squeezed vacuum.

10.8 What is meant by *amplitude squeezing*? Discuss how this kind of squeezing can be realized using self-phase modulation in combination with an optical filter.

10.9 Discuss how four-wave mixing inside an optical fiber can be employed for creating entangled photon pairs in the wavelength regime near 1500 nm. What is the source of quantum correlation between the signal and idler photons forming a pair?

10.10 Use the solution given in Eqs. (10.2.4) and (10.2.5) and calculate the correlation parameter defined in Eq. (10.4.2). You may consult Ref. [112].

10.11 Discuss a technique capable of creating polarization entanglement between the photon pairs created through spontaneous FWM inside an optical fiber.

10.12 Explain what is meant by time-bin entanglement. Sketch an experimental setup for creating this type of entanglement.

References

[1] P. D. Drummond and S. J. Carter, *J. Opt. Soc. Am. B* **4**, 1565 (1987).

[2] Y. Lai and H. A. Haus, *Phys. Rev. A* **40**, 844 (1989); **40**, 854 (1989).

[3] P. D. Drummond and J. F. Corney, *J. Opt. Soc. Am. B* **18**, 139 (2001).

[4] J. F. Corney and P. D. Drummond, *J. Opt. Soc. Am. B* **18**, 153 (2001).

[5] L. Mandel and E. Wolf, *Optical Coherence and Quantum Optics* (Cambridge University Press, New York, 1995).

[6] G. P. Agrawal, *Nonlinear Fiber Optics*, 4th ed. (Academic Press, Boston, 2007).

[7] M. Kitagawa and Y. Yamamoto, *Phys. Rev. A* **34**, 3974 (1986).

[8] E. M. Wright, *J. Opt. Soc. Am. B* **7**, 1142 (1990).

[9] K. J. Blow, R. Loudon, and S. J. D. Phoenix, *J. Opt. Soc. Am. B* **8**, 1750 (1991).

[10] F. X. Kärtner, L. Joneckis, and H. A. Haus, *Quantum Optics* **6**, 379 (1992).

[11] L. G. Joneckis and J. H. Shapiro, *J. Opt. Soc. Am. B* **10**, 1102 (1993).

[12] K. J. Blow, R. Loudon, and S. J. D. Phoenix, *J. Mod. Opt.* **40**, 2515 (1993).

[13] I. Abram and I. Cohen, *J. Mod. Opt.* **41**, 847 (1994).

[14] L. Boivin, F. X. Kärtner, and H. A. Haus, *Phys. Rev. Lett.* **73**, 240 (1994).

[15] F. X. Kärtner, D. J. Dougherty, H. A. Haus, and E. P. Ippen, *J. Opt. Soc. Am. B* **11**, 1267 (1994).

[16] P. L. Voss, K. G. Köprülü, and P. Kumar, *J. Opt. Soc. Am. B* **23**, 598 (2006).

[17] H. P. Thacker, *Rev. Mod. Phys.* **53**, 253 (1981).

[18] H. A. Haus, K. Watanabe, and Y. Yamamoto, *J. Opt. Soc. Am. B* **6**, 113 (1989).

[19] S. J. Carter, *Phys. Rev. A* **51**, 3274 (1995).

[20] H. A. Haus and Y. Lai, *J. Opt. Soc. Am. B* **7**, 386 (1990).

[21] M. D. Levenson, R. M. Shelby, A. Aspect, M. Reid, and D. F. Walls, *Phys. Rev. A* **32**, 1550 (1985).

[22] R. Loudon and P. L. Knight, *J. Mod. Opt.* **34**, 709 (1987).

[23] D. F. Walls and G. J. Milburn, *Quantum Optics* (Springer, New York, 1994).

[24] A. Sizmann and G. Leuchs, in *Progress in Optics*, Vol. 39, E. Wolf, Ed. (Elsevier, Boston, 1999), Chap. 5.

[25] H.-A. Bachor and T. C. Ralph, *A Guide to Experiments in Quantum Optics*, 2nd ed. (Wiley, Hoboken, NJ, 2003).

[26] P. D. Drummond and Z. Ficek, *Quantum Squeezing* (Springer, New York, 2004).

[27] R. M. Shelby, M. D. Levenson, S. H. Perlmutter, R. G. De Voe and D. F. Walls, *Phys. Rev. Lett.* **57**, 691 (1986).

[28] R. M. Shelby, M. D. Levenson, and P. W. Bayer, *Phys. Rev. B* **31**, 5244 (1985).

[29] B. Schumaker, S. H. Perlmutter, R. M. Shelby, and M. D. Levenson, *Phys. Rev. Lett.* **58**, 357 (1987).

[30] P. D. Drummond, S. J. Carter, and R. M. Shelby, *Opt. Lett.* **14**, 373 (1989).

[31] M. Shirasaki and H. A. Haus, *J. Opt. Soc. Am. B* **7**, 30 (1990).

[32] M. Rosenbluh and R. M. Shelby, *Phys. Rev. Lett.* **66**, 153 (1991).

[33] K. Bergman and H. A. Haus, *Opt. Lett.* **16**, 663 (1991).

[34] K. Bergman, C R. Doerr, H. A. Haus, and M. Shirasaki, *Opt. Lett.* **18**, 643 (1993).

[35] K. Bergman, H. A. Haus, E. P. Ippen, and M. Shirasaki, *Opt. Lett.* **19**, 290 (1994).

[36] M. Margalit, C. X. Yu, E. P. Ippen, and H. A. Haus, *Opt. Express* **2**, 72 (1998).

[37] S. R. Friberg, S. Machida, M. J. Werner, A. Levanon, and T. Mukai, *Phys. Rev. Lett.* **77**, 3775 (1996).

[38] M. J. Werner, *Phys. Rev. A* **54**, 2567 (1996).

[39] M. J. Werner and S. R. Friberg, *Phys. Rev. Lett.* **79**, 4143 (1997)

[40] A. Mecozzi and P. Kumar, *Opt. Lett.* **22**, 1232 (1997).

[41] D. Levandovsky, M. Vasilyev, and P. Kumar, *Opt. Lett.* **24**, 43 (1999).

[42] S. Spälter, N. Korolkova, F. König, A. Sizmann, and G. Leuchs, *Phys. Rev. Lett.* **81**, 786 (1998).

[43] E. Schmidt, L. Knöll, D. G. Welsch, M. Zielonka, F. König, and A. Sizmann, *Phys. Rev. Lett.* **85**, 3801 (2000).

[44] T. Opatrný, N. Korolkova, and G. Leuchs, *Phys. Rev. A* **66**, 053813 (2002).

[45] R.-K. Lee, Y. Lai, and Y. S. Kivshar, *Phys. Rev. A* **71**, 013816 (2005).

[46] S. Spälter, M. Burk, U. Strössner, M. Böhm, A. Sizmann, and G. Leuchs, *Europhys. Lett.* **38**, 335 (1997).

[47] S. Spälter, M. Burk, U. Strössner, A. Sizmann, and G. Leuchs, *Opt. Express* **2**, 77 (1998).

[48] F. König, S. Spälter, I. L. Shumay, A. Sizmann, T. Fauster, and G. Leuchs, *J. Mod. Opt.* **45**, 2425 (1998).

[49] S. Schmitt, J. Ficker, M. Wolff, F. König, A. Sizmann, and G. Leuchs, *Phys. Rev. Lett.* **81**, 2446 (1998).

[50] D. Krylov and K. Bergman, *Opt. Lett.* **23**, 1390 (1998).

[51] D. Krylov, K. Bergman, and Y. Lai, *Opt. Lett.* **24**, 774 (1999).

[52] M. Fiorentino, J. E. Sharping, P. Kumar, D. Levandovsky, and M. Vasilyev, *Phys. Rev. A* **64**, 031801 (2001).

[53] S. Lorentz, C. Silberhorn, N. Korolkova, R. S. Windeler, and G. Leuchs, *Appl. Phys. B* **73**, 855 (2001).

[54] M. Fiorentino, J. E. Sharping, P. Kumar, and A. Porzio, *Opt. Express* **10**, 128 (2002).

[55] M. Fiorentino, J. E. Sharping, P. Kumar, A. Porzio, and R. S. Windeler, *Opt. Lett.* **27**, 649 (2002).

[56] M. Meissner, C. Marquardt, J. Heersink, T. Gaber, A. Wietfeld, G. Leuchs, and U. L. Andersen *J. Opt. B* **6**, S652 (2004).

[57] H. H. Ritze and A. Bandilla, *Opt. Commun.* **29**, 126 (1979).

[58] D. Levandovsky, M. Vasilyev, and P. Kumar, *Opt. Lett.* **24**, 984 (1999).

[59] K. Hirosawa, H. Furumochi, A. Tada, and F. Kannari, *Phys. Rev. Lett.* **94**, 203601 (2005).

[60] A. Tada, K. Hirosawa, F. Kannari, M. Takeoka, and M. Sasaki, *J. Opt. Soc. Am. B* **24**, 691 (2007).

[61] A. S. Chirkin, A. A. Orlov, and D. Y. Paraschuk, *Quantum Electron.*, **23**, 870 (1993).

[62] N. V. Korolkova and A. S. Chirkin, *J. Mod. Opt.* **43**, 869 (1996).

[63] J. Hald, J. L. Sørensen, C. Schori, and E. S. Polzik, *J. Mod. Opt.* **47**, 2599 (2001).

[64] N. Korolkova, G. Leuchs, R. Loudon, T. C. Ralph, and C. Silberhorn, *Phys. Rev. A* **65**, 052306 (2002).

[65] W. P. Bowen, R. Schnabel, H.-A. Bachor, and P. K. Lam, *Phys. Rev. Lett.* **88**, 093601 (2002).

[66] W. P. Bowen, N. Treps, R. Schnabel, and P. K. Lam, *Phys. Rev. Lett.* **89**, 253601 (2002).

[67] J. Heersink, T. Gaber, S. Lorenz, O. Glöckl, N. Korolkova, and G. Leuchs, *Phys. Rev. A* **68**, 013815 (2002).

[68] J. Heersink, V. Josse, G. Leuchs, and U. L. Andersen, *Opt. Lett.* **30**, 1192 (2005).

[69] F. Popescu, *J. Opt. B* **7**, 70 (2005).

[70] J. F. Corney, P. D. Drummond, J. Heersink, V. Josse, G. Leuchs, and U. L. Andersen, *Phys. Rev. Lett.* **97**, 023606 (2006).

[71] V. Scarani, S. Iblisdir, N. Gisin, and A. Acín, *Rev. Mod. Phys.* **77**, 1225 (2005).

[72] V. B. Braginsky, V. I. Vorontsov, and K. S. Thorne, *Science* **209**, 547 (1980).

[73] M. D. Levenson, R. M. Shelby, M. Reid, and D. F. Walls, *Phys. Rev. Lett.* **57**, 2473 (1986).

[74] H A. Bachor, M. D. Levenson, D. F. Walls, S. H. Perlmutter, and R. M. Shelby, *Phys. Rev. A* **38**, 180 (1988).

[75] S. R. Friberg, S. Machida, and Y. Yamamoto, *Phys. Rev. Lett.* **69**, 3165 (1992).

[76] P. D. Drummond, J. Breslin, and R. M. Shelby, *Phys. Rev. Lett.* **73**, 2837 (1994).

[77] J. M. Courty, S. Spälter, F. König, A. Sizmann, and G. Leuchs, *Phys. Rev. A* **58**, 1501 (1998).

[78] S. R. Friberg, T. Mukai, and S. Machida, *Phys. Rev. Lett.* **84**, 59 (2000).

[79] F. König, B. Buchler, T. Rechtenwald, G. Leuchs, and A. Sizmann, *Phys. Rev. A* **66**, 043810 (2002).

[80] A. La Porta, R. E. Slusher, and B. Yurke, *Phys. Rev. Lett.* **62**, 28 (1989).

[81] S. F. Pereira, Z. Y. Ou, and H. J. Kimble, *Phys. Rev. Lett.* **72**, 214 (1994).

[82] K. Bencheikh, J. A. Levenson, P. Grangier, and O. Lopez, *Phys. Rev. Lett.* **75**, 3422 (1995).

[83] R. Bruckmeier, K. Schneider, S. Schiller, and J. Mlynek, *Phys. Rev. Lett.* **78**, 1243 (1997).

[84] R. Filip, *Phys. Rev. A* **67**, 042111 (2003).

[85] U. L. Andersen, O. Glöckl, S. Lorenz, G. Leuchs, and R. Filip, *Phys. Rev. Lett.* **93**, 100403 (2004).

[86] Y. Aharonov and M. S. Zubairy, *Science* **307**, 875 (2005).

[87] R. Clifton, J. Butterfield, and H. Halvorson, *Quantum Entanglements: Selected Papers* (Oxford University Press, New York, 2005).

[88] I. Bengtsson and K. Zyczkowski, *Geometry of Quantum States: An Introduction to Quantum Entanglement* (Cambridge University Press, New York, 2006).

[89] V. Vedral, *Introduction to Quantum Information Science* (Oxford University Press, New York, 2007).

[90] N. Gişin, G. G. Ribordy, W. Tittel, and H. Zbinden, *Rev. Mod. Phys.* **74**, 145 (2002).

[91] M. Le Bellac, *A Short Introduction to Quantum Information and Quantum Computation* (Cambridge University Press, New York, 2006).

[92] G. Van Assche, *Quantum Cryptography and Secret-Key Distillation* (Cambridge University Press, New York, 2006).

[93] D. Bouwmeester, A. K. Ekert, and A. Zeilinger, *The Physics of Quantum Information: Quantum Cryptography, Quantum Teleportation, and Quantum Computation* (Springer, New York, 2007).

[94] S. J. Freedman and J. F. Clauser, *Phys. Rev. Lett.* **28**, 938 (1978).

[95] D. N. Klyshko, *Photons and Nolinear Optics* (Gordon and Breach, New York, 1988).

[96] D. C. Burnham and D. L. Weinberg, *Phys. Rev. Lett.* **25**, 84 (1970).

[97] Z. Y. Ou and L. Mandel, *Phys. Rev. Lett.* **61**, 50 (1988).

[98] P. G. Kwiat, K. Mattle, H. Weinfurter, and A. Zeilinger, *Phys. Rev. Lett.* **75**, 4337 (1995).

[99] L. J. Wang, C. K. Hong, and S. R. Friberg, *J. Opt. B* **3**, 346 (2001).

[100] J. E. Sharping, M. Fiorentino, and P. Kumar, *Opt. Lett.* **26**, 367 (2001).

[101] M. Fiorentino, P. L. Voss, J. E. Sharping, and P. Kumar, *IEEE Photon. Technol. Lett.* **14**, 983 (2002).

[102] X. Li, J. Chen, P. Voss, J. Sharping, and P. Kumar, *Opt. Express* **12**, 3737 (2004).

[103] H. Takesue and K. Inoue, *Opt. Express* **13**, 7832 (2005).

[104] J. E. Sharping, J. Chen, X. Li, P. Kumar, and R. S. Windeler, *Opt. Express* **12**, 3086 (2004).

[105] J. G. Rarity, J. Fulconis, J. Duligall, W. J. Wadsworth, and P. St. J. Russell, *Opt. Express* **13**, 534 (2005).

[106] J. Fan, A. Dogariu, and L. J. Wang, *Opt. Lett.* **30**, 1530 (2005).

[107] J. Fan and A. Migdall, *Opt. Express* **13**, 5777 (2005).

[108] J. Fulconis, O. Alibart, W. J. Wadsworth, P. St. J. Russell, and J. G. Rarity, *Opt. Express* **13**, 7572 (2005).

[109] J. Fan, A. Migdall, and L. J. Wang, *Opt. Lett.* **30**, 3368 (2005).

[110] P. L. Voss and P. Kumar, *J. Opt. B* **6**, S762 (2004).

[111] Q. Lin, F. Yaman, and G. P. Agrawal, *Opt. Lett.* **31**, 1286 (2006).

[112] Q. Lin, F. Yaman, and G. P. Agrawal, *Phys. Rev. A* **75**, 023803 (2007).

[113] P. G. Kwiat, E. Waks, A. G. White, I. Appelbaum, and P. H. Eberhard, *Phys. Rev. A* **60**, 773 (1999).

[114] A. Yoshizawa and H. Tsuchida, *Appl. Phys. Lett.* **85**, 2457 (2004).

[115] F. König, E. J. Mason, F. N. C. Wong, and A. Albota, *Phys. Rev. A* **71**, 033805 (2005).

[116] H. Takesue and K. Inoue, *Phys. Rev. A* **70**, 031802 (2004).

[117] X. Li, P. L. Voss, J. E. Sharping, and P. Kumar, *Phys. Rev. Lett.* **94**, 053601 (2005).

[118] X. Li, P. L. Voss, J. Chen, J. E. Sharping, and P. Kumar, *Opt. Lett.* **30**, 1201 (2005).

[119] K. F. Lee, J. Chen, C. Liang, X. Li, P. L. Voss, and P. Kumar, *Opt. Lett.* **31**, 1905 (2006).

[120] J. Chen, K. F. Lee, C. Liang, and P. Kumar, *Opt. Lett.* **31**, 2798 (2006).

[121] X. Li, C. Liang, K. F. Lee, J. Chen, P. L. Voss, and P. Kumar, *Phys. Rev. A* **73**, 052301 (2006).

[122] C. Liang, K. F. Lee, T. Levin, J. Chen, and P. Kumar, *Opt. Express* **14**, 6936 (2006).

[123] J. F. Clauser, M. A. Horne, A. Shimony, and R. A. Holt, *Phys. Rev. Lett.* **23**, 880 (1969).

[124] J. Brendel, N. Gisin, W. Tittel, and H. Zbinden, *Phys. Rev. Lett.* **82**, 2594 (1999).

[125] I. Marcikic, H. de Reidmatten, W. Tittel, H. Zbinden, M. Legre, and N. Gisin, *Phys. Rev. Lett.* **93**, 180502 (2004).

[126] H. Takesue and K. Inoue, *Phys. Rev. A* **72**, 041804 (2005).

[127] H. Takesue, *Opt. Express* **14**, 3453 (2006).

[128] C. Silberhorn, P. K. Lam, O. Weiss, F. König, N. Korolkova, and G. Leuchs, *Phys. Rev. Lett.* **86**, 4267 (2001).

[129] P. van Loock, *Fortschritte Phys.* **50** 1177 (2002).

[130] C. Silberhorn, T. C. Ralph, N. Ltkenhaus, and G. Leuchs, *Phys. Rev. Lett.* **89**, 167901 (2002).

[131] O. Glöckl, S. Lorenz, C. Marquardt, J. Heersink, M. Brownnutt, C. Silberhorn, Q. Pan, P. van Loock, N. Korolkova, and G. Leuchs, *Phys. Rev. A* **68**, 012319 (2003).

[132] O. Glöckl, J. Heersink, N. Korolkova, G. Leuchs, and S. Lorenz, *J. Opt. B* **5**, S492 (2003).

[133] O. Glöckl, U. L. Andersen, and G. Leuchs, *Phys. Rev. A* **73**, 012306 (2006).

[134] S. Lorenz, J. Rigas, M. Heid, U. L. Andersen, N. Lütkenhaus, and G. Leuchs, *Phys. Rev. A* **74**, 042326 (2006).

[135] A. Einstein, B. Podolsky, and N. Rosen, *Phys. Rev.* **47**, 777 (1935).

[136] J. W. Pan, D. Bouwmeester, H. Weinfurter, and A. Zeilinger, *Phys. Rev. Lett.* **80**, 3891 (1998).

[137] R. E. S. Polkinghorne and T. C. Ralph, *Phys. Rev. Lett.* **83**, 2095 (1999).

[138] Y. H. Kim, S. P. Kulik, and Y. Shih, *Phys. Rev. Lett.* **86**, 1370 (2001).

[139] A. Poppe, A. Fedrizzi, R. Ursin, H. Böhm, T. Lörunser, O. Maurhardt, M. Peev, M. Suda, C. Kurtsiefer, H. Weinfurter, T. Jennewein, and A. Zeilinger, *Opt. Express* **12**, 3865 (2004).

[140] K. Gordon, V. Fernandez, G. Buller, I. Rech, S. Cova, and P. Townsend, *Opt. Express* **13**, 3015 (2005).

[141] K. Inoue, *IEEE J. Sel. Topics Quantum Electron.* **12**, 888 (2006).

[142] E. Diamanti, H. Takesue, C. Langrock, M. M. Fejer, and Y. Yamamoto, *Opt. Express* **14**, 13073 (2006).

[143] A. Soujaeff, T. Nishioka, T. Hasegawa, S. Takeuchi, T. Tsurumaru, K. Sasaki, and M. Matsui, *Opt. Express* **15**, 726 (2007).

[144] N. J. Cerf and P. Grangier, *J. Opt. Soc. Am. B* **24**, 324 (2007).

[145] N. Gisin and R. Thew, *Nature Photon.* **1**, 165 (2007).

[146] D. Bruss, G. Erdélyi, T. Meyer, T. Riege, and J. Rothe, *ACM Computing Surveys* **39**, Article 6 (2007).

[147] Companies involved include BBN Technologies, MagiQ Technologies, and id Quantique.

[148] A. K. Ekert, *Phys. Rev. Lett.* **67**, 661 (1991).

[149] G. Brassard, N. Lütkenhaus, T. Mor, and B. C. Sanders, *Phys. Rev. Lett.* **85**, 1320 (2000).

[150] Y. Zhao, B. Qi, X. Ma, H.-K. Lo and L. Qian, *Phys. Rev. Lett.* **96**, 070502 (2006).

[151] T. Jennewein, C. Simon, G. Weihs, H. Weinfurter, and A. Zeilinger, *Phys. Rev. Lett.* **84**, 4729 (2000).

[152] D. S. Naik, C. G. Peterson, A. G. White, A. J. Berglund, and P. G. Kwiat, *Phys. Rev. Lett.* **84**, 4733 (2000).

[153] W. Tittel, J. Brendel, H. Zbinden, and N. Gisin, *Phys. Rev. Lett.* **84**, 4737 (2000).

[154] P. A. Hiskett, D. Rosenberg, C. G. Peterson, R. J. Hughes, S. Nam, A. E. Lita, A. J. Miller, and J. E. Nordholt, *New J. Phys.* **8**, 193 (2006).

[155] D. Rosenberg, J. W. Harrington, P. R. Rice, P. A. Hiskett, C. G. Peterson, R. J. Hughes, A. E. Lita, S. W. Nam, and J. E. Nordholt, *Phys. Rev. Lett.* **98**, 010503 (2007).

[156] H. Takesue, S. W. Nam, Q. Zhang, R. H. Hadfield, T. Honjo, K. Tamaki, and Y. Yamamoto, *Nature Photon.* **1**, 343 (2007).

[157] K. Inoue and T. Honjo, *Phys. Rev. A* **71**, 042305 (2005).

Appendix A

Acronyms

Each scientific field has its own jargon, and the field of nonlinear fiber optics is no exception. Although an attempt was made to avoid extensive use of acronyms, many still appear throughout the book. Each acronym is defined the first time it appears in a chapter so that the reader does not have to search the entire text to find its meaning. As a further help, all acronyms are listed here in alphabetical order.

AM	amplitude modulation
APD	avalanche photodiode
ASE	amplified spontaneous emission
ASK	amplitude-shift keying
AWG	arrayed-waveguide grating
BPF	band-pass filter
CARS	coherent anti-Stokes Raman scattering
CPA	chirped-pulse amplification
CSRZ	carrier-suppressed return-to-zero
CW	continuous wave
DBR	distributed Bragg reflector
DCF	dispersion-compensating fiber
DDF	dispersion-decreasing fiber
DFB	distributed feedback
DM	dispersion managed
DPSK	differential phase-shift keying
DSF	dispersion-shifted fiber
EDFA	erbium-doped fiber amplifier
EDFL	erbium-doped fiber laser
FM	frequency modulation
FOPA	fiber-optic parametric amplifier
FROG	frequency-resolved optical gating
FWHM	full width at half maximum
FWM	four-wave mixing
GVD	group-velocity dispersion

ITU   International Telecommunication Union
MOPA   master oscillator–power amplifier
MZI   Mach–Zehnder interferometer
NLS   nonlinear Schrödinger
NALM   nonlinear amplifying-loop mirror
NOLM   nonlinear optical-loop mirror
NRZ   nonreturn-to-zero
OCT   optical coherence tomography
OOK   on–off keying
OTDM   optical time-domain multiplexing
PBS   polarizing beam splitter
PCF   photonic crystal fiber
PMF   polarization-maintaining fiber
QND   quantum nondemolition
RIFS   Raman-induced frequency shift
RIN   relative intensity noise
RMS   root-mean-square
RZ   return-to-zero
SBS   stimulated Brillouin scattering
SIT   self-induced transparency
SLALOM   semiconductor laser amplifier in a loop mirror
SNR   signal-to-noise ratio
SOA   semiconductor optical amplifier
SOP   state of polarization
SPM   self-phase modulation
SpRS   spontaneous Raman scattering
SQL   standard quantum limit
SRS   stimulated Raman scattering
TOAD   terahertz optical asymmetric demultiplexer
TOD   third-order dispersion
UNI   ultrafast nonlinear interferometric
WDM   wavelength-division multiplexing
XPM   cross-phase modulation

Index

absorption coefficient, 142, 211, 228
acoustic wave, 310, 459
add–drop multiplexer, 87, 376
air–silica interface, 402
Airy formula, 102
amplification
    chirped-pulse, 273–278, 431
    distributed, 148
    lumped, 149, 328–330
    parametric, 119, 120, 358, 360
    periodic, 149
    phase-sensitive, 389
    pulse, 158–169, 173
    Raman, *see* Raman amplification
    ultrashort pulse, 166
amplified spontaneous emission, 138, 141, 142,
        152, 305
amplifier
    cascaded, 305
    chain of, 305
    distributed-gain, 142
    fiber, *see* fiber amplifier
    in-line, 313
    lumped, 150, 319
    optical, 304, 317
    parametric, 169, 277, 291, 360, 385, 416,
        469, 470
    phase-sensitive, 469
    Raman, *see* Raman amplifier
    semiconductor optical, *see* semiconduc-
        tor optical amplifier
    unsaturated regime of, 134
    Yb-doped fiber, 164
anti-resonance condition, 108
anti-Stokes wave, 420, 476
apodization technique, 18
atomic clock, 427, 432
autocorrelation trace, 61, 110, 217, 257, 259,
        264, 277, 280, 441
avalanche photodiode, 476, 485

back-action evading measurement, 471, 474
balanced detection, 424, 474
bandwidth
    amplifier, 135
    Brillouin-gain, 191, 310
    detector, 137
    filter, 305, 464
    laser, 191
    Raman-gain, 311
BB84 protocol, 489
beat length, 60, 88, 234
Bell inequality, 475, 481, 482
Bessel function, 56, 59
biological imaging, 420–427
birefringence, 37, 87, 88, 215, 278, 289, 402
    built-in, 219
    circular, 358, 375
    linear, 219, 235, 375
    nonlinear, 217, 263
    residual, 464
bistable switching, 23
    XPM-induced, 38
bit-error rate, 310, 319
Bloch equations, 145
Bloch wave, 26, 27, 30
Boltzmann constant, 477
Bose distribution, 455
Bragg condition, 2, 5, 11, 150
Bragg diffraction, 2, 151, 230
Bragg grating, *see* grating
Bragg reflector, 212, 234, 260
Bragg soliton, 29–33
Bragg wavelength, 7, 9, 34, 40, 194, 195, 270,
        272
Brillouin crosstalk, *see* crosstalk
Brillouin gain, 191, 309
Brillouin scattering, 2, 199, 308–311
    control of, 310
    guided acoustic-wave, 459, 464, 469, 473
    spontaneous, 308

stimulated, 105, 189, 191, 198, 308, 459
thermal, 459
threshold of, 308
Brillouin shift, 308

capillary-stacking method, 401, 435
carbon nanotube, 213
carrier-envelope offset frequency, 429, 432
carrier-envelope phase, 429
CARS technique, 421, 422
cavity
    all-fiber, 183
    design of, 182
    dispersion-managed, 218, 232
    Fabry–Perot, 18, 182, 184, 186, 199, 206,
        212, 218, 224, 433
    figure-8, 183
    loss in, 182
    ring, 183, 186, 192, 206, 218
    sigma-shape, 207, 221
chalcogenide glass, 273, 404
chalcohalide glass, 65
chaos, 26, 106, 107, 110, 196
chirp, 333
    amplifier-induced, 290
    dispersion-induced, 247
    fiber-induced, 253
    grating-induced, 271
    GVD-induced, 248
    linear, 162–164, 253, 278
    negative, 289
    nonlinear, 256, 271, 273
    positive, 250, 289
    random, 255
    SPM-induced, 253, 254, 262, 278, 290,
        315, 324, 380
    XPM-induced, 36, 286, 288, 319, 341
chirp parameter, 289, 315
chirped-pulse amplification, *see* amplification
coherence degradation, 357
coherence length, 424
coherence time, 423, 484
coherent state, 458, 460
coincidence fringe, 481
commutation relation, 451, 458, 459
compression factor, 248, 253, 254, 256, 257,
    259, 262, 263, 270, 272, 280, 282,
    290, 294
compressor
    Bragg-grating, 268

cascaded, 264
compact, 271
design of, 252
four-stage, 264
grating-fiber, 249–261, 264, 291
optimization of, 262
pedestal in, 262
soliton-effect, 249, 261–268, 286, 289
two-stage, 257, 260, 264, 289
confinement factor, 436
continuum radiation, *see* radiation
conversion efficiency, 189, 360, 362, 385
correlation function, 454
coupled-mode equations, 86
    frequency-domain, 11, 56
    linear, 13
    nonlinear, 12, 20, 26, 62, 271, 272
    time-domain, 12, 56
coupled-mode theory, 11, 19, 55
coupler
    3-dB, 59
    active, 83
    asymmetric, 55, 57, 72, 81
    asymmetric dual-core, 86
    birefringent, 87
    coupled-mode equations for, 55
    directional, 102, 124, 224
    fiber, 182, 304, 363, 364, 388
    fused, 54
    grating-assisted, 85, 87
    intermodal dispersion in, 60
    linear theory of, 57
    multicore, 91
    paired solitons in, 76
    PCF, 90
    power transfer in, 58
    pulse propagation in, 70
    quasi-CW switching in, 62
    resonant, 124
    star, 89
    supermodes of, 60, 66, 68
    switching in, 64
    symmetric, 55, 57, 58, 62
    three-core, 92
    transfer matrix for, 59
    vector solitons in, 88
    WDM, 182, 223
coupling coefficient, 12, 18, 56, 59
    effective, 58
    frequency dependence of, 61

nonuniform, 42
    periodic modulation of, 85
coupling length, 58–61, 80, 89
coupling loss, *see* loss
    coupling, 206
cross state, 59, 63
cross-correlation, 213, 268
cross-gain saturation, 222, 223
cross-phase modulation, 12, 57, 115, 222, 256,
        273, 286, 317–321, 338–341, 353,
        363–368, 373, 417, 463
    intrachannel, 339–341
    QND measurement with, 474
crosstalk
    Brillouin-induced, 308–311
    FWM-induced, 321–324
    interchannel, 152
    Raman-induced, 311–314
    XPM-induced, 317–321
CSRZ format, 376
cubic phase distortion, 257

damage threshold, 261
decoherence, 484
decoy-state protocol, 490
delay-difference model, 197
delayed Raman response, 454
demultiplexer
    add–drop, 87
    terahertz optical asymmetric, 365
demultiplexing, 304, 371–376
dielectric coating, 182, 260
difference-frequency generation, 352
differential phase-shift keying, 320, 363, 376,
        490
diffraction length, 439
diffraction-limited beam, 190
digital logic, 118
dipole relaxation time, 134, 139, 156, 158
directional coupler, *see* coupler
dispersion
    anomalous, 159, 357, 389, 409, 418, 427,
        462
    chromatic, 424
    comblike, 283–285
    fourth-order, 282, 416
    grating-induced, 14
    group-velocity, 60, 324–327, 364, 426
    higher-order, 412, 414
    intermodal, 61, 78

limitations of, 306
    management of, 278
    material, 14, 405
    microstructured fiber, 402
    normal, 161, 164, 327, 331, 357, 415,
        426, 467
    photonic crystal fiber, 402
    polarization-mode, 61, 116, 172, 365
    second-order, *see* group-velocity disper-
        sion
    tailoring of, 426
    third-order, *see* third-order dispersion
    waveguide, 14, 405, 436
dispersion compensation, 15, 42, 121, 222, 270,
        314
dispersion curve, 403
    linear, 21
    nonlinear, 20
dispersion fluctuations, 416
dispersion length, 30, 60, 71, 80, 153, 328,
        338, 383, 437
    effective, 251
dispersion management, 115, 208, 209, 213,
        267, 293, 306, 308, 313, 316, 324,
        330–334
dispersion map, 320
    periodic, 330–334
dispersion parameter, 270, 405
dispersion relation, 13, 69, 149
dispersion slope, 362
dispersion-compensating fiber, *see* fiber
dispersion-decreasing fiber, *see* fiber
dispersive delay line, 221, 247, 257, 267, 268,
        272, 274, 276, 291
dispersive waves, 231, 327, 328, 338, 412
distributed amplification, 144, 148, 304, 306,
        329, 332, 343
distributed feedback, 5
double-pass configuration, 249, 252, 258, 277
DPSK format, 320, 363, 376, 387
dual-pump configuration, 361, 480

EDFA
    absorption spectrum of, 139
    C-band, 140
    cascaded, 140
    energy-levels of, 138
    gain flattening, 140
    gain spectrum of, 138
    L-band, 140

noise in, 143
pumping of, 138
rate-equation model for, 141
transient grating in, 45
effective cladding index, 403
effective core radius, 403
effective fiber length, 191
effective mode area, 12, 57, 170, 191, 261, 275, 309, 320, 357, 402, 404, 436
effective mode index, 41, 194
effective refractive index, 400, 403
electromagnetically induced transparency, 440
electron-beam lithography, 7
elliptic function, 62, 75
energy enhancement factor, 328
entanglement
    continuous-variable, 485–488
    EPR, 486
    four-partite, 488
    photn-pair, 475–480
    photon-pair, 486
    polarization, 480–485, 487, 489
    quantum, 475–488
    teleportation of, 488
    time-bin, 484
    verification of, 487
entanglement swapping, 488
EPR entanglement, *see* entanglement
erbium-doped fiber amplifier, *see* EDFA
error function, 354
etalon, 187, 192, 207
Euler–Lagrange equation, 73
evanescent wave coupling, 55
excited-state absorption, 141, 181, 188, 196
extrusion technique, 402
eye diagram, 353

Fabry–Perot cavity, *see* cavity
Fabry–Perot resonator, 101–105
    finesse of, 104
    free spectral range of, 103
    transmittivity of, 102
Faraday mirror, 207, 219, 483
Faraday rotator, 207, 219, 221, 277, 388
fast axis, 68, 116, 236, 365, 373, 482
feedback loop, 217, 434
fiber
    birefringent, 37, 60, 68, 87, 287, 482
    bismuth-oxide, 357, 375, 383, 404
    chalcogenide, 29, 270, 383

chalcohalide, 65
coiled, 189
dispersion-compensating, 152, 208, 213, 267, 276, 307, 320
dispersion-decreasing, 115, 173, 208, 264, 279–282, 330, 419
dispersion-flattened, 265, 374, 428, 432
dispersion-shifted, 115, 215, 259, 277, 280, 287, 309, 353, 477, 482
double-clad, 168, 188, 200, 219, 275, 278
dual-core, 54, 61, 64, 86, 89, 122, 224, 293
dye-doped, 65
endlessly single-mode, 404
erbium-doped, 138
fluorozirconate, 188
graded-index, 186
helical-core, 190
highly nonlinear, 209, 225, 261, 265, 267, 283, 355–357, 360, 362, 374, 379, 382, 385, 400–442
holey, 282, 400
hollow, 261, 288
hydrogen-soaked, 4
isotropic, 463
large-mode-area, 190, 261, 275, 276
lead-silicate, 404
microstructured, 167, 214, 217, 261, 267, 270, 276, 282, 357, 400–407, 411, 414, 418, 424, 430, 469, 478
multicore, 88, 294
multimode, 189, 276
nonsilica, 383
photonic bandgap, 265, 276, 401, 434–442
photonic crystal, 89, 158, 167, 200, 209, 261, 276, 278, 362, 375, 400–406
polarization-maintaining, 37, 87, 110, 113, 116, 207, 209, 219, 260, 356, 365, 366, 372, 385, 463, 468, 474
precompensation, 342
rare-earth-doped, 132, 180
reverse-dispersion, 268
semiconductor-doped, 65
spun, 375
standard, 115, 283, 306, 334
tight bending of, 276
transparent, 144
fiber amplifier, 132–173
    adiabatic amplification in, 159

bandwidth of, 135
basic concepts of, 132–137
cladding-pumped, 275
distributed, 144, 148, 159
erbium-doped, 138–144
gain of, 134
modulation instability in, 148–152
noise in, 136, 143
pulses in, 145, 158–169, 173
pumping of, 133
Sagnac loop with, 114
solitons in, 153–169
Yb-doped, 191, 200, 276, 278
fiber array, 91–94, 294
fiber coupler, *see* coupler
fiber dispersion, *see* dispersion
fiber laser
cavity of, 182
chaotic, 196, 197
cladding-pumped, 198, 221
coupled-cavity, 193, 224
DBR, 194
DFB, 193
distributed-feedback, 193
double-clad, 188
dual-frequency, 195, 283
erbium-doped, 180, 191–196, 425, 431, 432
figure-8, 215, 229, 431
line width of, 187, 192
mode-locked, 205–227, 419, 425, 431
modulation-instability, 227
multiwavelength, 194
Nd-doped, 180, 181, 186–188, 198, 221
output power of, 185
polarization effects in, 234
pumping of, 181
Q-switched, 198–201
self-pulsing in, 196
spectrum of, 230
stretched-pulse, 219
threshold condition for, 184, 185
threshold of, 184
timing jitter in, 208
Tm-doped, 181
tuning of, 187, 192
up-conversion, 181
Yb-doped, 188–191, 198, 225, 410, 416, 425
fiber resonator, 101–110

filter
acousto-optic, 140
add–drop, 86, 124
bandpass, 266, 289, 336, 380, 422, 465, 467, 477
birefringent, 187, 419
brick-wall, 466
comb, 195
Fabry–Perot, 209, 217, 227
grating-based, 16
guiding, 224
high-pass, 469
interference, 424
low-pass, 474
Mach–Zehnder, 122
narrowband, 224
notch, 356, 357
optical, 120, 140, 152, 208, 224, 305, 336, 378, 380, 384, 389, 392, 418, 464, 473
rocking, 88
sliding-frequency, 336
transfer function of, 305
tunable, 140, 383
WDM, 86
finesse, 102, 104
finite-element method, 402
fluorescence time, 134, 143, 147, 223
Fokker–Planck equation, 455
four-wave mixing, 2, 45, 119, 151, 169, 210, 227, 231, 256, 321–324, 338, 358, 368, 413–417
cascaded, 386, 414
dissipative, 225
efficiency of, 321
entanglement though, 475–484
intrachannel, 339, 341
resonant, 152
spontaneous, 459, 475
squeezing through, 458, 460
Fox–Smith resonator, 184
free-spectral range, 103, 207
frequency chirp, 154, 164, 168, 249, 289
frequency comb, 427–434
frequency jitter, 434
frequency-resolved optical gating, 164, 268
FROG trace, 164, 172, 268

gain bandwidth, 134, 186, 201
gain clamping, 143

gain dispersion, 155, 160, 165
gain flattening, 140
gain saturation, 134, 135, 185, 274, 290, 365, 386
gain spectrum
    amplifier, 138
    Brillouin, 191
    broadening of, 139
    flattening of, 140
    Raman, 80, 171, 191, 407, 455
gain switching, 116, 289
gain-recovery time, 366
gap soliton, 35
    coupled, 37
    Raman, 33
Gaussian statistics, 305
Geiger mode, 477, 485
ghost pulse, 340, 342
Ginzburg–Landau equation, 146, 148, 153, 156, 160, 203, 227, 229, 236, 291, 304
    generalized, 165
    quintic, 153
Gires–Tournois interferometer, 252, 260
Gordon–Haus jitter, *see* timing jitter
grating
    apodized, 18, 270, 272
    arrayed waveguide, 419, 477, 484
    Bragg, 4, 17, 83, 86, 102, 124, 143, 182, 187, 193, 224, 268, 311, 434, 484
    bulk, 274
    chalcogenide, 273
    chirped, 7, 15, 41, 194, 221, 224, 225, 260, 270, 274, 290
    couplers with, 85
    diffraction, 1
    dispersion relation for, 13
    dispersive, 187
    dynamic, 44
    external, 192
    fabrication of, 4–10
    fiber, 1, 187, 193, 224, 252, 268, 311, 357, 383, 391, 477
    group velocity in, 21
    index, 1, 150, 231
    linear properties of, 13
    long-period, 9, 39, 85, 140, 150, 412
    modulation instability in, 24
    nonlinear switching in, 33
    nonlinear-index, 152, 230
    nonlinearly chirped, 290

    nonuniform, 40
    optical bistability in, 22
    phase-shifted, 23, 40, 194
    polarization effects in, 37
    reduced speed in, 15
    sampled, 43, 194, 225, 391
    semiconductor, 34
    solitons in, 29
    stop band of, 14, 311
    superimposed, 225
    superstructure, 43
    transient, 44
    tunable, 357
grating pair, 250–252, 268, 274, 278, 288
    optimum separation of, 254
grazing angle, 251
group velocity, 12, 15, 20, 27, 57, 103, 266, 338, 429, 482
group-velocity dispersion, 12, 57, 103, 203
group-velocity mismatch, 116, 219, 223, 286, 313, 319, 320, 354, 365, 374, 416
GVD
    anomalous, 15, 25, 64, 70, 82, 109, 150, 159, 206, 213, 217, 218, 232, 249, 260, 262, 268, 290, 316, 323, 324, 382, 415, 426, 431
    average, 218
    grating-induced, 15, 24, 268, 269
    mismatch of, 83
    normal, 26, 64, 69, 81, 83, 108, 110, 150, 161, 213, 218, 232, 249, 268, 289, 291, 316, 324, 382, 407, 415, 431, 468
GVD parameter, 106, 324–327
    effective, 27, 251
    grating, 27

Hamiltonian, 73, 74, 452
Hankel function, 151
Hartree approximation, 455
Heisenberg uncertainty principle, 456, 471
helical core, 190
Helmholtz equation, 55
Hermite–Gauss function, 204, 331
heterodyne detection, 137, 430, 458
highly nonlinear fibers, *see* fiber
holographic technique, 5, 41
homodyne detection, 458, 487
homogeneous broadening, 133, 139
hydrogen soaking, 4, 9

hysteresis, 22, 105

idler wave, 359, 362, 414, 459, 476
inhomogeneous broadening, 139, 145, 157
intensity discriminator, 263
interferometer
    delay, 393
    Fabry–Perot, 101–110, 362
    Gires–Tournois, 252
    Mach–Zehnder, 121–125, 383, 462, 468
    Mach–Zender, 389
    Michelson, 125
    polarization-discriminating, 366
    Sagnac, 111–121, 353, 363, 385, 389, 463
    ultrafast nonlinear, 366
International Telecommunication Union, 418
inverse scattering method, 29, 30, 76, 153, 254, 325, 327, 408, 455
isolator, 144, 183, 217, 310, 388
    polarization-sensitive, 215
    polarizing, 217, 236
ITU grid, 419

Jacobi elliptic function, 62, 75

Kerr effect, 171
Kerr nonlinearity, 34, 90, 363, 410, 440
Kerr shutter, 217, 357, 375, 385
Kramers–Kronig relation, 3, 45

Lagrangian, 73
Langevin equation, 455
Langevin noise, 304
laser
    argon-ion, 2, 4, 191
    color-center, 110, 113, 263, 329
    DBR, 289
    DFB, 11, 23, 34, 116, 118, 193–195, 280, 418
    dye, 418
    excimer, 6
    external-cavity, 440
    fiber, *see* fiber laser
    figure-8, 215
    GaAs, 188, 191
    gain-switched, 419
    He–Ne, 116
    instabilities in, 196
    line width of, 194

    microchip, 200, 272
    mode-locked, 28, 64, 107
    Nd:glass, 9, 426
    Nd:YAG, 64, 107, 113, 118, 192, 258, 431
    Nd:YLF, 28, 32, 269
    Q-switched, 28, 107
    semiconductor, 23, 34, 138, 182, 186, 188, 192, 215, 280, 289, 417
        gain-switched, 281
    surface-emitting, 289
    threshold of, 184
    Ti:sapphire, 9, 105, 107, 108, 215, 260, 411, 413, 421, 423, 428, 430
laser ablation, 439
lightwave systems, 150
    coherent, 309, 320
    dispersion-managed, 306
    long-haul, 304, 315
    loss-managed, 304
    periodically amplified, 316
    pseudo-linear, 338–343
    terrestrial, 306
    undersea, 306
linear stability analysis, 24, 68
local oscillator, 459, 461, 463, 464
localized function approach, 402
logic gates, 37, 72, 118
longitudinal-mode spacing, 186, 201, 204, 208
loop mirror
    fiber, 112, 117
    parametric, 119
Lorentzian spectrum, 134
loss
    bending, 189
    cavity, 182, 184, 186, 203, 227
    compensation of, 304
    confinement, 403
    coupling, 490
    diffraction, 182, 252
    fiber, 304
    filter-induced, 465
    grating pair, 252
    insertion, 270
    intensity-dependent, 229
    internal, 184
    microstructured fiber, 402
    mode, 189
    nonlinear, 190
    pump, 185

spectral, 465
wavelength-selective, 192
loss management, 328–330

Mach–Zehnder interferometer, 121–125, 140,
         192, 473, 490
    applications of, 124
    asymmetric, 121, 485, 487
    chain of, 124
    nonlinear switching in, 121
    symmetric, 122
    unbalanced, 468
Markovian process, 305
master oscillator, 195
Maxwell–Bloch equations, 145, 156, 168, 209
metrology, 427–434
Michelson interferometer, 125, 287, 423
    nonlinear effects in, 125
microbending, 86
micromachining, 439
microscopy
    CARS, 420
    near-field, 420
microstructured fibers, *see* fiber
mirror
    amplifying-loop, 115, 183, 215
    Bragg, 198, 260
    chirped, 260
    dielectric, 182, 186
    fiber-loop, 111, 143, 183, 214, 292, 363
    loop, 365
    moving, 218
    nonlinear optical loop, 338, 353
    saturable absorber, 433
mode
    cladding, 39
    fundamental, 189
    higher-order, 168, 189
    longitudinal, 103
    temporal, 327
mode locking
    active, 202–205
    additive-pulse, 126, 211, 215, 224, 229
    AM, 202
    colliding-pulse, 225
    FM, 202
    harmonic, 205–210
    hybrid, 221–227
    interferometric, 126, 215
    master equation of, 203

passive, 85, 117, 125, 210–220
physics of, 201
regenerative, 208, 209
saturable-absorber, 211–214, 227
XPM-induced, 222
modulation
    amplitude, 419
    phase, 419
    synchronous, 337, 340
    synchronous phase, 338
modulation instability, 24, 67, 107, 148–152,
         266, 316, 317, 323, 357
    cavity-induced, 108
    experiments on, 28
    gain peak of, 227
    gain spectrum of, 25, 69, 108, 149, 152
    grating-induced, 272
    induced, 70, 151, 152, 266
    noise amplification by, 151
    self-induced, 227
    sidebands of, 150
    spontaneous, 70
modulator
    acousto-optic, 198, 199, 205, 224
    amplitude, 203, 205, 221, 337
    electro-absorption, 392
    electro-optic, 205
    external, 306
    FM, 225
    intracavity, 225
    $LiNbO_3$, 124, 206–208, 221, 277, 282,
         337, 428
    Mach–Zehnder, 209, 417
    phase, 205, 221, 288, 310, 337, 390
    spatial light, 422
molecular-beam epitaxy, 212
moment method, 407
MOPA configuration, 193, 195, 200
multiphoton ionization, 9
multiple-scale method, 26
multipole method, 402

NLS equation, 30, 81, 106, 145, 150, 204, 252,
         271, 279, 314, 325–327
    coupled, 67, 71, 91, 107, 286, 318, 353
    discrete, 93
    effective, 26, 272
    generalized, 109, 114, 256, 265, 407, 438,
         454
    grating, 26

quantum, 451–457, 462, 466, 471, 472
  standard, 314
  stochastic, 455
no-cloning theorem, 471
noise
  additive, 455
  amplification of, 151
  amplifier, 143
  amplitude, 389, 464, 467, 469, 486
  ASE, 143, 305
  background, 156
  broadband, 151, 230, 313, 317
  intensity, 434
  multiplicative, 455
  phase, 389, 393, 473
  pump-laser, 434
  quantum, 456–471
  shot, 136, 137, 457, 464, 465, 473, 486
  signal–spontaneous, 137
  spontaneous-emission, 137, 305
  squeezing of, 463
  white, 137, 305
noise figure, 121, 136, 143, 144, 366
noise operator, 455
nonlinear amplifying-loop mirror, 215
nonlinear dynamics, 106
nonlinear effects
  intrachannel, 339
nonlinear fiber-loop mirror, 111–121, 214, 463, 476
nonlinear length, 31, 150, 338, 381, 408, 437, 467
nonlinear map, 106
nonlinear optical loop mirror, *see* mirror
nonlinear parameter, 57, 272, 363, 383, 402, 404, 408, 436, 442
nonlinear phase shift, *see* phase shift
nonlinear polarization rotation, 217, 222, 229, 357
nonlinear pulse shaping, 117
nonlinear response function, 79, 170, 410, 453, 454
nonlinear Schrödinger equation, *see* NLS equation
nonlinear switching, *see* switching
normal ordering, 453
NRZ format, 150, 303, 316, 318, 376
NRZ-to-RZ conversion, 377
numerical aperture, 188–190

on–off keying, 309, 393
optical amplifier, *see* amplifier
optical bistability, 22, 26, 34, 104–105
optical clock, 118, 372, 384
optical coherence tomography, 423–427
optical filter, *see* filter
optical isolator, 105
optical pumping, 133
optical push broom, 36
optical sampling, 378
optical soliton, *see* soliton
optical switching, *see* switching
optical wave breaking, 161, 291

parabolic-gain approximation, 168
parametric amplification, *see* amplification
parametric amplifier, *see* amplifier
parametric gain, 359, 378, 459
parametric oscillator, 120, 413–417
period doubling, 26, 107, 110
period-doubling route, 107, 197
periodic boundary condition, 91, 332
periodic poling, 352
phase conjugation, 119, 121, 314
phase locking, 232
phase mask, 7, 9
phase matching, 323, 459, 477
  quasi-, 352
  SPM-mediated, 152
phase modulation, 310, 423
  cross, *see* cross-phase modulation
  self, *see* self-phase modulation
phase shift
  collision-induced, 472, 473
  differential, 366
  nonlinear, 105, 126, 318, 358, 363, 365, 453, 460, 462
  SPM-induced, 63, 104, 112, 122, 124, 151, 218, 291, 314, 315, 388, 442, 460, 461, 468
  vacuum-induced, 453
  XPM-induced, 112, 116, 118, 123, 218, 223, 318, 320, 353, 357, 365, 368, 385, 463
phase-conjugate mirror, 120
phase-locking technique, 207
phase-mask interferometer, 8
phase-matching condition, 2, 103, 119, 151, 231, 359, 379, 414, 479, 483
phase-sensitive detection, 458, 459

phase-space trajectory, 74
photon echo, 158
photon flux, 451, 453, 455
photon-number-splitting attack, 489
photon-pair generation, 475–484
photonic bandgap, 14, 44, 158, 401, 434–442
photonic crystal fibers, 400–442
    air-core, 437
    bandwidth of, 438
    fluid-filled, 439
    hollow-core, 434–442
    soliton formation in, 437
photonic nanowire, 267
photosensitivity, 3, 4
piezoelectric transducer, 208
PMMA cladding, 65
Poincaré sphere, 66, 67, 69, 469
Poisson distribution, 457
polarization controller, 183, 217, 368
polarization instability, 37, 236
polarization scrambler, 375
polarization-diversity scheme, 361, 375, 481, 482
polarization-maintaining loop, 221
polarization-mode dispersion, *see* dispersion
poling period, 433
population inversion, 133, 184
population-inversion factor, 137
positive-P representation, 455, 466
power penalty
    FWM-induced, 321
    Raman-induced, 312
    XPM-induced, 319, 321
power-conversion efficiency, 138
power-transfer function, 385, 388, 392
preform, 88, 190, 401
prism pair, 252, 261, 272
pseudo-random bit pattern, 341, 342, 360
pulse
    amplitude-squeezed, 485
    bell-shaped, 65
    chirped, 277, 331
    clock, 376, 391
    control, 366, 368
    entangled, 487
    Gaussian, 28, 163, 327, 331
    ghost, *see* ghost pulse
    hyperbolic secant, 326, 331
    parabolic, 162, 163, 171, 265, 278, 285
    Q-switched, 198–201, 272, 418, 439, 441

    Raman, 170
    self-similar, 161, 164, 171
    shadow, 340
    square-shaped, 65
    stretched, 274, 276, 278
    super-Gaussian, 316
    transform-limited, 217, 257, 289
pulse broadening
    dispersion-induced, 64, 119, 270, 306, 438, 455, 468
    loss-induced, 328
    SPM-enhanced, 316
pulse compression, 42, 160, 247–294
    amplifier-induced, 290
    Bragg solitons for, 272
    dispersion-decreasing fiber for, 279
    experiments on, 256, 263
    grating pair for, 250–252
    grating-based, 268
    interferometer-based, 292
    limitations of, 255
    pedestal-free, 268
    physical mechanism behind, 247
    quality of, 253
    soliton-based, 261, 441
    XPM-induced, 286
pulse-to-pulse fluctuations, 267
pump-probe delay, 286
pumping
    bidirectional, 138
pumping scheme, 181, 184
    backward, 138, 170
    bidirectional, 170, 194
    efficiency of, 138
    forward, 172
    four-level, 133
    synchronous, 223
    three-level, 133
push-broom effect, 36, 273

Q factor, 383
Q-switching technique, 198–201
QND measurement, 471, 473
quality factor, 253, 262, 268
quantum applications, 450–490
quantum computing, 475, 484, 488
quantum correlation, 476, 481, 484, 488
quantum cryptography, 475, 484, 488, 490
quantum dots, 214
quantum efficiency, 137

quantum erasing, 474
quantum key distribution, 488–490
quantum nondemolition scheme, 471, 475
quantum teleportation, 475, 484
quantum-well layer, 212, 228
quasi-periodic route, 197
quasi-phase matching, 352

radiation
    Cherenkov, 412, 418
    continuum, 78, 117, 231
    dispersive, 85
    infrared, 9
    nonsolitonic, 412
    soliton, 457
    ultraviolet, 6, 9
Raman amplification, 140, 170, 217, 265, 287,
    329
    distributed, 282, 305, 330
Raman amplifier, 169–173, 277, 285, 291, 311
Raman crosstalk, *see* crosstalk
Raman gain, 80, 170, 191, 225, 277, 304, 357,
    385, 389, 407, 476
Raman response function, 80, 171, 410, 454
Raman scattering, 311–314
    coherent anti-Stokes, 413, 420, 439
    intrapulse, 78, 80, 109, 114, 165, 229,
        265, 266, 268, 280, 289, 407, 408,
        418, 467, 469
    spontaneous, 454, 455, 466, 475–480
    stimulated, 33, 144, 158, 169, 189, 190,
        199, 227, 256, 258, 259, 271, 287,
        311, 338, 439, 441
    transient, 441
Raman shift, 170, 422, 441
Raman soliton, 409, 411, 418
Raman threshold, 190, 256, 311, 439, 441
Raman–Nath scattering, 8
Raman-induced frequency shift, 165, 168, 210,
    266, 279–281, 407–413, 438
rare earths, 132, 180
rate-equation approximation, 156
rate-equation model, 141, 196
Rayleigh range, 439
Rayleigh scattering, 198, 199
reactive-ion etching, 7
recirculating loop, 118, 329, 334
regenerator, 304, 306
    2R, 380–390
    3R, 390–393

all-optical, 380–393
    DPSK-Signal, 387
    FWM-based, 385
    non-soliton, 382
    soliton, 382
    SPM-based, 380–385
relative intensity noise, 193
relaxation oscillations, 196
repeater spacing, 304
resolution
    longitudinal, 424, 425, 427
    optical coherence tomography, 425
    temporal, 379
    ultrahigh, 424
ring resonator, 101–110, 125, 416
    modulation instability in, 107
Rowland ghost gap, 44
RZ format, 303, 376
RZ-to-NRZ conversion, 377

Sagnac interferometer, 111–121, 214, 293, 462,
    468, 476, 481
    applications of, 117
    balanced, 117, 463
    FWM in, 119
    nonlinear, 469
    nonlinear switching in, 112
    polarization-maintaining, 118
    switching characteristics of, 113
    transmittivity of, 111
    unbalanced, 114, 468
    wavelength conversion using, 117
    XPM in, 115
Sagnac loop, 111–121, 183, 229, 293, 365,
    389, 468
    asymmetric, 114, 120, 385, 486
    dispersion-imbalanced, 115
    SOA in, 365
    symmetric, 363
sampling oscilloscope, 378
saturable absorber, 85, 117, 198, 211, 222, 227,
    274, 338, 389, 392, 433
    carbon nanotubes based, 213
    fast, 211, 215
    InGaAs, 225
    pulse shortening in, 211
    quantum-dots based, 214
    quantum-well, 228
    semiconductor, 221
    sluggish response of, 117

superlattice, 212, 221
saturation energy, 143, 147, 228, 366
saturation power, 45, 134, 141, 185, 211
　　output, 135
SBS, *see* Brillouin scattering
　　cascaded, 199
　　dynamics of, 199
　　suppression of, 360
　　threshold of, 105, 191, 196, 199, 357,
　　　　360, 363
scanning electron microscope, 401
Schrödinger equation, 455
　　nonlinear, *see* NLS equation
　　standard, 451
second-harmonic generation, 288, 430
self-channeling, 440
self-focusing threshold, 9
self-frequency shift, 229
self-induced transparency, 157
self-phase modulation, 12, 57, 148, 203, 283,
　　　　314–317, 324–327, 338, 339, 364,
　　　　417, 428
　　quantum theory of, 452–454
　　squeezing through, 460–469
self-pulsing, 70, 193, 196, 199
self-referencing technique, 430, 432
self-similarity, 162, 163, 171, 284, 291
self-steepening, 78, 165, 265
semiconductor optical amplifier, 115, 121, 223,
　　　　290, 365, 366, 379
shot noise, *see* noise
sideband instability, 151
sigma configuration, 207, 221, 227
signal-to-noise ratio, 136, 304, 310, 317, 319,
　　　　360, 366, 384
similariton, 162, 172, 278, 284, 285
　　collision of, 172
single-mode condition, 404
single-photon detector, 490
single-photon source, 489
slope efficiency, 186, 190–192
slow axis, 68, 116, 236, 365, 373, 482
slow-light effect, 440
slowly-varying-envelope approximation, 145
soliton
　　adiabatic amplification of, 215
　　amplifier, 153–158
　　auto, 153, 154, 160, 213
　　Bragg, 29, 46, 87, 158, 272
　　bright, 156, 327

chirped, 154, 157, 160, 228
collision of, 31, 472
coupler, 90
dark, 30, 156, 173, 227
discrete, 93
dispersion-managed, 330–334
distributed amplification of, 330
fiber, 157
fission of, 80, 267, 268, 409, 469
fundamental, 30, 72, 79, 158, 159, 204,
　　　206, 215, 216, 279, 325, 407, 437,
　　　455, 462, 464, 469
gap, 30, 83
Gaussian shape for, 331
grating, 29
higher-order, 30, 160, 262, 266, 272, 287,
　　　290, 294, 325, 408, 409, 418, 441,
　　　466, 469
in-phase, 82
interaction of, 160, 215
loss-managed, 328–330
Maxwell–Bloch, 156
order of, 279, 286, 408, 455, 465
out-of-phase, 82
paired, 76
path-averaged, 328
periodic amplification of, 328–330
polarization-locked, 235
properties of, 325–327
pulse compression using, 261
QND measurements on, 471, 473
quantum, 455–457
radiation from, 78
Raman, 267, 407, 469
second-order, 33, 72
self-frequency shift of, 266, 407
SIT, 157, 158
spatial, 90, 93
squeezed, 120, 463
stability of, 156
switching of, 72, 74
third-order, 325
vector, 88, 235
soliton gas, 110
soliton pair, 78
　　asymmetric, 77
　　dark, 78
　　stability of, 77
　　symmetric, 77
soliton period, 281, 467

soliton systems
    amplifier spacing for, 328–330
    dispersion management for, 330–334
    jitter control in, 336
    timing jitter in, 334
soliton trapping, 166
spatial coherence, 7
spectral broadening
    SPM-induced, 261, 315, 378, 380, 413, 428, 438, 467
    XPM-induced, 376, 384, 474
spectral correlation, 465
spectral filtering, 465, 473
spectral hole-burning, 139
spectral inversion, 121, 314
spectral slicing, 418
spectral-window method, 256, 258
spectroscopy
    nonlinear, 420–423
    pump-probe, 420
    Raman, 420
    three-level, 440
split-step Fourier method, 71, 159, 163, 168, 315
SPM-induced phase shift, *see* phase shift
SPM-induced switching, *see* switching
spontaneous emission, 137, 141, 147, 151, 211, 229, 304, 366
    amplified, 117, 120, 305
    noise induced by, 305
spontaneous parametric down-conversion, 475
spontaneous-emission factor, 137, 143, 305
squeezed vacuum, 470
squeezing, 120, 457–471
    amplitude, 464–469, 486
    four-mode, 460
    FWM-induced, 458, 460
    photon-number, 464–469
    physics behind, 457
    polarization, 470, 487
    quadrature, 457–464
    SPM-induced, 460–471
    vacuum, 462
SRS, *see* Raman scattering
standard quantum limit, 461, 465, 469, 487
Stark splitting, 139
stimulated Brillouin scattering, *see* Brillouin scattering
stimulated emission, 133, 141

stimulated Raman scattering, *see* Raman scattering
stitching errors, 8
Stokes parameters, 66, 487
Stokes vector, 66, 469
Stokes wave, 158, 190, 191, 199, 308, 311, 421, 439, 441, 476
stop band, 34, 269
    edge of, 15, 16, 18, 27, 269, 271, 272
strain, sinusoidal, 310
streak camera, 290
subcellular structure, 423, 424
sum-frequency generation, 172
supercontinuum, 90, 198, 267, 268, 469
    applications of, 417–434
    generation of, 418
    microstructured fibers for, 418
    octave-spanning, 432
superlattice, 30, 212, 221
superluminescent diodes, 423
supermodes, 60, 66, 68, 79, 90
susceptibility
    dopant, 146
    second-order, 352, 475
    third-order, 475
switch
    interferometeric, 363–368
    polarization-discriminating, 366
    Sagnac-loop, 363
    SLALOM, 365
    TOAD, 365
    ultrafast, 363–368
    UNI, 366
switching
    applications of, 371–380
    bistable, 33
    contrast of, 64, 85
    FWM-based, 368
    high-contrast, 64
    low threshold, 23
    nonlinear, 33, 79, 89, 90, 92, 112, 122, 293
    observation of, 64
    packet, 363, 368
    power required for, 115
    pulse, 71
    quasi-CW, 62, 115
    soliton, 72, 74, 84, 113
    speed of, 364, 366
    SPM-induced, 33, 36, 113

threshold of, 65, 84, 114
XPM-induced, 36, 115, 117, 125, 363–368
switching window, 364, 374, 377
symmetry-breaking bifurcation, 75, 77
synchronous pumping, 107, 120

Talbot effect, 225
tapering technique, 405
Taylor series, 14, 56, 61, 146, 231, 250, 360
thermal phonons, 455
third harmonic generation, 440
third-order dispersion, 78, 109, 165, 231, 251, 260, 265, 267, 271, 281, 307, 360, 405, 415, 441
third-order nonlinearity, 454
Thirring model, 29, 30
three-level system, 440
three-wave mixing, 288
time standard, 427
time-bandwidth product, 206
time-division multiplexing, 281, 289
timing jitter, 119, 208, 209, 319, 334–434
control of, 336
Gordon–Haus, 334, 336
XPM-induced, 340
total internal reflection, 434
transition cross section, 133, 141, 146
transmitter, 304, 308
tunable delay line, 413, 432
two-level absorber, 157
two-photon absorption, 3, 65, 146, 155, 160, 203, 227, 270, 304, 422
two-photon fluorescence, 91, 422
two-photon fringe visibility, 482
two-photon interference, 484, 485
two-wave mixing, 45

V parameter, 59, 189, 307, 403
vacuum fluctuations, 453, 458
vacuum state, 457, 460, 463, 476
variational method, 73, 82, 315, 331

walk-off effect, 116, 123, 219, 286, 287, 319, 357, 364, 373, 377, 416
walk-off length, 286, 354
waveguide
array of, 88
$LiNbO_3$, 352, 433
nonlinear, 327, 440
periodically poled, 433
planar, 85, 88
silica, 485
temporal, 327
wavelength converter, 117, 223, 352–363
dual-pump, 361
FWM-based, 358
XPM-based, 353, 355
wavelength multicasting, 362, 368
wavelength tuning, 406–417
wavelength-division multiplexing, 86, 102, 139, 303, 418
WDM systems, 102, 139, 150, 152, 304, 308, 311, 322
dense, 418
sources for, 419
transmitter for, 418

XPM parameter, 57
XPM-induced phase shift, *see* phase shift
XPM-induced switching, *see* switching
XPM-induced wavelength shift, 392

zero-dispersion wavelength, 15, 82, 113, 116, 147, 165, 226, 259, 261, 265, 267, 287, 306, 317, 322, 353, 356, 365, 374, 386, 405, 426, 436, 463, 478

Printed and bound by CPI Group (UK) Ltd, Croydon, CR0 4YY

03/10/2024

01040414-0009